IEE CIRCUITS, DEVICES AND SYSTEMS SERIES 13

Series Editors: Dr D. G. Haigh
Dr R. S. Soin
Dr J. Wood

RFIC and MMIC design and technology

Other volumes in the Circuits, Devices and Systems series:

Volume 1	**GaAs technology and its impact on circuits and systems** D. G. Haigh and J. Everard (Editors)
Volume 2	**Analogue IC design: the current-mode approach** C. Toumazou, F. J. Lidgey and D. G. Haigh (Editors)
Volume 3	**Analogue-digital ASICs** R. S. Soin, F. Maloberti and J. Franca (Editors)
Volume 4	**Algorithmic and knowledge-based CAD for VLSI** G. E. Taylor and G. Russell (Editors)
Volume 5	**Switched-currents: an analogue technique for digital technology** C. Toumazou, J. B. Hughes and N. C. Battersby (Editors)
Volume 6	**High frequency circuits** F. Nibier and co-authors
Volume 7	**MMIC decisn** I. D. Robertson (Editor)
Volume 8	**Low-power HF microelectronics** G. A. S. Machado (Editor)
Volume 9	**VLSI testing: digital and mixed analogue/digital techniques** S. L. Hurst
Volume 10	**Distributed feedback semiconductor lasers** J. Carroll, J. Whiteaway and R. Plumb
Volume 11	**Selected topics in advanced solid state and fibre optic sensors** S. M. Vaezi-Nejad (Editor)
Volume 12	**Strained silicon heterostructures: materials and devices** C. K. Maiti, N. B. Chakrabarti and S. K. Ray (Editors)

RFIC and MMIC design and technology

Edited by
I. D. Robertson and
S. Lucyszyn

The Institution of Electrical Engineers

Published by: The Institution of Electrical Engineers, London,
United Kingdom

The Institution of Electrical Engineers,
Michael Faraday House,
Six Hills Way, Stevenage,
Herts. SG1 2AY, United Kingdom

British Library Cataloguing in Publication Data

Robertson, I. D. (Ian D.)
RFIC and MMIC design and technology.
–(IEE circuits, devices and systems series; no. 13)
1. Microwave integrated circuits–Design and construction
I. Title. II. Lucyszyn, S. III. Institution of Electrical Engineers
621.3' 8132

ISBN 0 85296 786 1

Printed in England by Antony Rowe Ltd, Chippenham, Wiltshire

Contents

Preface

RFIC and MMIC technology provides the core components for a wide range of microwave and millimetre-wave communications, radar and sensing systems. The technology can be said to be in its third "era": In the first era, the GaAs MESFET was used exclusively and many key circuit techniques were pioneered as designers adapted to the constraints of chip size by making use of the low parasitics of components and the ability to use many active devices. In the second era, GaAs and InP heterojunction devices (HEMTs and HBTs) were improved rapidly and enabled circuits to operate to well over 100 GHz, with new performance benchmarks being set at nearly every workshop and conference. The first book, *MMIC Design*, was written at an early stage of the second era, before active circuits operating above 30 GHz became commonplace.

This revised book is published at a time when the third era is well underway: on the one hand, HEMT technology has matured and GaAs MMICs operating to 100 GHz have become standard products; on the other, the advance of silicon and SiGe technology, combined with an unrelenting drive for lower component costs for mass consumer applications, has shifted the emphasis dramatically toward silicon technology. So, whereas the first edition was dedicated to GaAs, with a single chapter devoted to silicon millimetre-wave circuits, this one has been brought thoroughly up to date and finds silicon technology infused throughout.

The aim of this book is to give an introduction to the technology and bring together all the key circuit techniques into a single volume. It is intended for final year undergraduates and Masters-level students, as well as being a source of reference for engineers and researchers in the field. Chapter 1 gives and introduction to RFIC and MMIC technology, describing the applications, technology and circuit implementation techniques. Chapter 2 describes the different active devices in great detail and makes performance comparisons. Chapter 3 gives an extensive account of the passive components – both lumped and distributed – that are used. Chapter 4 describes the state-of-the-art in CAD techniques, with an emphasis on how to make the best use of commercial tools. Chapter 5 describes the principles of amplifier design, starting with the basics and leading up to some very advanced concepts. Chapter 6 describes oscillator design, and Chapter 7 gives a comprehensive account of the diverse techniques used for monolithic mixer design. Chapter 8 describes the circuit techniques used for the design of switches and variable attenuators. Chapter 9 covers the design of both digital and analogue phase shifters. The design of integrated antennas has developed into a subject in its own right, and Chapter 10 gives a thorough account of the many reported techniques. Chapter 11 brings together many of the circuit techniques from the preceding chapters, and looks at how complete monolithic transceivers can be realised. Finally, Chapter 12 describes the techniques used for making accurate on-wafer S-parameter measurements.

Ian D. Robertson and Stepan Lucyszyn

Foreword

Shortly after the publication of the first edition of this book, Jim Turner, in many ways the 'father' of the GaAs MMIC industry, sadly died. He is missed at Caswell and on the world scene and was active in device technology until the end. If Jim was the 'father' then Caswell was the 'cradle' of the technology, and it has been my privilege to work at that site since 1968, observing the developments and hopefully playing some small role in their growth. Since Jim wrote the foreword to the first edition of this book, the industry has continued to mature and expand at very much the rate he predicted.

The initial demonstration of the GaAs FET took place in the early 1960s, and the technologies that evolved from that primitive start now pervade many walks of life. Most people will own, have seen or used a system that depends critically upon GaAs-based devices for its functionality – whether they know it or not! Mobile phones, satellite TV receivers, communications links, high speed optical fibre links, radars... the list is almost endless. It is indeed a long way from the first small scraps of shiny black stuff, 40 years ago, to the 6in diameter wafers now entering production.

The jibe from the silicon industry that 'GaAs is the technology of the future, it always has been and always will be' is manifestly no longer true. Worldwide sales of GaAs components are now around $2 billion and growing at a compound annual rate of ~ 40%. For the first time in the history of GaAs devices, demand outstripped supply at the end of the 20th century, stimulating many suppliers to install increased capacity. Things have come a long way since the late 1980's, when several organisations ceased trading as they had invested in volume too early in the growth curve.

Developments in silicon and SiGe technology have advanced the performance of those technologies, and they are now capable of meeting the demands of some applications that could only be satisfied by GaAs. This is no threat to the viability of the GaAs industry as newer applications are now opening up in the millimetre-wave spectrum, and the long predicted exploitation of this frequency band finally seems poised to happen. LMDS (20–40 GHz) and automotive cruise control (76 GHz) are but two applications that are driving the technology forwards. The pseudomorphic HEMT, an exotic research device in the 1980s, is now in routine volume production and rapidly overtaking the MESFET as the key device. Further developments in HBTs, InP HEMTs, metamorphic HEMTs, etc., will provide all of the circuit functions that the systems designer will demand for many years to come. There is no sign that the generic technologies evolved from the GaAs MMIC will run out of steam for the foreseeable future, and we can be sure that other developments, currently being researched in some laboratory, will mature to fill any need that may eventually arise.

This book has been essential reading since its first publication in 1995, and it has been thoroughly revised and updated to take the reader through the technology, circuit design and application of the devices. We can look forward to many other revisions (I hope) of this valuable work, and we can be sure that the future will be even more wonderful, sophisticated and fascinating than we can possibly imagine. Good reading!

Fred Myers
Marconi Caswell

Contributors

Ian D. Robertson, University of Surrey, UK

Stepan Lucyszyn, Imperial College London, UK

Iain Thayne, Khaled Elgaid, Gary Ternent, University of Glasgow, UK

Christos E. Chrisostomidis, Greek Army, Greece

Douglas S. McPherson, Quake Technologies Inc., Canada

Mike W. Geen, Marconi Caswell, UK

K. K. Michael Cheng, Chinese University of Hong Kong, China

Jai S. Joshi, European Space Agency (ESTEC), The Netherlands

Steve J. Nightingale, ERA Technology, UK

Kian Sen Ang, DSO Laboratories, Singapore

Vincent F. Fusco, The Queen's University of Belfast, UK

Acknowledgements

The editors would first like to thank all the authors for their individual contributions and acknowledge the support of the University of Surrey and Imperial College London, for making it possible to complete this project. We would also like to acknowledge the financial support of the UK Engineering and Physical Sciences Research Council (EPSRC), which has underpinned our research activities for more than 15 years. The financial and in-kind support from the following companies is also appreciated:- The Asahi Chemical Company (Japan), EONCOM (Korea), RF Hitec (Korea), Ericsson (Sweden), European Space Agency (The Netherlands), COM-DEV (Europe), BAE Systems, BT, QinetiQ, Marconi Caswell, Matra Marconi Space and Thorn-EMI.

Much of the material presented in this book is the result of work carried out by research students and assistants over many years and we would like to thank the following people:- M. S. Aftanasar, K. S. Ang, A. Ashtiani, A. H. Baree, M. Bokatius, M. Chongcheawchamnan, C. Chrisostomidis, S. Economides, M. Gillick, S. Goasguen, T. Gokdemir, U. Karacaoglu, D. S. McPherson, R. Mellie, S. Miya, S. Nam, C. Y. Ng, C. Oztek-Yerli, G. Passiopoulos, D. Sanchez-Hernandez, F. Saunders, N. Siripon, G. W. Dahlmann.

In many cases we have greatly benefited from the experience and opinions of other researchers in the field and we would like to thank the following people for their indirect contributions to this book: M. Alderstein, F. Ali, A. Barnes, E. Bayar, J. S. Bharj, M. Brookbanks, S. Cornelius, L. Devlin, I. G. Eddison, H. J. Finlay, G. Green, R. H. Jansen, S. Marsh, F. Myers, S. E. Schwarz, B. Y. Song, J. Swift, M. J. Underhill and E. M. Yeatman.

On a personal note, Ian D. Robertson would like to thank C. W. Turner for sparking off his interest in MMICs during the Microwave Engineering undergraduate course (1983-84) at King's College London and also A. H. Aghvami for his support and encouragement as PhD supervisor and mentor for many years. The support of M. J. Kelly and B. L. Weiss, since the move to the University of Surrey in 1998, is acknowledged.

Stepan Lucyszyn would like to acknowledge H. C. Reader at the University of Stellenbosch and C. Stewart at Cascade Microtech Europe, for their assistance with the material in Chapter 12. On a personal note, he would like to thank his wife Rayna for her patience. Finally, the encouragement and support given by all the members of the Optical and Semiconductor Devices Group, since starting at Imperial College London, is gratefully appreciated.

Iain Thayne, Khaled Elgaid and Gary Ternent gratefully acknowledge the support of both EPSRC and the Department of Electronics and Electrical Engineering at the University of Glasgow during the preparation of their chapter.

Jai S. Joshi would like to acknowledge Philips Microwave for the use of their switch FET model and data for the FET switch performance.

Chapter 1

Introduction

I. D. Robertson

1.1 Introduction

A monolithic microwave integrated circuit (MMIC) is a microwave circuit in which the active and passive components are fabricated on the same semiconductor substrate. The frequency of operation can range from 1 GHz to well over 100 GHz, and a number of different technologies and circuit approaches can be used. The additional term 'monolithic' is necessary to distinguish them from the established microwave integrated circuit (MIC), which is a hybrid comprising a number of discrete active devices and passive components integrated onto a common substrate using solder or conductive epoxy adhesive. These became known as integrated circuits because the alternative is to employ hollow metal waveguides, where the inclusion of active devices is a matter requiring considerable mechanical design and workshop machining.

Strictly speaking, monolithic circuits operating above 30 GHz should perhaps be called monolithic millimetre-wave integrated circuits, but a number of alternative abbreviations have been used (such as MMWIC or M^3IC), and there doesn't seem to be general agreement as yet. Further confusion arises with the term 'RFIC', which is used by some to mean circuits operating in the low GHz range (most often using silicon technology), and by others as a vague general term for RF, microwave or millimetre-wave monolithic circuits. In this book the abbreviation MMIC is used almost exclusively, and it should be understood to include millimetre-wave circuits as well. All these terms may become redundant as the current trend is towards systems-on-a-chip, where the goal is an integrated circuit combining digital and analogue functions with micromechanical components, optical components and radio transceivers. So, a future edition of this book may have to be called something like 'Designing the RF part of a system-on-a-chip'.

Figure 1.1 shows a three-stage MMIC low-noise amplifier developed at Caswell for a satellite application (the chip size is $3.5 \times 1.5\ mm^2$). The input and output are at the left and right of the chip, respectively, and the pads are in a ground-signal-ground pattern to enable RF-on-wafer (RFOW) measurements to be made. This circuit contains three transistors, nine capacitors, ten spiral inductors, seven resistors, and a number of microstrip interconnections.

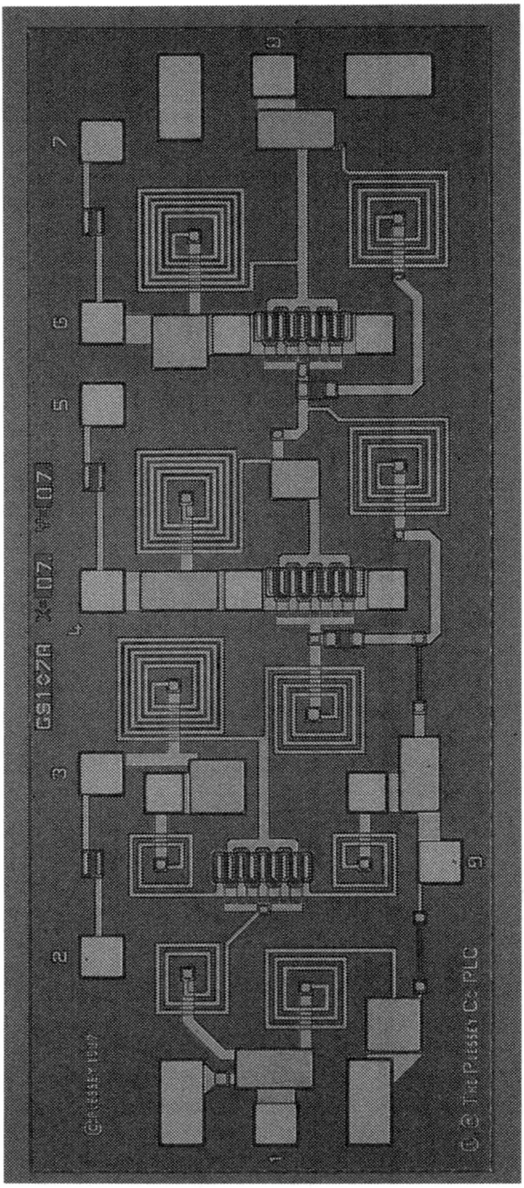

Figure 1.1 A three-stage low-noise MMIC amplifier developed at Caswell for a satellite receiver (Microphotograph courtesy of Marconi Caswell Ltd)

This heavy usage of passive components is necessary because matching networks are essential in order to achieve usable gain from the transistors at microwave frequencies. A further feature of design at microwave frequencies is that the interconnecting tracks have a major effect on the amplifier, and both the interconnections and any discontinuities (bends and T-junctions, for example) must be modelled as microstrip elements using microwave CAD. Whilst MMICs may have rather lower apparent complexity than other integrated circuits, they do offer the highest frequencies of operation. This means that MMICs have a special range of applications and that MMIC design is very different to conventional VLSI design, in which CAD offers a high degree of layout automation. The important transmission-line nature of interconnections on an MMIC requires far more involvement from the designer in the layout process.

The purpose of this chapter is to provide a general background in the history, applications and technology of MMICs. The advantages and disadvantages of MMICs compared with hybrid microwave integrated circuits are discussed in detail. The large range of available device technologies is introduced, and the major applications that have driven the development of MMIC technology are described.

1.2 A brief history of MMIC technology

Gallium arsenide (GaAs) has been used extensively in the development of MMICs because of its suitability for both high frequency transistors and low loss passive components. The technique that made the fabrication of stable high resistivity GaAs material possible was the development of the liquid-encapsulated Czochralski (LEC) method of growing GaAs ingots [1]. Much of the early research into the growth of GaAs using a boron trioxide encapsulant was carried out at the Royal Radar Establishment at Malvern [2], and this method is still in use today. The horizontal Bridgman method, whilst producing low dislocation densities, did not directly yield the high resistivity required to make the GaAs a suitable substrate for microwave components. The groundbreaking research into the fabrication of transistors using GaAs was first carried out by Jim Turner at Plessey Research (Caswell), where the earlier disappointing experiences with the GaAs bipolar transistor led, in 1962, to interest in the GaAs field effect transistor [3]. The first devices had a 24 μm gate length and produced power gain in the VHF band. In 1967, a 4 μm gate length device was fabricated, producing 10 dB gain at 1 GHz, and this became the first commercial device, called the GAT1, marketed by Plessey Optoelectronics and Microwave Ltd. The development of electron beam lithography allowed the first 1 μm device to be produced in 1971, giving gain at 10 GHz, and this was later marketed as the GAT3. In these first ten years, interest in the GaAs MESFET was confined to Caswell and IBM (Zürich).

Along with this activity to produce high performance microwave transistors, the concept of integrating semiconductor devices with microstrip circuitry was already being investigated. The first GaAs MMICs were simple circuits incorporating diodes and microstrip lines, and results were reported by Texas Instruments in 1968: Mao, Jones, and Vendelin [4] presented results for a single-ended 94 GHz mixer employing Schottky barrier diodes, and the design (but not measured performance) of a monolithic Gunn diode oscillator and frequency multiplier were described. In the same issue, Mehal and Wacker [5] presented the design of a 94 GHz balanced mixer and a 30 GHz Gunn oscillator. However, the transistor-based MMIC was probably first conceived by a silicon IC designer at Caswell, called Michael Gay, who created the layout of a 5 GHz receiver chip employing MESFETs with capacitors, inductors, and resistors, as shown in Figure 1.2. Whilst it was never realised, the idea led to the fabrication at Caswell of the world's first GaAs FET MMIC, reported by Pengelly and Turner [6], and shown in Figure 1.3; this was a single-stage X-band amplifier based on the GAT3 device, but with integrated loop inductors and interdigital capacitors. Later, Joshi *et al.* [7] presented results of the first FET oscillator MMIC, operating in J-band.

After these early successes, interest grew tremendously around the world and by 1979 the IEEE had established a symposium dedicated to GaAs IC developments. Thousands of papers have been published since then, many resulting from advances in the device technology, which have yielded higher frequencies and better performance. However, there are a number of papers in which really important new techniques were presented; often these papers attracted worldwide interest and their contributions became real milestones in the history of MMIC development. Some of these particularly noteworthy papers are introduced now, with apologies in advance to those whose work may have been missed. In 1981, Hornbuckle and Van Tuyl [8] presented results for direct-coupled amplifiers using transistors and level-shifting diodes exclusively. In 1982, Jamison *et al.* [9] presented the design of MMIC amplifiers using planar spiral transformers, and later the technique was extended to realise receiver chips with exceptionally high packing density [10]. For MMIC amplifier design the parallel [11] and series feedback [12] techniques were demonstrated, and both techniques showed the advantage of the well-controlled parasitics inherent in MMICs.

In power amplifier design, the thermal and matching problems encountered with MMICs were largely overcome with the introduction of the cluster matching technique described by Pavlidis [13]. For broad-band amplifiers, the travelling-wave (or distributed) amplifier technique was revisited, making full use of the high performance of the MESFET combined with the low parasitics of MMICs: in 1982, Strid and Gleeson [14] presented a particularly important paper describing a DC-12 GHz distributed amplifier along with some of the first ever results for RF-on-wafer (RFOW) measurements. For many years various companies have competed in the distributed amplifier performance stakes. However, the small size of the circuit reported by Kennan *et al.* [15] is still remarkable, with the coupling within the inductors actually being used to improve the performance. For high power the distributed amplifier has always been at a disadvantage, but the capacitively-coupled distributed amplifier technique presented by Ayasli *et al.* [16] in 1984 showed that the power limitations were not insurmountable.

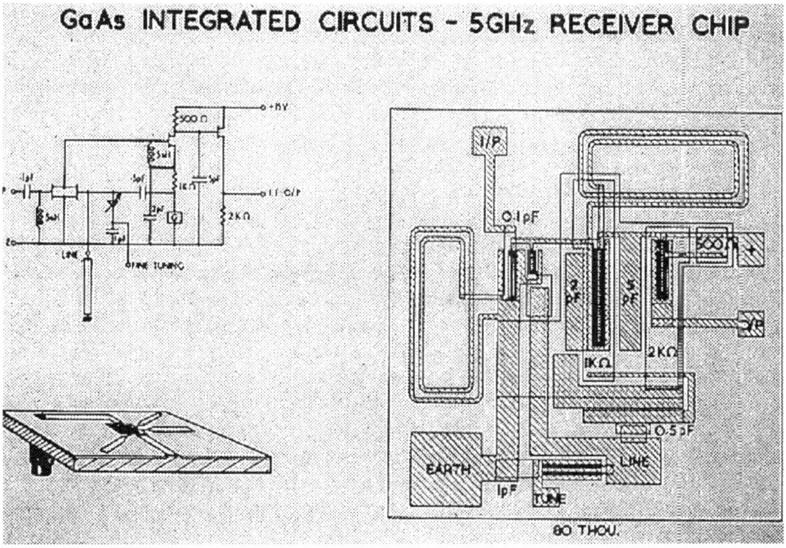

Figure 1.2 Pioneering MMIC layout in 1969
(Courtesy of Marconi Caswell Ltd)

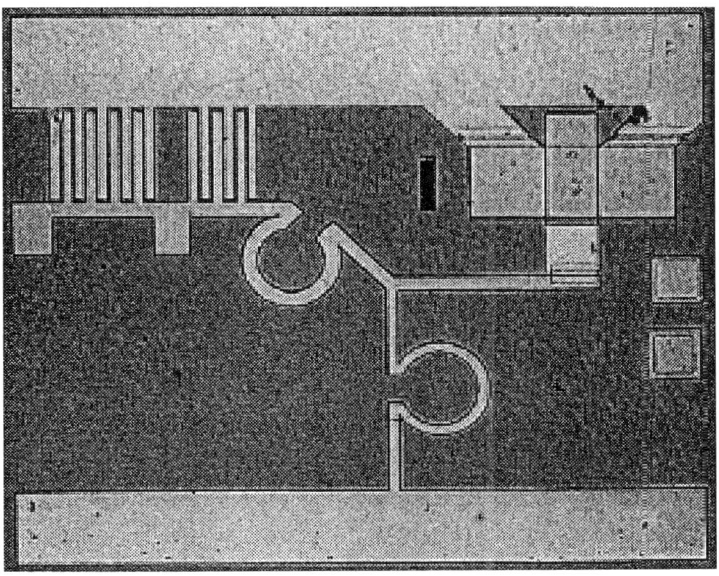

Figure 1.3 The world's first MMIC amplifier
(Courtesy of Marconi Caswell Ltd)

Distributed amplifiers entered a new era when HEMTs became available, and the standards were redefined by the 5-100 GHz bandwidth reported in 1990 by Majidy-Ahy *et al.* [17], for an InP MMIC distributed amplifier using coplanar waveguides. The distributed amplification technique received widespread interest for other circuit functions, and the broadband mixers and active baluns demonstrated by Pavio [18,19] received considerable acclaim. In Europe there was tremendous interest in satellite TV receivers, and the first 12 GHz MMIC receiver chips [20,21] were important milestones.

On-chip antennas were first demonstrated in diode mixer receivers [22], since their size dictates that they are mostly used in the millimetre-wave band. Now that millimetre-wave HEMT technology has matured, on-chip antenna elements are of even greater interest, with wafer-scale active phased-arrays having been proposed. In order to miniaturise circuits, active techniques have always been of interest; the active filter employing FETs in a transversal distributed amplifier topology was presented by Schindler *et al.* [23], and the active inductor was presented by Hara *et al.* [24], both in 1989. Hirota *et al.* [25] demonstrated that 'uniplanar' MMICs employing coplanar waveguide and slotline offered scope for a huge range of novel circuits, and Hiraoka *et al.* [26] demonstrated one of the first multi-layer circuits employing the thin-film microstrip technique. These concepts were further developed, leading to 'three-dimensional' MMIC technology [27].

As far as new circuit design techniques are concerned, the 1980s must surely represent the heyday for the MMIC designer. In the 1990s there was very rapid development of new devices, with HEMT amplifiers passing the 100 GHz barrier [28–30], GaAs HBTs providing new levels of high power performance [31], and SiGe technology being developed [32,33] to start challenging the dominance of III-V materials, even in the millimetre-wave range. Another major development in the 1990's was the increasing emphasis on the civil applications of MMIC technology in communications, and the first signs of the 'wireless revolution' as low cost MMIC transceivers made new systems possible [34,35]. Hence, designers were often employing established circuit ideas but making far more use of CAD tools in order to realise complex multifunction MMICs, whilst miniaturising the circuit area for high volume manufacture.

Commercial electromagnetic simulators became widely available in the late 1990s, and have greatly improved the designer's ability to model circuits with greater confidence. As we move into the new millennium, it is clear that systems integration is a major driving force, as pressures on manufacturing costs push for product solutions with ever lower component counts. Similarly, in communications applications especially, there is a great trend towards the integration of digital and analogue functions, with techniques such as software radio gaining popularity. The integration of such diverse functions has given silicon technology a tremendous boost for applications in the low GHz range, where so many wireless systems operate.

1.3 Advantages and disadvantages of MMICs

The advantages and disadvantages of MMICs, when compared with hybrid MICs, are listed in Table 1.1.

Table 1.1 Advantages and disadvantages of MMICs

MMICs	*Hybrid MICs*
Cheap in large quantities; especially economical for complex circuits	Simple circuits can be cheaper; automatic assembly is possible
Very good reproducibility	Poor reproducibility due to device placement and bond-wires
Small and light	Compact multilayer substrates with embedded passives now available
Reliable	Hybrids are mostly 'glued' together and so reliability suffers
Less parasitics – more bandwidth and higher frequencies	The best transistors are always available for LNAs and PAs
Space is at a premium; the circuit must be made as small as possible	Substrate is cheap, which allows microstrip to be used abundantly
Very limited choice of component	A vast selection of devices and components is available
Long turn around time (3 months)	Can be very fast (1 week), making multiple iterations possible
Very expensive to start up	Very little capital equipment is required

1.3.1 Cost

The low cost advantage stems from the simple fact that a single wafer can produce upwards of 1000 working amplifiers (for example), each with very similar performance and requiring no hand tuning and minimal assembly work. This argument holds as long as the circuit is fairly complicated and requires a large number of components. However, in many cases it is possible to use packaged transistors soldered onto a cheap substrate. Then the most expensive components are the transistors; the passive components such as filters and matching networks cost next to nothing, and the assembly is straightforward with packaged transistors. Even with bare chip devices, automated assembly can be used to good effect. For a circuit using a few transistors with a large amount of passive circuitry, it would be quite difficult to reduce the cost of the circuit by using MMICs: by including the passive circuitry on the GaAs the cost is almost certain to increase. Table 1.2 shows the approximate chip fabrication cost against chip size for a high yield MESFET process using ion-implantation. This represents prices in the year 2000 for high volume manufacture with 6in wafers, calculated from my own estimates after discussion with various foundries. As the chip size increases you get less chips from the wafer *and* lower yield; the cost rises rapidly as a result. However, the yield decrease would not be as bad if the circuit had a lot of passive components, so it is impossible to generalise too much.

What the table does show is that for a high volume application, such as DBS (satellite TV distribution), a MMIC must have high packing density to be competitive. A chip size of 2 mm² is really the maximum that should be considered for this function. At the other end of the scale, if you could design a single chip phased-array radar module in a 10 × 10 mm² chip, you would still get keen customers, even though each chip costs $156. In highly competitive commercial applications, such as personal communications and wireless LANs, there is continual pressure for chip costs to be reduced, and this makes the use of ever larger wafers advantageous if the necessary capital investment can be justified. This again gives silicon technology an advantage, since 12 in wafers are already widely available.

Table 1.2 Chip cost against size

Chip size (mm^2)	Typical yield (%)	Working circuits per 6" wafer	Bare chip cost ($) at $5k per wafer
1 × 1	80	12800	0.4
2 × 2	70	2800	1.8
5 × 5	45	288	17
7 × 7	30	98	51
10 × 10	20	32	156

1.3.2 Performance

Most MMIC devices have to be tailored to volume production and tend not to give state-of-the-art performance. This can be a serious problem for LNA and PA design, where performance is of prime concern. For the very best noise figure and power efficiency it is often necessary to use discrete transistors before and after the MMIC(s). With a hybrid MIC the designer can choose the best transistor for the job, and the transistors may be from different manufacturers. Discrete transistors can use the shortest gate length and optimum active layer structure for their application, regardless of other requirements, because yield is much less important. Using the same fabrication techniques for a complex MMIC would result in a very poor yield because of the multiplicative effect of the yields of individual transistors on the same chip. For low density circuits this is not so much of a problem, and so millimetre-wave circuits using HEMTs should give good performance and yield.

Special devices, such as Gunn diodes, PIN diode switches and hyperabrupt varactor diodes, are rarely incorporated into MMIC processes, and this can introduce further compromises for the MMIC designer: the FET switch is a poor substitute for the PIN diode, and the HEMT millimetre-wave oscillator will have a low output power compared with a Gunn diode. Most of these compromises can be absorbed into the specifications of the system design; good communications between the circuit designer and systems designer are very beneficial to the final product.

Table 1.3 Major MMIC manufacturers

GaAs		Silicon
Agilent	Motorola	Agilent
Alcatel-Telettra	NDI	Analog Devices
Alenia	NEC	Atmel
Alpha Industries	Northrop Grumman	Conexant
Anadigics	OKI	Harris
Celeritek	OMMIC	IBM
Conexant	Raytheon	Infineon
EiC Corp.	RF Micro Devices	Maxim
Filtronic	Samsung	Motorola
Fujitsu	Sanders	National Semiconductor
GaAsTEK	Sharp	NEC
Hexawave	Sony	Philips
Hitachi	Stanford Microdevices	Qualcomm
Honeywell	Toshiba	RF Micro Devices
Infineon	Triquint	SGS-Thomson
M/A-COM	Velocium (TRW)	Texas Instruments
Marconi Caswell	UMS	
Matsushita	Eoncom	
Mitsubishi	WIN	

1.3.3 Investment required

Setting up a MMIC fabrication facility for mass production is prohibitively expensive. There was a time when most major microwave companies had their own facilities as a matter of course, in order to demonstrate their capability. When the profitable applications were not forthcoming, there was a period of regrouping and consolidation. There are still over 35 major foundries/suppliers of GaAs MMICs, and there is a rapidly growing number of companies in the silicon RFIC/MMIC business. Some of the major GaAs and silicon RFIC and MMIC manufacturers are listed in Table 1.3.

Given that foundries are now widely available for MMIC fabrication, what is to prevent the entrepreneur from starting a profitable business? A major investment is required first of all in CAD facilities, since foundries can offer RFOW measurements. The task of designing a competitively priced circuit should not be underestimated, and extensive CAD facilities along with experienced designers are essential. The 'wireless revolution' offers considerable rewards to the company that succeeds, but the competition is fierce, not only between companies, but also between different technologies within the same company.

In contrast, developing hybrid circuits using packaged transistors can be carried out with a PC-based packaged like Genesys™ or MMICAD™, and with little more than an etching tank and soldering iron. A simple hybrid MIC could be laid out, constructed and tested in one day. Because the substrate is cheap and plentiful, the designer doesn't have to spend time squeezing the circuit into the smallest possible space. The prototype can be tweaked with a scalpel and conducting paint, or bond-wires and tuning pads, and then the final design completed. The capital investment required and development times are both considerably less for MICs, with obvious benefits.

1.3.4 Reproducibility

Reproducibility is excellent for MMICs because the active and passive components are produced by the same well-controlled fabrication steps, using the same photolithographic masks. Furthermore, since the active and passive component variations are related to the same physical parameters (such as the sheet resistance of the active layer), there is considerable scope for employing design-centring methods for maximising reproducibility and yield. In comparison, the hybrid MIC suffers from device placement and wire-bonding variations from circuit to circuit. Most hybrid circuits will require some form of hand tuning of the response, and this is very uneconomical for large volume manufacture.

In millimetre-wave design, the ability to control interconnection parasitics is a major advantage. The ease with which 30 GHz-plus monolithic circuits can be manufactured has made these frequencies available for widespread commercial applications, whereas this frequency resource was limited previously to expensive military and research activities. Most conventional millimetre-wave subsystems employ waveguide components, which are assembled and tuned individually by hand, making them very expensive. Microstrip-based MICs are exceedingly difficult to construct above 30 GHz; finding a suitable substrate and making a good launch to microstrip at 30 GHz are difficult enough, and then you have to overcome the huge parasitics of components such as chip capacitors and resistors, as well as the large reactances presented by the bond-wires. With experience and careful design, one can overcome these parasitics to construct a circuit in the research laboratory, but in large volume manufacture the parasitics would be very hard to control.

1.3.5 Reliability

MMICs are more reliable than hybrid circuits, as long as the fabrication process is carefully controlled and qualified. In contrast, the mechanical limitations of chip attachment and wire-bonding become all too apparent when a hybrid circuit is subjected to temperature cycling, shock and vibration.

1.3.6 Size and mass

MMICs are very compact, and this can be a significant advantage. In many commercial applications there is an increasing need to miniaturise the microwave circuits, with smart cards and radio transceivers for PCMCIA slots (in notebook computers) being good examples. In active phased-array antennas, each face of an array may have many hundreds of modules, and MMICs are essential in order to minimise the antenna size and mass.

1.4 Applications

The major applications of MMICs are shown in Table 1.4. Military and space applications have been a major driving force behind MMIC technology. The electronically-steered phased-array antenna is probably the single most important application for MMICs in these applications. This type of antenna is simply not practical with hybrid technology, due to the size and mass of the individual transmitter-receiver modules.

Table 1.4 Applications of MMICs

Military	Space	Civil
Phased-array radar	Communications satellites	Satellite TVRO
		VSAT earth terminals
Electronic warfare	Synthetic aperture radar	Mobile phones
		LOS communications
Smart munitions	Radiometers	Wireless LANs
		Fibre-optic systems
Synthetic aperture radar	Astronomy	Global positioning (GPS)
		Smart cards
Decoys	Low earth orbit mobile systems	Bluetooth™ and similar wireless systems
		LMDS and the wireless local loop
Passive millimetre-wave imaging	Steerable phased-array antennas	High speed internet access
		Autonomous cruise control
Altimeters	Remote sensing	Automatic tolling
		Search and rescue transponders
Instrumentation		Security scanners
		RFID and tagging
Remote sensing		Medical systems

Since the requirements of adaptive phased arrays have been so beneficial to MMIC development, it is important to understand them: Figure 1.4 compares a conventional satellite transmitter with a solid-state transmitter using adaptive beam-forming. The conventional transmitter uses a single high power travelling-wave tube amplifier (TWTA) transmitting through a beam-forming network (BFN) into the individual antenna feeds. This system has been used successfully on many major communications satellites. However, the TWTA is a vacuum tube device and as such has limited reliability and lifetime, requires high voltage supplies and is rather non-linear. The BFN is required to shape the transmitted beam so that coverage of the ground can be carefully designed to make optimum use of the satellite's precious transmitter power. However, since the BFN may be handling many hundreds of watts, is it impractical to employ electronic control of the BFN. The BFN is usually a mass of carefully designed waveguide 'plumbing'.

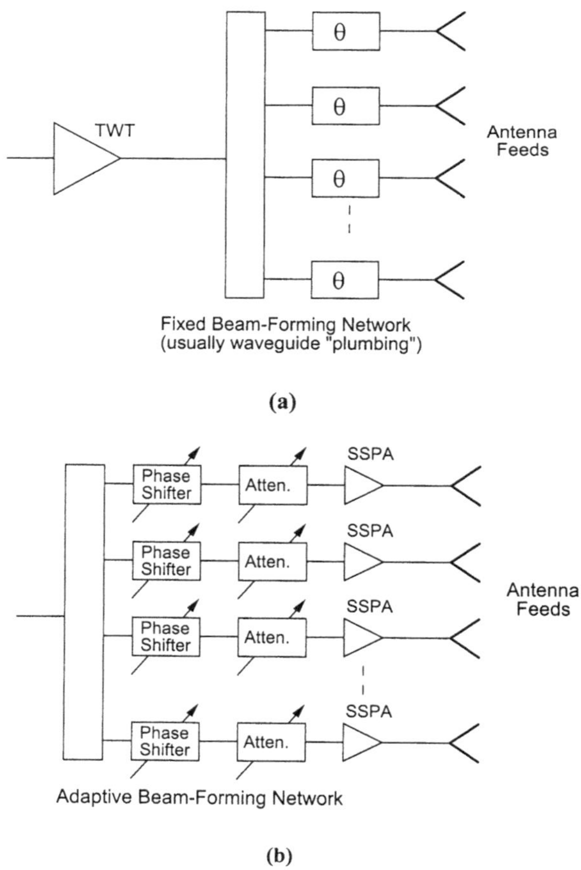

Figure 1.4 Phased-array antennas: (a) conventional, using a travelling-wave tube amplifier and (b) adaptive, using solid-state power amplifiers

In the solid-state transmitter, power amplification is distributed around the antenna. Hence, the individual power amplifiers are only 20 W or so, and solid-state power amplifiers (SSPAs) can readily be used. This means an improved lifetime and reliability, low voltage supplies and excellent linearity. In addition, there is clearly a large degree of redundancy in the arrangement. Since each antenna feed is now driven by a separate amplifier, the phase and amplitude of the signal can be individually controlled with electronic attenuators and phase shifters placed before the power amplifier. This means that the antenna beam can be steered electronically. In a satellite payload, this gives the operator the freedom to reconfigure the coverage patterns to match demand whilst the satellite is in operation. In a radar system, it means that the beam can be steered electronically very quickly, and can track fast-moving and multiple targets in a fashion which is impossible with a traditional rotating antenna. Since an adaptive phased-array antenna may have hundreds of transmitter-receiver modules, MMICs are essential because they give excellent reproducibility with small size and mass.

In addition to the military and space industries' drive for phased-array components, the satellite TV (DBS-direct broadcast by satellite) receiver (TVRO) provided an important 'carrot' for the development of MMIC technology. This high volume consumer application requires a fairly simple frequency downconverter. In the US the received frequency is in C-band at around 4 GHz, and in Europe it is in X-band at 11 to 12 GHz. The 'low noise block' (LNB) mounted at the antenna feed has to downconvert the received signal to the 1 to 2 GHz region, so that the signal can be fed through cheap coaxial cable into the indoor receiver/decoder unit. It was rather disappointing for the MMIC enthusiast that the first systems designed for the mass consumer market did not use any MMICs, and even today, the ones that do employ an MMIC are far from being the totally integrated single-chip receivers that were dreamed of in the pioneering days. To see why this is so, consider the simplified LNB block diagram shown in Figure 1.5. This is a very straightforward frequency downconverter. However, the noise figure of the LNA has to be as low as possible, and certainly less than 1 dB. The lower the noise figure, the better the TV picture or the smaller the dish. However, the first MMIC satellite TV receiver chips [20,21] were not able to achieve a low enough noise figure because the MESFET was not good enough, and because of the high on-chip matching network losses. The bandpass filter is required to give 50 dB or so of image rejection, and a filter of the required selectivity could not be integrated onto a chip by using lumped elements (too lossy) or microstrip (too large). The mixer was, until quite recently, a difficult component to realise on MMICs because traditional designs are so incompatible with the planar MMIC. The oscillator design is quite easy, but requires an off-chip dielectric resonator. Bringing all these points together, the MMIC single-chip satellite TV receiver actually needs an off-chip discrete LNA stage, an off-chip image filter, and an off-chip dielectric resonator, and on top of this the IF stages are cheaper in silicon. Hence, the MMIC has found it surprisingly hard to make inroads into this market, although eventually it did so to good effect. Nevertheless, the discrete HEMT front end stage feeding a MESFET-based MMIC appears to be preferred to a complete HEMT-based MMIC with low noise figure. This reflects the high costs of HEMT manufacture compared with ion-implanted MESFETs.

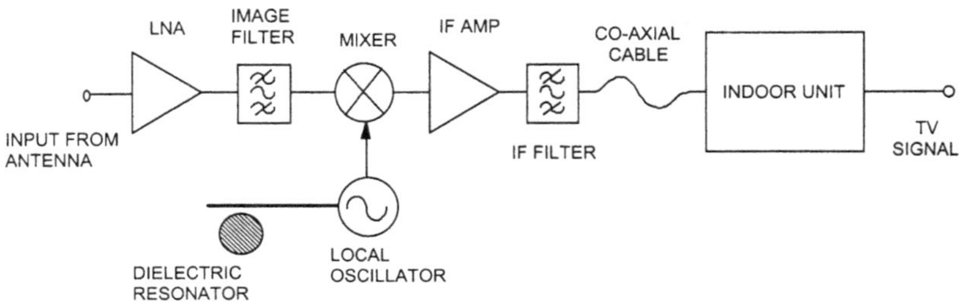

Figure 1.5 Block diagram of a satellite TV receiver's low noise block (LNB)

In the communications field, MMICs are applicable to fibre optic systems and to satellite, fixed-terrestrial and mobile radio systems. MMICs are an important part of the 'wireless revolution'. Mobile communications and high speed internet access systems are important consumer applications that make extensive use of RFIC and MMIC technology. As traffic builds up, and as new high data-rate services are envisaged, there are already major projects underway in Europe to develop MMICs for microcellular systems operating at 62.5/63.5 GHz [36]. In addition, wireless data communications is seen as a major growth area as personal digital assistants (PDAs) and notebook computers become commonplace, and there is a large market for RFIC/MMIC transceivers. In-building wireless LANs are increasingly being promoted [37], with frequencies of 2.4, 5.2, 17, 18 and 60 GHz being proposed for new systems. The rationale behind this new application of microwaves is that the transmitter-receivers are so much cheaper nowadays that the wireless system is an economical alternative to the extensive routing and laying of coaxial or fibre optic cables (although these are still used as the backbone of the network in most cases).

The microwave or millimetre-wave local multipoint distribution system (MVDS or LMDS) is a combined fibre optic and microwave distribution system, in which the enormous expense and effort required to connect homes with fibre optic cable is overcome by distributing the signals over the last mile or so by a short radio link, as shown in Figure 1.6. This system was extensively developed by British Telecom (BT), who ran a 29 GHz trial service at Saxmundham in the UK [38]. The combined cable and radio system is ideal for rural areas which are not economically viable for purely cable systems. The work at BT was extensively concerned with the development of an MMIC receiver for a low cost subscriber unit [39]. Wireless systems offering broadband communications to the home are rapidly being installed around the world, using a number of frequency allocations in the 24 to 42 GHz range [40–42].

In automotive applications [43], autonomous cruise control (ACC) is already available as an option for a number of luxury cars, and it is expected that these units will become standard in all new cars. The benefit of such technology is that it can contribute to the free movement of traffic, helping to reduce congestion. A wide range of MMICs and multi-chip modules has been reported for this 76.5 GHz application [44–47]. Further development of intelligent systems for traffic use will include more advanced radar sensors and imaging systems, for collision avoidance, and the widespread introduction of automatic tolling systems which require communication between the vehicle and numerous roadside units.

In the UK, the RTA ('Road Traffic Advisor') consortium [48] has already set up a trial system along a total of 350 km of the M4, M25 and M23 motorways.

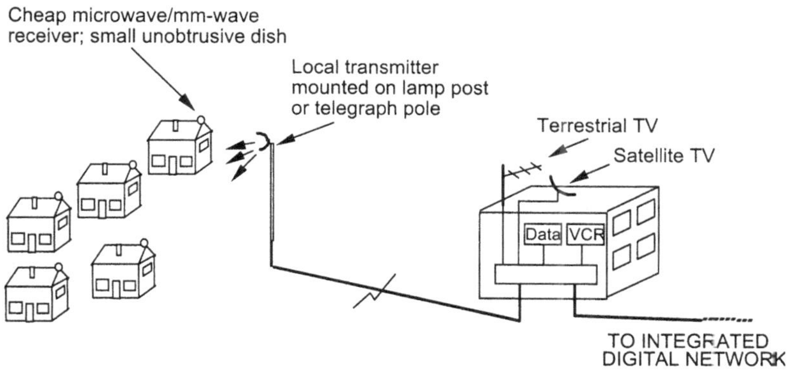

Figure 1.6 The millimetre-wave local multipoint distribution system (LMDS)

1.5 Active device technologies

GaAs has two key advantages over silicon for MMICs: (1) GaAs has higher saturated electron drift velocity and low-field mobility than silicon, resulting in faster devices and (2) GaAs can readily be made with high resistivity, making it a suitable substrate for microwave passive components. As a result, GaAs completely dominated the first 15 years of MMIC development, and even now the vast majority of MMICs above a few GHz are GaAs based. Nevertheless, improvements to the silicon bipolar device (such as narrow emitters and shallow bases) have enabled conventional silicon bipolar devices to achieve an f_t of over 25 GHz, and silicon dominates most applications up to 5 GHz or so. The competition between GaAs and silicon has become more fierce as a result of important developments in heterojunction devices with both materials. The main technologies which are used for RFICs and MMICs include:

1. GaAs MESFET
2. GaAs HEMT (conventional and pseudomorphic)
3. GaAs HBT
4. Silicon bipolar
5. Silicon-germanium HBT
6. Silicon CMOS
7. Silicon BiCMOS
8. InP HEMT
9. InP HBT
10. IMPATT, PIN, RTD, Gunn and Schottky diodes.

In addition, a number of new technologies have been developed, including LDMOS, GaN and SiC-based devices for high power applications [49–53], and

SiGe HMOST (or HFET) devices for low noise millimetre-wave operation on silicon [54].

The GaAs MESFET was the first microwave transistor and remains an important workhorse for many MMIC designs. It is easily fabricated using ion-implantation for high volume applications, and has good noise figure and output power performance. Most GaAs foundries offer 0.5 μm gate length MESFET processes, which are useful for circuits operating up to around 20 GHz, and circuits operating to over 30 GHz have been reported with shorter gate lengths. Foundries generally offer a selection of different device types including general purpose, low DC power, high RF power, switching, and enhancement-mode/depletion-mode devices. Using selective ion-implantation, it is often possible to use different types in the same circuit.

The GaAs-based pseudomorphic HEMT (high electron mobility transistor) offers a considerable increase in transconductance over the MESFET and, with gate lengths as short as 0.1 μm, circuits operating to well over 100 GHz have been reported. However, these high performance devices are less suited to high volume manufacture than the MESFET; the intricate material layers must be fabricated with either molecular beam epitaxy (MBE) or metal organic chemical vapour deposition (MOCVD), both of which are time consuming. The gates must be defined with electron beam lithography, which is rather slow, and for good noise figures the mushroom gate is needed, which must be produced using a precise multi-layer photoresist technique. PHEMT foundries commonly offer 0.25 μm gate-length devices, and 0.15 μm devices are available from some foundries. InP HEMTs have produced the absolute best high frequency performance, and monolithic amplifiers with 80 nm gate-length devices have been demonstrated at 200 GHz [55–57]. Power amplifiers with 1 W output power in V-band [58], and distributed amplifiers with 1 to 157 GHz bandwidth [59] have also been realised in InP technology. The fragility of InP wafers causes significant difficulties in mass production, however.

The GaAs heterojunction bipolar transistor (HBT) has significant advantages over the HEMT because it is a vertical structure. The critical device dimensions are defined by the material growth and doping rather than by lithography. An HBT with an emitter width of several microns can offer good microwave performance. The HBT is intrinsically a very high-gain device, but tends to suffer from large parasitic resistances and capacitances; it can offer exceedingly high power-density and efficiency. The high base resistance of HBTs means that their noise figure is considerably higher than that of HEMTs, but the device is continually being improved. A growing number of manufacturers are offering HBT foundry processes, and already some very competitive products are on offer, particularly low supply voltage components for mobile handsets.

Silicon bipolar technology has advanced tremendously in recent years and the homojunction device has already achieved an f_t of over 25 GHz [60]. In addition, the SiGe HBT [61–63] has been reported to have an f_{max} of over 160 GHz [64], and a noise figure of less than 1dB at 10 GHz has been achieved [65]. These technologies offer a major threat to the dominance of GaAs in MMIC development. Silicon bipolar devices can be integrated readily with conventional VLSI circuitry based on CMOS [66]. The combined BiCMOS processes can be expensive, but offer the ideal solution of true single chip integration in consumer products such as wireless transceivers, where digital/analogue circuit integration is more and more important [67]. For applications up to a few GHz the CMOS devices themselves have improved considerably, and microwave circuits in 'RF CMOS' technology have been reported [68]. The major advantage of silicon technology is that it is very mature and 12 in wafers are readily available, whilst GaAs MMIC fabrication is limited to 6 in at the moment. The traditional problem of realising good passive components on silicon substrates has been addressed with a wide range of solutions using high resistivity substrates, multilayer techniques or micromachining [69–74].

GaAs technology still has a major advantage in areas where high performance is paramount. However, if GaAs and silicon vendors are offering a chip with the same functionality, then it is the lowest price that determines the choice made by the customer. Ultimately, larger silicon wafers give a very valuable cost advantage, which is hard for the GaAs manufacturer to contend with. The solution is to use GaAs where its performance advantages are really needed, and not try to go head-to-head with silicon products when either technology will suffice.

For millimetre-wave applications the picture today is still very attractive for the III-V industry, with HEMT technology having matured and having no real equal in performance outside of the research laboratory. Nevertheless, there remains the fact that simple transmitter-receivers can be realised [73,74] using IMPATT or Gunn diode oscillators and Schottky barrier diode receivers, although they may have limited performance (particularly noise figure) and functionality.

1.6 Design approaches

As well as the different active device technologies, in general there are three quite distinct circuit design approaches, the choice of which largely depends on the operating frequency of the MMIC. The three techniques can be classified as 'all-transistor' techniques, lumped-element techniques and distributed techniques. There is inevitably some overlap of the useful frequency range of application for each approach, and the techniques may often be blended together in the same design. The class of distributed techniques can be further sub-divided depending on whether microstrip, coplanar waveguide (CPW) or micromachined transmission-lines are employed. Figure 1.7 shows the approximate frequency range to which each approach is applicable.

Circuits using entirely active techniques (e.g. as in [8]) tend to use small device peripheries, so that the resulting small input and output capacitances do not unduly affect the performance of circuits such as operational amplifiers. This approach is usable up to at least 5 GHz, and such high frequency of operation is achieved largely because of the low capacitance of a microwave transistor, rather than through the use of classical microwave design techniques. Nevertheless, the design of these 'conventional' circuits at such high frequencies does require tremendous design skill and experience. In the silicon industry, this sort of all-transistor design experience is widely available, but the GaAs industry has naturally tended to have microwave design engineers. It is partly because of this that, until recently, entirely active techniques were used relatively little in MMICs. In addition, most of the original interest in MMICs has been for space and military applications, where DC power consumption has been of considerable importance, making active techniques less attractive. However, the major advantage of active techniques is their high packing density, which leads to competitively priced products. Hence, in recent years there has been far more interest in these techniques for microwave applications such as mobile telephones operating at 1.8 GHz and wireless LANs operating at 2.4 GHz. Part of the reason for this has been the availability of high performance RF CMOS, silicon bipolar and GaAs HBT technology. These complementary or bipolar devices seem much more amenable to complex active circuits than MESFETs and HEMTs, and part of the reason for this must be the relatively poor control of the FET's DC characteristics, which has also limited their application to digital circuits. Figure 1.8 shows an example of an all-transistor MMIC: this is a narrowband filter using high-Q tuneable active inductors [75].

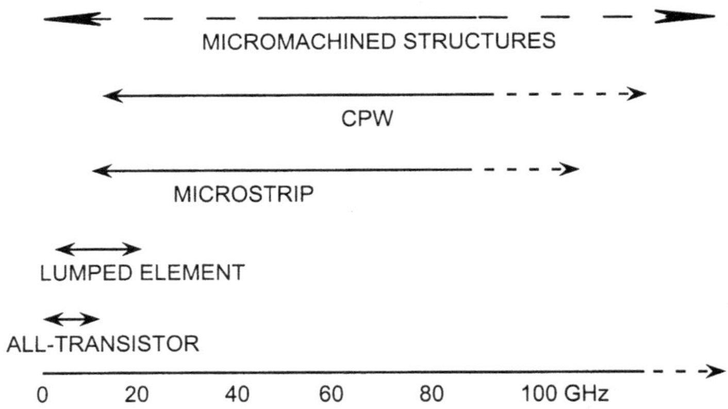

Figure 1.7 Typical frequency range of operation for various approaches to MMIC design

For higher operating frequencies, the transistor input and output capacitance must be accounted for. Lumped-element matching networks (using spiral inductors and overlay capacitors) provide the best solution at frequencies below 20 GHz. Figure 1.9 shows an example of an MMIC employing lumped-elements; the chip is a 1 to 2 GHz feedback amplifier. This could be considered to be the classical type of MMIC, with the spiral inductors being the visually striking trademark of MMIC technology. The principle advantage of this approach is that it offers high performance in terms of noise figure and power handling, when compared to an all-transistor design. Hence, it is well suited to applications such as satellite payloads and phased-array antennas, where high performance and low DC power consumption are very important [76].

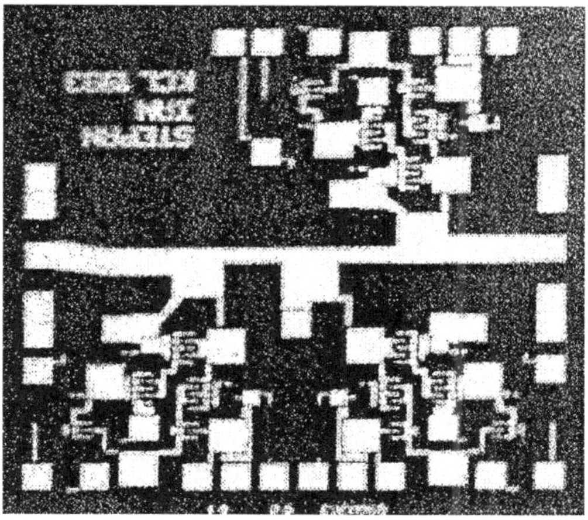

Figure 1.8 Microphotograph of an 'all-transistor' circuit: a MMIC filter using three active inductors

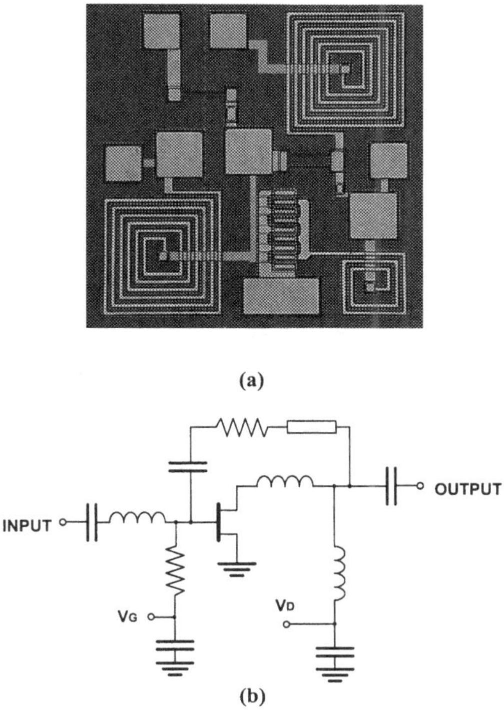

Figure 1.9 A lumped-element 1 to 2 GHz MMIC amplifier: (a) photograph and (b) circuit diagram

At frequencies higher than 20 GHz, the spiral inductors will be beyond their useful frequency range because of their own self-resonance, and so distributed elements are used. MMICs operating above 20 GHz will tend to use stubs and transformers for matching. The distributed elements can be realised in a number of transmission-line media, with microstrip and CPW being by far the most common. CPW has been extensively used for millimetre-wave circuits, its chief advantages being low dispersion and the ability to provide low inductance grounding of devices. With distributed matching techniques, the highest operating frequency of an MMIC is limited only by the maximum frequency at which the active devices still have usable available gain. The lowest frequency of operation for distributed circuits is determined by the chip size, since the physical length of matching elements is too great at frequencies below approximately 5 GHz. Figure 1.10 shows an example of a microstrip circuit employing distributed matching elements: it is a 35 GHz receiver [77]. Microstrip lines are used throughout to realise matching networks, couplers and a band-pass filter.

Microstrip circuits require through-substrate via-hole technology for the ground connections, and this limits the final wafer thickness to, typically, less than 200 μm. This makes it relatively difficult to process large diameter wafers, which is one reason why coplanar waveguide (CPW) transmission-lines may be preferred. Figure 1.11 shows a Ku-band two-stage balanced amplifier realised with CPW transmission lines.

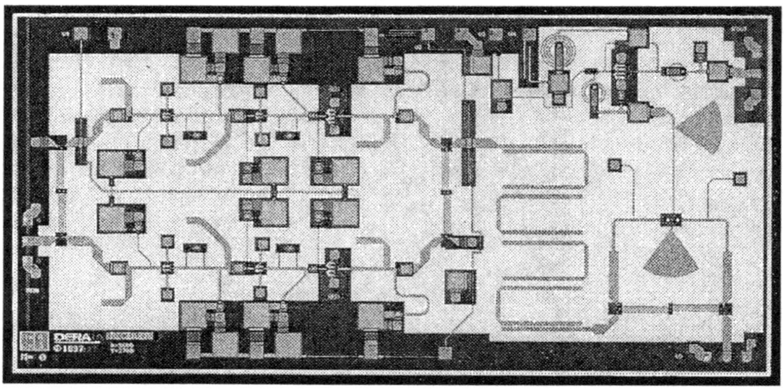

Figure 1.10 Microphotograph of a 35 GHz receiver chip using microstrip transmission-line elements (Published courtesy of QinetiQ: Crown copyright)

Both lumped and distributed components can be realised with multilayer technology [78]. Examples are miniature stacked spiral inductors and miniature thin-film microstrip (TFMS) transmission-lines. With these techniques, ground planes can be on the top-side of the chip; this is extremely desirable because wafer thinning to, say, 100 microns has become extremely difficult as wafer sizes have increased to 6 in. Multilayer passive components are described in Chapter 3. The high frequency limit is determined mostly by losses in the very small structures, but operation to over 60 GHz is possible for TFMS if a suitably low loss dielectric is used. Figure 1.12 shows the microphotograph of a uniplanar MMIC amplifier using TFMS lines.

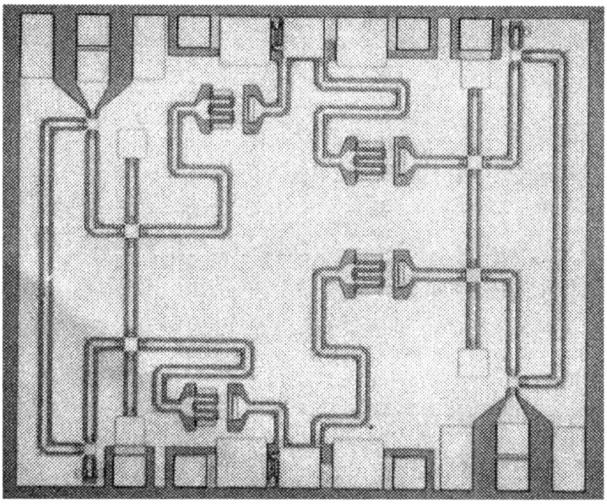

Figure 1.11 Microphotograph of a Ku-band two-stage balanced amplifier using coplanar waveguide (CPW) transmission lines

Beyond 100 GHz the losses of conventional planar transmission lines become very high, and it has been shown that various micromachined transmission-line media can be used. By making use of air as the principal dielectric, very low dielectric loss is achieved and the physical line lengths are longer. This is useful since line lengths otherwise start to become so small that circuit layout is difficult. Micromachined transmission-lines that have been proposed include CPW lines with etched grooves, microshield lines and rectangular waveguides [79,80].

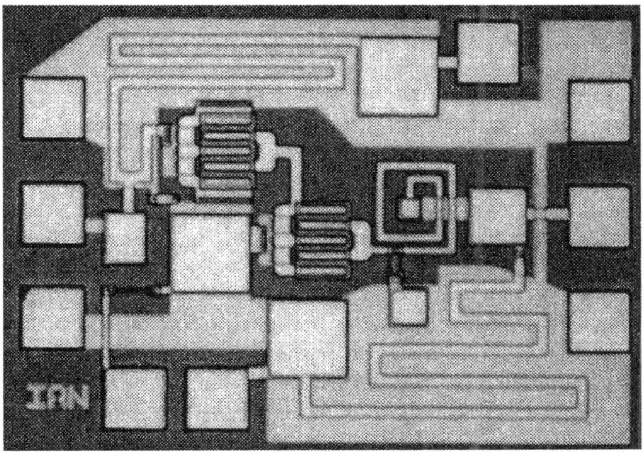

Figure 1.12 Microphotograph of a uniplanar MMIC amplifier using TFMS

1.7 Multi-chip module technology

Multi-chip module (MCM) technology offers an ideal solution for the manufacture of products for a huge variety of applications. The system-on-a-chip concept falls down in many cases when superior performance is offered by using a range of different implementation technologies. The MCM approach is to partition the system and realise each part with the best component for the task: CMOS for digital functions, PHEMT MMICs for millimetre-wave functions, and so forth, and then to combine the chips into a single module which contains all the interconnects, bias circuitry, filters and other passive components, and maybe a planar antenna. This compromise approach should be chosen wherever possible. It simply presents the most practical way of integrating a wide range of components into a single module. Single-chip integration is a fine ideal, and in many applications it is crucial (smart cards, for example), but compromises in performance, high costs of advanced processes, and lower yield achieved from such highly integrated systems present many challenges. In products, cost is of paramount importance, and if the MCM solution can be turned around faster and manufactured more cheaply then it is the best choice.

There are three main MCM technologies: MCM-D, which uses thin-film deposition techniques; MCM-C, which uses co-fired multilayer ceramic substrate technology; and MCM-L, which uses lamination technology. Broadly speaking, MCM-D uses normal spinning, deposition and photolithographic procedures to fabricate multilayer passive components and interconnects with small feature sizes [81]. For example, miniature TFMS lines could easily be employed, with 10 μm feature size, using the thin deposited dielectric layer(s) as the substrate. With suitable materials, MCM-D technology will offer the highest performance and is the best suited to flip-chip assembly. Figure 1.13 shows the photograph of a C-band module employing flip-chip silicon and GaAs MMICs for Hiperlan applications.

MCM-C technology employs ceramic materials in their un-fired ('green') state in the form of tapes or pastes. Multi-layer modules are formed from multiple layers of tapes, or multiple screen-prints of the pastes, and then the layers are all fired together. This co-fired ceramic technology has been used for many years in ceramic package manufacture. More recently, it has been developed further with new materials that can be fired at low temperature, and LTCC (low temperature co-fired ceramic) has attracted widespread interest for the manufacture of low-cost modules incorporating active circuits along with multilayer passive components. Via formation does, however, require mechanical punching or laser machining. An important development in this technology that overcomes this is the development of photo-imageable dielectrics and metal pastes. This makes it possible to approach the feature sizes of MCM-D technology whilst still using thick-film processing techniques, which are lower in cost. Materials that are available include the FODEL™ range from DuPont, and the KQ™ range of materials from Heraeus [82]. The KQ material has extremely low dielectric loss tangent and has been successfully used for 77 GHz filters for automotive applications.

MCM-L technology is essentially a very advanced PCB technology: multiple layers are formed by laminating successive layers, with photolithographic procedures being applied at each stage to define the metal patterns. Vias are formed by mechanical punching, drilling and routing, or laser machining, and 'microvias' are now in common usage. MCM-L technology is capable of making microstrip lines with approximately 50 μm line widths and gaps; this, allied to the use of Teflon™ material, which has a low dielectric loss tangent, means that MCM-L technology can be applied even to millimetre-wave modules. Its principal advantage is that very large boards can be processed (18 × 24 in^2 is common), and this results in extremely low costs. As a result, MCM-L technology is widely used for high volume consumer products in the < 20 GHz range.

As applications beyond 100 GHz develop, there is a need for transmission-line media with lower loss than can be achieved with microstrip or CPW. For example, power combining multiple chips can only be practical if the combiner has very low loss. For this reason, there is great interest in multi-chip modules employing rectangular waveguide or other low-loss micromachined transmission medium. Figure 1.14 illustrates the concept of how MMIC power amplifiers could be power combined in such a module [83]. Dielectric-filled waveguides were proposed many years ago by Hinken [84] as a low-cost circuit medium, and more recently such waveguides have been demonstrated at X-band using MCM technology [85]. To extend the technique to beyond 100 GHz requires a material with very low loss tangent [86]. The concept offers an attractive solution to the manufacture of low cost millimetre-wave modules incorporating both MMICs and high-Q passive components.

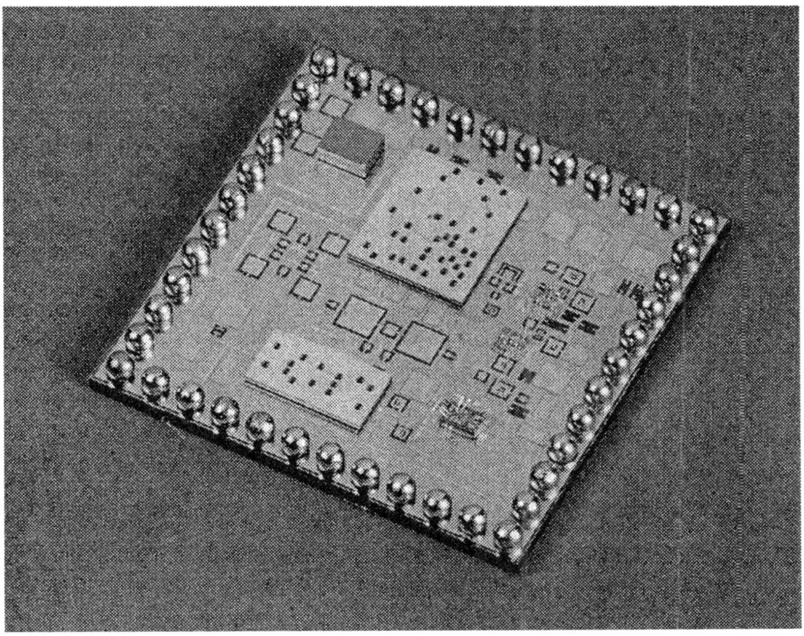

Figure 1.13 Hiperlan transceiver MCM (Courtesy of D. J. Pedder, Intarsia Corp.)

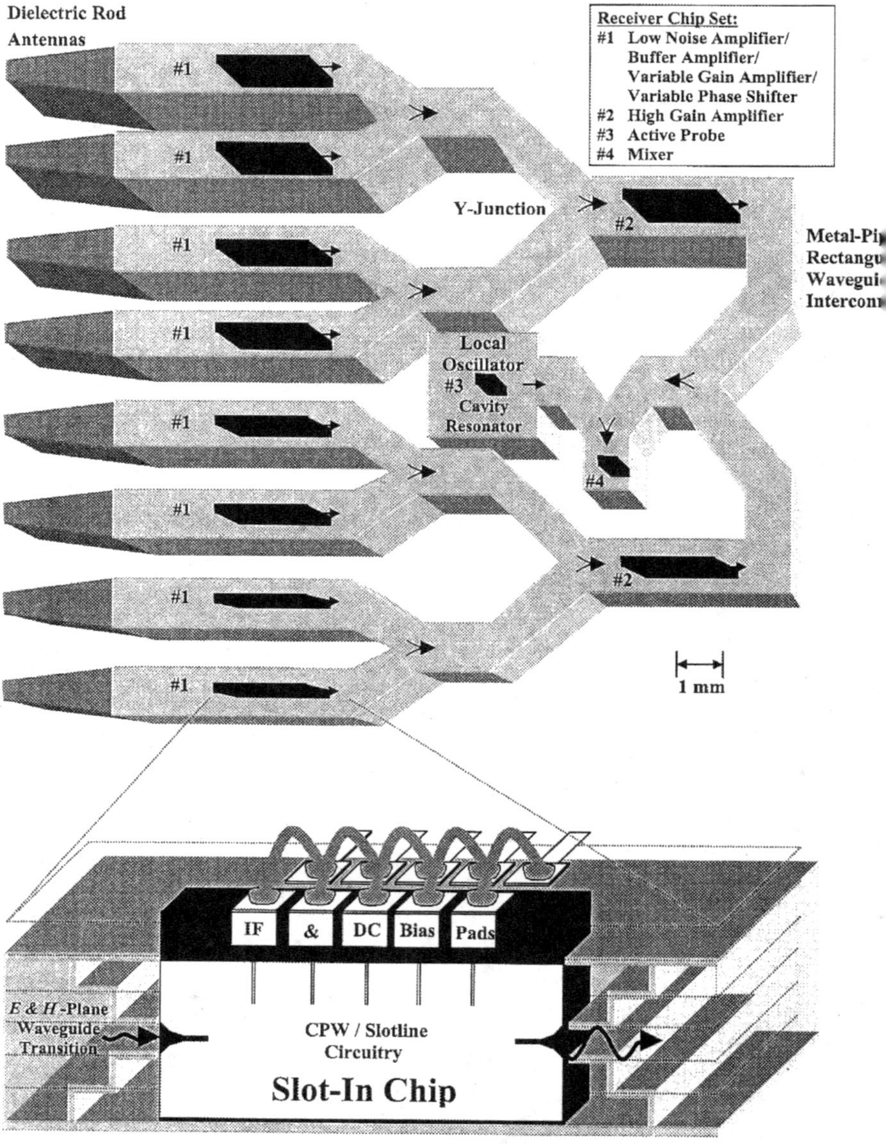

Figure 1.14 Futuristic illustration of 75–300 GHz MCM concept, based on rectangular waveguide interconnects

1.8 References

1 METZ, E. P. A., MILLER, R. C., and MAZELSKY, R.: 'A technique for pulling single crystals of volatile materials', *Journal of Applied Physics*, 1962, **33** (6), pp. 2016–2017

2 MULLIN, J. B., STRAUGHAN, B. W., and BRICKELL, W. S.: 'Liquid encapsulation techniques: The use of an inert liquid in suppressing dissociation during the melt-growth of InAs and GaAs crystals', *Journal of the Physics and Chemistry of Solids*, 1965, **26**, pp. 782–784

3 TURNER, J. A.: 'History of the GaAs FET at Caswell (1964-1985)'. *IEE colloquium on modelling, design, and application of MMICs*, London, 1994, digest no. 1994/092, pp. 1/1–1/3

4 MAO, S., JONES, S., and VENDELIN, G. D.: 'Millimeter-wave integrated circuits', *IEEE Journal of Solid-State Circuits*, 1968, **SC-3** (2) pp. 117–123

5 MEHAL, E. W., and WACKER, R. W.: 'GaAs integrated microwave circuits', *IEEE Journal of Solid State Circuits*, 1968, **SC-3** (2), pp. 113–116

6 PENGELLY, R. S., and TURNER, J. A.: 'Monolithic broadband GaAs FET amplifiers', *Electronics Letters*, 1976, **12** (10), pp. 251–252

7 JOSHI, J. S., COCKRILL, J. R., and TURNER, J. A.: 'Monolithic microwave GaAs FET oscilllators'. *IEEE GaAs IC Symposium*, Lake Tahoe, Nevada, USA, 1979

8 HORNBUCKLE, D. P. and VAN TUYL, R. L.: 'Monolithic GaAs direct-coupled amplifiers', *IEEE Transactions on Electron Devices*, 1981, **ED-28** (2), pp.175–182

9 JAMISON, S. A., PODELL, A., HELIX, M., NG, P., and CHAO, C.: 'Inductively coupled push-pull amplifiers for low cost monolithic microwave ICs'. *IEEE GaAs IC Symposium Digest*, 1982, pp. 91–93

10 FERGUSON, D., BAUHAHN, P., KEUPER, J., LOKKEN, R., CULP, J., CHAO, C., and PODELL, A.: 'Transformer coupled high-density circuit technique for MMIC'. *IEEE MTT-S International Microwave Symposium Digest*, 1984, pp. 34–36

11 RIGBY, P. N., SUFFOLK, J. R., and PENGELLY, R. S.: 'Broadband monolithic low noise feedback amplifiers', *IEEE Microwave and Millimeter-Wave Monolithic Circuits Symposium Digest*, 1983, pp. 71–75

12 LEHMANN, R. E., and HESTON, D. D.: 'X-band monolithic series feedback LNA'. *IEEE Microwave and Millimeter-Wave Monolithic Circuits Symposium Digest*, 1985, pp. 54–57

13 PAVLIDIS, D., ARCHAMBAULT, Y., EFTHIMEROU, M., KAMINSKY, D., BERT, A., and MAGARSHACK, J.: 'A new specifically monolithic approach to microwave power amplifier design'. *IEEE Microwave and Millimeter-Wave Monolithic Circuits Symposium Digest*, 1983, pp. 54–58

14 STRID, E. W., and GLEESON, K. R.: 'A DC-12 GHz monolithic GaAs FET distributed amplifier', *IEEE Transactions on Microwave Theory and Techniques*, 1982, **MTT-30** (7), pp. 969–975

15 KENNAN, W., ANDRADE, T., and HUANG, C.: 'A miniature 2 to 18 GHz monolithic GaAs distributed amplifier'. *IEEE Microwave and Millimeter-Wave Monolithic Circuits Symposium Digest*, 1984, pp. 41–44

16 AYASLI, Y., MILLER, S. W., MOZZI, R., and HAINES, L. K.: 'Capacitively coupled travelling-wave power amplifier'. *IEEE Microwave and Millimeter-Wave Monolithic Circuits Symposium Digest*, 1984, pp. 52–54

17 MAJIDI-AHY, R., NISHIMOTO, C. K., RIAZIAT, M. *et al.*: '5–100 GHz InP coplanar waveguide MMIC distributed amplifier', *IEEE Transactions on Microwave Theory and Techniques*, 1990, **MTT-38**, pp. 1986–1993.
18 HOWARD, T. S., and PAVIO, A. M.: 'A dual-gate 2-18 GHz monolithic FET distributed mixer', *IEEE Microwave and Millimeter-Wave Monolithic Circuits Symposium Digest*, 1987, pp. 27–30
19 PAVIO, A. M. *et al.*: 'Double balanced mixers using active and passive techniques', *IEEE Transactions on Microwave Theory and Techniques*, 1988, **MTT-36** (12), pp. 1948–1957
20 HARROP, P., LESARTRE, P., and COLLET, A.: 'GaAs integrated all-FET front end at 12 GHz', *IEEE GaAs IC Symposium Digest*, 1980, paper no.28
21 KERMARREC, C. et al.: 'The first GaAs fully integrated microwave receiver for DBS applications at 12 GHz', *Proceedings of 14th European Microwave Conference*, 1984, pp. 749–754
22 NIGHTINGALE, S. J. *et al.*: 'A 30 GHz monolithic single balanced mixer with integrated dipole receiving element', *IEEE Microwave and Millimeter-Wave Monolithic Circuits Symposium Digest*, 1985, pp. 74–77
23 SCHINDLER, M. J., and TAJIMA, Y.: 'A novel MMIC active filter with lumped and transversal elements', *IEEE Transactions on Microwave Theory and Techniques*, 1989, **MTT-37** (12), pp. 2148–2153
24 HARA, S., TOKUMITSU, T., and AIKAWA, M.: 'Lossless, broadband monolithic microwave active inductors', *IEEE MTT-S International Microwave Symposium Digest*, 1989, pp. 955–958
25 HIROTA, T., TARUSAWA, Y., and OGAWA, H.: 'Uniplanar MMIC hybrids—a proposed new MMIC structure', *IEEE Transactions on Microwave Theory and Techniques*, 1987, **MTT-35**, pp. 576–581
26 HIRAOKA, T., TOKUMITSU, T., and AKAIKE, M.: 'Very small wide-band MMIC magic-T's using microstrip lines on a thin dielectric film', *IEEE Transactions on Microwave Theory and Techniques*, 1989, **MTT-10**, p. 1569–1575
27 TOKUMITSU, T. *et al.*: '3-Dimensional MMIC technology – A possible solution to masterslice MMIC on GaAs and Si', *IEEE Microwave and Guided Wave Letters*, 1995, **5**, pp. 411–413
28 WANG, H., et al.: 'A high gain low noise 110 GHz monolithic two-stage amplifier', *IEEE MTT-S International Microwave Symposium Digest*, 1993, pp. 783–785
29 DUH, K. H. G. *et al.*: 'A super low-noise 0.1 μm T-Gate InAlAs-InGaAs-InP HEMT', *IEEE Microwave and Guided Wave Letters*, 1991, **1** (5), pp. 114–116
30 WANG, H. *et al.*: 'A monolithic W-band three-stage LNA using 0.1μm InAlAs/InGaAs/InP HEMT technology', *IEEE MTT-S International Microwave Symposium Digest*, 1993, pp. 519–522
31 SHIMURA, T. *et al.*: '1W Ku-band AlGaAs/GaAs power HBTs with 72% peak power added efficiency', *IEEE MTT-S International Microwave Symposium Digest*, 1994, pp. 687–690
32 PATTON, G. L. *et al.*: '75 GHz f_T SiGe-Base HBT's', *IEEE Electron Device Letters*, 1990, **11**, pp. 171
33 LUY, J. F., STROHM, K. M., and SASSE, E.: 'Si/SiGe MMIC technology', *IEEE MTT-S International Microwave Symposium Digest*, 1994, pp. 1755–1757

34 DEVLIN, L. M., BUCK, B. J, CLIFTON, J. C., DEARN, A. W., and LONG, A. P.: 'A 2.4 GHz single chip transceiver', *IEEE Microwave and Millimeter-Wave Monolithic Circuits Symposium Digest*, 1993, pp. 23–26
35 WILLIAMS, D. A.: 'A frequency hopping microwave radio system for local area network communication', *IEEE MTT-S International Microwave Symposium Digest*, 1993, pp. 685–690
36 CHELOUCHE, M., and PLATTNER, A.: 'Mobile broadband system (MBS): trends and impact on 60 GHz band MMIC development', *Electronics and Communication Engineering Journal*, 1993, pp. 187–197
37 van NEE, R., AWATER, G., MORIKURA, M., TAKANASHI, H., WEBSTER, M., and HALFORD, K. W.: 'New high-rate wireless LAN standards', *IEEE Communications Magazine*, 1999, **37** (12), pp.82–8.
38 PILGRIM, M., and SEARLE, R. P.: 'MM-wave direct-to-home multichannel TV delivery system', *IEEE MTT-S International Microwave Symposium Digest*, 1989, pp. 1095–1098
39 WILSON, P. G., and BARNES, B. C.: 'Millimetre-wave downconverter using monolithic technology for high volume application', *IEEE MTT-S International Microwave Symposium Digest*, 1989, pp. 1099–1102
40 SKORO J.: 'LMDS: broadband wireless access', *Scientific American*, 1999, **281** (4), pp.108–9.
41 McCOY, J.W., and LERUDE, D. G.: 'An overview of local multipoint communications systems', *Proceedings of 1997 International Conference on GaAs Manufacturing Technology*, 1997, pp.2–5.
42 EVANS, J.V.: 'Proposed U.S. global satellite systems operating at Ka-Band', *Proceedings of 1998 IEEE Aerospace Conference*, **4**, pp.525–37.
43 SUSSMAN, J. N.: 'Intelligent vehicle highway systems: challenge for the future', *IEEE MTT-S International Microwave Symposium Digest*, 1993, pp. 101–104
44 RAFFAELLI, L., STEWART, E., QUIMBY, R., BORELLI, J., GESSBERGER, A., and PALMIERI, D.: 'A low cost 77 GHz monolithic transmitter for automotive collision avoidance systems', *IEEE Microwave and Millimeter-Wave Monolithic Circuits Symposium Digest*, 1993, pp. 63–66
45 SIWERIS, H.J., WERTHOF, A., TISCHER, H. *et al.*: 'Low cost GaAs PHEMT MMICs for millimeter-wave sensor applications', *Proceedings of 1998 Radio Frequency Integrated Circuits (RFIC) Symposium*, pp. 283–286.
46 SHIMURA, T., KAWASAKI, Y., OHASHI, Y. *et al.*: '76GHz flip-chip MMICs for automotive radars', *Proceedings of 1998 Radio Frequency Integrated Circuits (RFIC) Symposium*, pp. 25–28
47 WOLLITZER, M., BUECHLER, J., LUY, J.-F. *et al.*: 'Multifunctional radar sensor for automotive application', *IEEE Transactions on Microwave Theory and Techniques*, 1998, **46** (5), pp.701–708
48 CARTER, T. and CARTER. A.: 'Clear road ahead? intelligent transport systems', *IEE Review*, 2000, **46** (2), pp.9–14.
49 EASTMAN, L. F.: 'AlGaN/GaN microwave power HEMT's', *IEEE GaAs IC Symposium Digest*, 1999, pp.15–18.
50 MICOVIC, M., NGUYEN, N. X., JANKE, P. *et al.*: 'GaN/AlGaN high electron mobility transistors with f_t of 110 GHz', *Electronics Letters*, 2000, **36** (4), pp.358–9.

51 WU, Y-F., KELLER, B.P., KELLER, S. *et al.*: 'GaN-based FETs for microwave power amplification', *IEICE Transactions on Electronics*, 1999, **E82-C** (11), pp.1895–905.
52 PALMOUR, J.W., ALLEN, S.T., SADLER, R.A., PRIBBLE, W.L., SHEPPARD, S.T., and CARTER, C.H. Jr.: 'Progress in SiC materials and microwave devices', *1999 GaAs MANTECH Conference Digest of Papers*, pp. 241–4
53 TREW, R.J.: 'Experimental and simulated results of SiC microwave power MESFETs', *Physica Status Solidi A*, 1997, **162** (1), pp.409–19
54 KONIG, U.: 'Progress in SiGe heterostructure devices', *IEE Colloquium on Advances in Semiconductor Devices* (ref. no.1999/025), 1999, pp.6/1–6.
55 KOK, Y.L., WANG, H., HUANG, T.W. *et al.*: '160–190-GHz monolithic low-noise amplifiers', *IEEE Microwave and Guided Wave Letters*, 1999, **9** (8), pp. 311–13
56 MATLOUBIAN, M.: 'Advances in millimeter-wave FET MMIC technology', *Proceedings of 1999 IEEE Radio Frequency Integrated Circuits Symposium*, Anaheim, CA, USA, 1999, pp.141–144
57 MATLOUBIAN, M.: 'Advances in millimeter-wave HEMTs', *Compound Semiconductor*, 1999, **5** (8), pp.24–6
58 CHEN, Y.C., INGRAM, D.L., YAMAUCHI, D. *et al.*: 'A single chip 1-W InP HEMT V-band module', *1999 GaAs Symposium Digest*, pp.149–52
59 AGARWAL, B., SCHMITZ, A.E.. BROWN. J.J. *et al.*: '112-GHz, 157-GHz, and 180-GHz InP HEMT traveling-wave amplifiers', *IEEE Transactions on Microwave Theory and Techniques*, 1998, **46** (12), pp.2553–9
60 FENK, J.: 'Highly integrated RF–IC's for GSM and DECT systems – A status review', *IEEE Transactions on Microwave Theory and Techniques*, 1997, **45** (12), pp.2531–2539
61 RUSSER, P.: 'Si and SiGe millimeter-wave integrated circuits', *IEEE Transactions on Microwave Theory and Techniques*, 1998, **46** (5), pp.590–603
62 LARSON, L.E., and DELANEY, M.J.: 'Applications of Si/SiGe technology for high-speed communications systems', *IEEE MTT-S International Microwave Symposium Digest*, 1999, **3**, pp.1077–80
63 RAINER G., OTZFRIED, F., BEIßWANGER, S.G. *et al.*: 'RFIC's for mobile communication systems using SiGe bipolar technology', *IEEE Transactions on Microwave Theory and Techniques*, 1998, **46** (5), pp.661–668
64 SCHUPPEN, A., and ERBEN, U.: 'Enhanced SiGe heterojunction bipolar transistors with 160 GHz F_{max}', IEDM, Washington DC, 1995, pp. 743–746
65 SCHUMACHER, H., ERBEN, U., and GRUHLE, A.: 'Low-noise performance of SiGe heterojunction bipolar transistors', *IEEE MTT-S International Microwave Symposium Digest*, 1994, pp. 1167–1170
66 AHLGREN, D.C.. KING, N., FREEMAN, G., GROVES, R., and SUBBANNA, S.: 'SiGe BiCMOS technology for RF device and design applications', *1999 IEEE Radio and Wireless Conference Digest* , pp.281–4
67 STRANGE, S., and STRANGE, J.: 'Integration trends with radio applications', *Proceedings of 1998 Radio Frequency Integrated Circuits (RFIC) Symposium*, pp. 83–86
68 KIM, C.S., PARK, M., KIM, C.H. *et al.*: 'A fully integrated 1.9-GHz CMOS low-noise amplifier', *IEEE Microwave and Guided Wave Letters*, 1998, **8** (8), pp. 293–295

69 NISHIKAWA, K., TOYODA, I., KAMOGAWA, K., and TOKUMITSU, T.: 'Three-dimensional silicon MMIC's operating up to K-band', *IEEE Transactions on Microwave Theory and Techniques,* 1998, **46** (5), pp. 677–684
70 LU, L.-H., RIEH, J.-S., BHATTACHARYA, P. *et al.*: 'K-band Si/SiGe HBT MMIC amplifiers using lumped passive components with a micromachined structure', *Proceedings of 1998 Radio Frequency Integrated Circuits (RFIC) Symposium*, pp.17–20
71 REYES, A. C., GHAZALY, S. M. E., and DORN, S. J.: 'Coplanar waveguides and microwave inductors on silicon substrates', *IEEE Transactions on Microwave Theory and Techniques*, 1995, **43**, pp. 2016–2022
72 BURGHARTZ, J. N., SOYUER, M., and JENKINS, K. A.: 'Integrated RF components in a SiGe bipolar technology', *IEEE Journal of Solid-State Circuits*, 1997, **32**, pp. 1440–1445
73 RUSSER, P.: 'Silicon monolithic millimetre-wave integrated circuits', *Proceedings of 21st European Microwave Conference*, 1991, pp. 55–71
74 BÜCHLER, J., KASPER, E., LUY, J. F., RUSSER, P., and STROHM, K. M.: 'Silicon millimeter-wave circuits for receivers and transmitters', *IEEE Microwave and Millimeter-Wave Monolithic Circuits Symposium Digest*, 1988, pp. 67–70
75 LUCYSZYN, S., and ROBERTSON, I. D.: 'Monolithic narrow-band filter using ultrahigh-Q tunable active inductors', *IEEE Transactions on Microwave Theory and Techniques*, 1994, **MTT-42**, pp. 2617–2622
76 BAYAR, E., ROBERTSON, I. D., and AGHVAMI, A. H.: 'L-band GaAs MMIC amplifier design and test', *ESA Workshop on MMICs for Space Applications*, ESTEC, Noordwijk, Netherlands, 1990.
77 MUNDAY, P. D., BARNES, A. R., MOORE, M. T., BALL, G., and BROWN, A. L.: 'A 35 to 40 GHz image rejection receiver front end', *IEE Colloquium on Millimetre-Wave Circuits and Technology for Commercial Applications*, 1999, pp. 2/1–6
78 Nishikawa, K., *et al.*: 'Millimeter-wave three-dimensional masterslice MMICs', *Proceedings of 1998 Radio Frequency Integrated Circuits (RFIC) Symposium*, pp. 239–242
79 LUBECKE, V. M., MIZUNO, K., and REBEIZ, G.M.: 'Micromachining for Terahertz Applications', *IEEE Transactions on Microwave Theory and Techniques*, 1998, **46** (11), pp.1821–1831
80 BLONDY, P., BROWN, A. R., CROS, D., and REBEIZ, G.M.: 'Low-loss micromachined filters for millimeter-wave communication systems', *IEEE Transactions on Microwave Theory and Techniques*, 1998, **46** (12), pp 2283–2288
81 ARNOLD, R.G., and PEDDER, D. J.: 'Microwave characterization of microstrip lines and spiral inductors in MCM-D technology', *IEEE Transactions on Components, Hybrids and Manufacturing Technology*, **15** (6), pp. 1038–1045
82 WOOD, J., SABO, C., LONDON, A., and BARNWELL P.: 'Processing of photoimageable thick film materials', *Proceedings of the SPIE – the International Society for Optical Engineering*, 1999, **3582**, pp.783–8
83 LUCYSZYN, S. *et al.*: 'Terahertz multi-chip module (T-MCM) technology for the 21st Century?', *IEE Colloquium on Multi-Chip Modules and RFICs*, London, 1998, pp. 6/1–8

84 HINKEN J.H., 'Waveguides become integrated circuits in new space and cost saving method', *Microwave Systems News*, 1983, **13**, pp. 106-118

85 STEVENS, D., GIPPRICH, J.: 'Microwave characterization and modeling of multilayered cofired ceramic waveguides', *International Journal of Microcircuits and Electronic Packaging*, 1999, **22** (1), pp.43–8

86 LUCYSZYN, S., BUDIMIR, D., WANG, Q. H., and ROBERTSON, I. D.: 'Design of compact monolithic dielectric-filled metal-pipe rectangular waveguides for millimetre-wave applications', *IEE Proceedings H (Microwaves, Antennas and Propagation)*, 1996, **143** (5), pp. 451–453

Chapter 2

Devices and fabrication technology

I. Thayne, K. Elgaid and G. Ternent

2.1 Introduction

Advances in integrated circuit technology are the key to opening and fully exploiting new market opportunities, as exemplified recently by the massive expansion of the mobile phone sector in the telecommunications arena. The success of mobile communications is fundamentally underpinned by advances in semiconductor manufacturing technology. These have delivered key high performance components in large volume and with high yield, resulting in low unit costs. Today, the RF circuit designer has a greater choice in technology than ever before with the availability of internal corporate and external foundries offering silicon and III-V semiconductor bipolar and field effect device processes. The choice of technology ultimately depends on the application area and the component specifications. In this chapter, the wide range of active and passive device technologies utilised in contemporary MMIC realisation are reviewed, and key RF performance parameters such as bandwidth, gain, noise, power handling and linearity are compared.

2.2 Substrates and technologies

Before discussing active and passive MMIC components, it is important to mention the properties of the semiconductor substrate on which the circuits will be produced. Until relatively recently, III-V semiconductors (e.g. GaAs) were the only viable option for the realisation of RF MMIC components, as only III-V devices could deliver the performance required for efficient system operation at RF frequencies. Continual development of silicon bipolar transistor technology has changed this scenario dramatically, such that silicon-based RF MMICs are now widely available. These advanced silicon technologies are produced on wafer diameters of up to 300 mm and benefit from mature, high yield, processing methodologies. This combination results in economies of scale, which make advanced silicon technologies highly attractive for high volume, low unit cost market sectors, such as in the massively expanding areas of RF chipsets for wireless systems. Compatibility with traditional silicon technology generally requires the use of low resistivity substrates, however, which leads to major issues associated with parasitic capacitance; this can influence active device performance and also seriously compromise the loss and Q-factor of passive components, such as transmission lines and inductors.

In contrast, III-V substrates are semi-insulating and so suffer none of the parasitic capacitance, loss or Q-factor limitations mentioned above. Furthermore, 150 mm diameter GaAs substrates are now being used in a production environment, and when combined with a simpler process flow and lower mask count, this means that GaAs technologies can offer similar economies of scale to advanced silicon technologies. The detailed vertical architecture of the various active devices used in current MMIC production will be described later. In defining a vertical architecture, processes such as ion implantation, chemical vapour deposition and molecular beam epitaxy are use to engineer dopant and material composition profiles with resolutions of atomic monolayer precision [1].

In ion implantation, the substrate is irradiated with a beam of ions of given density and energy. The density determines the concentration of implanted species, whilst the energy defines the depth of the implantation. After implantation, the substrate is annealed at high temperature (around 1000° C) to 'activate' the implant, i.e. to melt and recrystallise the semiconductor substrate incorporating the implanted material into the substrate. This technique is most often used in silicon-based technology to form doped contact regions within the device; for example, in defining the emitter, base and collector regions in the silicon bipolar transistor (described in detail in Section 2.4).

Chemical vapour deposition and molecular beam epitaxy are material growth techniques in which layers of material are deposited on the surface. These techniques allow the growth of high quality III-V and silicon single crystal semiconductor substrates with abrupt changes in material composition and doping concentration. Chemical vapour deposition (CVD) is used in the production of SiGe bipolar transistors, in which the base region of a conventional bipolar transistor is formed from a layer of SiGe. Molecular beam epitaxy (MBE) is used in the growth of the high electron mobility transistors (HEMTs). In contemporary devices, the abrupt changes in material composition and the precise placement of the dopant atoms are a requirement for optimal device operation.

The realisation of an MMIC is a process in which the final circuit elements are constructed by a combination of substrate patterning followed by selective implantation, etching, and metallisation [2]. The detailed process flow is manufacturer specific, but to give some indication of the likely steps involved, the following section shows schematic cross-sections of a III-V HEMT coplanar waveguide MMIC, a III-V heterojunction bipolar transistor microstrip MMIC and a Si bipolar MMIC process, respectively, together with descriptions of the likely process flows.

2.2.1 III-V HEMT coplanar waveguide MMICs

In a III-V HEMT process, the starting semiconductor wafer generally comprises a semi-insulating GaAs or InP substrate, on which are epitaxially grown a number of thin semiconductor layers of highly controlled composition, thickness and doping, which together define the transistor vertical architecture. Full details of typical HEMT vertical architectures are described later in this chapter.

A generic HEMT coplanar waveguide (CPW) MMIC is shown schematically in Figure 2.1. This technology requires at least nine process levels, starting with the definition of the HEMT active area by selectively etching the substrate to leave islands (or mesas) of the vertical epitaxial structure that will eventually form the transistors of the MMIC. Following this isolation etching process, the 1st level metal to form the lower plate of metal-insulator-metal (MIM) capacitors is patterned and deposited. Thereafter, the dielectric for the MIM capacitors is deposited and patterned before defining the HEMT ohmic contacts. As will be shown later, relatively large value semiconductor resistors can also be formed using the combined processes of isolation and ohmic contact formation. Next, the gates of the HEMT are patterned — usually a very shallow etch is performed after gate definition, prior to metallisation. This gate recess etch is a key part of the transistor formation, as it plays a major role in defining the threshold voltage and breakdown characteristics of the device. Following gate metallisation, thin film metal resistors are patterned and defined before the whole circuit is interconnected using thick 2nd level metal. The 2nd level metal is used to form the coplanar waveguide transmission lines, small value interdigitated capacitors and the top plate of MIM capacitors.

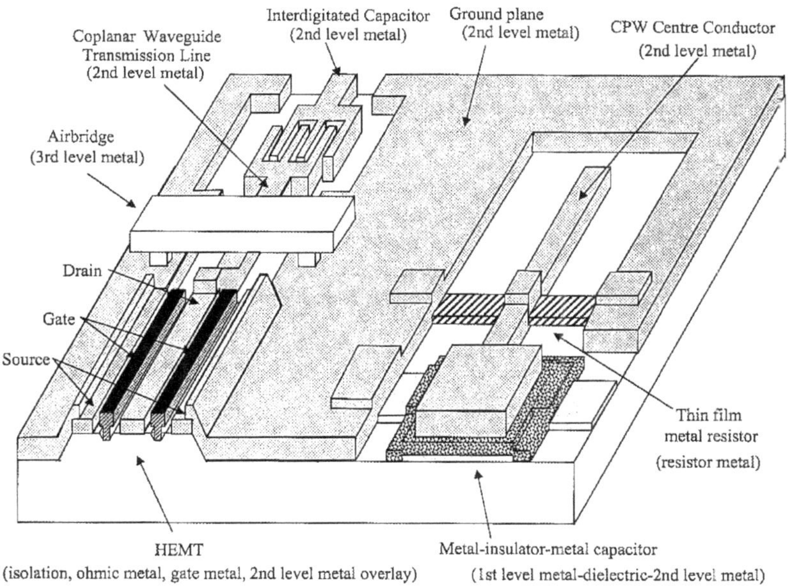

Figure 2.1 Schematic of III-V HEMT coplanar waveguide MMIC

Finally, in a CPW topology, airbridges can be formed by electroplating, to ensure that the generation of unwanted modes is suppressed at discontinuities in the circuit. This 3rd level metal is usually formed in a two-step process, in which the airbridge posts are first formed before defining the airbridge spans. In many cases, the use of CPW removes the need to thin the wafer and to fabricate through-substrate via-holes, which are needed for microstrip-based MMICs, as described in Section 2.2.2. In practical applications, the front face of the MMIC will be passivated with a protective coating of silicon nitride (Si_3N_4) or polyimide before being tested, diced and packaged.

2.2.2 III-V HBT microstrip MMICs

As in the case of the III-V HEMT MMIC process, outlined in Section 2.2.1, the starting semiconductor wafer of the III-V heterojunction bipolar transistor (HBT) microstrip technology is based on a semi-insulating GaAs or InP substrate, on which are grown a number of thin layers of semiconducting material, whose individual properties are tailored to optimise device performance. Details of typical HBT vertical architectures are described later in this chapter. In contrast to the coplanar waveguide technology, the microstrip MMIC process flow requires wafer thinning and backside wafer processing to enable the formation of via-holes, which are needed to ground the active and passive structures realised on the front-face of the substrate. A cross-sectional schematic of a generic HBT microstrip MMIC is shown in Figure. 2.2. This type of HBT technology requires up to 15 process levels, starting with repeated patterning, metal deposition and substrate etching to define the emitter, base and collector contacts. The main challenge at this stage is to accurately, uniformly and repeatably control the etching step to the base region when defining the base contact, as the base layer in the vertical architecture is typically less than 20 nm thick and buried several hundred nanometres below the wafer surface. Semiconductor resistors can be defined during the above processing stages by contacting regions of the collector layer in the vertical architecture.

Having defined the active devices, thin film metal resistors can next be realised, if required, using an identical process to that described in Section 2.2.1. Next, the 1st level metal is defined and deposited. This layer is used to form the lower plate of metal-insulator-metal capacitors, can be used for interconnects, and acts as bondpad metallisation, which can be contacted using via-holes to the ground plane on the backside of the wafer. The process flow continues with the deposition and patterning of Si_3N_4 to form the insulator layer of metal-insulator-metal capacitors. Subsequently, a thick layer of polyimide is spun-on and patterned. This interlayer dielectric allows the formation of metal cross-overs, and can also be used to form MIM capacitors with lower values than those produced using the Si_3N_4 layer, as the polyimide is generally much thicker and has a lower relative permittivity. The final front-face processing step involves definition and deposition of thick 2nd level metal, which is used for the formation of low-loss microstrip transmission lines, spiral inductors and the top plate of MIM capacitors. Next, the wafer, originally of thickness in the range 400–600 μm (depending on its diameter) is thinned to around 100 μm. Via-holes to connect to ground contacts on the front-face are then formed by etching through the wafer. Finally, a thick groundplane metallisation is formed on the back of the wafer, usually by electroplating.

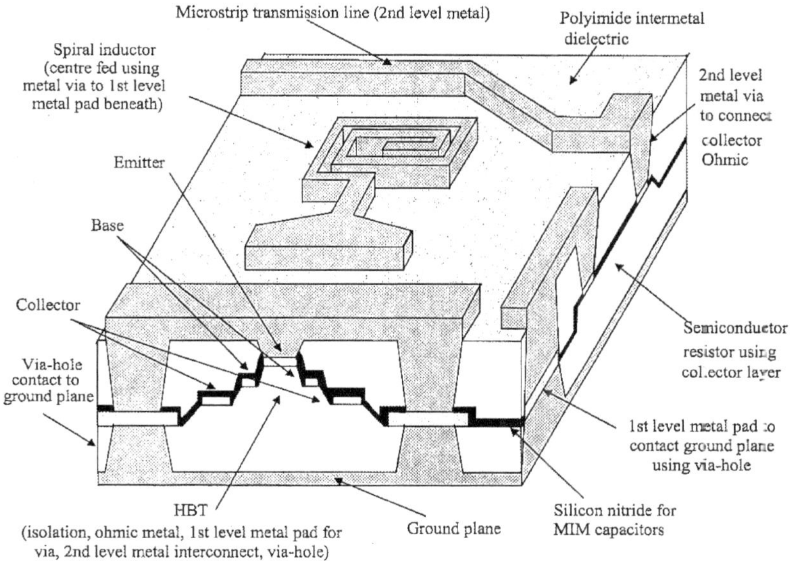

Figure 2.2 Schematic of a III-V HBT microstrip MMIC technology

From the above, it is clear that microstrip technology is more complex than the coplanar waveguide MMIC process flow, yet microstrip technology has, until recently, been the dominant MMIC technology. This is partly historic, as the wide range of active and passive models, originally developed for hybrid microstrip circuit design, are easily scaled for MMIC design. In contrast, models for coplanar waveguide elements have traditionally been rather scarce, due to the complexities and lack of scalability when describing the airbridge properties, which are so crucial to coplanar waveguide technology. This situation is beginning to change, however, as more powerful electromagnetic simulators make the analysis and performance prediction of coplanar waveguide components more straightforward.

2.2.3 Silicon bipolar transistor MMICs

Silicon bipolar technology has continued to evolve, to address the needs of contemporary analogue and digital applications. Most recently, the development has been strongly driven to meeting the requirements of the wealth of high volume consumer electronics applications that have begun to appear in the low GHz frequency range, such as RF chipsets for mobile phones and Bluetooth™ technologies. As will be shown in Section 2.4, advanced silicon transistor technology is capable of providing the performance required for these applications, but, as mentioned previously, one of the major hurdles to applying silicon technology in these areas is overcoming the challenges imposed by the conducting silicon substrate, which can result in significant device capacitance, high loss in transmission lines and low Q passive components. It is testament to the determination of the silicon bipolar technologist that many of these limitations have been overcome by the development of highly advanced, complex process flows. Besides optimising transistor design to minimise intrinsic capacitance, major breakthroughs have been the development of multilayer metal technologies, which both physically separate and screen critical elements from the conducting substrate, and the introduction of low permittivity insulating layers to reduce interconnect capacitance.

A schematic cross-section of a contemporary RF silicon bipolar MMIC technology is shown in Figure 2.3. Full details of this type of process, which can run to some 40 levels, (compared to the 9 levels of a III-V coplanar technology and 15 levels for a III-V microstrip process) is beyond the scope of this text, but includes shallow and deep trench isolation, multiple implantations for contact formation, at least two polysilicon levels for contact and resistor formation, SiO_2 layers for isolation, often at least two low permittivity dielectric layers, three or four levels of metallisation and complex via intermetal studs to interconnect between metal layers. Figure 2.4 is a scanning electron micrograph image through a contemporary silicon bipolar RF MMIC which clearly shows three levels of metallisation and stud-interconnect technology.

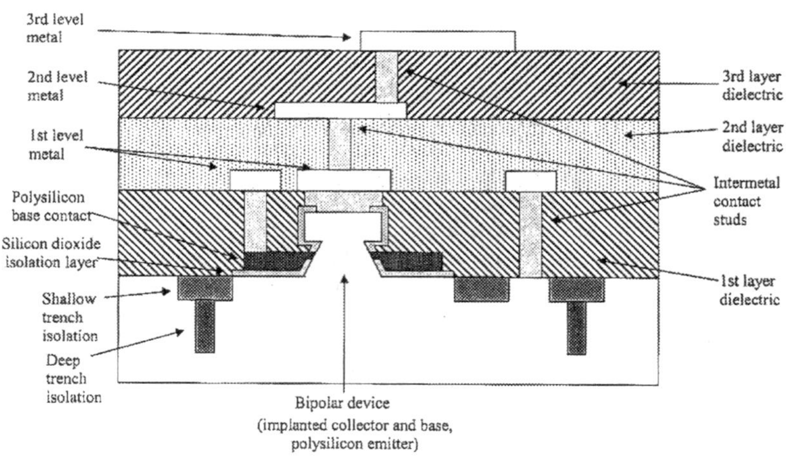

Figure 2.3 Schematic cross-section of a silicon RFIC technology

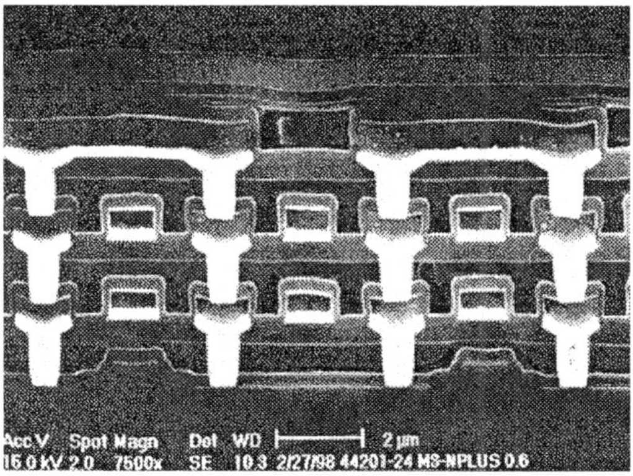

Figure 2.4 Scanning electron micrograph of advanced silicon RFIC technology (Courtesy of and copyright to Philips Semiconductors)

The long history of exceptional success, within the silicon industry, in overcoming technological challenges has been key to developing the types of advanced bipolar technology described above. This type of technology is also integrable with CMOS, and so offers a powerful combination of intrinsic RF capability together with massive signal processing capability — all on the same chip. It is fair to say that present III-V technology cannot compete with this level of complexity. However, the strengths of the III-V technologies are in their intrinsic levels of performance at RF, which are superior to silicon bipolar technologies, and the suitability for integration of optoelectronic devices. Most recently, the introduction of SiGe heterojunction bipolar technologies has narrowed the performance gap with III-V components, and currently, a fierce battle for market share in the low GHz chipset arena is raging. It would seem that the diverse range of microwave and millimetre-wave applications will be sufficiently large that both silicon and III-V based MMIC technologies will succeed in the coming years, but the diversity of technology available to the MMIC designer is now greater than it has ever been in the past.

Having reviewed the types of technology confronting today's MMIC designer, the remainder of this chapter will focus on describing the various passive and active elements, and their derivatives, in more detail.

2.3 Passive lumped-elements

In an MMIC, circuit functionality is achieved by integrating passive and active elements, such as resistors, capacitors, inductors and transistors. These elements can be interconnected using transmission lines. This section focuses on the realisation of lumped-element passive components, highlighting typical structures, together with an insight into the typical element values, and an overview of the role parasitic effects have on passive lumped-element performance at microwave frequencies.

2.3.1 Resistors

Resistors can be produced using either deposited thin metal films, such as NiCr or TaN, by using doped regions of the semiconductor substrate, or layers of polysilicon in the case of silicon RF processes. In all cases, the resistor value is determined by the geometry of the material, as shown in Figure 2.5.

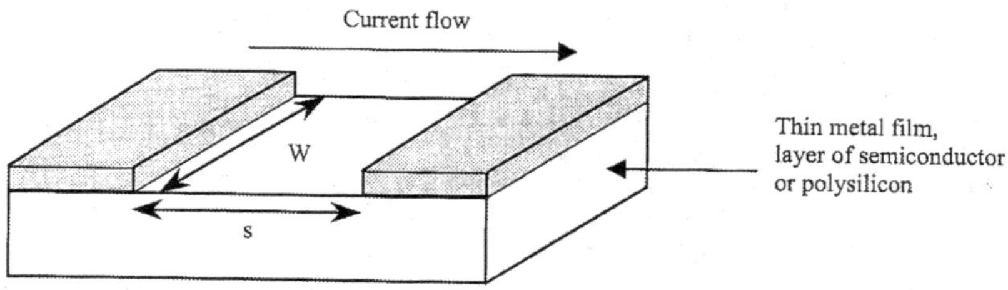

Figure 2.5 Schematic of a generic resistor

The resistance is given by:

$$R = R_{sh}\frac{s}{W} + 2R_c \tag{2.1}$$

where R_{sh} is the sheet resistance of the metal film or doped semiconductor region
s is the separation between ohmic contacts (defining the length of the resistor parallel to current flow)
W is the width of the resistor in the direction perpendicular to current flow
R_c is the resistance of the contact at either end of the structure

In the case of semiconductor resistors, a linear current voltage relationship only exists for electric field strengths of less than 1 $kVcm^{-1}$. To avoid non-linearity, the resistor length should be chosen so that the field across the structure is less than 1 $kVcm^{-1}$, and thus a resistor that is required to drop 2 V should be at least 20 μm long. The choice of resistor structure (thin film or semiconductor) will depend on the desired value. Thin metal films typically have a sheet resistance in the range 10–100 Ω/square and thus are useful for relatively low values of resistor (up to a few hundred ohms). Larger resistor values (e.g. for biasing circuitry) can be realised using layers in the substrate's vertical architecture. Typical sheet resistances of such semiconductor layers are in the range 200 to 600 Ω/square, and so are well suited to realising resistances up to a few kilohms. A simple RF equivalent circuit model for a resistor is shown in Figure 2.6. In this model, C represents the capacitance (in the 1-20 fF range, providing parasitic capacitances are minimised) across the resistor, whilst the series inductance (of up to 0.5 nH per mm, dependent on the detailed cross-section of the resistor) arises from the finite geometry of the structure.

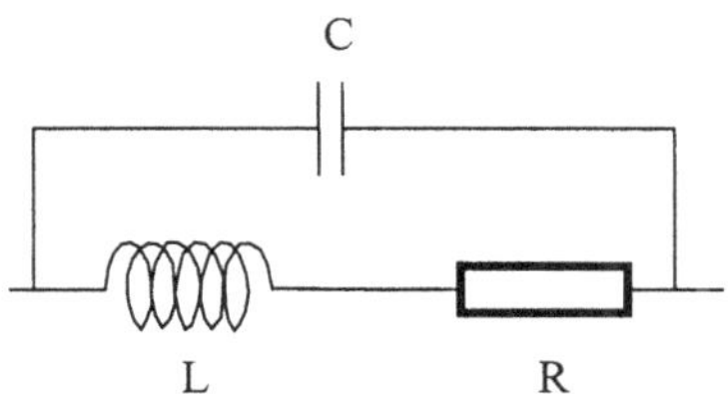

Figure 2.6 Lumped-element equivalent circuit of a resistor

2.3.2 Capacitors

2.3.2.1 Metal-insulator-metal (MIM) capacitors

These capacitors are formed by sandwiching a layer of dielectric (usually silicon nitride, polyimide, silicon dioxide or a combination of these) between two metal plates, as shown in Figure 2.7. Typical fabrication geometries and dielectric thicknesses mean that MIM structures are generally used for capacitance values in the range 0.1-20 pF.

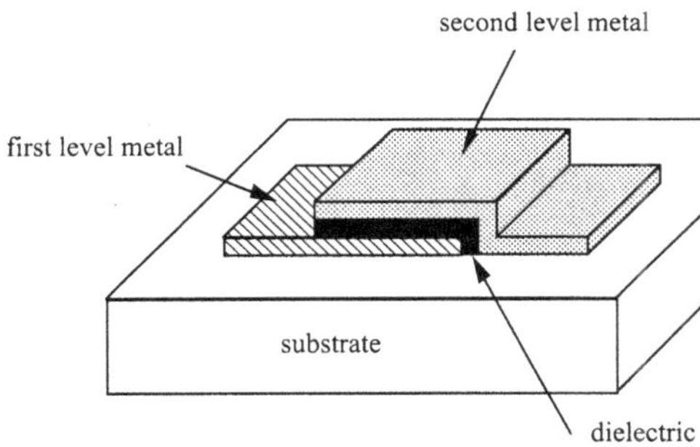

Figure 2.7 Schematic of a metal-insulator-metal capacitor

For a top-plate area A, dielectric thickness d and relative permittivity ε_r, the structure is, to first order, a parallel plate configuration, whose capacitance C is given by:-

$$C = \frac{A\varepsilon_o\varepsilon_r}{d} \tag{2.2}$$

where ε_o is the free space permittivity

At RF, the parasitic inductance and resistance of the MIM capacitor can be represented in the equivalent circuit model of Figure. 2.8.

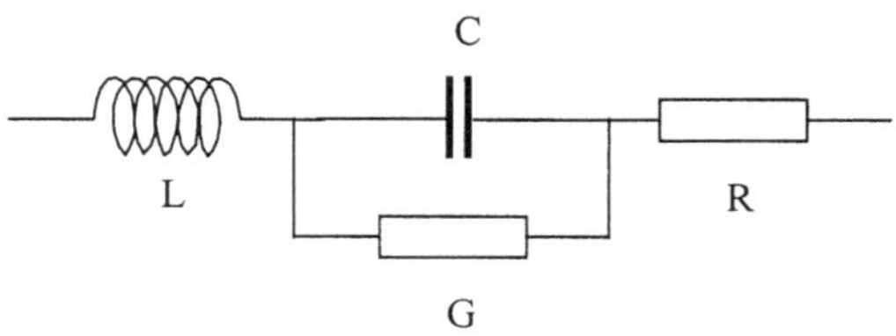

Figure 2.8 Lumped-element equivalent circuit of a capacitor

The conductance, G, associated with leakage currents and dielectric losses of the insulating layer can be neglected in most cases. The series resistance is determined by the resistivity of the first and second level metals, scaled by the geometry of the capacitor; typical values are in the range 0.5–2 Ω, depending on the layout. The parasitic inductance can be estimated by considering the capacitor to be similar to a length of microstrip line, and that the current is continuous through the dielectric with a return path via the groundplane of the circuit. In this situation, the design equations for microstrip lines discussed in Chapter 3 can be used to deduce an inductance per unit length of around 0.8 pH/μm, depending on the capacitor geometry and substrate thickness. Typical geometries for MIM capacitors have plate areas ranging from 10 μm^2 to 200 μm^2, with dielectric thicknesses ranging from 150 nm for Si_3N_4 to 10 μm for polyimide, resulting in capacitance values in the range 0.1 pF to 50 pF.

2.3.2.2 Interdigitated capacitors

Figure 2.9*a* shows an interdigitated capacitor comprising a number of thin parallel conducting strips of length l, interlinked alternately to give a total structure width w. The capacitance of this structure may be expressed as:

$$C(pF) = \frac{\varepsilon_r + 1}{w} l[(N-3)A_1 + A_2] \tag{2.3}$$

where A_1 and A_2 in pF are given by $8.85 \times 10^{-12} w$ and $9.92 \times 10^{-12} w$, respectively, with w in cm in both cases. This approximation is valid for large substrate thickness h, in relation to the strip spacing s ($h/s > 100$). Figure 2.9*b* compares the measured and calculated capacitance for interdigitated capacitors structures of various finger lengths l of width 2 μm and finger separation 1 μm.

The lumped-element equivalent circuit of Figure 2.8 can also be used to describe the interdigitated capacitor. The series resistance can be calculated from the resistivity of the metal used to form the capacitor, together with the detailed capacitor geometry, i.e. the number of parallel fingers, their length and width. The parasitic inductance can be estimated along similar lines to that described above for the MIM capacitor. Current flow can be envisaged as continuous from one set of interdigitated fingers to the other, and thus the structure's inductance can be estimated using the microstrip design equations of Chapter 3, by assuming the line length is equal to the interdigitated finger length l, with a line width of w. For typical geometries, this would result in parasitic inductances in the range 20–50 pH. Typical fabrication geometries and tolerances result in interdigitated structures being used to realise capacitor values in the 10–100 fF range. Outside this range, parasitic contributions tend to dominate the structure's performance.

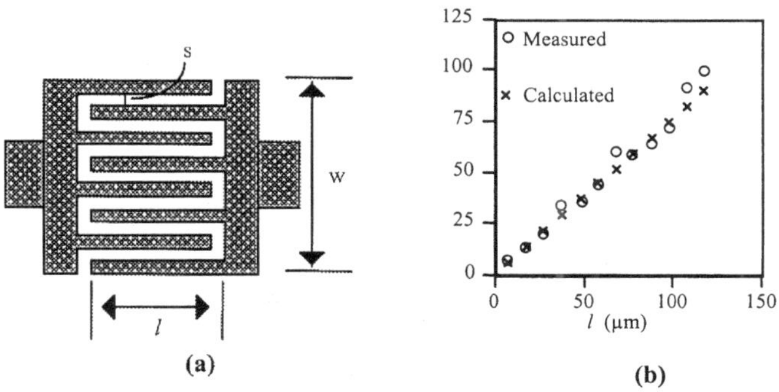

Figure 2.9 Interdigitated capacitor: (a) layout and (b) comparison between the measured and calculated capacitance as a function of the finger length (l)

2.3.3 Spiral inductors

The spiral inductor is a crucial feature in a range of MMICs, particularly for low GHz applications such as Bluetooth™ and mobile phone chip sets, where it can be used as a compact matching element. The requirements for inductors in MMICs are:

1. large inductance with small area
2. high resonance frequency (f_R)
3. high Q-factor.

Recently, massive investment and progress has been made to produce high performance inductors in silicon-based MMIC technology, due to the rapid explosion of low GHz, high volume applications. The silicon-based solutions, which will be summarised below, utilise highly sophisticated and complex processing strategies to produce inductor structures with performance comparable to that of III-V technologies. This in turn has driven III-V based inductor technologies to even higher levels of performance.

2.3.3.1 Spiral inductors in III-V technology

The realisation of a spiral inductor using III-V MMIC technology requires a minimum of two layers of metal, together with an airbridge or a dielectric via to establish a connection to the centre of the inductor. Spiral inductors can be realised in circular, octagonal or square shapes. Given their complex nature, theoretical descriptions of the inductance of spiral structures give estimates to an accuracy of only around 20%, and for this reason, an experimental approach to characterising these structures is usually taken. In this methodology, a range of structures with various geometries are produced and characterised over the frequency range of applicability, and fitted to a lumped-element equivalent circuit for the spiral inductor, an example of which is shown in Figure 2.10. From this, general trends on the impact of geometry on inductance, resistance and capacitance can be extracted.

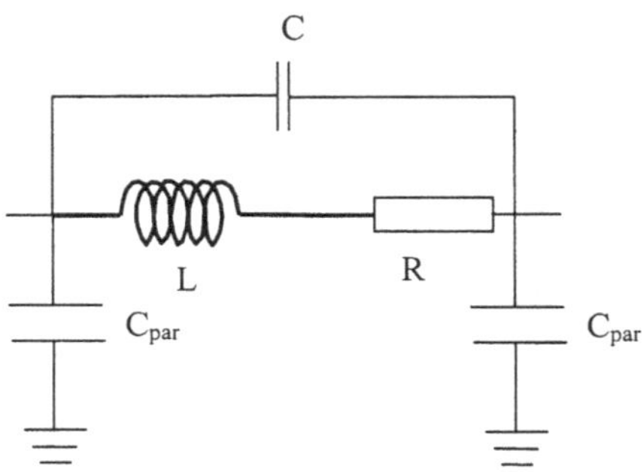

Figure 2.10 RF equivalent circuit of a spiral inductor

The simplest form of spiral inductor can be realised using first layer metal deposited directly on the semiconductor surface. However, this structure suffers from relatively high inter-turn capacitance through the substrate, which reduces both the Q-factor and self-resonant frequency of the inductor. To overcome this limitation, the inductor can be realised on a thick layer of dielectric, such as silicon nitride or polyimide. Both these dielectric materials have lower relative permittivity than III-V semiconductors, and thus, the inter-turn capacitance can be reduced. Alternatively, if the MMIC technology includes an air-bridge process, the spiral inductor can be suspended in air, as shown in Figure 2.11, to further increase the Q-factor by reducing the parasitic capacitance. In these ways, inductance values in the range 0.1 nH to 10 nH can be realised without seriously compromising the self-resonant frequency (limited by capacitive inter-turn coupling) and Q.

Figure 2.11 Spiral inductor using air-bridge technology

The 3.5 turn inductor of Figure 2.11 has an inductance of 2.3 nH and a Q-factor of 25 at a resonant frequency of 11 GHz. Inductance values up to 25 nH can be produced using stacked spiral approaches. This is most easily realised in processes incorporating thick dielectric layers, but have also been demonstrated using airbridge technologies. The negative impact of stacked spiral structures is that the increased interturn capacitance reduces the self-resonant frequency. As an example, a stacked version of the above 3.5 turn planar inductor, although having an inductance of around 9–10 nH, would typically have a self-resonant frequency of around 6–7 GHz.

2.3.3.2 Spiral inductors on low resistivity (CMOS grade) silicon

Spiral inductors on low resistivity (CMOS grade) silicon have been investigated since 1990. The development of wireless communications in the low-GHz range and the improved high frequency operation of silicon based devices have provided a motivation for developing passive RF components for silicon MMICs. The low resistivity silicon substrate causes substantial transmission-line losses and large parasitic capacitance at high frequency and so limits both the Q-factor and self-resonant frequency of spiral inductors. To overcome these limitations, a number of techniques have been developed to improve the quality of spiral inductors fabricated on a low resistivity silicon substrate, including:

1. multiple metal layers
2. bulk micromachining to remove resistive material under inductors
3. spin-on dielectrics
4. patterned polysilicon groundplanes.

The basic design of spiral inductors on silicon is very similar to that on a semi-insulating III-V substrate, in that the conductor width, spacing and number of turns are designed for a given inductance. In general, increasing L will decrease Q and resonant frequency, f_R, due to the introduction of additional parasitic capacitance and resistance. The equivalent circuit representation of a spiral inductor on silicon is similar to that on a III-V substrate, except for the addition of the spiral–substrate capacitance and substrate resistance.

Initial attempts to fabricate spiral inductors using a standard silicon process employed a single layer spiral on 1.7 μm of SiO_2 on CMOS grade silicon, with 1.8 μm aluminium metallisation. Using today's standard technology, multiple metal layers have been used to improve on inductor performance. By using up to five levels of metallisation, each separated by a thick layer of oxide, the inductor is separated further from the substrate, decreasing substrate loss. In addition, the shunt-connected levels of metal reduce resistive loss in the spiral. Further reduction in metal losses can be achieved using copper or aluminium/copper metallisation. Table 2.1 summarises the performance of state-of-the-art inductors.

It has also been shown that by etching the silicon substrate in spaces between turns increases inductor performance, but this deviates somewhat from standard silicon processing. Other novel techniques using standard silicon processing have been used to further improve inductor quality. A wide range of these techniques is described in Chapter 3.

Table 2.1 Summary of inductor performance on CMOS grade silicon.

L (nH)	Q	f_Q (GHz)	f_R (GHz)	Technique
16	4.3	0.8	2.9	Standard Si process, 2 level inductor, Al conductors
10	3	0.9	2.47	Single layer of SiO_2, standard process
10	17	2	11	100 µm layer of polyimide, 1 level of Au metallisation
9	9.5	1.8	9	Standard silicon process, single Al metal on layer 3
7.5	5	2	6.8	Al on 5.2 µm oxide with no ground shield
7.4	6.8	2	3.6	Al on 5.2 µm oxide with polysilicon ground shield
1.95	9.3	4	20	Standard Si process with 3 layers AlCu
1.9	8	4.1	9.7	Single layer of SiO_2, standard process
1.45	24	2.3	24	Standard Si process with 5 levels of Al
1.35	18	3.7		Standard Si process with 1 level of Cu
1.2	15	6	25	7.5 µm layer of polyimide with 1 level of Au

Patterned ground shields, placed directly on the substrate underneath the spiral, have been shown to shield the electric field from the substrate, reducing energy loss and increasing Q of a given spiral inductor by 10-33%. Figure 2.12 shows a cross-section of an inductor with patterned polysilicon ground-plane using a standard three-layer silicon process. Alternative dielectrics, such as spin-on polyimide and hydrogen silsequioxane (HSQ), with Au metallisation, have been introduced recently with performance comparable to III-V technologies. Circuit demonstrators, such as VCOs and LC bandpass filters, have been fabricated using multiple layers of metal, patterned ground shields and centre-tapped designs, all aimed at maximising inductor performance. It seems inevitable that multilayer metal and dielectric processes, patterned ground shields and novel inductor structures will proliferate in the future.

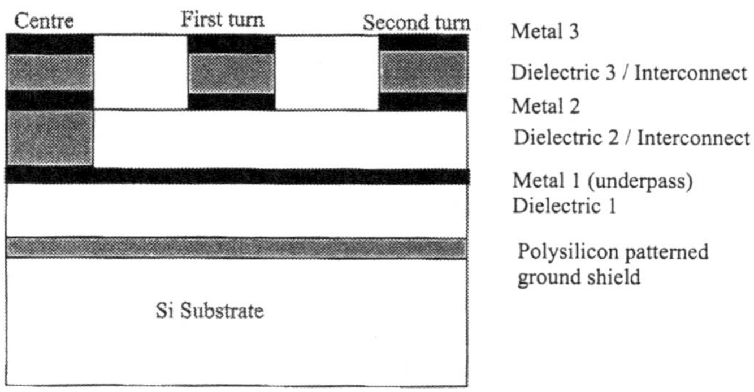

Figure 2.12 Two-level shunt connected spiral inductor with patterned polysilicon ground plane using a standard three-layer silicon process

2.4 Bipolar transistors

More than at any time in the past, the MMIC designer of today has a wide choice of active device technologies when designing a circuit. Bipolar and field effect transistor technologies, based on Si, SiGe, GaAs and InP, are available for MMIC production using both open foundry and internal corporate processes. In the following sections, the current range of available active device technologies and key parameters of device operation are presented, enabling a performance comparison to be made.

In many applications, bipolar devices have performance advantages over field effect devices, such as HEMTs: thus, the MMIC designer should be aware of the relative merits of competing technologies, in order to make an informed choice of process for a given application. The financial value of the low GHz markets has resulted in a major push in the development of advanced silicon bipolar processes. The raw performance of current silicon bipolar transistors is adequate to meet the specification requirements of many of the new wireless applications, and, when combined with the high levels of integration, yield and economies of scale, present a strong challenge to III-V technologies. The recent emergence of SiGe heterojunction bipolar MMIC technology has resulted in even more fierce competition in the new wireless market sectors. The early years of the new millennium will be a fascinating time, as the bipolar and field effect device technologies battle for supremacy and incorporation into the ever-increasing array of consumer electronics products which depend on MMICs for their operation.

This section begins with an overview of bipolar transistor operation, highlighting key aspects of the device that influence performance, before summarising the range of bipolar technologies currently available.

2.4.1 Overview of bipolar transistor operation

Since its invention in 1948, the bipolar transistor has been a cornerstone of modern electronics. Even today, in its highly evolved form, this device looks set to play an important role in GHz frequency MMICs. The bipolar transistor is a three-terminal device, with emitter, base and collector contacts as shown diagrammatically in Figure 2.13, and schematically in cross-section in Figure 2.14.

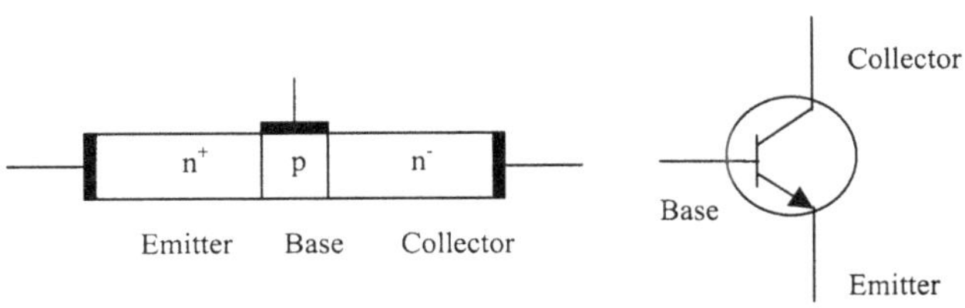

Figure 2.13 Diagrammatic representation of a bipolar transistor

The emitter, base and collector regions have a *p-n-p* or *n-p-n* doping strategy, and the device comprises two back-to-back *p-n* junctions. The key aspect of the bipolar transistor is that the two *p-n* junctions interact, as the base region is designed to be much smaller than the diffusion length of minority carriers in the base, which are electrons in an *n-p-n* device and holes in a *p-n-p* transistor. In general, the doping levels in the three regions of the bipolar device are designed so that the emitter is more heavily doped than the base, which is in turn more heavily doped than the collector. An idealised doping profile schematic is shown in Figure 2.15. The choice of doping strategy and the underlying principles of bipolar device operation can be understood with reference to the band diagram of Figure 2.16.

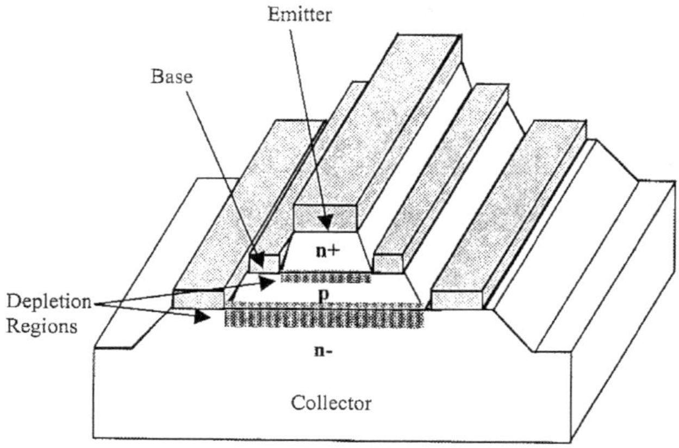

Figure 2.14 Cross-sectional schematic of a bipolar transistor

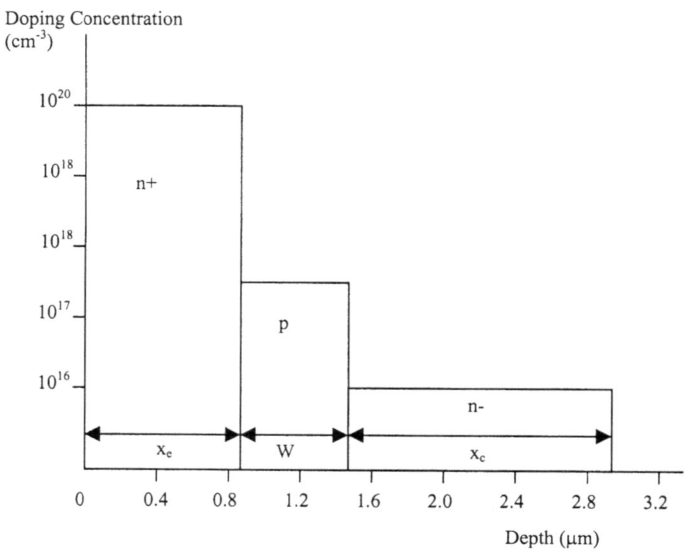

Figure 2.15 Idealised doping strategy of a bipolar transistor

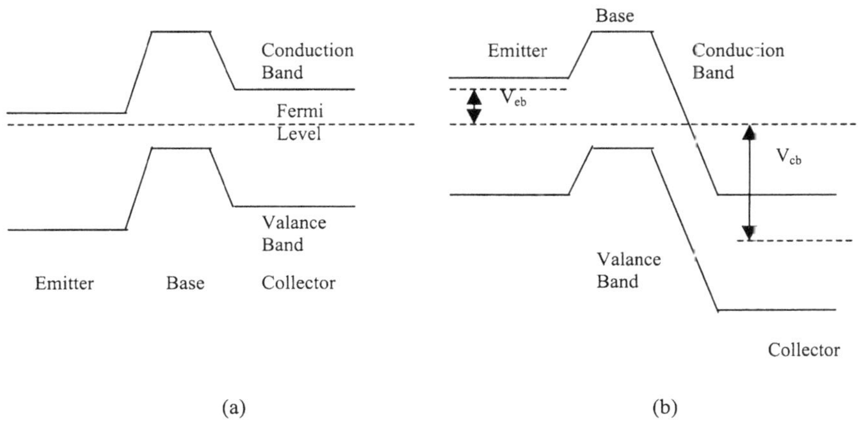

Figure 2.16 Energy band diagrams of an npn *bipolar transistor (a) in equilibrium and (b) under typical operating conditions*

Under typical operating conditions, the emitter–base junction is forward biased, whilst the collector–base junction is reverse biased. In this configuration, electrons will be injected from the emitter into the base region and will diffuse towards the collector. As the base is designed to be thin compared with the minority carrier diffusion length, nearly all of the injected electrons reach the collector–base junction, where they are rapidly swept into the collector contact by the strong electric field in the reverse-biased collector–base junction. Thus, the emitter and collector currents are practically equal. Quantitatively, this operation can be described as follows.

For the *n-p-n* device of Figure 2.16, at the forward-biased emitter–base junction, the electron concentration in the base is:

$$n_{be} \equiv n_b(x_e) = n_{bo}e^{\frac{V_{be}}{V_{th}}} >> n_{bo} \tag{2.4}$$

where n_{bo} is the equilibrium concentration of electrons in the base, V_{be} is the base–emitter voltage and V_{th} is the thermal voltage, (k_BT/q). As V_{be} is positive, n_{be} is very much greater than n_{bo}.

Similarly, at the reverse-biased collector–base junction, the electron concentration in the base is:-

$$n_{bc} \equiv n_b(W + x_e) = n_{bo}e^{\frac{V_{cb}}{V_{th}}} << n_{bo} \tag{2.5}$$

As V_{cb} is negative, n_{bc} is very much less than n_{bo}.

As there is an electron concentration gradient across the base region, a diffusion current exists. Given that the base is much thinner than the diffusion length, carrier recombination in the base is very small and so, to first order, the electron concentration varies linearly across the base. Also, as $n_{be} >> n_{bc}$, the linear variation of electron concentration across the base of width W can be written as:

$$n_b(x) = n_{be} \frac{W - x + x_e}{W} \tag{2.6}$$

The electron diffusion current across the base can be written as:

$$I_{nb} = qD_n \frac{\partial n_b}{\partial x} A \tag{2.7}$$

where, q is the electron charge, D_n is the electron diffusion coefficient and A is the cross-sectional area of the device.

Substituting eqn. (2.4) into eqn. (2.6), then differentiating in eqn. (2.7), gives:

$$I_e \cong I_c \cong I_{nb} = \frac{qD_n n_{bo} A e^{\frac{qV_{be}}{k_B T}}}{W} \tag{2.8}$$

Thus, a change in the emitter–base voltage leads to an exponential change in the emitter and collector currents. Although electron recombination in the base is small, and therefore the emitter and collector currents are roughly equal, a base current will flow as a result of holes from the base being injected into the emitter across the forward-biased emitter–base junction. The analysis of this situation is very similar to that above, where the base current is determined by the hole diffusion current flowing due to the concentration gradient of holes in the n-doped emitter region of the device. Assuming that the emitter width, x_e, of Figure 2.15 is much less than the hole diffusion length (which is the case for the vertical architecture of practical devices), the base current can be written as:

$$I_b = \frac{qD_p p_{eo} A e^{\frac{qV_{be}}{k_B T}}}{x_e} \tag{2.9}$$

where D_p is the hole diffusion coefficient and p_{eo} is the equilibrium concentration of holes in the emitter.

Comparing rough orders of magnitude in eqns. (2.8) and (2.9), W and x_e are of comparable value. However, the emitter is much more heavily doped than the base, and thus p_{eo} is much smaller than n_{bo}. Therefore, the base current is much smaller than the emitter and collector currents.

2.4.1.1 DC operation

Armed with the above equations, two scenarios can be envisaged whereby the bipolar device can be configured to operate as an amplifier. In the first situation, a small voltage is applied between the emitter and base, whilst connecting a load between the collector and base; this common-base configuration is shown in Figure 2.17. With a suitable load, the voltage developed at the output can be larger than the emitter–base input voltage, so this configuration has voltage gain. In addition, the input power is proportional to I_eV_{be}, whilst the output power is related to I_cV_{load}. As V_{load} can be larger than V_{be}, and I_e is roughly equal to I_c, the circuit also provides power gain.

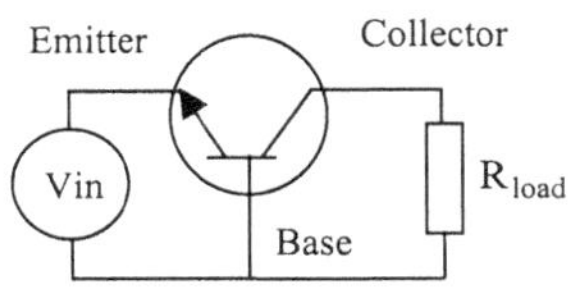

Figure 2.17 Common-base bipolar transistor amplifier

The second configuration is shown in Figure 2.18. This is the common-emitter topology, which can be used as a current amplifier. The input current is the base current, given by eqn. (2.9). The output current is the collector current, given by eqn. (2.8). The ratio is the current gain, $\beta = I_c/I_b \approx I_e/I_b$, which can be written as:

$$\beta \approx \frac{D_n N_{de} x_e}{D_p N_{ab} W} \tag{2.10}$$

where N_{de} and N_{ab} are the *n*- and *p*-doping concentrations in the emitter and base, respectively.

As mentioned above, the emitter is much more heavily doped than the base, and thus the current gain is much greater than unity. Also, it is worth noting that $D_n/D_p = \mu_n/\mu_p$, where μ_n and μ_p are the electron and hole mobilities, respectively. In most semiconductors, the electron mobility is higher than the hole mobility, and thus *n-p-n* bipolar devices are generally used in preference to *p-n-p* devices to achieve higher levels of current gain.

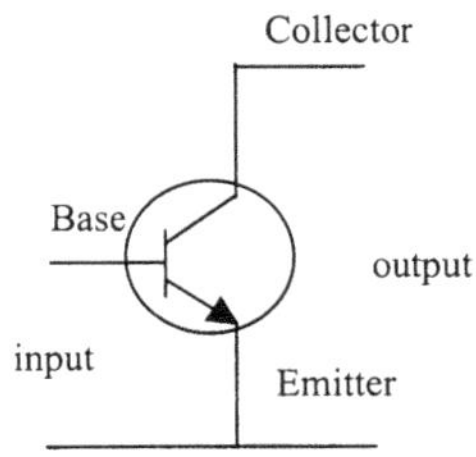

Figure 2.18 Common-emitter configuration of a bipolar transistor

Eqn. (2.10) shows that to obtain a large current gain:

1. electron mobility should be increased
2. doping concentration in the emitter should be maximised
3. base doping should be minimised
4. base width should be reduced.

However, additional considerations have to be made in optimising the device's vertical architecture. Firstly, increasing the emitter doping results in modifications of the material properties of the emitter. In particular, the effective bandgap of the layer is reduced. In silicon, the bandgap reduction is given by:

$$\Delta E_g = \frac{3q^2}{16\pi\varepsilon_s}\left(\frac{q^2 N_d}{\varepsilon_s k_B T}\right)^{\frac{1}{2}} \tag{2.11}$$

This bandgap reduction results in an increase in the concentration of holes injected into the emitter, and thus, the band-gap narrowing reduces the maximum current exponentially gain to:

$$\beta \approx \frac{D_n N_{de} x_e}{D_p N_{ab} W} e^{\left(\frac{-\Delta E_g}{k_B T}\right)} \tag{2.12}$$

Secondly, reducing the base doping and width result in an increase in the base resistance, which will limit the high frequency performance of the device. This will be discussed later.

The above analysis has shown that the emitter, base and collector currents flowing in a bipolar device are determined by the voltages applied across the transistor junctions. This analysis allows the device's family of current–voltage (I–V) curves to be constructed. Figure 2.19 shows the common-base output characteristics, in which the collector current is plotted as a function of collector–base voltage, for various emitter currents. The collector current is virtually identical to the emitter current, even down to zero collector–base voltage, where any excess electrons in the base are extracted by the collector. To reduce the collector current to zero, it is necessary to forward bias the collector–base junction to balance the electron concentrations at the emitter–base and collector–base junctions. In this way, there will be no electron diffusion current through the base.

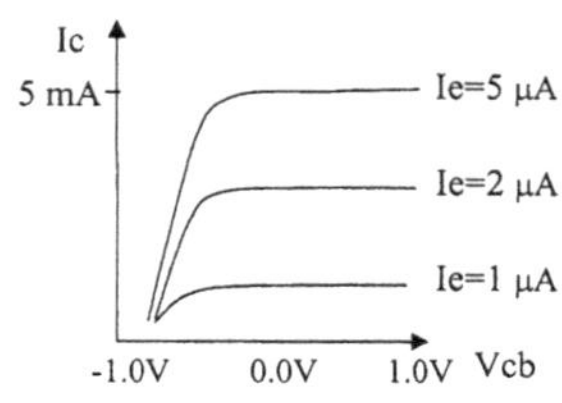

Figure 2.19 Common-base characteristics of a bipolar transistor

Figure 2.20 shows the common-emitter output characteristics, where the collector current is plotted as a function of collector–emitter voltage for various base currents. In this configuration, a certain forward base–emitter voltage is required to give a constant base current. For low collector–emitter voltages, therefore, the collector–base junction may actually be forward-biased, limiting the collector current as described above. Increasing the collector–emitter voltage to the point that the collector–base junction becomes reverse-biased, results in a rapid increase in collector current. Increasing the emitter–collector voltage beyond this situation, a preliminary inspection of eqn. (2.10) would suggest a constant collector current dependent on the current gain of the device. However, closer consideration of eqn. (2.10) shows that the device β is dependent on the width of the base, W. This value is actually the effective base width, which is governed by the depletion regions formed at the emitter–base and collector–base junctions. Whilst the emitter–base junction width remains essentially constant due to the fixed base current flowing, the collector–base junction width changes with collector voltage. Thus, β is bias dependent and gives rise to a finite device output conductance, as shown in Figure 2.20.

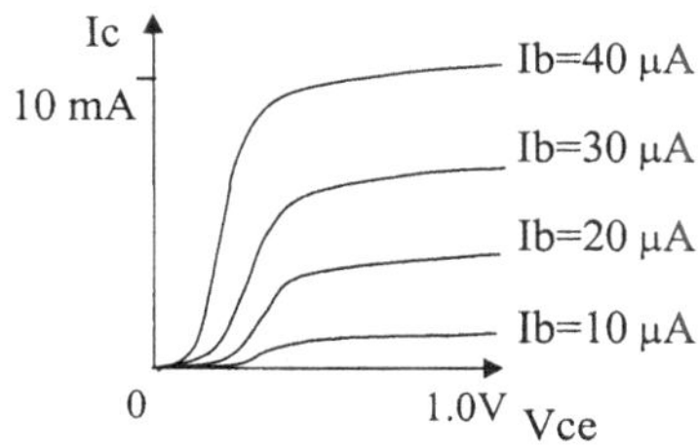

Figure 2.20 Common-emitter characteristics of a bipolar transistor

In the common-emitter configuration, the transconductance g_m, defined as $g_m = \partial I_c / \partial V_{be}$, can achieve large values due to the exponential dependence of the collector current on emitter–base voltage described by eqn. (2.8). Differentiating eqn. (2.8) leads to $g_m = I_c q / k_B T$. The large magnitude of g_m is an important consideration in applications requiring large output voltage swing, such as in power amplifiers.

A final point before leaving the basic DC common-emitter configuration is that this condition requires only a single power supply polarity, which can be an important consideration in simplifying the design topology.

2.4.1.2 RF operation

In the preceding section, it was shown that the operation of the bipolar transistor is governed by the emitter, base and collector currents flowing in response to voltages applied across the transistor junctions. Both junctions have depletion regions associated with them, and thus capacitances, which influence the high frequency response of the device. A lumped-element equivalent circuit to describe a bipolar transistor operated in common-emitter configuration is shown in Figure 2.21. Using this equivalent circuit, a number of key RF figures-of-merit for device operation can be determined.

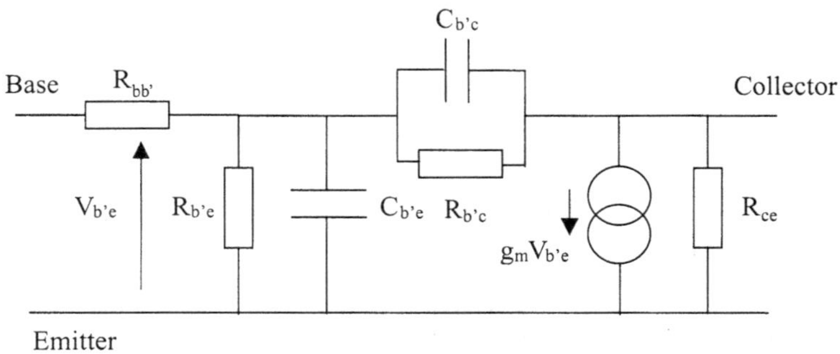

Figure 2.21 Simplified lumped-element equivalent circuit for a bipolar transistor

The f_T of a bipolar device is defined as the frequency at which the common-emitter short circuit current gain falls to unity. Using the simplified equivalent circuit of Figure 2.21, the intrinsic emitter-base voltage drop $V_{b'e}$ can be written as:

$$V_{b'e} = \frac{i_b}{g_{b'e} + j\omega(C_{b'e} + C_{b'c})} \tag{2.13}$$

where $g_{b'e}$ is the conductance of the forward-biased base–emitter junction, $C_{b'e}$ is the intrinsic base–emitter capacitance and $C_{b'c}$ is the intrinsic base–collector capacitance.

Hence, the frequency dependent short circuit current gain $\beta(\omega)$ can be written as:

$$\beta(\omega) = \frac{-i_{out}}{i_{in}} = \frac{g_m}{g_{b'e} + j\omega(C_{b'e} + C_{b'c})} \tag{2.14}$$

This can be written in terms of β, the DC current gain of the device:

$$\beta(\omega) = \frac{\beta}{1 + \dfrac{j\omega(C_{b'e} + C_{b'c})}{g_{b'e}}} \tag{2.15}$$

At frequency $f = f_T$, $\beta(\omega)$ falls to unity, and thus rearranging eqn. (2.15) gives:

$$1 + j\frac{2\pi f_T(C_{b'e} + C_{b'c})}{g_{b'e}} = \beta \tag{2.16}$$

Assuming β>> 1, this can be rewritten as:

$$f_T \cong \frac{\beta g_{b'e}}{2\pi(C_{b'e}+C_{b'c})} = \frac{g_m}{2\pi(C_{b'e}+C_{b'c})} \tag{2.17}$$

Essentially, this analysis defines f_T in terms of the delays associated with charging the depletion regions associated with the emitter–base and base–collector junctions. A more complete expression for f_T can be determined by including time constants associated with transiting the collector depletion region τ_{cT}, the collector charging time τ_c, the base transit time τ_b and charging of parasitic capacitances. In this situation, the short circuit current gain can be written as:

$$f_T = \frac{1}{2\pi\tau_{eff}} \tag{2.18}$$

where

$$\tau_{eff} = \tau_E + \tau_B + \tau_{cT} + \tau_c + \tau_{para} \tag{2.19}$$

τ_E has already been dealt with above, and can be written as:

$$\tau_E = \frac{(C_{b'e}+C_{b'c})}{g_m} = \left(\frac{k_B T}{qI_e}\right)(C_{b'e}+C_{b'c}) \tag{2.20}$$

The base charging time, τ_B, can be written as:

$$\tau_B = \frac{W^2}{\eta D_B} \tag{2.21}$$

where η, and thus the total delay, depends strongly on the doping strategy in the base.

The collector depletion region transit time, τ_{cT}, can be written as:

$$\tau_{cT} = \frac{x_{bc}}{v_{sat}} \tag{2.22}$$

where x_{bc} is the base–collector depletion region width and v_{sat} is the electron saturation velocity.

The collector charging time, τ_c, can be written as:

$$\tau_c = r_c C_c \tag{2.23}$$

where r_c is the series resistance of the collector and C_c is the collector capacitance.

Thus, for a large value of f_T, the device should have a large DC current gain, a small dynamic base resistance, small values of junction capacitance, and should be operated at high current levels. This in turn means that the device should have small cross-sectional area to minimise capacitance, a thin highly-doped base region and a large current gain.

The maximum available gain of a device is defined as the forward power gain with the optimal matching conditions at both input and output. The maximum frequency of operation, f_{max}, is the frequency at which the MAG falls to unity. For a bipolar device, represented by the lumped-element equivalent circuit of Figure 2.21, it can be shown that:

$$MAG = \frac{\left(\frac{f_T}{f^2}\right)}{8\pi C_c r_b} \tag{2.24}$$

and thus

$$f_{\max} = \left(\frac{f_T}{8\pi C_c r_b}\right)^{\frac{1}{2}} \tag{2.25}$$

Therefore, to have a large value of f_{max}, the device should have:

1. high f_T
2. small capacitance
3. small base resistance.

which, in turn, requires that the vertical architecture is designed to have:

1. large base doping concentration
2. small base width
3. large current gain.

2.4.2 Heterojunction bipolar transistors (HBTs)

As devices are aggressively scaled to reach ever higher operating frequencies, thinner base regions and higher doping levels are required, which result in higher electric fields across narrow depletion regions and give rise to significant leakage effects. As a result, conventional technologies to define such thin, highly doped structures are being seriously stressed. More specifically, a number of features desirable for the realisation of high frequency bipolar transistors are conflicting. As an example, increasing the base and emitter doping concentrations (to minimise the junction capacitance) actually reduces the device current gain, due to bandgap narrowing in the emitter and increased minority carrier injection from the base to the emitter. For these reasons, a great deal of attention has been focussed on more advanced vertical architectures, and consequently, on sophisticated technologies capable of delivering the required structures. Bipolar technologies have embraced heterojunction strategies to great effect

— the additional degree of freedom in engineering a device, having the ability to modify the bandgap of certain layers in the structure, can significantly enhance device performance. Heterojunctions have been used in III-V bipolar transistors for many years, and more recently they have been used to great effect in silicon bipolar technology [4].

In a heterojunction bipolar transistor (HBT), the emitter has a wider bandgap than the base. Most HBT structures are realised using epitaxial growth techniques such as CVD or MBE. The band diagram for such a device is shown in Figure 2.22. The bandgap discontinuity, $\Delta E_{bg} = \Delta E_c + \Delta E_v$, is split between the valance and conduction bands. The importance of this configuration can best be understood by recalling that the common-emitter current gain, β, can be written as:

$$\beta \approx \frac{D_n N_{de} x_e}{D_p N_{ab} W} e^{\left(\frac{-\Delta E_g}{k_B T}\right)} \tag{2.26}$$

In the case of the HBT, the bandgap narrowing is offset by the increase in the valance band offset ΔE_v, which presents an increased barrier to hole injection from the base into the emitter, resulting in an increase in the current gain. This increase in current gain can be used directly to enhance device performance, or can be traded against increasing the base doping to reduce the base resistance, thus improving f_{max}. As a further example, the conduction band offset at the emitter–base junction means that the doping concentration of the emitter could be reduced, even to a level below that of the base, without compromising device operation. In this way, the emitter–base capacitance can be reduced.

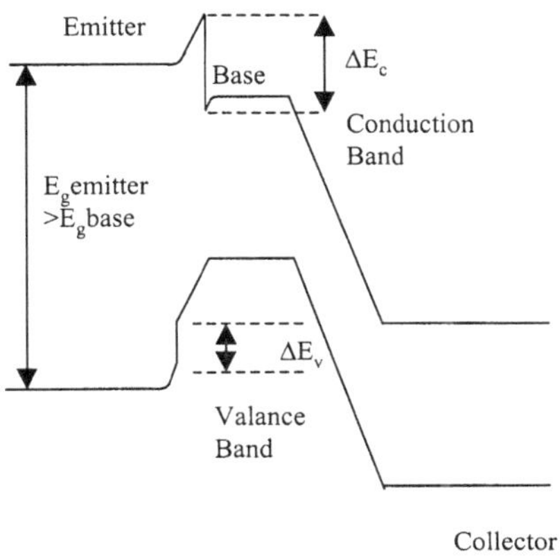

Figure 2.22 Band diagram of heterojunction bipolar transistor

2.4.3 Summary of contemporary bipolar technologies

Currently, four types of bipolar technology are available to the MMIC designer:

1. Silicon bipolar processes
2. SiGe heterojunction bipolar processes
3. GaAs-based heterojunction bipolar processes
4. InP-based heterojunction bipolar processes.

Typical device cross-sections and a brief explanation of the key features are now given for each technology.

2.4.3.1 Silicon bipolar processes

A number of manufacturers currently offer highly evolved silicon bipolar MMIC technologies, primarily targeted towards low GHz chipsets for wireless applications. The cross-section of such a silicon bipolar device is shown in Figure 2.23. The key feature of the advanced silicon bipolar technologies is the use of highly doped polysilicon layers, with steep doping gradients, to minimise the access resistance, junction capacitance and parasitic capacitance. In addition, the use of a buried layer to bring the collector contact to the top of the chip reduces parasitic collector–base capacitance, so improving the high frequency performance.

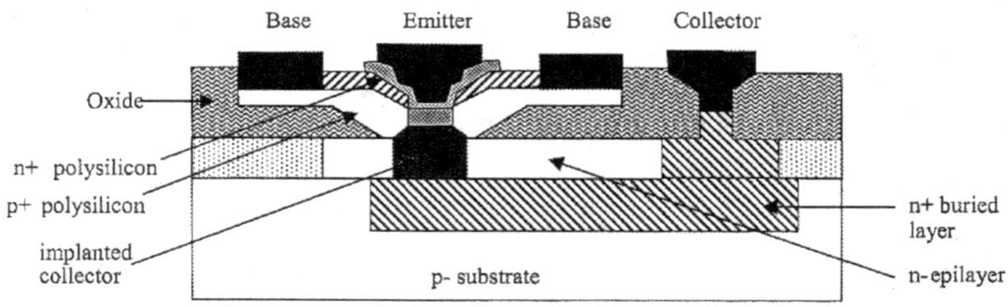

Figure 2.23 Cross-section of a silicon bipolar technology

2.4.3.2 SiGe heterojunction bipolar processes

SiGe HBT technologies have now matured and look set to capture a significant market share of the low GHz communications arena. SiGe HBTs utilise a layer of SiGe in the base of the transistor, grown by CVD. As SiGe has a smaller bandgap than silicon, the advantages of HBTs can be exploited in a traditional silicon environment. Currently, two base growth strategies are in production. In the first, the Ge concentration is graded across the width of the base, as shown in Figure 2.24. In addition to the Ge grading, a spike of B is incorporated in the base to supply the *p*-doping.

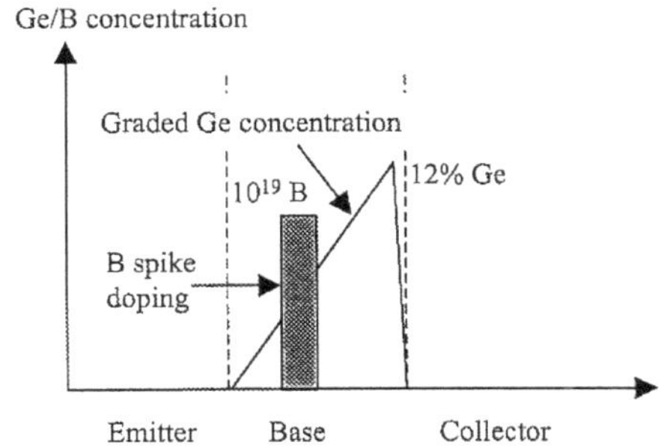

Figure 2.24 Doping strategy for SiGe HBTs

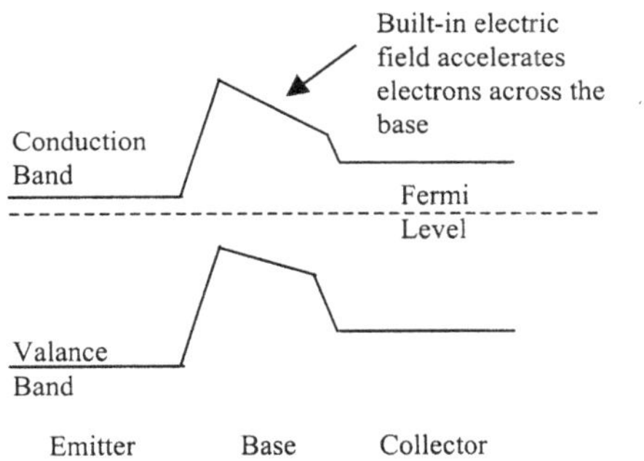

Figure 2.25 Band diagram of graded-base SiGe HBT

As SiGe has a bandgap that is lower than silicon, and increasing the germanium concentration reduces the bandgap further, grading the germanium concentration across the base results in the band diagram shown in Figure 2.25. The graded base structure introduces an additional built-in electric field that accelerates electrons through the base, reducing the transit time by around a factor of two (when compared with a conventional bipolar device). This results in an increase in device operating frequency. The maximum concentration of Ge that can be introduced into the base is limited due to the different size of Ge and Si atoms. For this reason, the SiGe alloy is strained when grown on the Si substrate. Provided that the strained SiGe thickness is kept below a critical value, the strain is accommodated in the layer.

In the second SiGe process, a 'box' profile of Ge in the base is used again, along with a spike of B doping. In both strategies, the addition of Ge in the base of the device reduces the transit time, whilst the higher base doping (made possible by the emitter–base heterojunction) improves the current gain and reduces the base resistance, which results in improved RF performance. The polysilicon and buried-layer collector strategies mentioned above are equally applicable to the SiGe HBT, and when combined with the intrinsic performance advantages offered by incorporating SiGe into the vertical architectures, result in a silicon-based RF technology with performance almost matching that of III-V devices. The high volume manufacturing capability of the silicon microelectronics industry, together with more subtle effects relating to device linearity, power handling and noise performance, ensure that SiGe HBT technology will play a major role in many applications. The details of the noise and power handling capability are discussed later in this chapter.

2.4.3.3 GaAs-based heterojunction bipolar technologies

III-V HBTs predominantly based on the GaAs/AlGaAs material system have been investigated for around twenty years. The technology has matured to the stage that a number of foundries currently offer high volume production GaAs-based HBTs processes [5]. At the time of writing in early 2001, III-V HBT technologies can be found in around 25% of all mobile phone handsets. More significantly, given the impending roll-out of 3rd generation UMTS mobile services, close to 80% of all CDMA handsets currently contain GaAs-based HBTs. Figure 2.26 shows the cross-section of an AlGaAs/GaAs HBT, in which the emitter is formed from *p+* AlGaAs, the base is *n* GaAs and the collector is either *p-* GaAs or *p-* AlGaAs — the complete structure is grown on a semi-insulating GaAs substrate. The use of a *p*-doped AlGaAs collector results in a double heterojunction device, which takes advantage of the wide band-gap collector to improve both the device breakdown and speed.

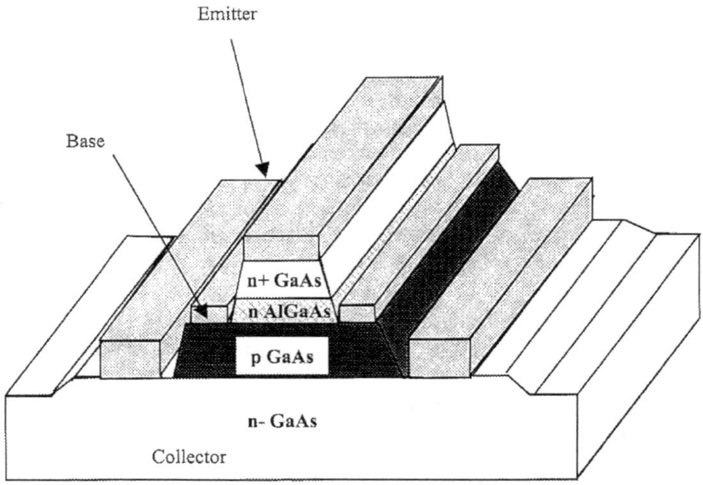

Figure 2.26 Cross-section of an AlGaAs/GaAs HBT

2.4.3.4 InP-based heterojunction bipolar technologies

A study of the transport properties of III-V semiconductors reveals that materials like $In_{0.53}GaAs$, which can be grown strain-free on InP substrates, have impressive low field mobility and high field saturation velocity figures-of-merit. For these reasons, it was widely predicted that InP-based bipolar devices would deliver the highest levels of performance, when compared with devices based on either the silicon or GaAs material systems. This has indeed proved to be the case, and, as will be shown in Section 2.6, InP-based HBTs currently offer some of the highest levels of RF performance available from a three-terminal device. InP-based HBTs have received considerable attention in recent years, as a result of the demand for electronics for 40 Gb/s fibre applications. In addition to the intrinsic high frequency performance capabilities of this technology, it is perceived that InP HBT technology may offer a route to monolithic integration of both electronic and long wavelength optical functionality on the same chip.

2.5 Field effect transistors

The GaAs metal–semiconductor field effect transistor (MESFET) was the first three-terminal active RF device. Since their introduction at Caswell, III-V field effect transistors have been a cornerstone of MMIC technology. Today, around 100 foundries worldwide offer III-V field effect transistor MMIC processes based on MESFET or HEMT (high electron mobility transistor) technology. At the time of writing, silicon CMOS technology is on the verge of becoming a viable competitor for low GHz RF applications, for the same reasons as Si bipolar technology described in the previous sections. This section on field effect transistors will, however, concentrate mostly on III-V Schottky gated devices (MESFETs and HEMTs), but will make comparisons with figures-of-merit available form contemporary silicon CMOS. It is likely that there will continue to be significant improvements in the RF performance of silicon CMOS. The relative importance of this technology in future RF applications should not be discounted by the designer, particularly if high functionality, low cost, high volume MMIC performance for low GHz applications is required.

This section begins with an overview of field effect transistor operation, leading to an identification of the aspects of the device that determine RF performance. The section ends with a summary of the various HEMT technologies currently available to the MMIC designer, highlighting the relative merits of each device type.

2.5.1 Overview of MESFET operation

The MESFET has source, drain and gate contacts as shown in Figure 2.27. The device consists of two ohmic contacts (the source and drain), between which the third electrode (the gate) is aligned. The gate forms a rectifying metal–semiconductor Schottky barrier contact.

In operation, current flows from source to drain under the gate of the device, via a conducting channel. Thus, current flow is predominantly parallel to the surface of the device, in contrast to the bipolar device, in which transport is perpendicular to the surface. The MESFET is a unipolar device; in other words, current is carried by only electrons or holes. This is again in contrast to the bipolar device, which relies on transport of both electrons and holes for its function. In general, due to the inferior mobility of holes, current in MESFETs is carried by electrons, and so the device cap and channel are doped *n*-type.

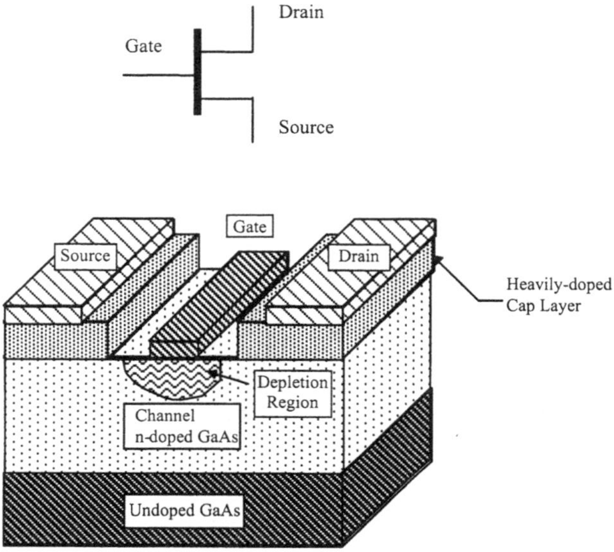

Figure 2.27 Cross-section of a MESFET device

In its simplest configuration, the MESFET vertical architecture comprises three layers. Uppermost is a heavily doped cap layer, to facilitate the formation of low resistance source and drain contacts. Below this is a lower doped, thicker channel layer, via which current flows through the device. Finally, above the semi-insulating substrate is a high resistance undoped buffer layer, whose function is to confine the carriers to the device channel. Under typical operating conditions, the source contact of the MESFET is grounded. Assuming *n*-channel operation (the current is carried by electrons), the drain contact is positively biased with respect to the source. The Schottky gate contact can be either positively (forward) or negatively (reverse) biased with respect to the source.

The simplest way to envisage MESFET operation is to independently consider the operation of the gateless MESFET and Schottky diode structures, shown schematically in Figures 2.28 and 2.30, respectively, and then to combine these situations in the three terminal MESFET configuration. The gateless MESFET of Figure 2.28 is identical to the semiconductor resistor described in Section 2.3.

The source and drain contacts are of width W and separated by a distance s. A positive voltage, V_{ds}, is applied to the drain and current I_{ds} flows. To simplify the analysis, assume the source and drain contacts are sitting directly on the channel layer of thickness h and doping concentration N_d.

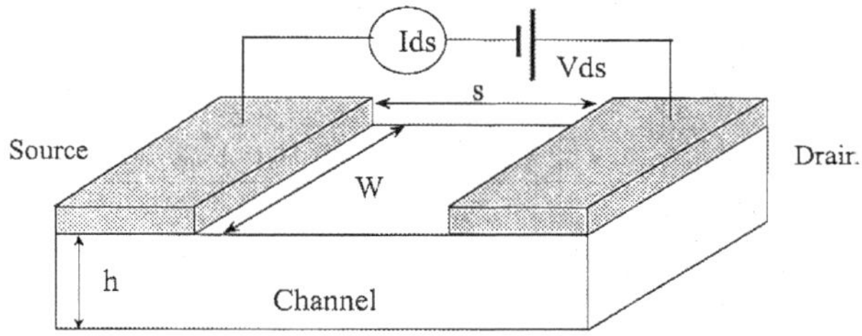

Figure 2.28 Schematic of gateless MESFET

The current flowing from source to drain *Ids* can be written as:

$$I_{ds} = qN_d hWv \tag{2.27}$$

where q is the electron charge and v is the velocity of electrons in the channel. The electron velocity can be written as:

$$v = \mu E \tag{2.28}$$

where μ is the electron mobility and E is the electric field, $E = V_{ds}/s$. Thus, eqn. (2.27) can be rewritten as:

$$V_{ds} = I_{ds} \frac{s}{W} R_{sh} \tag{2.29}$$

where $$R_{sh} = \frac{1}{qN_d h\mu} \tag{2.30}$$

In other words, a linear current–voltage relationship exists, the details of which depend on the vertical architecture (channel thickness, doping concentration and mobility — parameters outside the control of the MMIC designer) and the device geometry (contact width and separation). Eqn. (2.29) can be depicted graphically as in Figure 2.29 — the slope of the I_{ds}/V_{ds} plot is related to R_{sh}. The linear relationship does not hold beyond electric fields of around 3 kV/cm, as the carrier velocity saturates due to scattering processes in the device channel, the details of which are beyond the scope of this text. The macroscopic effect of velocity saturation is that the current saturates for electric fields of greater than 0.3 V/μm, as shown in Figure 2.29.

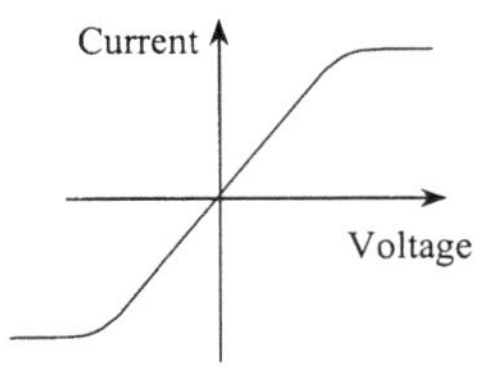

Figure 2.29 Current–voltage relationship of a gateless MESFET

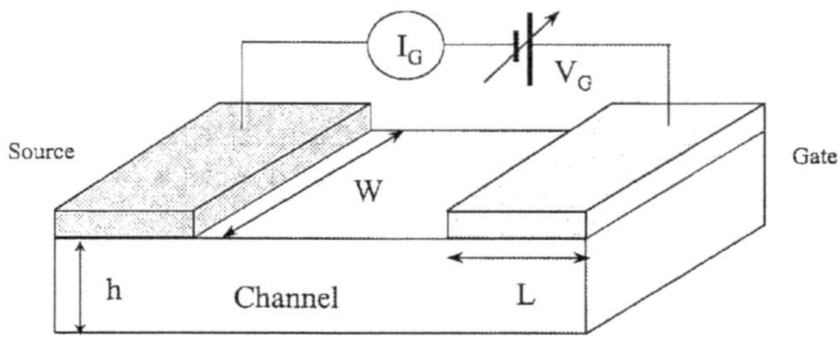

Figure 2.30 Schematic of Schottky diode

The Schottky gate/source contact configuration, shown in Figure 2.30, can be considered in a similar fashion. The gate current I_G can be written as:-

$$I_G = A^* T^2 LW e^{(\frac{-qV_{bi}}{kT})} e^{(\frac{qV_G}{nkT})} \tag{2.31}$$

where A^* is the effective Richardson constant (8.8 $\mathrm{AcmK^{-1}}$ at 300 K)
T is the absolute temperature
L is the Schottky contact length
W is the Schottky contact width
V_{bi} is the built in voltage on the diode
K is Boltzmann's constant (1.38×10^{-23} J/K)
N is the diode ideality factor (around 1.1-1.2).

Examination of eqn. (2.31) shows that if V_G is negative, both exponential terms are very small, and thus the gate current is small. For V_G positive but less than V_{bi}, the exponential terms tend to cancel each other, again leading to small values of gate current. However, when V_G is more positive than V_{bi}, the current starts to increase exponentially. To ensure small values of gate current, voltages of less than the built-in voltage must be applied. This will not necessarily always be the case, particularly in power amplifier applications, and thus, large variations in the device input impedance may have to be considered in the design of such components.

In the situation where the applied voltage is less than the built-in voltage, the lack of current flow results in the formation of a depletion region below the gate, whose thickness, a, is determined by the equation:

$$a = \left(\frac{2\varepsilon_0 \varepsilon_r (V_{bi} - V_G)}{qN_d} \right)^{\frac{1}{2}} \qquad (V_g < V_{bi}) \tag{2.32}$$

where ε_o and ε_r are the free space and semiconductor permittivities, respectively.

Eqn. (2.32) shows that the depletion region size a can be varied by changing the voltage applied to the gate. A large negative gate voltage will result in a large depletion region.

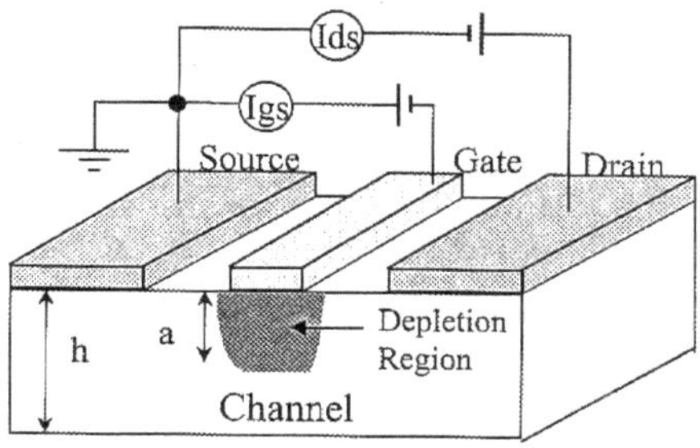

Figure 2.31 Schematic of a MESFET

MESFET operation can now be understood by combining the gateless FET and Schottky diode structures together, as shown in Figure 2.31. For zero gate-source voltage, assume that the channel thickness and doping concentration are such that a depletion region of thickness a ($< h$) is formed. Then, at low drain-source voltages, the drain-source current will be:

$$I_{ds} = qN_d(h-a)W\mu\frac{V_{ds}}{s} \tag{2.33}$$

From eqn. (2.32), it is clear than making V_{gs} more negative increases the size of the depletion region, decreasing the drain-source current. The drain-source current, therefore, depends on the gate-source voltage for a given drain-source voltage, as shown in Figure 2.32. At a certain negative value of gate-source voltage, the depletion region thickness will extend to the full channel depth, preventing drain-source current flow. This is termed the 'pinch-off' voltage of the device, V_p. From eqn. (2.32), the pinch-off voltage is:

$$V_p = \frac{qN_d a^2}{2\varepsilon_o\varepsilon_r} - V_{bi} \tag{2.34}$$

At high drain voltages, the drain current saturates due to the field dependent velocity saturation mentioned before. In this situation, the drain current can be written as:

$$I_{ds} = qN_d(h-a)v_{eff}W \tag{2.35}$$

where v_{eff} is the velocity of carriers in the device channel.

The application of negative gate voltage results in a higher electric field for a given drain bias. Thus, current saturation occurs at lower drain voltages as the gate bias is made more negative. As a result, the $I_{ds}(V_{ds}, V_{gs})$ characteristics can be split into two gate and drain bias zones — the linear and saturation regions shown in Figure 2.32.

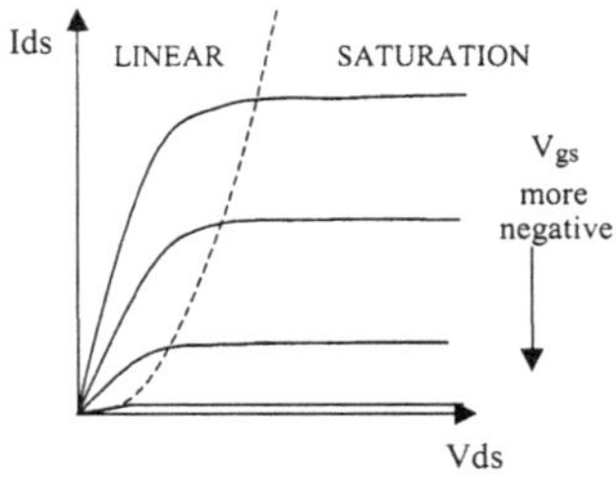

Figure 2.32 Output characteristic of a GaAs MESFET

2.5.1.1 MESFET DC operation

Figure 2.33 shows the experimental characteristics of a GaAs MESFET, which follows the analysis presented above. The main difference between the experimental and simplified DC characteristics presented above is the finite slope of the I_{ds} curves in the saturation region. A number of factors contribute to this finite output conductance, g_{ds}, defined as:

$$g_{ds} = \left.\frac{\partial I_{ds}}{\partial V_{ds}}\right|_{V_{gs}=\text{constant}} \tag{2.36}$$

These factors include charge injection from the channel to the undoped buffer and conduction mechanisms via surface and channel-substrate states — mechanisms beyond the scope of this text.

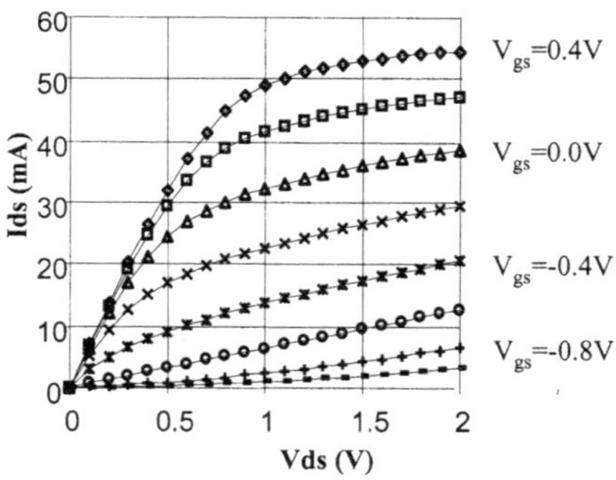

Figure 2.33 Output characteristics of a MESFET

The gain mechanism of a FET is embodied in the transconductance (g_m), defined as the modulation of channel current (I_{ds}) by the gate voltage (V_{gs}) with a fixed drain voltage (V_{ds}):

$$g_m = \left.\frac{\partial I_{ds}}{\partial V_{gs}}\right|_{V_{ds}=\text{constant}} \tag{2.37}$$

In the simple approximation described above, the transconductance can be obtained by substituting the expression for the depletion region thickness of eqn. (2.32) into the drain current equation (2.33), and differentiating with respect to V_{gs}. This gives:

$$g_m = \frac{\varepsilon_o \varepsilon_r v_{eff} W}{a} \tag{2.38}$$

A large transconductance requires high carrier velocity in the device channel, which is separated from the gate by a thin depletion region. Increasing the depletion region thickness, as a result of applying a large negative gate bias, for example, will reduce the transconductance. Further, for a given carrier velocity, the transconductance can be enhanced by increasing the channel doping concentration to reduce the depletion region thickness. The negative impact of this is that the breakdown voltage will be compromised.

The voltage gain (A_V), which is directly related to power gain, is defined as:

$$A_v = \frac{\partial V_{ds}}{\partial V_{gs}} = \frac{g_m}{g_{ds}} \tag{2.39}$$

2.5.1.2 MESFET RF operation

In MMIC design, it is the bias-dependent equivalent circuit of the MESFET, deduced from RF measurements, that concerns the design engineer. The DC bias conditions are important only to establish the values of the equivalent circuit elements. Figure 2.34 shows a schematic cross-sectional diagram of the GaAs MESFET, onto which an electrical network has been superimposed. Each element is intended to represent the electrical character of a particular region of the device. For example, the metallurgical resistance associated with the gate is represented by the lumped-element R_g. Figure 2.35 shows the basic equivalent circuit required to describe the electrical behaviour of the physical structure of the FET at microwave frequencies. The region within the dashed box is defined as the intrinsic region of the MESFET, which is physically situated below the gate

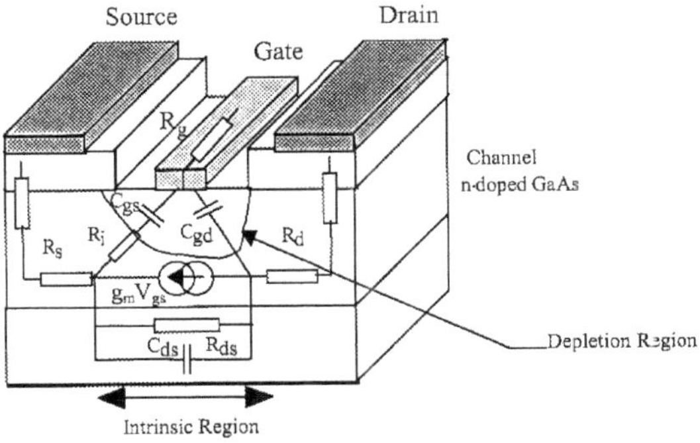

Figure 2.34 Cross-section of MESFET with superimposed lumped-element equivalent circuit

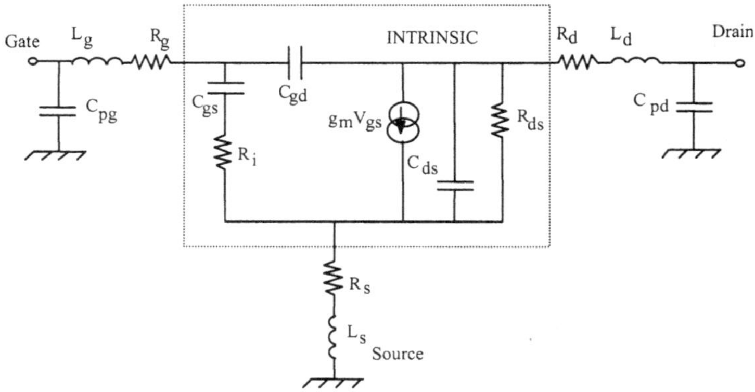

Figure 2.35 Lumped-element equivalent circuit of a MESFET

The gate-source capacitance, C_{gs}, represents the charge in the depletion region under the gate at the source side of the device. The gate-drain capacitance, C_{gd}, is the capacitance associated with the charge in the depletion region at the drain edge of the space-charge layer, generated and modulated by the gate-drain voltage. Together, C_{gs} and C_{gd} comprise the total gate capacitance, which can, to first order, be described as a parallel plate geometry, whose capacitance can be written as:

$$C_g = \frac{\varepsilon_o \varepsilon_r W L_g}{a} \tag{2.40}$$

where L_g is the device gate length.

The channel resistance or input resistance R_i (in series with C_{gs}) represents the resistance of the semiconductor underneath the gate. The current generator $g_m V_{gs}$ represents the gain of the device, where g_m is the intrinsic transconductance and V_{gs} is the voltage dropped across the gate capacitance. R_{ds} represents the finite output resistance of the device. The output resistance at microwave frequencies is usually lower than the DC value, as a result of dispersion, which is mainly due to the charge exchange with deep levels located at the interface of the device channel and buffer. The drain-source capacitance, C_{ds}, in parallel with R_{ds}, arises from capacitive coupling between the doped regions of the drain and source, separated by the depletion region.

The extrinsic elements of the equivalent circuit are shown outside the dashed line in Figure 2.35. The series parasitic resistances, comprising the source resistance R_s and the drain resistance R_d, each have two components: the contact resistance component (R_c), which arises from the ohmic contact to the device channel, and the bulk resistance of the semiconductor in the source-gate (or drain-gate) access regions. The gate resistance R_g is determined by the number of gate fingers (m), the gate cross-sectional area (gate length (L_g)$\times$gate height (y))

and its width (W). Multi-finger gates are connected in parallel, resulting in a gate resistance reduced by a factor of m^2:

$$R_g = \frac{\rho W}{3m^2 y L_g} \tag{2.41}$$

where ρ is the resistivity of the gate metal.

The RF gate resistance is distributed, resulting in the factor of 3 in the denominator of the expression. The gate, drain and source parasitic inductances, L_g, L_d and L_s, arise from feed pads of the electrodes. The parasitic geometrical capacitances, C_{pg} and C_{pd}, caused by the electric field distributions between metallic contacts in the MESFET, are determined by the layout of the device.

The cut-off frequency, f_T, is the frequency at which the short circuit current gain, $i_{out}/i_{in} = i_d/i_g$, falls to unity. Using the intrinsic part of the equivalent circuit model of Figure 2.35, with the output shorted, $V_{out} = 0$ (drain is shorted to source):

$$\frac{i_d}{i_g} \approx \frac{g_m}{j\omega(C_{gs} + C_{gd})} \tag{2.42}$$

At $f = f_T$, $\frac{i_d}{i_g} = 1$, therefore:

$$f_T = \frac{g_m}{2\pi(C_{gs} + C_{gd})} \tag{2.43}$$

Substituting eqns. (2.38) and (2.40) for g_m and $(C_{gs}+C_{gd})$, respectively, the cut-off frequency can be rewritten as:

$$f_T = \frac{v_{eff}}{2\pi L_g} \tag{2.44}$$

Taking into consideration the parasitic resistances of the equivalent circuit shown in Figure 2.35, f_T can be written as:

$$f_T = \frac{\mathrm{g_m}}{2\pi\left[\left(\mathrm{C_{gs}} + \mathrm{C_{gd}}\right)\left(1 + \frac{\mathrm{R_s} + \mathrm{R_d}}{\mathrm{R_{ds}}}\right) + \mathrm{g_m C_{gd}}\left(\mathrm{R_s} + \mathrm{R_d}\right)\right]} \tag{2.45}$$

Therefore, for large f_T, gate capacitance and parasitic resistances should be minimised, whilst the transconductance should be maximised. This requires:

1. a high effective velocity in the channel
2. a short gate length
3. well engineered, low parasitic resistance technology.

The maximum available gain (MAG) is defined as the maximum power gain that can be achieved with conjugate input and output matching. The maximum frequency of oscillation (f_{max}) is defined as the frequency at which the MAG falls to unity. Using the equivalent circuit shown in Figure 2.35, it can be shown that:

$$\mathrm{MAG} = \frac{\left(f_T / f \right)^2}{4\left(\dfrac{R_g + R_i + R_s}{R_{ds}} \right) + 4\pi f_T C_{gd} \left(2R_g + R_i + R_s \right)} \tag{2.46}$$

and thus:

$$f_{\max} = \frac{f_T}{2\left(\dfrac{R_g + R_i + R_s}{R_{ds}} + \left(2\pi f_T R_g C_{gd} \right) \right)^{\frac{1}{2}}} \tag{2.47}$$

To optimise the MAG and f_{max}, many of the requirements for high f_T also have to be satisfied, but in addition, the gate resistance must also be minimised. This can be achieved using a multiple gate finger approach, as eqn. (2.41) showed that the gate resistance decreased as m^2, where m is the number of parallel gate fingers. Driving multiple parallel gate fingers can, however, result in a significant increase in inter-electrode capacitance, which limits the device f_T and f_{max}. Thus, the more common solution is to reduce the gate resistance by forming a T-gate: the cross-section of such a gate is shown in Figure 2.36. This structure combines a small footprint, to produce a short gate-length device, whilst simultaneously having a large cross-sectional area to minimise the gate resistance.

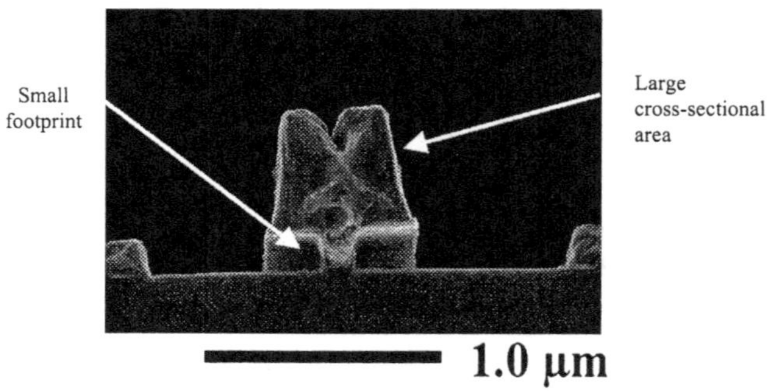

Figure 2.36 Cross-section of a 0.12 μm T-gate

2.5.2 The high electron mobility transistor (HEMT)

The high electron mobility transistor (HEMT) is a field effect device whose operation is very similar to that of the MESFET. The difference is that the HEMT utilises a vertical architecture, which incorporates layers of different materials that are chosen to form a device channel in which the electrons are physically separated from their parent donors. This leads to large electron mobilities, as ionised impurity scattering is reduced. In addition, a large electron concentration is formed at a well-defined plane, close to the surface of the material structure.

In the HEMT, current flows from the source to the drain under the gate contact via the high mobility channel. This results in improved high frequency noise and gain characteristics compared to conventional GaAs MESFETs. The following explanation of device operation is equally applicable to any HEMT, but it is easiest to explain by considering a specific example: the AlGaAs/GaAs HEMT — the first device of this type to be demonstrated, as the lattice constants of AlGaAs and GaAs are almost identical, which simplifies the complexities of the growth of high quality single-crystal AlGaAs/GaAs heterostructures. The cross-section of an AlGaAs/GaAs HEMT is shown in Figure 2.37, together with the conduction band profile.

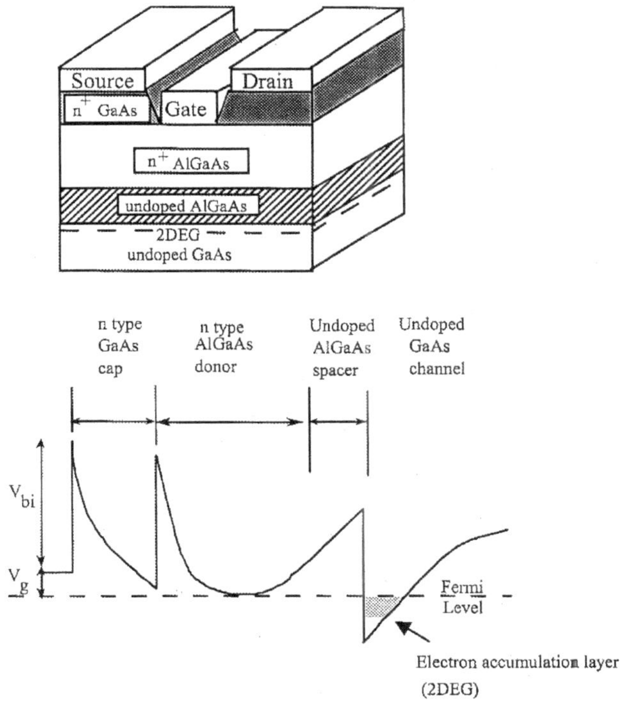

Figure 2.37 Cross-section of AlGaAs/GaAs HEMT and its conduction band diagram

The doped GaAs cap facilitates the formation of low resistance ohmic contacts to the device. As in the case of the GaAs MESFET, the gate is placed between the source and drain in a recess trench, to reduce the access resistance and to improve the vertical scaling of the device. The *n*-doped AlGaAs layer provides the electrons for the device channel, whilst the undoped AlGaAs spacer layer is introduced to improve the channel mobility by spatially separating the donor atoms from the electron accumulation layer. Both AlGaAs layers have aluminium concentrations in the range 20–30%, depending on the particular application. The aluminium concentration determines the size of the conduction band discontinuity at the AlGaAs/GaAs interface, which in turn controls the concentration of the electron accumulation layer in the GaAs channel. The larger the aluminium concentration, the larger the channel concentration. Aluminium concentrations in the range 20–30% balance channel concentration with mobility, which decreases with aluminium concentration due to random alloy scattering. The high mobility channel formed in the undoped GaAs channel layer results from the fact that AlGaAs has a larger bandgap than GaAs, and thus the electrons in the doped AlGaAs donor layer naturally accumulate at the AlGaAs/GaAs interface as it is energetically favourable for them to reside in the GaAs channel layer.

It is worth noting that the electric field in the undoped GaAs channel, caused by the electron accumulation layer, is very strong (~10^7 V/m), and so confines the electrons in a very narrow quasi-triangular notch within 15–20 nm of the interface. This dimension is commensurate with the electron wavelength, resulting in quantisation of the electron momentum in the direction perpendicular to the interface. The electrons are free to move in the plane parallel to the interface, however, and so are two-dimensional in nature. For this reason, the electron accumulation layer is usually known as a two-dimensional electron gas (2DEG).

The current passing from source to drain through the 2DEG is determined by the sheet electron concentration beneath the gate contact, which can be modulated by varying the bias on the Schottky contact. In essence, the gate can be considered as a capacitor whose plates are formed by the gate metal and the 2DEG. The depleted AlGaAs donor and spacer layers form the dielectric of the capacitor. In the HEMT, the gate capacitance remains relatively constant with changing gate bias, as the 'plate' separation is fixed by the thickness of the AlGaAs donor and spacer layers. Current modulation in the HEMT occurs as charge is added to or removed from the 2DEG in response to variations in voltage applied to the gate following the classic '$Q = CV$' capacitor equation, where Q is the charge in the 2DEG, C is the gate capacitance and V is the applied gate voltage.

By sufficiently reverse biasing the gate, the 2DEG can be completely annihilated, resulting in no current flow from source to drain. In this situation, the device is pinched-off, and this condition defines the device threshold voltage. If the gate is forward biased sufficiently, a channel of undepleted carriers will be established in the doped AlGaAs layer. In this case, there are two paths by which current can travel from source to drain:

1. via the 2DEG
2. through the doped, undepleted AlGaAs via the 'parallel conduction' layer.

Operating in this mode tends to result in reduced performance, as the advantages of the high mobility device channel are diluted by carrier transport through the doped AlGaAs donor layer, which has a lower electron mobility than GaAs doped at a similar level.

2.5.2.1 HEMT DC operation

The output characteristic of a HEMT, shown in Figure 2.38, is very similar to that of a GaAs MESFET, in that it has linear and saturated regions. For an identical geometry device, the knee voltage for device saturation occurs at lower values of drain-source bias in a HEMT, due to the higher electron mobility. This also reduces the device access resistance, improving the extrinsic transconductance. In the linear region, the drain-source current of a HEMT can be written as:

$$I_{ds} = qn_{2D}\mu W \frac{V_{ds}}{s} \tag{2.48}$$

where
- n_{2D} is the sheet channel concentration under the gate (cm^{-2})
- μ is the 2DEG mobility
- W is the device width
- V_{ds} is the applied drain source voltage
- s is the source drain contact separation.

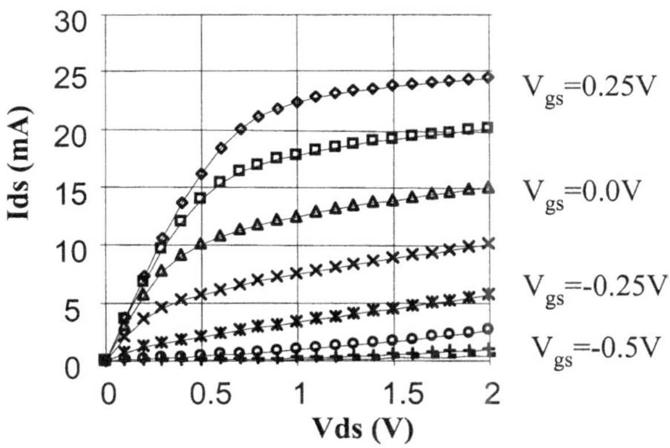

Figure 2.38 Output characteristic of a HEMT

In the saturation region, the drain-source current can be written as:

$$I_{ds} = qn_{2D}v_{eff}W \tag{2.49}$$

where v_{eff} is the effective channel velocity.

The simple charge control model described above can be used to derive an expression for the intrinsic transconductance of a HEMT. The total charge under the gate can be written as:

$$WLqn_{2D} = \frac{WL\varepsilon_o\varepsilon_r}{h}V_{gs} \tag{2.50}$$

where L is the device gate length
$\varepsilon_o\varepsilon_r$ is the effective permittivity of AlGaAs donor & spacer layers
h is the combined thickness of the AlGaAs donor & spacer layers
V_{gs} is the applied gate-source voltage.

Differentiating eqns. (2.49) and (2.50) yields:

$$g_m = \frac{\partial I_{ds}}{\partial V_{gs}} = \frac{\varepsilon_o \varepsilon_r v_{eff}}{h} W \tag{2.51}$$

Therefore, the transconductance of a HEMT depends on the gate-channel separation h and the effective carrier velocity v_{eff}.

2.5.2.2 HEMT RF behaviour

The RF operation of a HEMT is identical to that of the MESFET, as described above in Section 2.5.1.2. The expressions derived for RF figures-of-merit (f_T, f_{max} and MAG) are equally valid for a HEMT, and show that, as in the case of the MESFET, the device transconductance should be maximised whilst minimising the gate capacitance through the use of a short gate length and reducing the parasitic source and drain resistance.

2.5.3 Summary of contemporary HEMT technologies

The preceding discussion on HEMT operation was based on the AlGaAs/GaAs structure, which was the first HEMT device to be demonstrated. Whilst still available in a number of foundries, this baseline HEMT technology has been superseded by a range of devices with more complex vertical architectures, each designed to improve the channel mobility and effective velocity, thus reducing the device parasitic resistance whilst simultaneously offering enhanced intrinsic high frequency performance. These technologies are:

1. pseudomorphic GaAs HEMTs
2. lattice matched and pseudomorphic InP HEMTs
3. metamorphic GaAs HEMTs.

As in Section 2.4.3, a brief description of each of these technologies is now given.

2.5.3.1 Pseudomorphic GaAs HEMTs

The GaAs pseudomorphic HEMT (pHEMT) is currently the most commonly available HEMT technology [6]. The term 'pseudomorphic' arises from the fact that the device channel is formed from InGaAs, with an indium concentration in the range 20–30%, as shown in Figure 2.39. As a result, the lattice constant of the channel is larger than the GaAs substrate, buffer and cap layers and AlGaAs donor and spacer layers, which results in the formation of a strained channel. If the channel layer is kept sufficiently thin, however, the strain can be accommodated in the crystal without the introduction of defects, and the channel takes on the lattice constant of the rest of the structure – i.e. the channel is pseudomorphic. The addition of indium increases the channel mobility and, for concentrations in the range 20–30%, also slightly increases the effective velocity, both of these factors improving the high frequency performance of the device. Furthermore, inspection of the band diagram of Figure 2.39 reveals that the conduction band discontinuity at the AlGaAs/InGaAs interface is larger than in the AlGaAs/GaAs case, as InGaAs has a smaller bandgap than GaAs. As a result, the sheet channel concentration in the GaAs pHEMT is larger, giving greater drive current capability. In addition, the electrons in the device channel are confined at the backside by the conduction band step at the InGaAs/GaAs buffer interface. This potential step results in improved charge control, which reduces output conductance and improves the sharpness of the pinch-off characteristics of the device. pHEMT wafers are currently available in high volumes with diameters up to 150 mm, and so offer excellent economies of scale. This, combined with their excellent RF performance, makes this technology attractive for a number of mobile and broadband wireless applications.

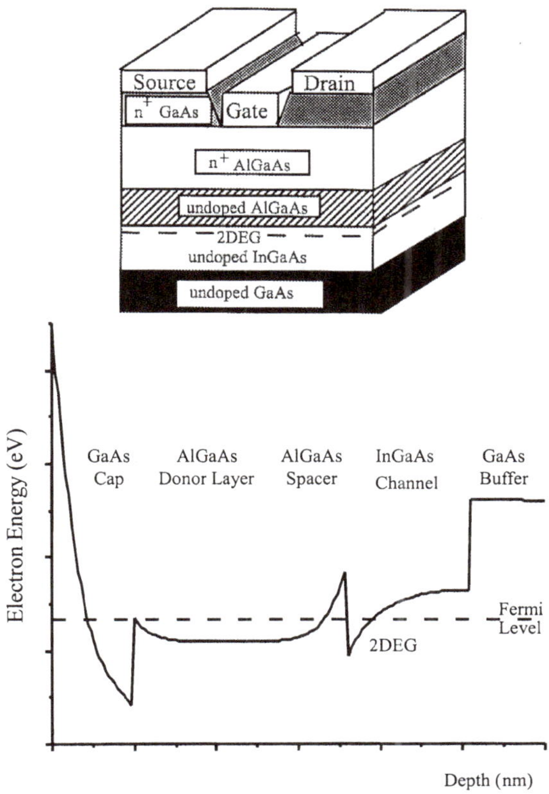

Figure 2.39 Cross-section of a GaAs pHEMT grown on a GaAs substrate and its conduction band diagram

2.5.3.2 Double heterojunction HEMTs

To increase the power handling capability of a HEMT, an additional supply of doping can be introduced below the device channel as shown in Figure 2.40. In essence, this can result in a doubling of the sheet carrier concentration in the channel, with a commensurate increase in the drain source current for a given device width. In addition, the device access resistance is decreased, due to a reduction in the sheet resistance, which results in a lowering of drain voltage required to drive the device into saturation. For these reasons, the double heterojunction HEMT is commonly employed for power amplifier applications. Careful consideration of the vertical architecture design is required in this type of structure to ensure that nonlinearities in the gate bias dependent device transconductance and gate capacitance, arising from the way in which the 2DEG channels are modulated, are minimised for optimal intermodulation performance.

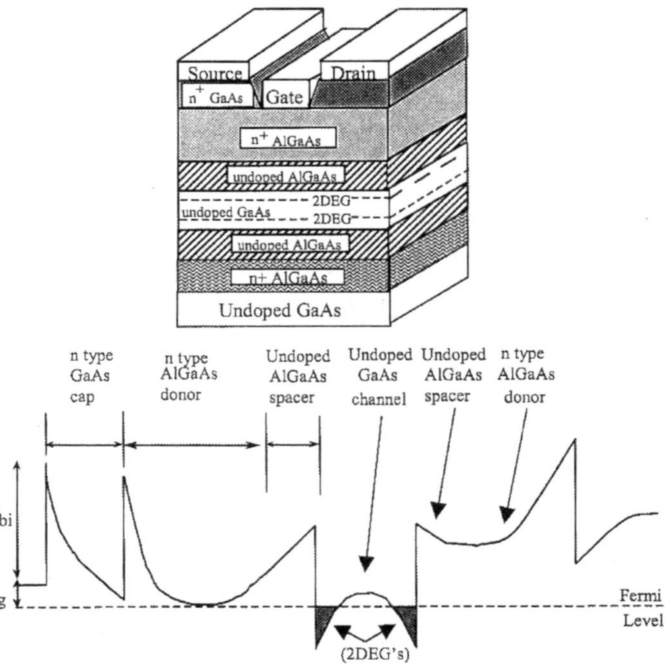

Figure 2.40 Cross-section of a double-heterojunction GaAs pHEMT and its conduction band diagram

2.5.3.3 Lattice matched and pseudomorphic InP HEMTs

As mentioned above, adding indium to the channel of a HEMT increases the conduction band offset at the spacer/channel interface, due to the smaller bandgap of InGaAs, which in turn increases the sheet electron concentration in the channel. Furthermore, due to details of the band structure, the effective velocity and mobility of carriers increases with indium concentration. Thus, for very high frequency operation, large indium concentrations in the device channel are required. Regrettably, due to the mismatch in lattice constant, it has not, until recently, been possible to grow high quality films of InGaAs on GaAs substrates with indium concentrations beyond 35%. However, InP has a larger lattice constant than GaAs, and so completely unstrained HEMTs can be grown on an InP substrate, with a channel indium concentration of 53%. This high indium channel concentration device is known as a lattice matched InP HEMT and has significantly improved transconductance and f_T, when compared with GaAs pHEMTs, due to enhancement in the channel velocity and mobility. The buffer, spacer and donor layers in lattice matched InP HEMTs are formed from InAlAs with 52% indium concentration, which has exactly the same lattice constant as both InP and $In_{0.53}GaAs$. The negative side of this device structure, whose cross-section is shown in Figure 2.41, is the relatively small bandgap of the $In_{0.53}GaAs$ channel layer, which compromises the breakdown voltage. Further improvements in transconductance and f_T can be made by increasing the channel's indium concentration, forming pseudomorphic InP HEMTs. For channel indium concentrations much beyond 65%, however, breakdown effects so seriously compromise device performance as to render them useless in practical applications.

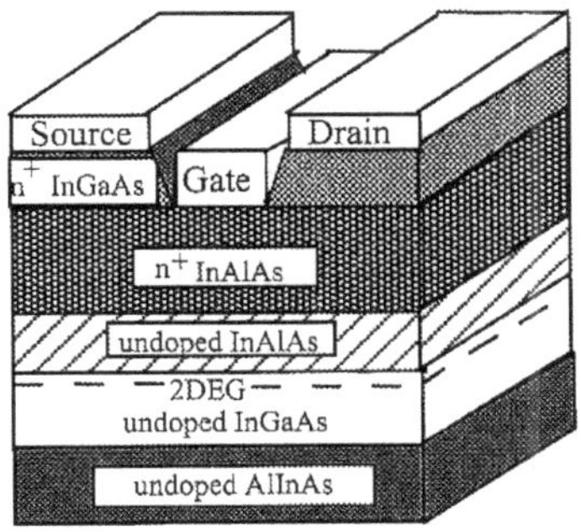

Figure 2.41 Cross-section of InP HEMT grown on an InP substrate

2.5.3.4 Metamorphic GaAs HEMTs

Whilst InP substrate based devices deliver impressive performance at high frequencies, these technologies do suffer from a number of drawbacks. InP substrates are expensive and brittle: although the substrate represents a small fraction of the overall cost of MMIC fabrication, the mechanical properties of InP make handling difficult, which can reduce yield, particularly in backface processing. Furthermore, owing to the maturity of GaAs crystal growth and processing, InP substrates are generally of poorer quality in terms of defect density and morphology, which can further reduce device and circuit yield. Finally, InP substrates are not currently available in diameters exceeding 100 mm, thus precluding the economies of scale possible with 150 mm diameter GaAs wafers.

For these reasons, the use of a GaAs substrate is advantageous in MMIC production, but clearly it would be desirable if the resulting devices and circuits had 'InP-like' performance. It is not possible to directly grow high indium concentration channel layers (for example, $In_{0.53}GaAs$) on a GaAs substrate, as the lattice constant mismatch is too great to be accommodated without the introduction of severe defects and dislocations in the crystal lattice. However, the use of a buffer layer below the channel, in which the indium concentration is either linearly graded or increased in a series of steps, does allow the growth of high quality, single-crystal channel layers with high indium concentrations on a GaAs substrate. As the channel lattice constant and the substrate lattice constant are different, devices produced on these layer structures are known as 'metamorphic' HEMTs. The cross-section of a generic metamorphic GaAs HEMT is shown in Figure 2.42.

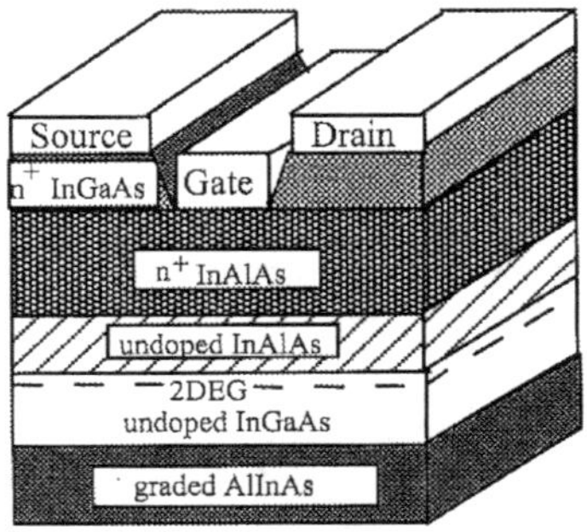

Figure 2.42 Cross-section of metamorphic GaAs HEMT grown on a GaAs substrate

At the time of writing, metamorphic GaAs HEMTs are close to foundry release, with two types of technology likely to be offered:

1. channel In concentrations of >53%, for high gain, low noise applications
2. channel In concentrations in the 30–40% range for power amplifiers.

2.6 Comparison of bipolar and field effect devices

Having described the operation and variety of the bipolar and field effect device technologies currently available to the MMIC designer, their relative RF performance characteristics are now described. The following sections compare and contrast the gain, noise, power handling and linearity capabilities of each of the technologies described above.

2.6.1 f_T and f_{max} of bipolar and field effect devices

As mentioned in Sections 2.4 and 2.5, the realisation of high f_T, f_{max} and MAG in bipolar and field effect devices requires the simultaneous optimisation of numerous parameters in both the device vertical architecture and process technology. These requirements are summarised in Table 2.2.

Table 2.2 Requirements and implications for both transistor families

Technology	Generic requirement	Implication
Bipolar	Small base and emitter capacitance	Small emitter geometry
	Large current gain	Thin, highly doped base
	Small base resistance	Heterojunction technology
	Low parasitic resistance and capacitance	High mobility and velocity materials
FET	Small gate capacitance	Small footprint T-gate structure
	Large transconductance	Device channel close to surface
	Small gate resistance	Heterojunction technology
	Low parasitic resistance and capacitance	High mobility and velocity materials

Translating the observations of Table 2.2 into practical devices with the yield, volume, reliability and cost requirements of contemporary MMIC applications has been, and will continue to be, the challenge faced by silicon and III-V technologists. The current status of the various technologies is summarised in Table 2.3.

Table 2.3 Comparison of f_T and f_{max} for various transistor technologies

Technology	Minimum feature size	f_T	f_{max}
Si bipolar	0.5 μm	50 GHz	50 GHz
SiGe HBT	0.8 μm	130 GHz	160 GHz
GaAs HBT	1.0 μm	180 GHz	280 GHz
InP HBT	1.0 μm	228 GHz	270 GHz
GaAs MESFET	0.2 μm	80 GHz	120 GHz
GaAs pHEMT	0.12 μm	120 GHz	200 GHz
InP HEMT	0.12 μm	250 GHz	350 GHz
GaAs mHEMT	0.12 μm	225 GHz	250 GHz

It is clear that III-V devices outperform silicon technologies, with InP-based HBT and short gate-length HEMT processes offering the highest frequency operation, in keeping with the generic requirements summarised in Table 2.2. The feature size requirements of HEMT technologies impact both the process cost and yield. Currently, the majority of 0.15 μm gate-length HEMTs are realised using electron beam lithography. This is a relatively slow writing process, when compared with an optical stepper based technology, which would be used for high volume manufacture of bipolar devices with minimum emitter feature sizes of around 1 μm. Furthermore, with the low process yields of short gate-length HEMT technologies, it is difficult to realise high volume, low cost LSI components such as (de)-multiplexers for 40 Gb/s optical fibre applications, which can be more readily produced using bipolar technologies.

2.6.2 Noise performance of bipolar and field effect devices

The broadband noise performance of a three-terminal device can be split into three zones as shown in Figure 2.43. At low frequency, the noise is dominated by trapping effects in the bulk material, at junction interfaces and at the surface of the substrate. These effects combine to give a $1/f$ relation to the noise power spectral density. As current transport in a bipolar device is perpendicular to the surface and all junctions are buried, it can be envisaged that the 1/*f* noise corner frequency of bipolar devices should occur at relatively low frequency. In contrast, in field effect device technologies, current transport is parallel to and less than 100 nm from the material surface and a number of heterointerfaces. As a consequence, the 1/*f* corner frequency of MESFETs and HEMTs will be relatively high. Above the 1/*f* corner frequency, the noise is determined by the current shot noise, together with Johnson noise from resistors in the current path. In bipolar technologies, contributions to the Johnson noise arise from access resistance in the emitter and collector regions, the intrinsic resistance of the base and the dynamic resistance of the base–emitter and base–collector junctions. In MESFET and HEMT technologies, the source and drain access resistances and channel resistance contribute to the Johnson noise.

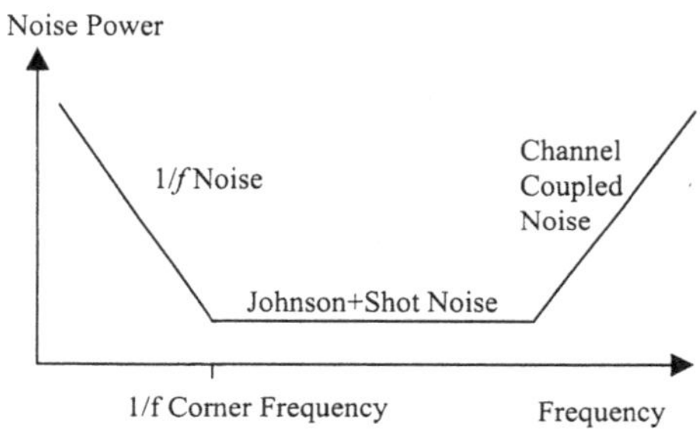

Figure 2.43 Frequency-dependent noise power density spectra of a three-terminal device

As the frequency increases, the device capacitances couple the combined shot and Johnson noise arising within the device to the input and output terminals. At the input, the coupled noise current combines with any base (gate) current and is dropped across the base (gate) resistance, giving rise to an input voltage noise source. This noise voltage is amplified in the device and combines with the noise voltage source arising from capacitive coupling to the output to produce an output noise signal. As the frequency increases, the capacitive coupling becomes more efficient and so the noise increases. From the foregoing, it should be clear that a low noise device should have low capacitance, small input resistance and a low internal resistance, to reduce the Johnson noise, which can be capacitively coupled to the input and output. In addition, the noise will be current-dependent, reflecting the shot noise contribution.

Table 2.4 compares the noise performance of contemporary bipolar and field effect technologies. As expected, the 1/*f* noise performance of bipolar devices is unsurpassable, making these technologies particularly attractive for the realisation of oscillator circuits having low phase noise. At high frequencies, HEMT technologies offer the best noise performance, due to the low internal device resistances and the key role that T-gate technology plays in producing a small-capacitance short gate with low resistance.

Table 2.4 Comparison of noise performance of bipolar and field effect transistor technologies

Technology	1/f corner frequency	Minimum noise figure	Associated gain
Si bipolar	< 1kHz	1.5 dB @ 2 GHz	21 dB @ 2 GHz
SiGe HBT	< 1kHz	0.65 dB @ 2GHz 3 dB @ 12 GHz	21 dB @ 2 GHz 5 dB @ 12 GHz
GaAs HBT	< 1kHz	0.4 dB @ 2 GHz	25 dB @ 2 GHz
GaAs MESFET	>10 MHz	0.8 dB @ 12 GHz	12 dB @ 12 GHz
GaAs pHEMT	>10 MHz	0.25 dB @ 2 GHz 1 dB @ 18 GHz	16 dB @ 4 GHz 10 dB @ 18 GHz
InP HEMT	>10 MHz	0.3 dB @ 18 GHz	17 dB @ 18 GHz
GaAs mHEMT		0.4 dB @ 18 GHz	11.5 dB @ 18 GHz

2.6.3 Power and linearity performance of bipolar and field effect devices

Power handling and linearity performance are important figures-of-merit for power amplifier design. The device power handling capability can be defined in terms of the maximum and minimum voltage and current levels that can be supported by the transistor. As shown in Figure 2.44, the voltage swing is determined by the values of V_{dsmin} and V_{dsmax}, which in turn are governed by the knee-voltage and breakdown voltage of the devices, respectively. From a device technology viewpoint, therefore, it is important to minimise the device access resistance, whilst maximising the breakdown voltage and simultaneously maximising the drive current capability for a given device geometry.

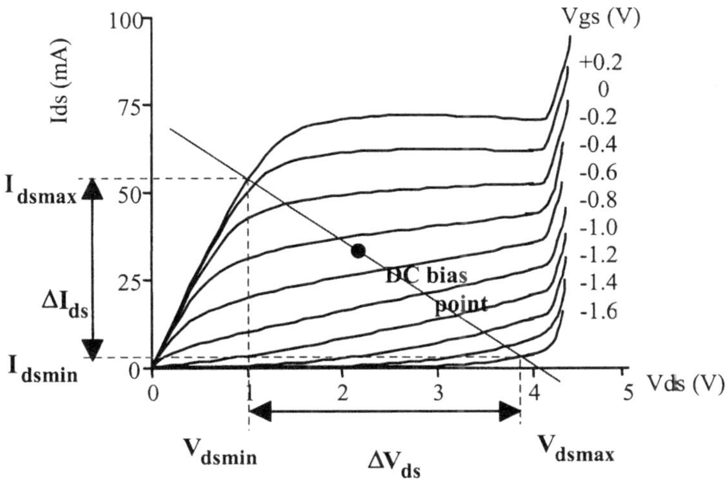

Figure 2.44 Maximum and minimum voltage and current levels to determine transistor power handling performance

The breakdown voltage is governed by the peak electric field in the device, which for a given operating voltage tends to be lower in bipolar technology than for a field effect device. Therefore, the breakdown voltage in bipolar devices tends to be superior, making bipolar devices attractive for high power applications. Despite this, the breakdown voltage of HEMT technologies can be engineered by utilising double gate recess regions and asymmetrical placement of the gate in the source-drain gap, both of which combine to reduce the peak electric field at the drain end of the gate. InP-based HEMTs tend to have inferior breakdown properties, due to the relatively narrow bandgap of the typical layers in such a device (e.g. $In_{0.53}GaAs$). As a result, impact ionisation effects in the device channel are particularly severe for InP-based HEMTs, limiting their usefulness in power amplifier applications.

The simultaneous delivery of high frequency performance and good breakdown characteristics is a particularly attractive possibility with metamorphic GaAs HEMTs, as this technology offers the full range of indium concentrations in the device channel. This freedom allows a trade-off to be made between RF gain and breakdown, which may be an important consideration in millimetre-wave applications such as broadband wireless data distribution and optoelectronic modulator drivers for 40/80 Gb/s applications.

Currently, linearity is a key power amplifier performance figure-of-merit, as the specification of 3rd generation CDMA mobile wireless systems puts severe limitations on the intermodulation performance and adjacent channel spectral regrowth of the transistor technologies from which amplifiers are designed. As an example of this, in contemporary CDMA systems, many power amplifiers are currently run well below saturation, simply to meet 3rd order intermodulation specifications. Similar situations arise in the design of mixers for 3rd generation applications.

Table 2.5 summarises the power handling and linearity performance of both bipolar and field effect technologies. Direct comparisons of both power handling and linearity are constrained by the DC operating conditions, which vary for the different technologies and the RF drive conditions for which the 3rd order intermodulation data is gathered; therefore, the data presented in Table 2.5 should be interpreted with some caution. Nevertheless, it is clear that both HBT and HEMT technologies are delivering impressive power and linearity performance, which the designer is now able to exploit.

Table 2.5 Comparison of power and linearity for various transistor technologies

Technology	Frequency	Gain	1 dB compression point	3rd order intercept point	Power added efficiency
Si bipolar	900 MHz	15 dB			71 %
SiGe HBT	900 MHz	12.5 dB		35 dBm	72 %
GaAs HBT	1.8 GHz	20 dB	0.25 W	42 dBm	>40 %
GaAs pHEMT	1.8 GHz		2 W	48 dBm	60 %

2.7 Summary

In this chapter, the wide range of active and passive device technologies utilised in contemporary MMIC realisation have been reviewed and key RF performance parameters, such as bandwidth, gain, noise, power handling and linearity have been compared. From the information presented in Section 2.6, it should be clear that no single technology simultaneously offers optimal performance, when considering key RF figures-of-merit. Therefore, the application specification, cost, yield and reliability will drive the final technology choice.

2.8 References

1 WILLIAMS, R. E.: 'Modern GaAs processing methods' (Artech House, 1990)

2 LADBROOKE, P. H.: 'MMIC design; GaAs FETs and HEMTs' (Artech House, 1989)

3 Proceedings of 1998 IEEE topical meeting on *Silicon Monolithic Integrated Circuits in RF Systems*. 1999

4 YUAN, J. S.: 'SiGe, GaAs and InP heterojunction bipolar transistors' (Wiley Series in Microwave and Optical Engineering, 1999)

5 KOBAYASHI, K. W. *et al.*: 'HBT IC technology for wireless applications beyond 2000'. *1997 IEEE GaAs IC symposium short course on IC Technology for Wireless Applications*, 1997

6 Ross, Svensson and Lugli (Eds): 'Proceedings of the NATO Advanced Study Institute on *Pseudomorphic HEMT technology and applications*'. (Kluwer Academic Publishers, 1996)

Chapter 3

Passive components

M. Gillick, I. D. Robertson and S. Lucyszyn

3.1 Introduction

It is the extensive use of passive components that makes MMIC design so different to conventional integrated circuit design and layout. In MMIC design, passive components are used for impedance matching, DC biasing, phase-shifting, filtering and many other functions. This chapter describes the key passive components that are used in MMIC design. These components include not only the basic lumped inductors, capacitors and resistors, but also a wide range of distributed transmission-line components. These transmission-line components include microstrip lines and elements such as bends and T-junctions, as well as standard building blocks like couplers and power splitters/combiners. Most often, these building blocks are realised in the microstrip transmission-line medium. However, other types of transmission line such as coplanar waveguide (CPW), slotline, and 'thin-film microstrip' are being used more and more. The principal advantages to be gained from using CPW are increased packing density of circuits and reduced dispersion for millimetre-wave operation. Transitions between the different transmission-line media can be used to realise components such as baluns. At the lower microwave frequencies transmission-line components are often too large to be practical. To overcome this, the standard microstrip components such as couplers and power splitters can be realised with lumped-element equivalent networks or by using the lumped-distributed miniaturisation technique. These approaches are an interesting example of how the precisely defined passive components available on MMICs can be used to develop new circuit techniques which overcome the fundamental size limitation not encountered with MIC circuits.

There are relatively few design equations in the description of the passive components since these are highly process-dependent. For a particular MMIC fabrication process, the manufacturer's design manual is the essential authority on the exact design rules for component layout as well as the appropriate design curves and equations. Nevertheless, in many cases the design manual is little more than a reference guide to the process data, and this chapter gives a great deal of background information which is essential for all but the most experienced designer.

3.2 Inductors

Depending on the inductance required, MMIC inductors can be realised either as straight narrow tracks (ribbon inductors), as single loop inductors, or as multi-turn spiral inductors. A microstrip ribbon inductor and its equivalent circuit are shown in Figure 3.1. For short lengths ($<\lambda_g/4$), the inductance and shunt end capacitances can be calculated from the following well known equations [1]:

$$L = \frac{Z_0}{2\pi f}\sin\left(\frac{2\pi l}{\lambda_g}\right) \tag{3.1}$$

and

$$C = \frac{1}{2\pi f Z_0}\tan\left(\frac{\pi l}{\lambda_g}\right) \tag{3.2}$$

A narrow track with high Z_0 is needed to achieve high inductance with low parasitic capacitance. However, in practice the choice of track width is determined by fabrication limits, by the DC current carrying capacity and by the high resistance of very narrow tracks. The track length is limited simply by the need to ensure a realistic and economical chip size. The ribbon inductor is thus limited to values of less than 1 nH, but is a relatively 'pure' inductor with low parasitics. Hence it is often used in distributed amplifiers where very large bandwidths are required. Distributed amplifiers are described in detail in Chapter 5.

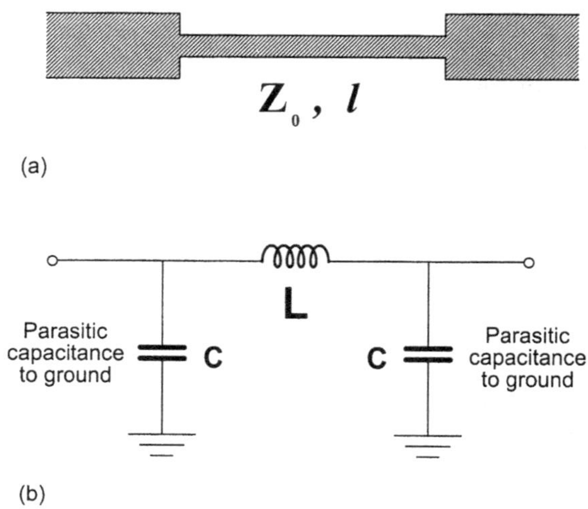

Figure 3.1 The ribbon inductor: (a) microstrip layout and (b) equivalent circuit

3.2.1 Loop inductors

Single loop inductors like the one shown in Figure 3.2 were used extensively in the pioneering days of MMICs. This is probably because the processing did not at first offer air-bridges for spiral inductors, and because the limited experience of coupled lines on GaAs made designers reluctant to use meandered ribbon inductors. It is fair to say that in recent years loop inductors have been used very little because of their inefficient use of chip area. A number of texts give useful design equations for loop inductors [2–4].

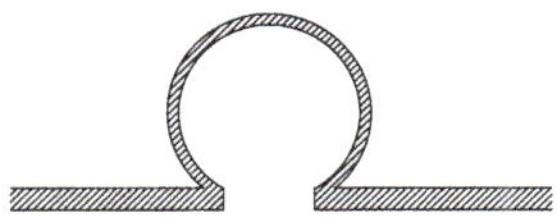

Figure 3.2 Layout of a single loop inductor

3.2.2 Spiral inductors

Spiral inductors are essential for values above approximately 1 nH. Comparing a spiral inductor with a meandered ribbon, one can see from Figure 3.3 that for the meandered inductor (*a*) the current in adjacent tracks is flowing in opposite directions, whereas for the spiral inductor (*b*) the current in adjacent tracks is in the same direction. The resulting mutual inductance yields a significant increase in the spiral inductor's overall inductance. In the meandered case, however, the inductance is reduced and the overall inductance is significantly less than that of a straight track equal to the unwound length. Since the DC resistance is the same, the spiral inductor thus has higher Q.

Figure 3.3 Large value inductance using (a) meandered track and (b) spiral

The drawback of the spiral is that the need to connect the centre turn back to the outside circuit dictates that either air-bridge crossovers or dielectric spaced underpasses must be used. There are a number of different solutions to this same fundamental connection problem, and these are illustrated in Figure 3.4. Note that square inductors are shown; rectangular ones could be used if layout constraints made them necessary, and (approximately) round ones would have slightly superior performance at the cost of greatly increased layout complexity and a less convenient shape for integration with other components. However, these alternative inductors would have to be characterised first.

Inductor type (*a*) has a single air-bridge span from the centre of the spiral to the outside. Whilst this type has been used successfully, it has the major drawback that the span of the air-bridge is perilously large, especially when the inductor has many turns. The maximum span of an air-bridge is limited by fabrication technology and reliability issues. It is difficult to ease the problem by having landing posts between the turns because adjacent tracks are very close. Hence, the more commonly accepted air-bridge solution is the type (*b*). In this type of spiral inductor the connection to the centre turn stays at the substrate level, and each turn of the inductor jumps over it using a short air-bridge. This type thus has more air-bridges, but they are each of minimum length so that reliability is improved. The third type of spiral inductor (*c*) actually forms the entire spiral out of air-bridges, which jump from landing post to landing post. The advantage of this structure is that since the track spends much of the time off the substrate surface, the parasitic capacitance between turns and to ground is reduced, giving a higher useful maximum operating frequency.

The alternative to using air-bridges is to employ a two-level metal process with a spacer dielectric. Polyimide is often chosen as it has a low ε_r and because thick films can be deposited very quickly. In this type of inductor (*d*), the turns are normally on the top metal layer above the dielectric so that the parasitic capacitance is reduced, and so that the turns can be plated up for less series resistance. The underpass is necessarily of a much thinner metal, and may thus be made quite wide in order to maintain a usable current carrying capacity and minimise series resistance. The connection between the upper metal and the underpass is made with a via in the dielectric(s). These vias should not be confused with the through-substrate vias used for grounding components.

On standard silicon substrates the spiral inductor has parasitics which are so severe that the inductor is virtually useless. Thus, there has been a great deal of activity to make improved inductors on silicon, using either multilayer techniques or micromachining, both of which are discussed in later sections.

3.2.3 Stacked spirals

Stacked spiral inductors consist of a pair of inter-wound spirals placed on separate metal layers. The major advantage is that the turns are much more tightly packed than normal photolithography and metal patterning would allow for a single layer spiral, and since they are separated vertically to some extent there is less capacitance between adjacent turns than there would normally be with such small gaps. However, since the dielectric spacer can only planarise the circuit to a certain extent, the lower metal thickness is limited to 1 μm or less, and this means that the stacked spiral has a high series resistance. The inductance per unit area of stacked spiral inductors is very high, but their resonant frequency is much lower, and so they are limited to frequencies below 5 GHz or so. They often find application as bias chokes, where high inductance is required, and in this application the high series resistance can actually be beneficial to the amplifier's stability, matching and bandwidth.

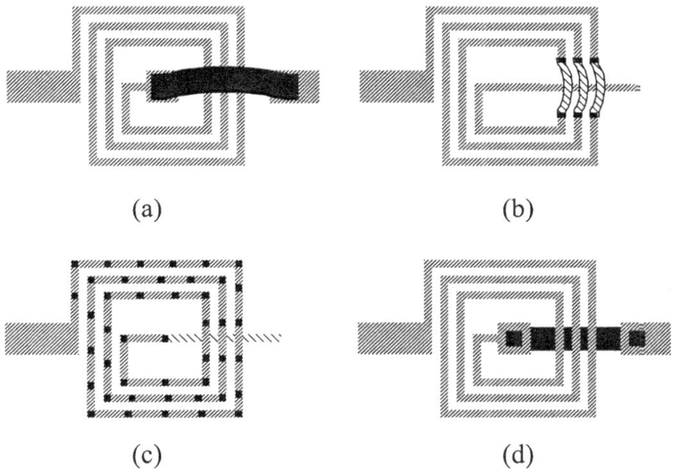

Figure 3.4 Spiral inductors: (a) single air-bridge, (b) air-bridges over an underpass, (c) formed entirely of air-bridges and (d) using two metal levels for an underpass

3.2.4 Spiral inductor models

The basic inductor equivalent circuit consists of a prime inductance along with its associated series resistance, inter-turn and crossover feedback capacitance, and some capacitance to ground. This lumped element model is shown in Figure 3.5. For silicon substrates it is necessary to add shunt resistance elements to ground, which are coupled via capacitors if there is an added dielectric layer or oxide. Direct calculations of prime inductance can be made using the classical formulas given by Grover [4], with appropriate correction factors [5]. It is more common, however, for the equivalent circuit data to be found for a number of different inductor geometries, and to use empirical curve-fitted expressions for the values of the equivalent circuit elements [6]. Any equivalent circuit using lumped elements can only accurately model the component when its dimensions are less than approximately $\lambda/20$. In practice, such lumped equivalent circuits are accurate to frequencies above the self-resonant frequency of the inductor, beyond which one could argue that the 'inductor' is no longer useful anyway. However, in a circuit such as a mixer, an inductor may be used for IF matching at 1 GHz in a circuit which has a 20 GHz RF input. In this case it is essential that the inductor is modelled properly at both the IF and at the RF frequency. In order to achieve an accurate model at such high frequencies the distributed nature of the inductor must be considered. In essence, this means treating the inductor as an interconnection of multiple coupled microstrip lines. For a square inductor it is surprisingly effective to consider each side of the inductor as a separate coupled-line problem, with special treatment for the corners [7]. A number of quite rigorous approaches have been developed for the modelling of spiral inductors [8]. The coupling between adjacent spiral inductors has received far less theoretical treatment, but some very useful practical investigations have been made [9]. In recent years, numerical electromagnetic field solutions have been reported [10], and this type of modelling is described further in Chapter 4. This type of modelling can be used in conjunction with measured inductor data to develop inductor models which are accurate and computationally efficient for incorporation into standard CAD packages. This approach allows more flexibility to the designer compared with the approach where a large number of standard inductor geometries are characterised by the foundry. Nevertheless, the foundry's own models should always be used in preference to those from any CAD package, unless that particular foundry has evaluated the alternative models for its own inductors.

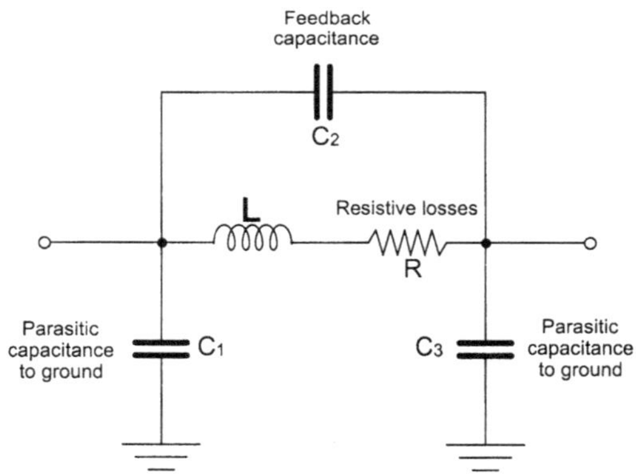

Figure 3.5 Inductor model

3.2.5 Planar spiral transformers

The planar spiral transformer was first employed on MMICs by Honeywell [11,12]. Figure 3.6 shows a transformer employing underpasses, although air-bridge techniques can also be employed. The transformer offers DC blocking, matching and DC bias injection functions in a very small size. As a result, transformer-based circuits can achieve exceptionally high packing density, and Pacific Monolithics made a successful business based on circuits using this technique. In particular, they sold the first commercially successful MMIC for C-band DBS TV receivers.

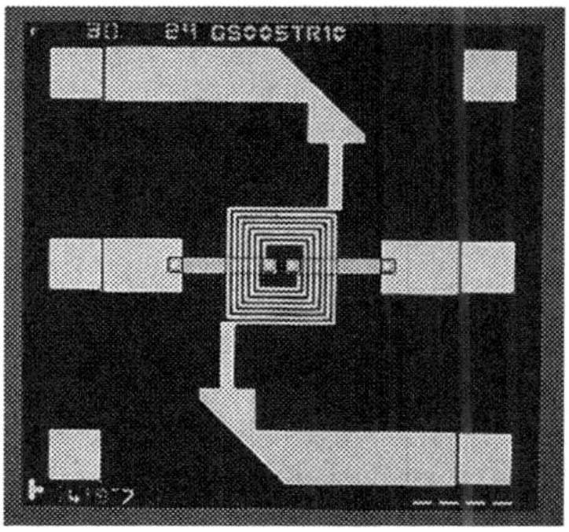

Figure 3.6 Photograph of a spiral transformer (Courtesy of Marconi Caswell Ltd)

The planar spiral transformer has a high capacitance between the two spirals, and this makes the transformer resonate as the capacitive coupling becomes dominant at higher frequencies. To minimise this parasitic capacitance whilst achieving a high mutual inductance, the turns need to be very narrow and close together. Typically a transformer would have 5 μm tracks and 5 μm gaps. Inevitably, the narrow tracks lead to high series resistance and so planar transformers are rather lossy. In a circuit such as an amplifier, matching can be achieved by resonating the self inductance with the transistor capacitance, rather than by employing a turns ratio for impedance transformation. Employing a different number of turns in the basic transformer does not necessarily give the desired effect, because the magnetic field of the larger coil is not tightly coupled to the smaller coil. The lumped element equivalent circuit is only valid to frequencies up to and just over the resonant frequency of the transformer.

The 'transformer' becomes a much more useful component if it is treated as a coupled transmission-line structure. The transformer of Figure 3.6 has a resonant frequency of approximately 6 GHz [13], and is only usable as a true lumped transformer up to 4 GHz or so. Normal centre-tapped transformer balun operation is very hard to achieve because of parasitic capacitance. However, the same structure can be used a quadrature backward-wave coupler with a centre frequency roughly corresponding to an unwound length of $\lambda_g/4$ [14]. A single 'tri-filar' transformer [15] or a pair of transformers [16] can be used to realise a Marchand balun, which offers broad-band balun operation. These baluns are described in Chapter 7.

3.3 Capacitors

Both overlay capacitors (metal-insulator-metal (MIM) capacitors) and interdigital capacitors can be used on MMICs. Interdigital capacitors can be used for values up to approximately 1 pF, above which their size and the resulting distributed effects prevent their use. Overlay capacitors are therefore essential for all but the simplest circuits, and are particularly important for DC blocking and decoupling, where large capacitor values are required.

3.3.1 Overlay capacitors

These consist of a metal-insulator-metal 'sandwich', with the most common insulators being silicon nitride, silicon dioxide, and polyimide. Silicon nitride is popular since it has a fairly high ε_r, and can also be used for passivating the exposed GaAs in the active devices. In processes which use polyimide as the spacer dielectric for spiral inductors, the polyimide can also be used to realise small value capacitors for applications such as matching networks and filter inter-resonator coupling, and they are significantly smaller than interdigital capacitors. The type of connection used from the capacitor to the rest of the circuit depends on whether air-bridges or a polyimide-based two metal level process are used. In a two metal level process the main microstrip circuitry is normally on the upper metal, and so the lower plate has to be brought up to the top metal through a via-hole in the dielectric, as shown in Figure 3.7*a*. Alternatively, with air-bridges, the upper plate is connected to the rest of the circuit with an air-bridge as shown in Figure 3.7*b*. This is preferable to the method shown in Figure 3.7*c*, because the sharp edge along the interconnection would make a short-circuit failure more likely and reduce the breakdown voltage rating of the capacitor. Since the width of air-bridges is limited by the need during fabrication to remove the temporary photoresist support, often several air-bridges would be put in parallel in order to achieve a low inductance connection, rather than using a single wide air-bridge.

(a)

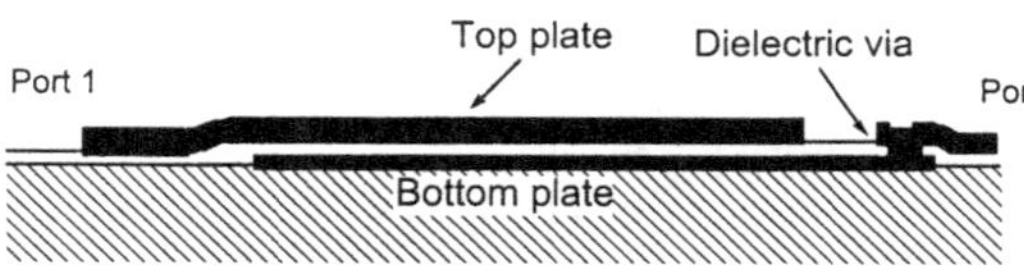

(b)

(c)

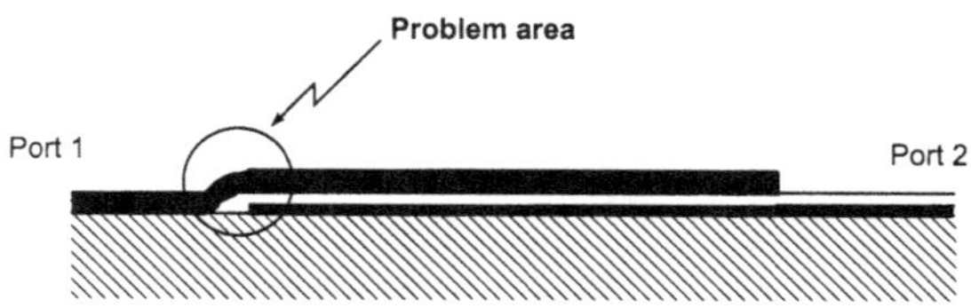

Figure 3.7 Overlay capacitors: (a) using a dielectric via (b) with an air-bridge and (c) without an air-bridge or spacer dielectric

Figure 3.8 shows the typical equivalent circuit model of a monolithic capacitor. On high resistivity substrates the major parasitic is often the shunt capacitance to ground, especially for larger capacitances. On standard silicon substrates there will be severe substrate parasitics. This simple model is not truly distributed and cannot represent real layout situations where asymmetrical or multiple connections are made to the capacitor. Process-oriented models have been developed which model the distributed nature accurately and model realistic layout scenarios [17].

Finally, it should be mentioned that the exact size of the fabricated capacitor may often be slightly different to the layout dimension, and this is a particularly important effect for small capacitors. The manufacturer's foundry design manual should be consulted to make sure that the correct allowances are made at the layout stage.

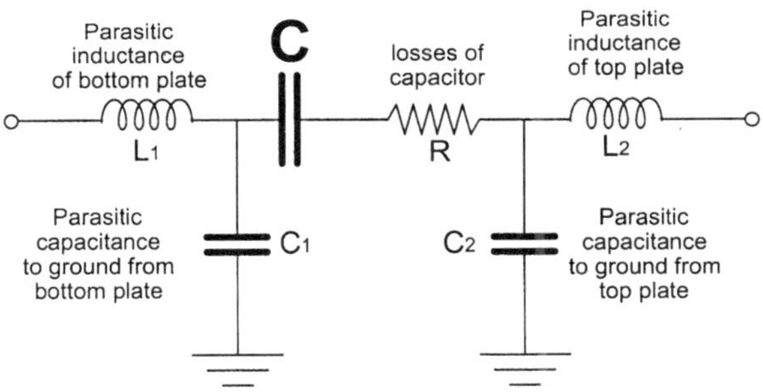

Figure 3.8 Capacitor model

3.3.2 Interdigital capacitors

These consist of a number of interleaved microstrip fingers coupled together as shown in Figure. 3.9. The maximum value of an interdigital capacitor is limited by its physical size, and its maximum usable operating frequency is limited by the distributed nature of the fingers. They certainly cannot be used for values above 1 pF, and even a 0.5 pF interdigital capacitor will measure approximately $400 \times 400\ \mu m^2$. Nevertheless, since interdigital capacitors do not use a dielectric film, their capacitance tolerance is very good and is limited only by the accuracy of the metal pattern definition. Hence they are ideal as tuning, coupling and matching elements, where small capacitor values are required and precise values are necessary.

Figure 3.9 Interdigital capacitor

3.4 Resistors

Resistors on MMICs can either use the doped semiconductor layer (mesa resistors or implanted planar resistors) or a deposited thin-film resistive layer. In either case, since the layer or film thickness is fixed it is very convenient to quote resistivity in terms of an ohms-per-square figure. Hence, the value of the resistor is chosen by selecting a suitable aspect ratio, as shown in Figure 3.10. The absolute size of the resistor is determined by the tolerance required and the expected power dissipation. The area of the ohmic contact pads is particularly important in determining the tolerance of the resistor, as their contact resistance is difficult to control precisely and must be minimised by making the pads suitably large. A practical limit is imposed by the higher parasitic capacitance of large pads and the resistor's physical size.

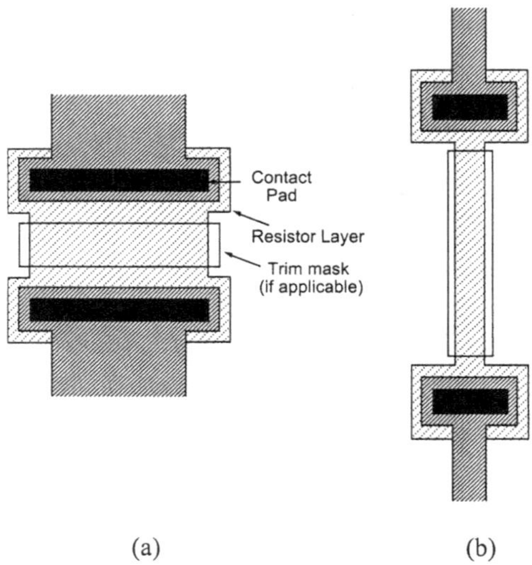

Figure 3.10 Resistor examples: (a) small value (≈50 Ω) and (b) large value (≈3000 Ω)

3.4.1 GaAs resistors

The term 'mesa' refers to the situation where the whole wafer is doped and subsequently selectively etched, so that active regions remain only where required and are left standing proud of the substrate surface. This is the case when epitaxial growth or blanket ion-implantation are used. However, selective ion-implantation or proton isolation are often used, and in this case the active regions remain flush with the wafer surface. Hence, resistors using the GaAs active layer can be either mesa resistors or implanted planar resistors. Figure 3.11 shows the typical form of a mesa resistor. The GaAs resistor relies on the linearity of the semiconductor's current-field characteristic at low electric field values. Hence, it is important to consider how much current is to be passed through the resistor. A design rule specifying the maximum electric field allowed in the resistor is normally given as a figure of volts-per-unit-length for a certain permissible percentage deviation from perfect linearity. If a resistor is used for the biasing of a FET gate (or bipolar transistor base) this does not present a problem, but for drain or collector biasing it is very easy to exceed the maximum ratings if care is not taken. Resistor biasing in any case greatly increases the total DC power consumption, and the operating temperature of the resistor should be considered. For mesa resistors, the orientation of the resistor must be considered when calculating the exact layout dimensions of the resistor. The crystallographic orientation means that the mesa slope along the edges will be either positive or negative, depending on whether the resistor is horizontal or vertical on the layout. This makes the effective width different and necessitates a small correction factor. For mesa resistors the active region may undergo an additional etching process (trimming), intended to increase the ohms-per-square figure and improve the tolerance on resistors. In this case the resistor will require an additional mask level to facilitate this etching without affecting the active devices.

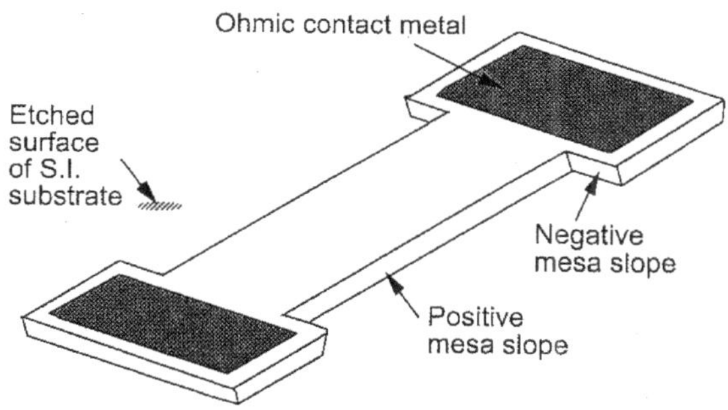

Figure 3.11 Mesa resistor view showing positive and negative mesa edges

3.4.2 Thin-film resistors

Sputtered thin-film resistors offer improved linearity and lower temperature coefficients compared with the mesa types. In addition, the Ω/square figure can be optimised for the circuit designer without any limitations imposed by the requirements of the active devices. The most commonly used materials are Tantalum Nitride, Cermet, and Nickel Chrome. Their temperature coefficient of resistance is less than one tenth that of GaAs, and sheet resistances of 50 Ω/square can be produced, which is convenient for the circuit designer.

3.4.3 Resistor models

Fortunately, the monolithic resistor is relatively free of parasitics. However, a frequency-dependent resistance results from the frequency dependence of the skin-depth. In all but the most broadband circuits this can be accounted for at the design stage. The parasitic shunt capacitance to ground of the resistor and its contact pads, and the small end-to-end feedback capacitance of the resistor, have to be included in any simulation. The typical equivalent circuit model of a resistor is shown in Figure 3.12. The distributed nature of the resistor is taken into account with the series inductance. For greater accuracy it may be necessary to model the resistor as a proper distributed RLC ladder, or as a lossy transmission line.

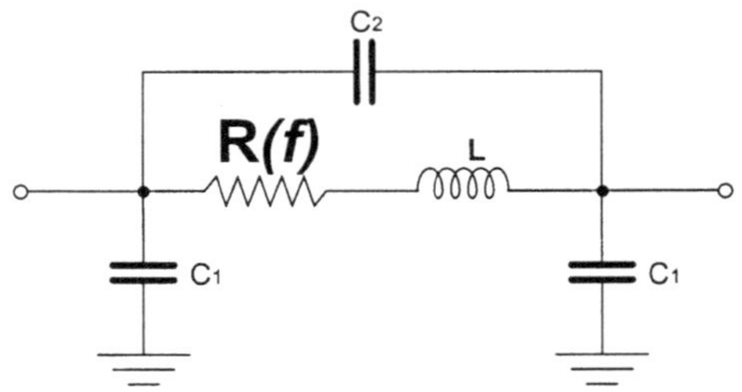

Figure 3.12 Resistor equivalent circuit

3.5 Via-holes and grounding

The through-substrate via-hole, shown schematically in Figure 3.13*a*, is an essential component for all but the most simple microwave circuits as it enables a low inductance ground to be established within the circuit. Without via-holes, devices would have to be placed very near the edge of the chip and grounded with either wrap-around metallisation or bond-wires, as illustrated in Figures 3.13*b* and *c*. These methods of grounding impose severe restrictions on the topology of the circuit. However, even when via-holes are used there are several pitfalls regarding grounding to be avoided at the design and layout stage. The most important principle to remember is that a mass of metal can never be assumed to be a 'ground bar'. The grounding connection must always be considered as a transmission-line element in its own right and modelled accordingly. Figure 3.14 shows a classic example of a circuit layout based on the 'ground bar' principle that may not work in practice. In Figure 3.14*a* the FETs of a two-stage amplifier share a common grounding pad, which is grounded with a via-hole. As the equivalent circuit shows, this grounding method provides a direct feedback path that is almost certain to cause instability. The solution in this case is to provide each FET with its own ground pad, as shown in Figure 3.14*b*. The source inductance is not removed, but there is no longer a feedback path.

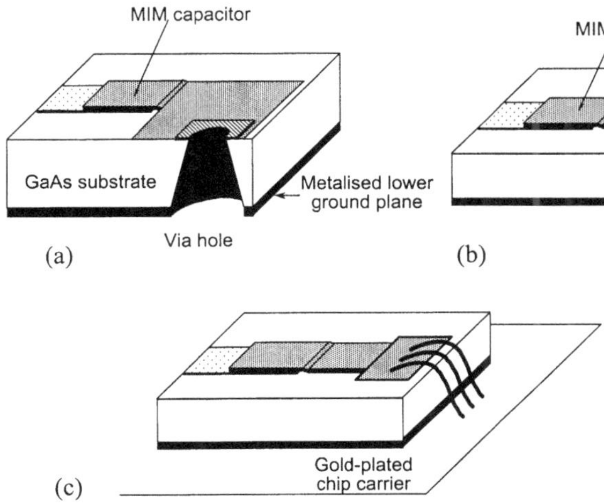

Figure 3.13 Grounding methods: (a) through-substrate via-holes, (b) wrap-around grounding and (c) bond-wires

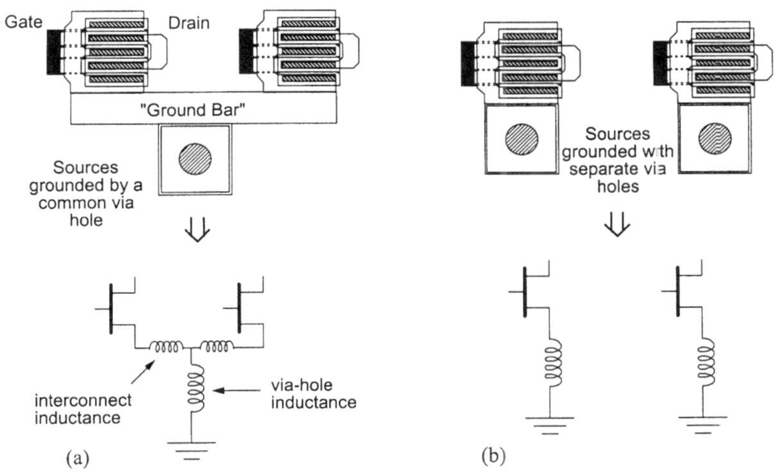

Figure 3.14 An example of improper grounding: (a) FETs with a common source ground pad, leading to unwanted feedback and (b) solution: separate source ground pads

A second example of the inappropriate use of a 'ground-bar' is shown in Figure 3.15. The ground bar is used as a 'neat' common ground for the DC bias decoupling capacitors in a two-stage amplifier. However, the distributed nature of the ground bar means that none of the decoupling capacitors are truly grounded, and so a dangerous feedback path is again created. In many cases, when two components share a single via-hole ground, even the inductance of the via-hole itself is enough to allow the two components to interact undesirably. This can normally be detected by proper modelling.

For MMIC design in the millimetre-wave range, even more subtle grounding effects have become apparent, such as the mutual coupling between via-holes placed in close proximity. In this situation, the inductance is much higher than expected as the current crowds together down one side of each via-hole. Hence, because grounding is so critical in millimetre-wave circuits, even the through-substrate via-hole ground may not offer low enough inductance. CPW offers a very attractive solution to this problem, and is becoming increasingly popular as the design tools improve.

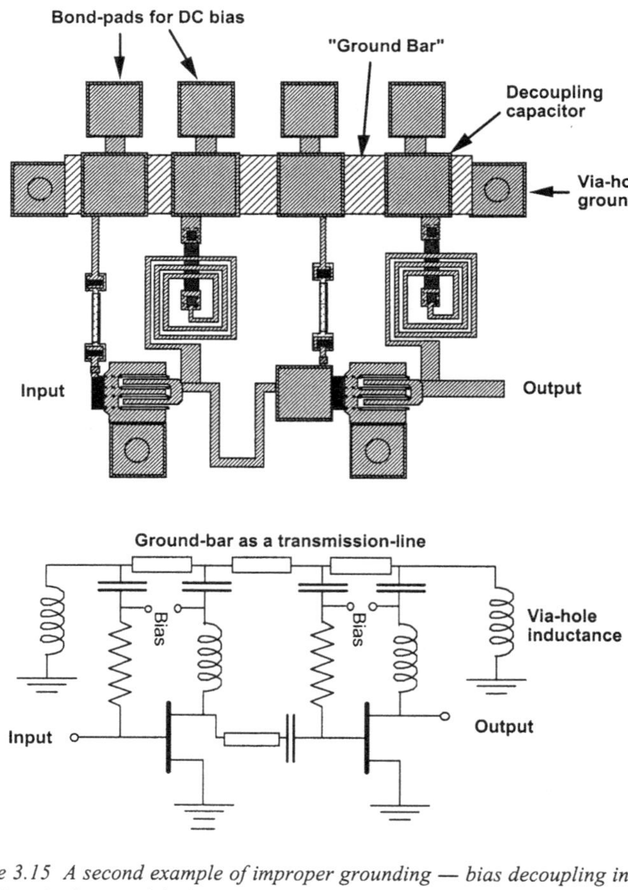

Figure 3.15 A second example of improper grounding — bias decoupling in a two-stage amplifier: (a) layout of the bias networks and (b) equivalent circuit

3.6 Microstrip components

For many years microwave CAD packages have offered a wide range of models for microstrip lines and discontinuities. However, most of these have been developed principally for MIC circuits on substrates such as alumina and Duroid™. MMICs have much smaller features than MICs, and it is important to ensure that the standard models are not used outside their range of validity. The models are further limited because MMICs often have extra dielectric passivation and spacer layers. Many CAD packages have addressed these limitations and now provide models that are suitable for MMIC microstrip elements. However, the problem remains that on MMICs the components are packed more tightly and the coupling between elements must be accounted for. As a result, it may be necessary to employ field-based simulation for some parts of a circuit layout. All these aspects of CAD and modelling are described in more detail in Chapter 4, but it is necessary to highlight the problem here before describing the use of transmission-line components in MMIC design.

3.6.1 Transmission losses

The total loss of a microstrip line is made up of conduction loss, substrate dielectric loss and radiation loss. With semi-insulating GaAs substrates the conduction losses predominate. With silicon, the substrate resistivity greatly affects the losses; most processes use low resistivity substrates and the substrate loss is enormous. Even if an insulating layer is used – either a thermal oxide (SOI technology) or a deposited thin film – the capacitive coupling through the layer at high frequencies means that there is little benefit to the microwave loss. The same is true of lumped elements. To achieve low loss with silicon substrates, it is necessary to use high resistivity substrates, which may be more expensive and impose some processing restrictions. Losses in microstrip lines on silicon substrates have received widespread attention [18–24], and it is found that conductor loss is dominant if the substrate resistivity is in the order of several thousand Ωcm. Figure 3.16 shows the simulated loss at 60 GHz vs. substrate resisitivity, for a 50 Ω line on a 200 μm silicon substrate. For these simulations (on *em*™) the conductors are lossless. It should be noted that the characteristic impedance of lossy lines is not purely real, and special techniques must be used for on-wafer calibrations and for accurate determination of the complex characteristic impedance [25,26].

At high frequencies, the current flowing in a conductor is mostly confined near the outer surface of the conductor. The skin depth, δ_s, of a conductor is defined as the distance in the conductor (along the direction normal to its surface) in which the current density drops to $1/e$ of its surface value. The current decays exponentially to a negligible value at a distance of about 4 to 5 skin depths. The conductivity of typical metals (e.g. gold for GaAs and InP processes and aluminium for silicon and SiGe processes) used in MMICs is high, although finite. For these metals, the skin depth at microwave frequencies is very small, as can be seen in Table 3.1. Most GaAs and InP foundries adopt two different metal layer thicknesses. The top-level metal, used to provide low ohmic-loss lines, has a thickness of the order of 3 μm, while the lower (under-pass) metals are only of the order of 0.5 μm thick.

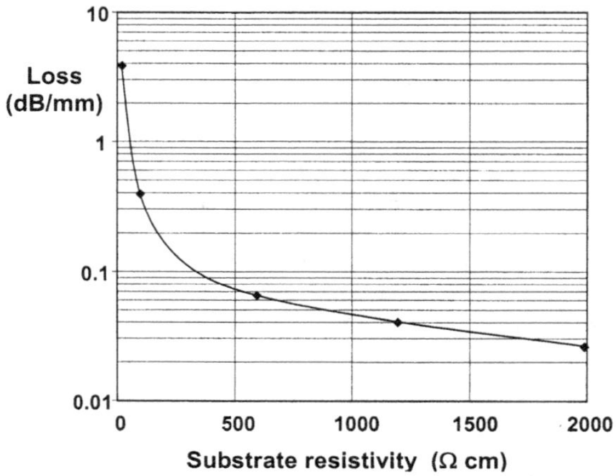

Figure 3.16 Simulated transmission loss at 60 GHz vs. substrate resistivity, for a 50 Ω microstrip line on 200 μm thick silicon (starting at 10 Ω cm, and with lossless conductors)

Table 3.1 Calculated skin-depths

	Gold		Aluminium	
Frequency (GHz)	δ_s (μm)	$5 \times \delta_s$ (μm)	δ_s (μm)	$5 \times \delta_s$ (μm)
1	2.49	12.43	2.67	13.37
5	1.11	5.56	1.20	5.98
10	0.79	3.93	0.85	4.23
30	0.46	2.27	0.49	2.44
60	0.32	1.60	0.35	1.73
77	0.28	1.42	0.30	1.52
110	0.24	1.19	0.26	1.28
140	0.21	1.05	0.23	1.13

3.6.2 Microstrip couplers and power splitters

Microstrip couplers and power splitters are used extensively in MMIC balanced amplifiers, power amplifiers, mixers and phase shifters. Many well-known texts give extensive accounts of the design of Lange, branch-line and rat-race couplers, and Wilkinson power splitter/combiners [1]. For completeness, these components are illustrated in Figures 3.17–3.20. Since they employ lines of length $\lambda_g/4$ or more, they tend to be used sparingly below 10 GHz. The Lange coupler is widely used since it readily provides a broadband 90° phase split. In comparison, the branch-line coupler has a narrow bandwidth and has an inconvenient shape, which is very wasteful of chip space. Nevertheless, the branch-line coupler has been used successfully, particularly in millimetre-wave circuits, where it can offer lower insertion loss than the Lange coupler which has rather narrow fingers in MMIC implementations. As a guide, on a 200 μm thick GaAs substrate the fingers of a Lange coupler are typically 20 μm wide with 10 μm gaps. In determining the exact layout dimensions, frequency dispersion and the effect of any dielectric passivation layers must be taken into account [17].

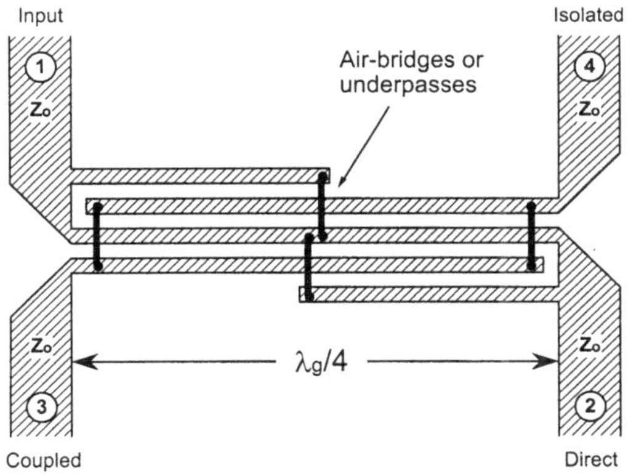

Figure 3.17 Layout of the Lange coupler

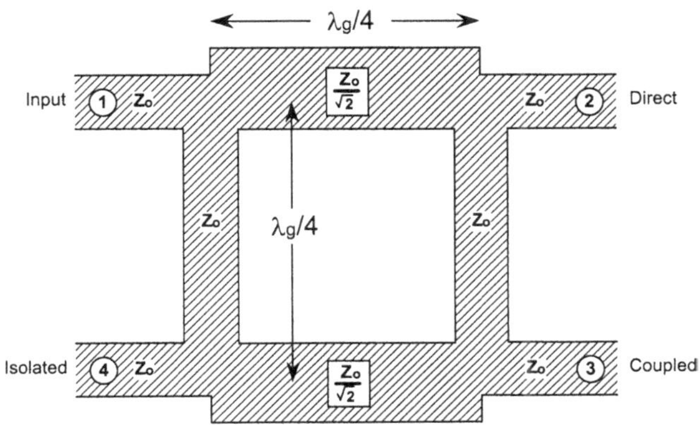

Figure 3.18 The microstrip branch-line coupler

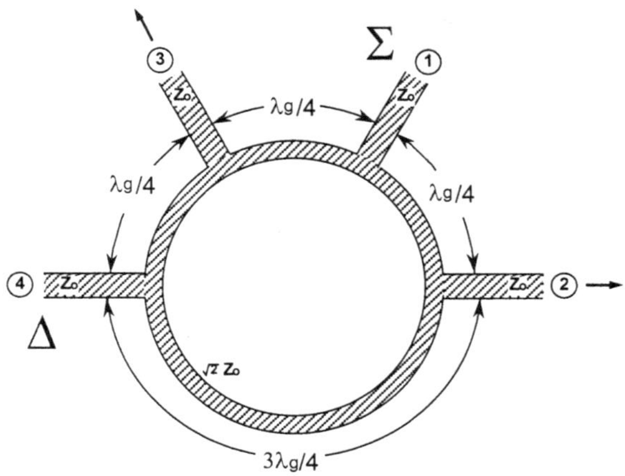

Figure 3.19 The microstrip rat-race coupler

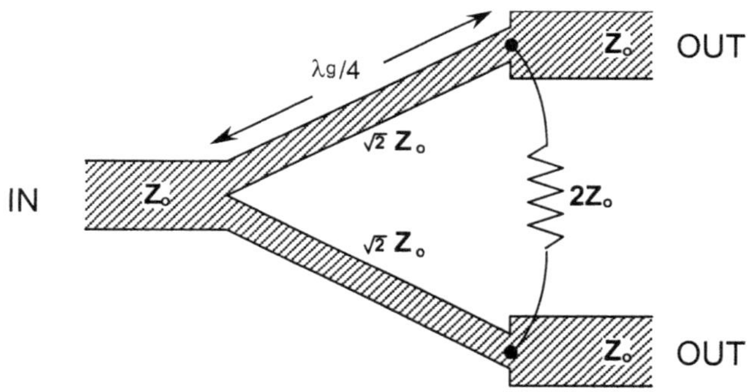

Figure 3.20 Microstrip layout of the Wilkinson power splitter

3.6.3 Special MMIC realisations

On a 200 μm GaAs substrate, the branch-line coupler, for example, measures approximately 6×6 mm^2 at 5 GHz. It is apparent, therefore, that the standard microstrip couplers and power splitters are not ideal for MMICs operating at frequencies lower than approximately 10 GHz, simply because of their size. Miniaturising these elements by using meandered microstrip lines can make a useful size reduction, but this miniaturisation is limited by the coupling between the lines, which will have a particularly undesirable effect on the phase balance of a coupler. There are other alternatives in this case: one could use active techniques, and many excellent techniques are described in Chapter 7. However, active techniques can impose severe penalties in circuit complexity, DC power consumption and noise figure. A more attractive alternative is to miniaturise the original passive component by employing lumped element equivalents [27,28] or the lumped-distributed technique [29].

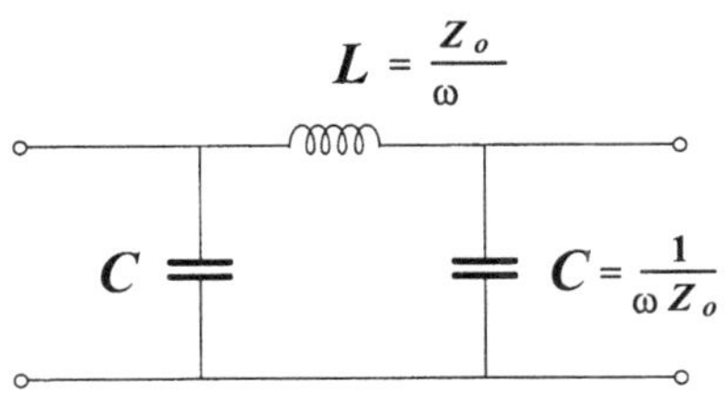

Figure 3.21 The lumped element equivalent of a quarter-wave transmission line

The inductor–capacitor π-network of Figure 3.21 can be used to replace a transmission line of characteristic impedance Z_o, which is a quarter wavelength (90°) long at a frequency f, if their values are calculated as follows:

$$L = \frac{Z_o}{2\pi f} \tag{3.3}$$

and

$$C = \frac{1}{2\pi f Z_o} \tag{3.4}$$

However, the structure is clearly a low-pass filter, and its cut-off frequency is given by:

$$f_c = \frac{1}{2\pi\sqrt{LC}} \tag{3.5}$$

As an example, the equivalent of a 50 Ω quarter-wavelength line at 10 GHz would require C = 0.32 pF and L = 0.8 nH. Although the lumped-element equivalent network cuts off quite rapidly after 10 GHz, in practice, components such as branch-line couplers and Wilkinson power splitters can achieve comparable performance in lumped element form to the transmission-line circuit. The major penalty with this technique is that the insertion loss will tend to be higher, especially at low frequency, when multi-turn spiral inductors are required. Figure 3.22 shows the lumped-element form of the Wilkinson power divider, and Figure 3.23 shows a lumped-element branch-line coupler, and these have both been used extensively in many MMIC designs.

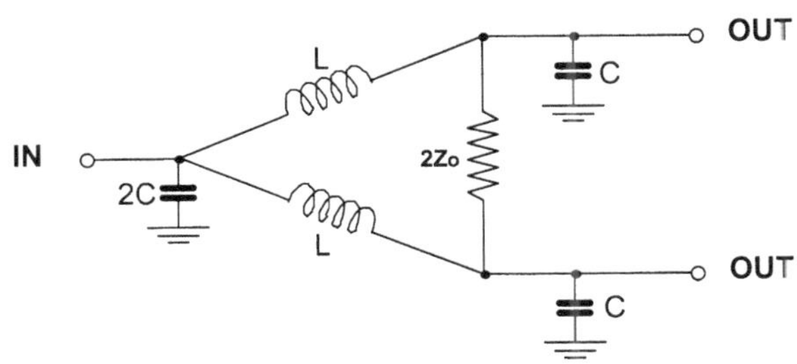

Figure 3.22 Lumped-element Wilkinson power divider

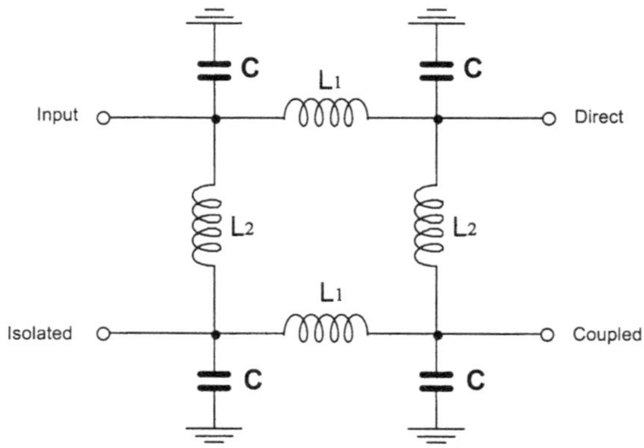

Figure 3.23 Lumped-element branch-line coupler

In the lumped-distributed or 'reduced-size' [29] technique, each quarter-wavelength line required in the conventional coupler or power splitter is replaced with an equivalent π-network, consisting of a short high-impedance line with a shunt capacitor at each end, as shown in Figure 3.24. If the characteristic impedance of the quarter-wavelength line to be substituted for is $Z_{\lambda/4}$, and the short line has electrical length θ, then at the design frequency f, the characteristic impedance of the short line is given by:

$$Z_o = \frac{Z_{\lambda/4}}{\sin(\theta)} \tag{3.6}$$

and the shunt capacitor value is given by:

$$C = \frac{\cos(\theta)}{2\pi f Z_{\lambda/4}} \tag{3.7}$$

From eqn. (3.6) it is evident that the minimum length of the short line is dictated by the maximum value of characteristic impedance that can be used, which is typically 90 to 100 Ω. Notice that when θ is 90° (a quarter wavelength), then $Z_o = Z_{\lambda/4}$ and $C = 0$. The lumped-distributed equivalence is exact at the design frequency, and in practice components such as branch-line couplers will have performance very similar to the standard version. A lumped-distributed branch-line coupler is shown in Figure 3.25.

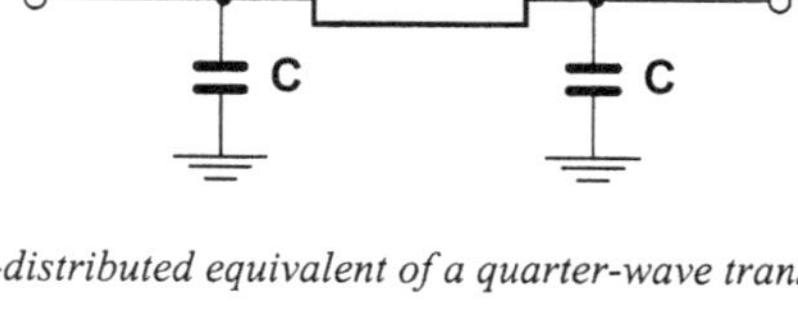

Figure 3.24 Lumped-distributed equivalent of a quarter-wave transmission line

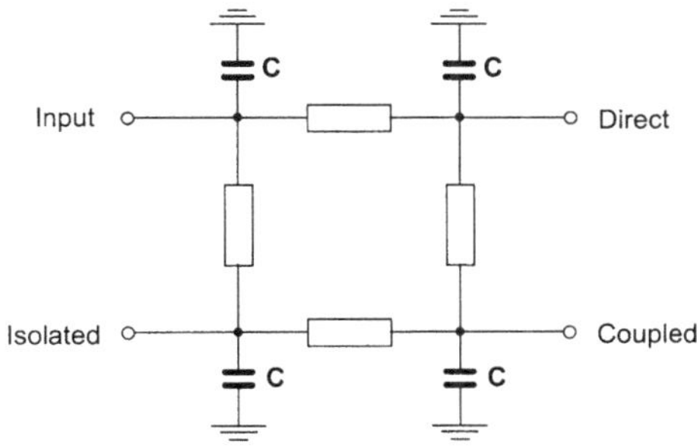

Figure 3.25 Lumped-distributed branch-line coupler

In microstrip form the lumped-distributed technique is not so attractive, since it requires extra via-hole grounds for the shunt capacitors. However, in uniplanar or CPW MMICs the lumped-distributed technique is very attractive, since the capacitors can readily be grounded using the ground-planes adjacent to the signal track.

3.7 Coplanar circuits

As a result of the effort to reduce chip sizes and increase operating frequency into the millimetre-wave band, there has been widespread development of circuits employing coplanar waveguide (CPW), slotline and coplanar strips (CPS). The major attraction of these 'uniplanar' techniques is that through-substrate via-holes are not required for grounding purposes. The structure of a CPW transmission line is shown in Figure 3.26*a*. It consists of a signal conductor placed between two ground planes, and since the dominant mode of CPW is quasi-TEM there is no low frequency cut-off. The principal advantages of CPW are as follows:

1. devices and components can be grounded without via-holes
2. it suffers from much less dispersion than microstrip, making it suitable for millimetre-wave circuits
3. a given characteristic impedance can be realised with almost any track width and gap combination
4. a considerable increase in packing density is possible because the ground planes provide shielding between adjacent CPW lines
5. with the back-face ground plane removed, lumped-elements exhibit less parasitic capacitance.

Previously, the lack of accurate design information and discontinuity models precluded CPW from being extensively used in MMICs. This situation has improved dramatically, however, and many CAD packages now have CPW elements in their libraries, as discussed in Chapter 4. These models include bends, steps, T-junctions and many other common elements. One of the major problems with CPW is that the mode of propagation can easily degenerate from quasi-TEM into a balanced coupled-slotline mode. This happens very often at discontinuities, but can be avoided by incorporating grounding straps between the ground planes, using either air-bridges or underpasses. A second reason why CPW has not been widely used is due to the belief that it has inherently higher conduction loss than microstrip. However, it has been shown that at millimetre-wave frequencies CPW can be equal to or better than microstrip when loss and dispersion on GaAs substrates are used as a basis for comparison. This comparison has shown that minimum loss for a given CPW cross-section occurs at about 60 Ω impedance, whereas the minimum loss for microstrip occurs at about 25 Ω. However, the physical sizes of these minimum loss microstrip lines are much larger than the CPW type.

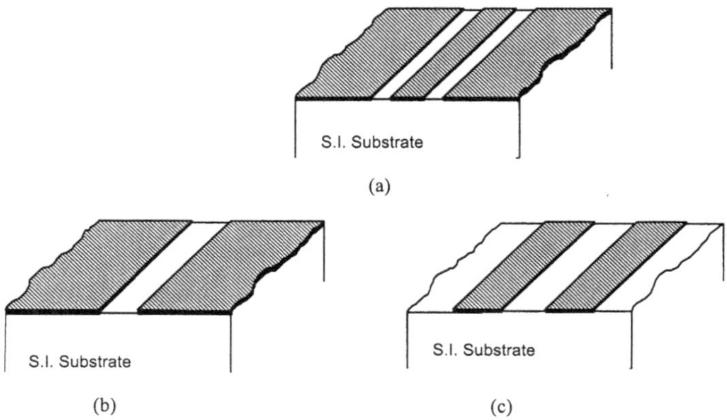

Figure 3.26 Transmission lines for uniplanar MMICs: (a) coplanar waveguide, (b) slotline and (c) coplanar strips

3.7.1 Slotline

Slotline consists of a pair of ground planes with a narrow slot between them, as shown in Figure 3.26*b*, and the signal propagates in a TE (transverse electric) mode. This means that slotline is not an ideal general purpose transmission-line medium, but this mode makes it very useful for circuits such as balanced mixers and amplifiers, where push–pull operation is required. Also, since CPW has the quasi-TEM mode and slotline has the TE mode, a number of useful hybrid junctions and transition circuits can be produced [30]. The most popular of these is the CPW-to-slotline transition, shown in Figure 3.27. The version shown is realised by employing a two metal layer structure, without the need for via-holes or air bridges. At the transition, the two CPW ground planes are connected with a thin metallised underpass. One of the CPW ground planes then becomes one of the slotline conductor planes. The CPW centre conductor extends over the underpass and is connected to the second slotline conductor. At high frequencies, this transition converts the quasi-TEM mode into the slotline's TE mode, yielding balun operation. This type of transition has demonstrated balun operation over more than two octaves of bandwidth, and has been used in miniaturised uniplanar mixers and amplifiers [31,32].

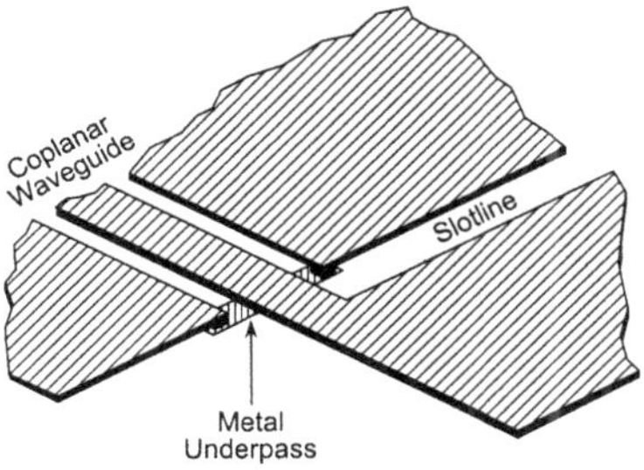

Figure 3.27 CPW-to-slotline transition

3.7.2 Coplanar strips

The coplanar strip (CPS) transmission medium consists of a pair of closely coupled parallel strips, as shown in Figure 3.26*c*. It is essentially a planar equivalent to the twisted pair or Lecher line. As a balanced transmission line, it is ideally suited to balanced mixers and push–pull amplifiers. However, the lack of design information has severely restricted its use, as well as the fact that a circuit's input and output signals are usually required eventually to be in an unbalanced medium such as a microstrip or co-axial line. However, as integration levels in subsystems increase this latter limitation is less severe, whereas the less critical grounding requirements make CPS an attractive medium.

3.7.3 CPW circuits

The main feature of CPW circuits is that grounding can be achieved directly on the front face without a via-hole. As well as simplifying the processing, this can also save space and improve performance. For example, in a transistor layout the area used up by the via-holes is often greater than that of the transistor itself. Furthermore, in CPW there is a considerable reduction in common-lead inductance, which is particularly important for millimetre-wave circuits. CPW distributed amplifiers operating to over 100 GHz have been reported [33]. To illustrate the compact nature of CPW circuits, Figure 3.28 shows a small part of a CPW amplifier layout. It shows a bias resistor, which is decoupled to ground, and part of a meandered CPW matching line. The decoupling capacitor is directly grounded simply by placing the lower plate directly under the ground plane; the meandered line can be folded tightly together because of the shielding effect of the intervening ground plane. Note again the use of ground straps to prevent unwanted propagation modes being excited at the bend discontinuities.

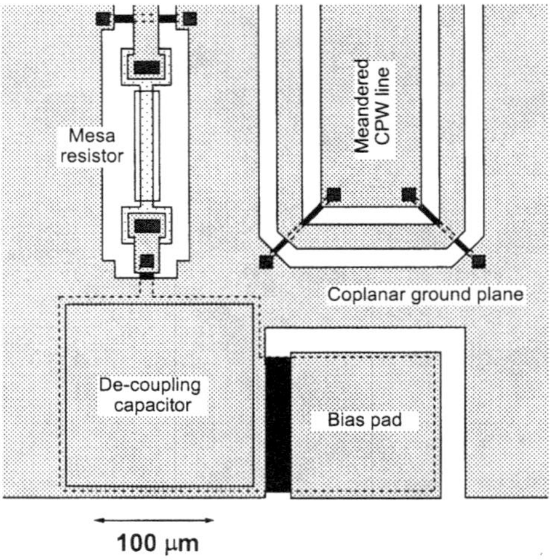

Figure 3.28 Part of a CPW amplifier layout

3.7.3.1 Lumped-distributed CPW circuits

The 'reduced-size' or 'lumped-distributed' technique [29] is ideal for CPW MMICs because the shunt capacitors can be implemented very easily. An impedance-transforming CPW branch-line coupler operating at 15.5 GHz is shown in Figure 3.29 [34] with an area of only 800 μm × 800 μm, which includes four sets of RFOW probe pads. The coupler's measured results showed an amplitude response of 4.7 ± 0.3 dB, an input and output return loss of better than 12 dB and 25 dB, respectively, and an isolation of better than 20 dB over the range 14-16 GHz. This coupler was applied to the design of a two-stage balanced amplifier [35].

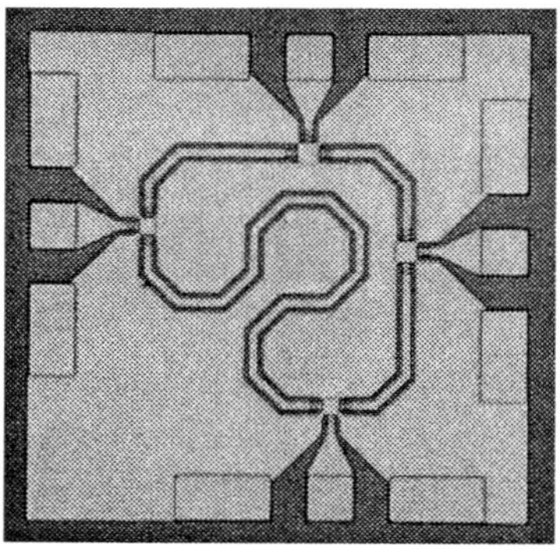

Figure 3.29 Photograph of a Ku-band CPW branch-line coupler

3.8 Multilayer techniques

The majority of MMIC designs and layouts are still based on the traditional microstrip design techniques developed for hybrid MICs. These techniques do not take full advantage of the high integration levels achievable with integrated circuit technology, and a typical MMIC has more area devoted to passive components than to the active devices. For this reason, Japanese industry has been pioneering ways of increasing levels of integration for many years [36–42], and these multilayer techniques have been adopted widely. In the late 1980's, a joint venture between ATR and NTT produced thin-film microstrip line technology [36]. Since then, ATR has gone on to develop this uniplanar technology further. Using gold and polyimide materials, thin-film guided-wave structures have been investigated in microstrip, inverted microstrip and stripline forms. This technology has been subsequently integrated with active devices to realise both microwave and mm-wave balanced amplifiers [37]. NTT has taken this concept much further by developing their own ultra-compact 3D passive circuit technology [38,39] and their 'Master-Slice' technology [40,41]. With the former, vertical gold walls and microwires have been constructed in air and also buried in a polyimide dielectric. With the Master-Slice technology, selected areas of an MMIC's active layer are covered by an upper ground plane. Passive circuits are then realised in a layer above this upper ground plane. This technique is then repeated so that a number of passive layers are stacked on top of one another. Another recent Japanese innovation, led by the Mitsubishi Electric Corporation, has been the buried microstrip line (BMSL) [42]. Here, a gold-plated ground trough is partially filled with a polyimide dielectric. The main signal track either sits on or is immersed within the dielectric. The dielectric-supported BMSL can be viewed as the multilayer counterpart to the micromachined air-filled MS-CPW line with lower shielding cavity. With its inherent high level of line-to-line isolation, the BMSL is ideal for realising high packing density microwave and mm-wave monolithic integrated circuits.

3.8.1 Thin-film microstrip

Thin-film microstrip (TFMS) lines are essentially highly miniaturised microstrip lines, fabricated on a thin dielectric film, as shown in Figure 3.30. A conventional 50 Ω MMIC microstrip line might be 80 to 150 μm wide; a TFMS line would typically be less than 10 μm wide. Hence, TFMS circuitry can give a dramatic reduction in circuit size, and because the ground plane is on the front face of the chip, via-hole grounds are not required. A disadvantage of TFMS is that the ε_r of common dielectrics is considerably lower than that of GaAs or silicon, and so the physical line lengths are longer. This compounds the problem of high transmission loss due to the high resistance of the very narrow signal conductors. However, by careful choice of dielectric material and thickness, the loss can be made acceptable, and NTT has demonstrated 50 Ω TFMS lines on a triple layer of polyimide with acceptable loss. TFMS lines have already been successfully applied to the miniaturisation of branch-line couplers and a range of novel circuits [43–45]. A 1 W X-band power amplifier has been demonstrated using short transformers with characteristic impedances as low as 4 Ω.

As well as basic TFMS lines, multilayer circuits may incorporate inverted TFMS lines, 'square-ax' coaxial lines, striplines, broadside coupling structures, multi-level spiral inductors and transformers, and low impedance CPW lines using centre conductor or ground plane underpasses [46,47]. Now that suitable CAD tools are available, these techniques are becoming very popular, both for GaAs and silicon MMICs.

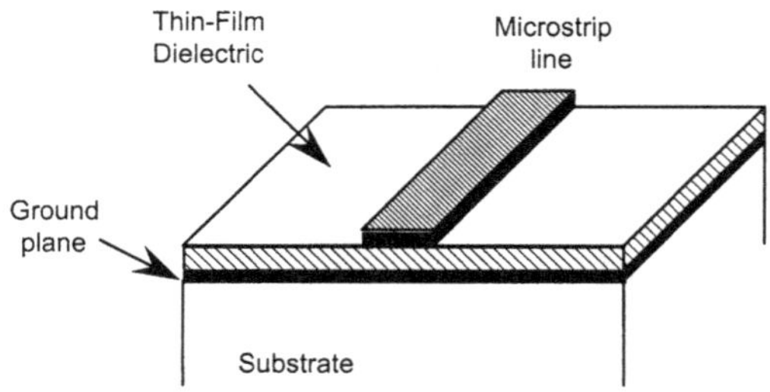

Figure 3.30 Thin-film microstrip transmission line medium

3.8.2 Multilayer directional couplers and baluns

The Lange coupler has an interdigitated structure in order to achieve the tight coupling required for a 3 dB directional coupler. The potential use instead of overlaid structures on MMICs was first recognised by Marczewski and Niemyjski [48]. With multilayer MMICs very tight coupling can be achieved more easily by employing one of a number of broadside or offset-broadside coupling structures. Among the advantages of using multilayer coupling structures are their compactness, tight coupling ability, flexibility, and also their ability to be folded or meandered. The first practical multilayer monolithic coupler [49] employed broad-side coupling, with one track directly above the other, and a photograph of this coupler is shown in Figure 3.31. However, this method leads to high insertion loss because the lower metal is normally very thin due to processing limitations. The microstrip overlaid coupler was later improved [50] by using offset broad-side coupling, which enables the lower metal track to be plated-up with the upper metal. When compared to a Lange coupler this coupler achieves the same bandwidth performance while using only half of the chip area. A CPW multilayer directional coupler has also been realised [51], operating in the frequency range 12–36 GHz with an equal power imbalance of less than 1 dB.

The use of multilayer structures for the realisation of monolithic Marchand baluns was first demonstrated by Pavio and Kikel [52,53]. The multilayer structure makes it possible to access the ground plane of the input unbalanced transmission line, without any complex suspended microstrip structure. Multilayer Marchand baluns have been widely developed [54], and various implementations are described in Chapter 7.

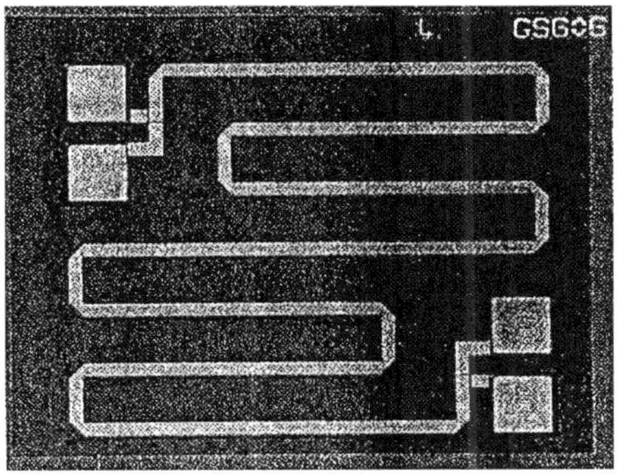

Figure 3.31 Photograph of the first monolithic multilayer coupler

3.9 Micromachined passive components

Micromachined silicon components, using selective crystallographic etching techniques, have been widely developed for high volume commercial markets such as air-bag sensors, displays, disk drives and print-heads. These miniature components are classified as micro-electromechanical systems (MEMS) or microsystems. For microwave circuits, MEMS technology has the important feature of being able to realise moving parts for switching, tuning and steering as well as structures using air as the main dielectric, leading to low loss. The application of the first on-chip submillimetre-wave micromachined structure was demonstrated in 1980 at the University of California. Here, a 2.5 THz tapered silicon rod antenna, with an integrated bismuth microbolometer detector, was realised on a thin SiO_2 membrane support [55]. In the late 1980s, the University of Michigan group followed this pioneering work with results for micromachined waveguide horns and arrays operating at 92, 240, 370, 802 GHz and 2.5 THz [56,57]. Moving micromechanical parts have a wide range of applications in microwave and mm-wave systems. Notable MEMS components include a microwave switch having a reported figure of merit of $f = 1/(2\pi R_{ON}C_{OFF}) = 2$ THz [58], a sub-mm-wave slide tuning mechanism at 620 GHz [59], a parallel-plate variable capacitor [60] and tuneable filters [61].

3.9.1 Micromachined inductors

Micromachining has been widely applied in various ways to the realisation of high-Q inductors on low resistivity silicon substrates. Figure 3.32 shows a selection of the techniques that have been reported in the literature. One of the most direct approaches is to remove the substrate from around the inductor [62–67] by using either surface machining and a membrane support, as shown in (a) and (b), or bulk backside machining, as shown in (c).

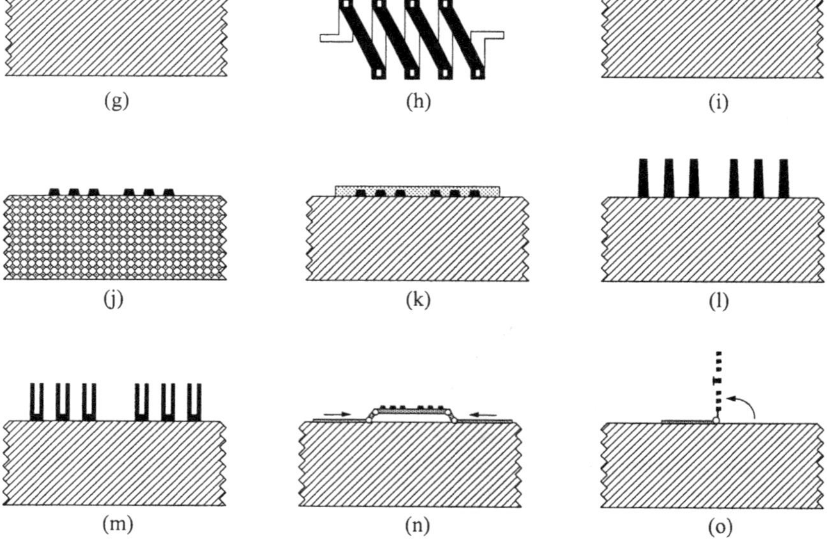

Figure 3.32 Some inductor structures for improved performance: (a) removal of substrate by surface micromachining, (b) cavity backed, (c) removal of substrate by bulk micromachining of back-face, (d) added dielectric layer, (e) thick multilayer conductor, (f) air-bridge inductor, (g) air gap stacked spiral inductor, (h) solenoidal structure, (i) optimised non-uniform width, (j) porous silicon, (k) deposited magnetic film, (l) thick conductors by LIGA (or 'microwires'), (m) U-shaped microwires, (n) levitation by linear actuators and (o) self-assembly by surface tension

Using multilayer techniques [68–73], the inductor can be mounted on top of a dielectric film, as shown in (d), and multiple metal layers can be paralleled up to form a single thick conductor track, as shown in (e). However, the capacitive coupling through the dielectric film to the lossy substrate limits the performance enhancement that can be obtained. Using air-bridge technology [74,75], the inductor tracks can be raised off the substrate, as shown in (f), and stacked air-gap inductors have been reported, as shown in (g). Air bridges and direct machining [76,77] have also been used to create solenoid-type structures, as shown in (h). The layout and structure of the inductor can be optimised using computer algorithms [78–80], leading to non-uniform inductors with wider tracks near the outer edges, as shown in (i). Porous silicon can be used as an alternative [81–83] and techniques for locally modifying the material have been reported. Thin magnetic films can be deposited [84], as shown in (k), and vertical conductor structures [39,85], as shown in (l) and (m), can be fabricated using LIGA or multilayer techniques.

Radically different approaches have been reported in recent years, including raising the inductor up off the substrate using linear actuators [86], as illustrated in (n), or using tensile stress in the metallisation to lift the conductor off the surface [87]. The approach pioneered at Imperial College, London, is to fabricate self-assembled vertical components, as shown in (o), using solder surface tension as the lifting force [88]. Figure 3.33 shows a microphotograph of one of these structures; the inductor, including air-bridge, was fabricated in the normal manner on the surface of the chip, and the whole structure lifted up by applying the correct heating profile to activate the solder hinges.

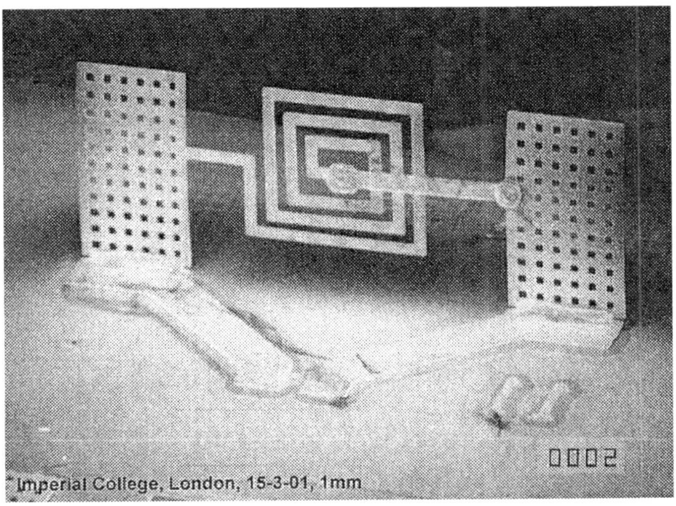

Figure 3.33 High-Q inductor fabricated by self-assembly
(Courtesy of E. M. Yeatman and G. W. Dahlmann, Imperial College, London)

3.9.2 Micromachined transmission lines

For silicon MMICs, microstrip and CPW lines have extremely high loss unless high-resisitivity silicon is used. Various multilayer techniques have been employed [89–92], mostly based around the concept of using an additional dielectric layer between the conductors and the substrate. However, at millimetre-wave frequencies, even GaAs transmission lines have considerable transmission loss, and there has thus been a great deal of interest in using micromachining as a means of realising low loss transmission lines. Figure 3.34 shows some of the most commonly-used structures. One of the simplest techniques is to employ CPW, and to remove the substrate from the vicinity of the slot by etching grooves [93–95], as shown in (a). Alternatively, the CPW can be mounted on a membrane [96–99], and the substrate removed, as shown in (b) to (f). Membrane-supported coplanar waveguide (MS-CPW) lines demonstrating bandwidths in excess of 1 THz with very low dispersion characteristics have been reported, along with developments such as MS-CPW lines with lower shielding cavities and MS-CPW lines with both lower and upper shielding cavities. Using such ultra-high performance microshield transmission lines, extremely low loss filters have been realised at various frequencies [100–102]. For example, one at 95 GHz had a measured insertion loss of 2.2 dB, and another at 250 GHz had a measured insertion loss of only 1.5 dB. Another form of CPW is one where the centre conductor is formed from air-bridges, so that the fields are mostly confined in air, as shown in (g).

Microstrip line losses can be reduced by mounting the conductor on a membrane, above a trench, as shown in (h), by removing the substrate from beneath the signal conductor using back-side micromachining, as shown in (i), or by using ultra-thick conductors fabricated using a LIGA process, as shown in (j). Finally, microstrip losses can be reduced by employing V-shaped conductors, as shown in (k) and (l), to address the problem of current crowding at the edges of the conductor.

3.9.2.1 Micromachined and multilayer rectangular waveguides

The metal-pipe rectangular waveguide (MPRWG) has always been the preferred choice for implementing a high performance guided-wave structure. MPRWGs have much lower conduction current densities, compared to conventional microstrip, CPW, slotlines or CPS lines. Therefore, their ohmic losses can be very low indeed. Moreover, because the propagating fields are contained within the enclosed structure, there is no leakage of electromagnetic energy. As a result, this type of guided-wave structure enjoys the benefits of having a high Q-factor and excellent line-to-line isolation. In electrical performance terms, the MPRWG can be considered as the ultimate guided wave structure, although it is physically large compared with planar media and has a restricted bandwidth.

The University of California first proposed the idea of implementing air-filled MPRWGs, using integrated circuit technology, back in 1980 [55]. More than a decade later, in collaboration with the University of Arizona, the California Institute of Technology demonstrated a W-band micromachined air-filled MPRWG [103]. By using a two-wafer sandwich approach, shown in Figure 3.34*m*, they achieved a measured level of insertion loss of only 0.04 dB/λg at 100 GHz. Six months later, the University of Bath reported the construction of a 600 GHz micromachined air-filled MPRWG structure, realised using thick photoresist as a former, as shown in Figure 3.34*n* [104]. A consortium of UK universities (the Terahertz Integrated Technology Initiative (TINTIN) project) have tried to develop this idea further, with the aim of realising an integrated 200 GHz downconverter and a 1.6 THz mixer, implemented using micromachined air-filled MPRWG technology [105]. With micromachined structures there are a number of practical difficulties in launching signals efficiently into hollow structures. To overcome this, TINTIN proposed a monolithic waveguide horn antenna to act as an interface between the air-filled MPRWG and free-space beams [106]. In its simplest from, this 'bespoke aerial' is a conventional *H*-plane sectoral horn antenna with a tapered slot in the upper wall. The slot reduces the -3 dB beamwidth and elevates the boresight of the antenna. TINTIN found that an exponentially tapered horn gives better simulated results with reduced-height waveguides.

One of the major drawbacks with air-filled MPRWG structures is that they become physically too large, and therefore too expensive, for monolithic applications below *circa* 200 GHz. This problem is further compounded by very poor yield, which excludes this technology from commercial exploitation [107]. Dielectric-filled MPRWGs, shown in Figure 3.34*o*, can solve this problem, and can be realised using multilayer processing [107–111] or micromachining. Compared to an air-filled waveguide, a MPRWG having a dielectric-filler with relative permittivity ε_r can achieve a reduction in the physical area by a factor of ε_r to 1. However, it is extremely important to choose a material with a low $\tan\delta$ in order to achieve low loss. The first experimental dielectric-filled MPRWG is shown in Figure 3.35, and was fabricated on a GaAs wafer [108,110]. The physical dimensions for the complete structure were 1×8 mm^2. The waveguide was designed to have a centre frequency of 105 GHz and to be measured using conventional on-wafer probing techniques. This robust multilayer technology is designed to be compatible with modern microelectronic manufacturing, while still offering a very high performance. To this end, the T-MCM technology, introduced in Chapter 1, operating in the 75 to 300 GHz region (i.e. well beyond the current *circa* 100 GHz limit, offered by flip-chip) could soon be a reality [111]. Figure 3.36 shows a photograph of an experimental MPRWG fabricated using advanced photoimageable thick-film materials, operating in the 60 to 90 GHz band [112]. This waveguide achieved a transmission loss of approximately 0.5 dB/mm, which is comparable to that achieved with microstrip lines on the same dielectric. By increasing the height of the waveguide, the loss can be considerably lower than that of planar transmission lines. The photograph clearly shows the CPW-to-TFMS-to-MPRWG transitions that are used to enable measurements to be made using RFOW probes. The transition can achieve 20 dB return loss over the entire waveguide band and requires no additional fabrication processing steps. Using this transition, the dielectric-filled MPRWG can be easily integrated with active devices and other standard MMIC components.

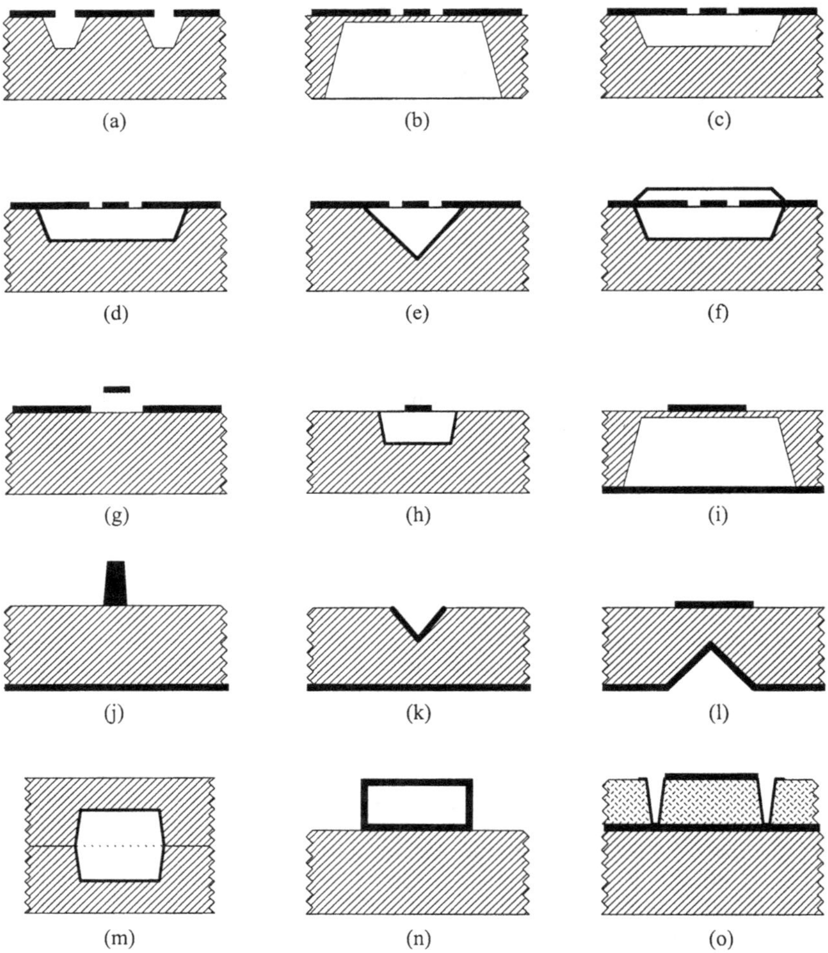

Figure 3.34 Some micromachined transmission-line structures: (a) CPW with grooves, (b) CPW with substrate removed by back-side bulk micromachining, (c) membrane-supported CPW, (d) microshield line, (e) V-shaped microshield line, (f) shielded microshield line, (g) CPW with air-bridge centre conductor, (h) membrane-supported microstrip, (i) suspended microstrip, (j) microstrip with ultra-thick conductor, (k) microstrip with V-shaped signal conductor, (l) microstrip with V-shaped ground plane, (m) air-filled waveguide using a wafer sandwich, (n) air-filled waveguide using thick photoresist former and (o) dielectric-filled waveguide

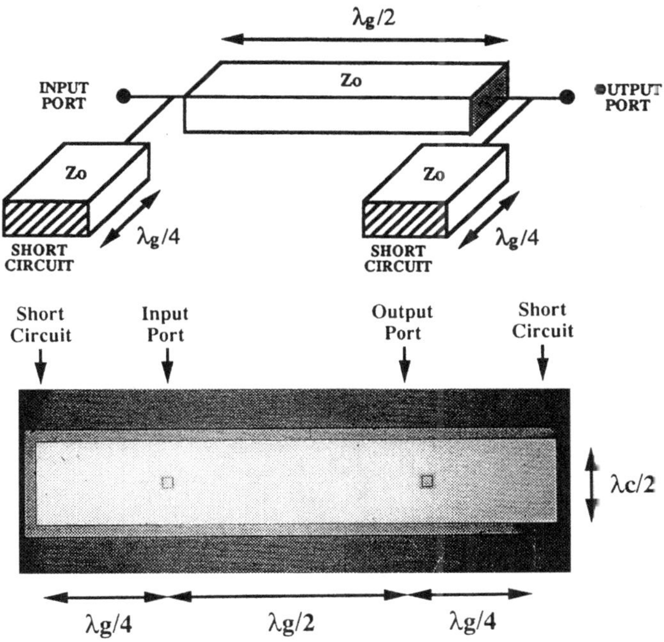

Figure 3.35 Schematic and microphotograph of the first dielectric-filled metal-pipe rectangular waveguide, having a design frequency of 105 GHz on GaAs

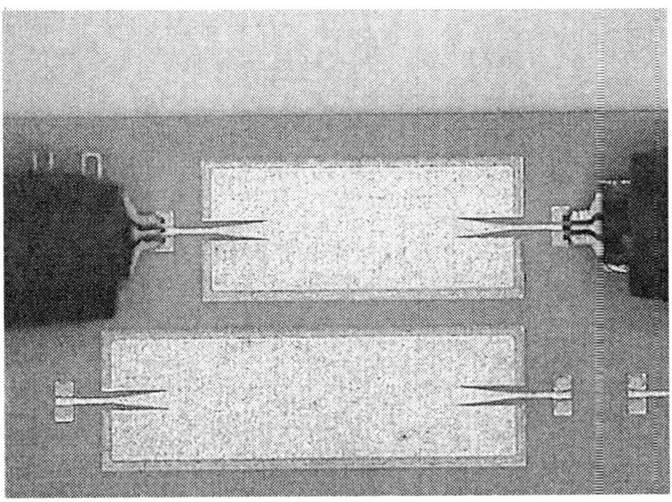

Figure 3.36 Photograph of thick-film MPRWG on wafer STD calibration standards

3.10 References

1 EDWARDS, T. C.: 'Foundations for microstrip circuit design' (John Wiley & Sons, New York, 1992, 2nd edn.)

2 PENGELLY, R. S.: 'Microwave field effect transistors- theory, design and applications' (Research Studies Press Ltd, England, 1986)

3 LADBROOKE, P. H.: 'MMIC design: GaAs FETs and HEMTs' (Artech House, London, 1989)

4 GROVER, F. W.: 'Inductance calculations' (Van Nostrand, Princeton, NJ, 1946)

5 GREENHOUSE, H. M.: 'Design of planar rectangular microelectronic inductors', *IEEE Transactions on Parts, Hybrids and Packaging*, 1974, **PHP-10** (2), pp. 101–109

6 PETTENPAUL, E., KAPUSTA, H., WEISGERBER, A., MAMPE, H., LUGINSLAND, J., and WOLFF, I.: 'CAD models of lumped elements on GaAs up to 18 GHz', *IEEE Transactions on Microwave Theory and Techniques*, 1988, **MTT-36**, (2), pp. 294–304

7 HILL, A., and TRIPATHI, V. K.: 'Analysis and modeling of coupled right angle microstrip bend discontinuities', *IEEE MTT-S International Microwave Symposium Digest*, 1989, pp. 1143–1146

8 SCHMÜCKLE, F. J.: 'The method of lines for the analysis of rectangular spiral inductors', *IEEE Transactions on Microwave Theory and Techniques*, 1993, **MTT-41** (6/7), pp. 1183–1186

9 HOWARD, G. E., DAI, J., CHOW, Y. L., and STUBBS, M.: 'The power transfer mechanism of MMIC spiral transformers and adjacent spiral inductors', *IEEE MTT-S International Microwave Symposium Digest*, 1989, pp. 1251–1254

10 BECKS, T., and WOLFF, I.: 'Analysis of 3-D metallization structures by a full-wave spectral domain technique', *IEEE Transactions on Microwave Theory and Techniques*, **MTT-40** (12), pp. 2219–2227

11 JAMISON, S. A., PODELL, A., HELIX, M., NG, P., and CHAO, C.: 'Inductively coupled push-pull amplifiers for low cost monolithic microwave Ics', *IEEE GaAs IC Symposium Digest*, 1982, pp. 91–93

12 FERGUSON, D., BAUHAHN, P., KEUPER, J. *et al.* 'Transformer coupled high-density circuit technique for MMIC', *IEEE MTT-S International Microwave Symposium Digest*, 1984, pp. 34–36

13 WIEMER, L., JANSEN, R. H., ROBERTSON, I. D., and SWIFT, J. B.: 'Computer simulation and experimental investigation of square spiral transformers for MMIC applications', *IEE Colloquium on Computer-Aided Design of Microwave Circuits*, Digest no. 99, 1985, pp. V1–5

14 ECONOMIDES, S. B., KHALID, A., REZAZADEH, A. A., and ROBERTSON, I. D.: 'A study of CPW spiral transformers using multilayer GaAs MMIC technology', *Proceedings of Microwaves and RF Conference*, 1996, pp. 420-425

15 BOULOUARD, A., and LE ROUZIC, M.: 'Analysis of rectangular spiral transformers for MMIC applications', *IEEE Transactions on Microwave Theory and Techniques*, **MTT-37**, 1989, pp. 1257–260

16 GOKDEMIR, T., ECONOMIDES, S., KHALID, A. H., REZAZADEH, A. A., and ROBERTSON, I. D.: 'Design of CPW baluns using spiral and overlaid couplers', *IEEE MTT-S International Micowave Symposium Digest*, 1997, pp. 401–404

17 JANSEN, R. H., ARNOLD, R. G., and EDDISON, I. G.: 'A comprehensive CAD approach to the design of MMICs up to millimetre-wave frequencies', *IEEE Transactions on Microwave Theory and Techniques*, 1988, **MTT-36**, pp. 208–219

18 BUECHLER, J., KASPER, E., RUSSER, P., and STROHM, K. M.: 'Silicon high-resistivity-substrate millimeter-wave technology', *IEEE Transactions on Electron Devices*, 1986, **ED-33** (12), pp. 2047–2052

19 HYLTIN, T. M.: 'Microstrip transmission on semiconductor dielectrics', *IEEE Transactions on Microwave Theory and Techniques*, 1965, **MTT-13**, pp. 777–781

20 ROSEN, A., CAULTON, M., STABILE, P. *et al.*: 'Silicon as a millimeter-wave monolithically integrated substrate - A new look', *RCA Review*, 1981, **42**, pp. 633–660

21 STABILE, P. J., and ROSEN, A.: 'A silicon technology for millimeter-wave monolithic circuits', *RCA Review*, 1984, **45**, pp. 587–605

22 STROHM, K. M., LUY, J.-F., KASPER, E., BUECHLER, J., and RUSSER, P.: 'Silicon technology for monolithic millimeter wave integrated circuits', *Mikrowellen & HF Magazin*, 1984, **14** (8), pp. 750–760

23 PONCHAK, G. E.: 'RF transmission lines on silicon substrates', *Proceedings of 29th European Microwave Conference*, Munich, 1999, **1**, pp.158–61

24 ZHENG, J., HAHM, Y-C., TRIPATHI, V. K., and WEISSHAAR, A.: 'CAD-oriented equivalent-circuit modeling of on-chip interconnects on lossy silicon substrate', *IEEE Transactions on Microwave Theory and Techniques*, 2000, **48** (9), pp.1443–51

25 WILLIAMS, D. F., ARZ, U., and GRABINSKI, H.: 'Accurate characteristic impedance measurement on silicon', *IEEE MTT-S International Microwave Symposium Digest*, 1998, **3**, pp.1917–20

26 WILLIAMS, D. F., and MARKS, R. B.: 'On-wafer impedance measurement on lossy substrates', *IEEE Microwave and Guided Wave Letters*, 1994, **4** (6), pp.175–6

27 GUPTA, R. K., and GETSINGER, W. J.: 'Quasi-lumped-element 3- and 4-port networks for MIC and MMIC applications', *IEEE MTT-S International Microwave Symposium Digest*, 1984, pp. 409–411

28 VOGEL, R. W.: 'Analysis and design of lumped- and lumped-distributed-element directional couplers for MIC and MMIC applications', *IEEE Transactions on Microwave Theory and Techniques*, 1992, **MTT-40**, pp. 253–262

29 HIROTA, T., MINAKAWA, A., and MURAGUCHI, M.: 'Reduced-size branch-line and rat-race hybrids for uniplanar MMICs', *IEEE Transactions on Microwave Theory and Techniques*, 1990, **MTT-38**, pp. 270–275

30 HIROTA, T., TARUSAWA, Y., and OGAWA, H.: 'Uniplanar MMIC hybrids-A proposed new MMIC structure', *IEEE Transactions on Microwave Theory and Techniques*, 1987, **MTT-35**, pp. 576–581

31 MURAGUCHI, M., HIROTA, T., MINAKAWA, A., IMAI, Y., ISHITSUKA, F., and OGAWA, H.: '26 GHz-band full MMIC transmitters and receivers using a uniplanar technique', *IEEE MTT-S International Microwave Symposium Digest*, 1990, pp. 873–876

32 TON, T. N., CHEN, T.H, DOW, G. S, NAKANO, K., LIU, L. C. T., and BERENZ, J.: 'A Q-band monolithic AlGaAs/GaAs HEMT CPW downconverter', *IEEE GaAs IC Symposium Digest*, 1990, pp. 185–188

33 MAJIDI-AHY, R., NISHIMOTO, C. K., RIAZIAT, M. *et al.*: '5-100 GHz InP coplanar waveguide MMIC distributed amplifier', *IEEE Transactions on Microwave Theory and Techniques*, 1990, **MTT-38**, pp. 1986–1993

34 GILLICK, M., ROBERTSON, I. D., and JOSHI, J. S.: 'Design and realisation of reduced-size impedance transforming uniplanar MMIC branchline coupler', *Electronics Letters*, 1992, **28** (16), pp. 1555–1557

35 GILLICK, M., ROBERTSON, I. D., and JOSHI, J. S.: '15 GHz balanced CPW amplifier using impedance-transforming lumped-distributed branch-line couplers', *IEE Proceedings — Microwaves, Antennas and Propagation*, 1994, **141**, pp. 241–245

36 HIRAOKA, T., TOKUMITSU, T., and AKAIKE, M.: 'Very small wide-band MMIC magic-T's using microstrip lines on a thin dielectric film', *IEEE Transactions on Microwave Theory and Techniques*, 1989, **MTT-10**, pp. 1569-1575

37 IMAOKA, T., BANBA, S., MINAKAWA, A., and IMAI, N.: 'Millimeter-wave wide-band amplifiers using multilayer MMIC technology', *IEEE Transactions on Microwave Theory and Techniques*, 1997, **45** (1), pp. 95–101

38 HIRANO, M., NISHIKAWA, K., TOYODA, I., AOYAMA, S., SUGITANI, S., and YAMASAKI, K.: 'Three-dimensional passive circuit technology for ultra-compact MMIC's', *IEEE Transactions on Microwave Theory and Techniques*, 1995, **43** (12), pp. 2845–2850

39 ONODERA, K., HIRANO, M., TOKUMITSU, M., TOYODA, I., NISHIKAWA, K., and TOKUMITSU, T.: 'Folded U-shape microwire technology for ultra-compact three-dimensional MMIC's', *IEEE Transactions on Microwave Theory and Techniques*, 1996, **44** (12), pp. 2347–2353

40 AIKAWA, M., TOKUMITSU, T., and NISHIKAWA, K.: 'Advanced MMIC technology for the next generation-3D MMICs and master-slice technology', *Proceedings of 26th European Microwave Conference*, 1996, pp. 748–753

41 TOYODA, I., TOKUMITSU, T., and AIKAWA, M.: 'Highly integrated three-dimensional MMIC single-chip receiver and transmitter', *IEEE Transactions on Microwave Theory and Techniques*, 1996, **44** (12), pp. 2340–2346

42 ISHIKAWA, T., and YAMASHITA, E.: 'Characterization of buried microstrip lines for constructing high-density microwave integrated circuits', *IEEE Transactions on Microwave Theory and Techniques*, 1996, **44**, pp. 840–847

43 GILLICK, M., and ROBERTSON, I. D.: 'An X-band monolithic power amplifier using low characteristic impedance thin-film microstrip transformers', *IEEE Microwave and Guided Wave Letters,* 1992, pp. 328–330

44 LUCYSZYN, S., ROBERTSON, I. D., and AGHVAMI, A. H: 'Monolithic analogue phase shifter and cascode FET amplifier using uniplanar techniques', *Proceedings of 24th European Microwave Conference*, Cannes, France, 1994, pp. 554–559

45 TOKUMITSU, T., HIRAOTA, T., NAKAMOTO, H., and TAKENAKA, T.: 'Multilayer MMIC using a 3 μm × 3-layer dielectric film structure', *IEEE MTT-S International Microwave Symposium Digest,* 1990, pp. 831–834

46 GOKDEMIR, T., KARACAOGLU, U., BUDIMIR, D., WANG, Q. H., REZAZADEH, A. A., and ROBERTSON, I. D.: 'A comparison of miniature transmission-line media for multi-level MMICs', *Proceedings of European Microwave Conference*, 1996

47 GILLICK, M., and ROBERTSON, I. D.: 'Ultra low impedance CPW transmission lines for multilayer MMICs', *IEEE MTT-S International Microwave Symposium Digest,* 1993, pp. 127–130

48 MARCZEWSKI, W., and NIEMYJSKI, W.: 'The overlapped microstrip for MICs and MMICs', *Proceedings of 14th European Microwave Conference*, 1984, Liege, Belgium, pp. 166–171

49 ROBERTSON, I. D., and AGHVAMI, A. H: 'Novel coupler for gallium arsenide monolithic microwave integrated circuit applications', *Electronics Letters*, 1988, **24** (25), 1577–1578

50 LUCYSZYN, S., and ROBERTSON, I. D.: 'An improved multi-layer quadrature coupler for MMICs', *Proceedings of 21st European Microwave Conference*, Stuttgart, 1991, pp. 1154–1158

51 GILLICK, M., ROBERTSON, I. D., and JOSHI, J. S.: 'A 12-36 GHz MMIC 3dB coplanar waveguide directional coupler', *Proceedings of 22nd European Microwave Conference*, Espoo, Finland, 1992, pp. 724–728

52 MARCHAND, N.: 'Transmission-line conversion transformers', *Electronics*, 1944, **17** (12), pp. 142–145

53 PAVIO, A. M., and KIKEL, A.: 'A monolithic or hybrid broadband compensated balun', *IEEE International Microwave Symposium Digest*, 1990, pp. 483–486

54 JANSEN, R. H., JOTZO, J., and ENGELS, R.: 'Improved compaction of multilayer MMIC/MCM baluns using lumped element compensation', *IEEE MTT-S International Microwave Symposium Digest*, 1997, **1**, 1997, pp.277–80

55 RUTLEDGE, D. B., SCHWARZ, S. E., HWANG, T. L., ANGELAKOS, D. J., MEI, K. K., and YOKOTA, S.: 'Antennas and waveguides for far-infrared integrated circuits', *IEEE Journal of Quantum Electronics*, 1980, **QE-16**, pp. 508–516

56 REBEIZ, G. M., REGEHR, W. G., RUTLEDGE, D. B, SAVAGE, R. L., and LUHMANN, N. C.: 'Submillimeter-wave antennas on thin membranes', *International Journal of Infrared and Millimeter Waves*, 1987, **8** (10), p. 1249–1255

57 REBEIZ, G. M.: 'Millimetre-wave and terahertz integrated circuit antennas', *Proceedings of the IEEE*, 1992, **80** (11), pp. 1748–1770
58 GOLDSMITH, C., LIN, T-H., POWERS, B., WU, W-R., and NORVELL, B.: 'Micromechanical membrane switches for microwave applications', *IEEE MTT-S International Microwave Symposium Digest*, 1995, pp. 91–94
59 LUBECKE, V. M., McGRATH, W. R., STIMSON, P. A., and RUTLEDGE, D. B.: 'Performance of micromechanical tuning elements in a 620-GHz monolithic integrated circuit', *Proceedings of 7th International Symposium on Space Terahertz Technology*, 1996
60 CHIAO, J-C., FU, Y., CHOUDHURY, D., and LIN, L-Y.: 'MEMS millimeterwave components', *IEEE MTT-S International Microwave Symposium Digest*, 1999, pp. 463–466
61 KIM, H-T., PARK, J-H., KIM, Y-K., and KWON, Y.: 'Millimeter-wave micromachined tunable filters', *IEEE MTT-S International Microwave Symposium Digest*, 1999, pp. 1235–1238
62 JIANG, H., WANG, Y., YEH, J-LA., and TIEN, N. C.: 'Fabrication of high-performance on-chip suspended spiral inductors by micromachining and electroless copper plating', *IEEE MTT-S International Microwave Symposium Digest*, 2000, **1**, pp. 279–82
63 LUM L-Hg., PONCHAK, G. E., BHATTACHARYA, P., and KATEHI, L. P. B.: 'High-Q X-band and K-band micromachined spiral inductors for use in Si-based integrated circuits', *IEEE Topical Meeting on Silicon Monolithic Integrated Circuits in RF Systems*, 2000, pp. 108–12
64 OZGUR, M., ZAGHLOUL, M. E., and GAITAN, M.: 'High Q backside micromachined CMOS inductors', *Proceedings of the 1999 IEEE International Symposium on Circuits and Systems*, **2**, pp. 577–80
65 RIBAS, R. P., LESCOT, J., LECLERCQ, J-L., KARAM, J. M., NDAGIJIMANA, F.: 'Micromachined microwave planar spiral inductors and transformers', *IEEE Transactions on Microwave Theory and Techniques*, 2000, **48** (8), pp. 1326–35
66 MIAO, J., HARTNAGEL, H. L., WILSON, R. J., and WEISS, B. L.: 'Improved free standing semi-insulating GaAs membranes for sensor applications', *Electronics Letters,* 1995, pp. 1047–1049
67 BEILENHOFF, K., MUTAMBA, C., PFEIFFER, J. *et al.*: 'III-V semiconductor structuring for mm and sub-mm waves'. Workshop Proceedings (M-FrW3) at the European Microwave Week, 1999
68 BERG, H., JACOBSSON, H., LEWIN, T., BERG, K., SCHUPPEN, A., and GEVORGIAN, S.: 'Low loss passive microwave components in standard silicon technology', *IEEE MTT-S International Microwave Symposium Digest*, 2000, **2**, pp. 1153–6
69 PARK, M., LEE, S., YU, H. K., KOO, J. G., and NAM, K. S.: 'High Q CMOS-compatible microwave inductors using double-metal interconnection silicon technology', *IEEE Microwave and Guided Wave Letters*, 1997, **7** (2), pp. 45–7
70 GOKDEMIR, T., KARACAOGLU, U., BUDIMIR, D. *et al.*: 'Multilayer passive components for uniplanar Si/SiGe MMICs', *IEEE Radio Frequency Integrated Circuits (RFIC) Symposium Digest*, 1997, pp. 233–6

71 KIM, B-K., KO, B-K., LEE, K., JEONG, J-W., LEE, K-S., and KIM, S-C.: 'Monolithic planar RF inductor and waveguide structures on silicon with performance comparable to those in GaAs MMIC', *IEEE International Electron Devices Meeting*, Technical Digest, 1995, pp. 717–20

72 KAMOGAWA, K., NISHIKAWA, K., TOKUMITSU, T., and TANAKA, M.: 'A novel high-Q inductor based on Si 3D MMIC technology and its application', *IEEE MTT-S International Microwave Symposium Digest*, 1999, **2**, pp. 489–92

73 SOYUER, M., BURGHARTZ, J. N., JENKINS, K. A., PONNAPALLI, S., EWEN, J. F., and PENCE, W. E.: 'Multilevel monolithic inductors in silicon technology, *Electronics Letters*, 1995, **31** (5), pp. 359–60

74 YOON, J-B., HAN, C-H., YOON, E., and KIM, C-K.: 'Monolithic high-Q overhang inductors fabricated on silicon and glass substrates', *Proceedings of IEEE International Electron Devices Meeting*, Washington, 1999, pp. 753–6

75 PARK, S. W., and SEO, K. S.: 'Air-gap stacked spiral inductor', *IEEE Microwave and Guided Wave Letters*, 1997, **7** (10), pp. 329-31

76 YOUNG, D. J., MALBA, V., OU, J-J., BERNHARDT, A. F., and BOSER, B. E.: 'Monolithic high performance three-dimensional coil inductors for wireless communication applications', *Proceedings of IEEE International Electron Devices Meeting*, 1997, pp. 67–70

77 PARK, J. Y., and ALLEN, M. G.: 'New micromachined inductors on silicon substrates', *IEEE Transactions on Magnetics*, 1999, **35** (5), pp. 3547–9

78 LOPEZ-VILLEGAS, J. M., SAMITIER, J., CANE, C., LOSANTOS, P., and BAUSELLS, J.: 'Improvement of the quality factor of RF integrated inductors by layout optimization', *IEEE Transactions on Microwave Theory and Techniques*, 2000, **48** (1), pp. 76–83

79 KOUTSOYANNOPOULOS, Y., PAPANANOS, Y., BANTAS, S., and ALEMANNI, C.: 'Novel Si integrated inductor and transformer structures for RF IC design', *Proceedings of the 1999 IEEE International Symposium on Circuits and Systems*, 1999, **2**, pp. 573–6

80 LONG, J. R., and COPELAND, M. A.: 'The modeling, characterization, and design of monolithic inductors for silicon RF IC's', *IEEE Journal of Solid-State Circuits*, 1997, **32** (3), pp. 357–69

81 YU, M-J., CHAN, Y-J., LAIH, L-H., and HONG, J-W.: 'Improved microwave performance of spiral inductors on Si substrates by chemically anodizing a porous silicon layer', *Microwave and Optical Technology Letters*, 2000, **26** (4), pp. 232–4

82 WELTY, R. J., PARK, S. H., ASBECK, P. M., DANCIL, K-PS., and SAILOR, M. J.: 'Porous silicon technology for RF integrated circuit applications', *Proceedings of IEEE Topical Meeting on Silicon Monolithic Integrated Circuits in RF Systems*, Digest of Papers, 1998, pp.160–3

83 NAM, C-M., and KWON, Y-S.: 'High-performance planar inductor on thick oxidized porous silicon (OPS) substrate', *IEEE Microwave and Guided Wave Letters*, 1997, **7** (8), pp. 236–8

84 YAMAGUCHI, M., BABA, M., SUEZAWA, K. *et al.*: 'Magnetic RF integrated thin-film inductors', *IEEE MTT-S International Microwave Symposium Digest*, 2000, **1**, pp. 205–8

85 GROVES, R., MALINOWSKI, J. VOLANT, R., and JADUS, D.: 'High Q inductors in a SiGe BiMOS process utilizing a thick metal process add-on module', *Proceedings of 1999 Bipolar/BiCMOS Circuits and Technology Meeting*, Minneapolis, 1999, pp. 149–52
86 FAN, L., CHEN, R. T., NESPOLA, A., and WU, M. C.: 'Universal MEMS platforms for passive RF components: suspended inductors and variable capacitors', *Proceedings of MEMS 98*: IEEE Eleventh Annual International Workshop on Micro Electromechanical Systems (An Investigation of Micro Structures, Sensors, Actuators, Machines and Systems), 1998, pp. 29–33
87 GAMMEL, P. L., BARBER, B. P., LUBECKE, V., BELK, N., and FREI, M. R.: 'Design, test, and simulation of self-assembled micromachined RF Inductors', *Proceedings of SPIE*, 1999, **3680** (1–2), pp. 582–91
88 DAHLMANN, G. W., and YEATMAN, E. M.: 'High Q microwave inductors on silicon by surface tension self-assembly', *Electronics Letters*, 2000, **36** (20), pp. 1707–8
89 PONCHAK, G. E., and KATEHI, L. P. B.: 'Measured attenuation of coplanar waveguide on CMOS grade silicon substrates with polyimide interface layer, *Electronics Letters*, 1998, **34** (13), pp. 1327–9
90 HENRI, H., GONZAGUE, S., MATTHIEU, V., ALAIN, C., and GILLES, D.: 'Ultra low loss transmission lines on low resistivity silicon substrate', *IEEE MTT-S International Microwave Symposium Digest*, 2000, **3**, pp. 1809–12
91 KLEVELAND, B., LEE, T. H., and WONG, S. S.: '50-GHz interconnect design in standard silicon technology', *IEEE MTT-S International Microwave Symposium Digest*, 1998, **3**, pp. 1913–16
92 WARNS, C., MENZEL, W., and SCHUMACHER, H.: 'Transmission lines and passive elements for multilayer coplanar circuits on silicon', *IEEE Transactions on Microwave Theory and Techniques*, 1998, **46** (5) pp.616–22
93 WILLIAMS, D. F., and SCHWARZ, S.E.: 'Reduction of Propagation Losses in Coplanar Waveguide', *IEEE International Microwave Symposium Digest*, 1984, pp. 453–454
94 OZGUR, M., ZAGHLOUL, M. E., and GAITAN, M.: 'Micromachined 28-GHz power divider in CMOS technology', *IEEE Microwave and Guided Wave Letters*, 2000, **10** (3), pp. 99–101
95 ISHII, H., SAHRI, N., NAGATSUMA, T. *et al.*: 'A new fabrication process for low-loss millimeter-wave transmission lines on silicon', *Japanese Journal of Applied Physics*, 2000, **39** (4B), pp. 1982–6
96 KATEHI, L. P. B., REBEIZ, G. M., WELLER, T. M. *et al.*: 'Micromachined circuits for millimeter- and sub-millimeter-wave applications', *IEEE Antennas and Propagation Magazine*, 1993, **35** (5), pp. 9–17
97 WELLER, T. M., KATEHI, L. P. B., and REBEIZ, G. M.: 'High performance microshield components', *IEEE Transactions on Microwave Theory and Techniques*, 1995, **43** (3), pp. 534–543
98 DRAYTON, R. F., and KATEHI, L. P. B.: 'Development of self-packaged high frequency circuits using micromachining techniques', *IEEE Transactions on Microwave Theory and Techniques*, 1995, **43** (9), pp. 2073–2080

99 CHI, C-Y., and REBEIZ, G. M.: 'Conductor-loss limited stripline resonator and filters', *IEEE Transactions on Microwave Theory and Techniques*, 1996, **44** (4), pp. 626–630

100 CHI, C-Y., and REBEIZ, G. M.: 'Planar microwave and millimetre-wave lumped elements and coupled-line filters using micro-machining techniques', *IEEE Transactions on Microwave Theory and Techniques*, 1995, **43** (4), pp. 730–738

101 WELLER, T. M., KATEHI, L. P. B., and REBEIZ, G. M.: 'A 250-GHz microshield bandpass filter', *IEEE Microwave and Guided Wave Letters*, 1995, **5** (5), pp. 153–155

102 ROBERTSON, S. V., KATEHI, L. P. B., and REBEIZ, G. M.: 'Micromachined W-band filters', *IEEE Transactions on Microwave Theory and Techniques*, 1996, **44** (4), pp. 598–606

103 McGRATH, W. R., WALKER, C., YAP, M., and TAI, Y-C.: 'Silicon micromachined waveguides for millimeter-wave and submillimeter-wave frequencies', *IEEE Microwave and Guided Wave Letters*, 1993, **3** (3), pp. 61-63

104 TREEN, A. S., and CRONIN, N. J.: 'Terahertz metal pipe waveguides', *18th International Conference Digest on Infrared and Millimetre Waves*, 1993, pp. 470–471

105 COLLINS, C. E., POLLARD, R. D., MILES, R. E. *et al.*: 'The TINTIN project and the need for integrated antennas at terahertz frequencies', *Proceedings of International Symposium on Antennas and Propagation (ISAP '96)*, Japan, 1996, pp. 1229–1232

106 COLLINS, C. E., DIGBY, J. W., POLLARD, R. D. *et al.*: 'W-band measurements of 100 μm height micro-machined air-filled rectangular waveguides', *IEEE MTT-S International Microwave Symposium Digest*, 1997, pp. 1439–1442

107 LUCYSZYN, S.: 'The future of on-chip terahertz metal-pipe rectangular waveguides implemented using micromachining and multilayer technologies', *IEE Colloquium Digest on Terahertz Technology and its Applications*, London, 1997, pp. 10/1–10

108 LUCYSZYN, S., WANG, Q. H., and ROBERTSON, I. D.: '0.1 THz rectangular waveguide on GaAs semi-insulating substrate', *Electronics Letters*, 1995, **31** (9), pp. 721–722

109 LUCYSZYN, S., BUDIMIR, D., and ROBERTSON, I. D.: 'Design of low-loss monolithic millimetre-wave filters using dielectric-filled metal-pipe rectangular waveguides', *Proceedings of ESA Workshop on Advanced CAD for Microwave Filters and Passive Devices*, ESTEC, Noordwijk, 1995, pp. 381–387

110 LUCYSZYN, S., BUDIMIR, D., WANG, Q. H., and ROBERTSON, I. D.: 'Design of compact monolithic dielectric-filled metal-pipe rectangular waveguides for millimetre-wave applications', *IEE Proceedings — Microwaves, Antennas and Propagation*, 1996, **143** (5), pp. 451–453

111 LUCYSZYN, S., SILVA, S. R. P., ROBERTSON, I. D., COLLIER, R. J., JASTRZEBSKI, A. K., THAYNE, I. G., and BEAUMONT, S. P.: 'Terahertz multi-chip module (T-MCM) technology for the 21st century', *IEE Colloquium Digest on Multi-Chip Modules and RFICs*, London, 1998, pp. 6/1–8

112 AFTANASAR, M. S., YOUNG, P. R., ROBERTSON, I. D, MINALGIENE, J., and LUCYSZYN, S.: 'Photoimageable thick-film millimetre-wave metal-pipe rectangular waveguides', *Electronics Letters*, 2001, **37** (18), pp. 1122–1123

Chapter 4

CAD techniques

D. S. McPherson and C. E. Chrisostomidis

4.1 Introduction

CAD techniques are now an integral part of the MMIC design cycle. Consequently, the architecture of most commercially available software products is based on a universal sequence of interrelated steps. Referring to Figure 4.1, the procedure begins with a set of specifications that must be achieved by the design. Using experience, reference material and theoretical principles, the designer interprets the specifications and proposes a preliminary design topology.

The preliminary design is then represented in the CAD package using library components and models. In the first design iteration, it is often prudent to simplify the circuit representation by using idealised components. This allows the designer to ascertain the worth of the proposed topology quickly. If the idealised design does not exceed the required specifications, it is almost certain that a practical realisation of the topology will fail. Whether in the first or subsequent design iterations, the cycle proceeds to simulation, using circuit and electromagnetic (EM) simulators. There are no clearly defined procedures as to when and how to use these. In some instances, there is a strong mutual dependence between the results of the two classes of simulators. As a result, the simulations are carried out concurrently. In other cases, there are distinctly separable portions of the MMIC that are more amenable to one class of simulation than another, where simulations can be carried out sequentially.

It is seldom the case in MMIC design that a preliminary design immediately yields optimal performance. The level of complexity, in even the simplest MMIC, is sufficiently high that most initial hand calculations cannot predict the net effect of all of the circuit's components. Circuit optimisation permits the designer to automatically make tradeoffs between these various effects. When investigating a circuit topology with ideal elements, the result of this optimisation should give an upper bound on its performance. If it is below or only slightly above the required specifications, the designer must re-evaluate the topology or even negotiate the specifications themselves. For later design iterations, the optimisation results represent the best performance achievable. If the designer meets the specifications, it is then possible to finalise the layout, check that it conforms to the design rules and submit it for fabrication.

It should be noted that even in the final stages of the cycle, redesign remains a possibility. A circuit design may meet all of the electrical specifications and dimensional constraints, but then fail for topological reasons. For example, when converting a schematic to a layout, some components may be too close together or even be forced to overlap. This can occur with the physical layout of feedback loops or when there is insufficient space to place all DC and RF probe pads. The placement of RF probe pads can also lead to problems with the design. Many on-wafer probe station measurements require that the input and output RF pads be in-line with each other and located at opposite sides (i.e. East and West) of the chip. At the same time, DC bias pads should be located on the orthogonal sides (i.e. North and South). In some circuit designs, this requirement is not easily met.

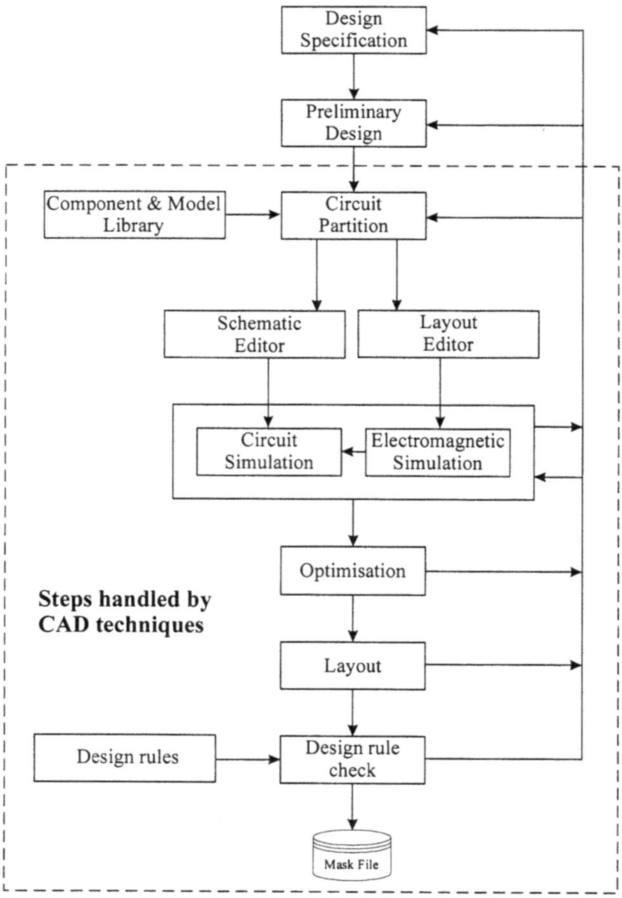

Figure 4.1 Complete MMIC design cycle

4.2 Integrated CAD design environment

Commercial CAD packages are more than just a simple tool for predicting the response of MMIC circuits. Increasingly, they provide an integrated design environment that manages all aspects of an MMIC design, from initial circuit specifications to layout design rule checking. Typically, this environment features a structured file management scheme, schematic editor, synthesis tools, component libraries, simulation engines with optimisation capability, output displays, layout editor and layout design rule checker. It may also contain a 2.5-D EM field solver, for simulating arbitrary single or multilayer planar structures. By providing the user with a complete set of tools, within a single package, the MMIC design cycle can be executed with far greater efficiency.

The philosophy in most commercial CAD packages is to divide the integrated environment into a series of programs or sub-packages, which can be interrelated as shown in Figure 4.2. The first sub-package is the schematic editor, where the topology of the circuit is represented in a graphical symbol format. To support the schematic editor, most commercial packages provide useful tools to help evaluate component values (e.g. for matching networks and filters) and physical parameters (e.g. for transmission lines). For any given design, there can be many individual schematics. Furthermore, they can be organised hierarchically, so that one schematic can contain links to another. This helps to manage complex designs and allows commonly used sub-circuits to be stored in libraries and inserted into subsequent designs, as required.

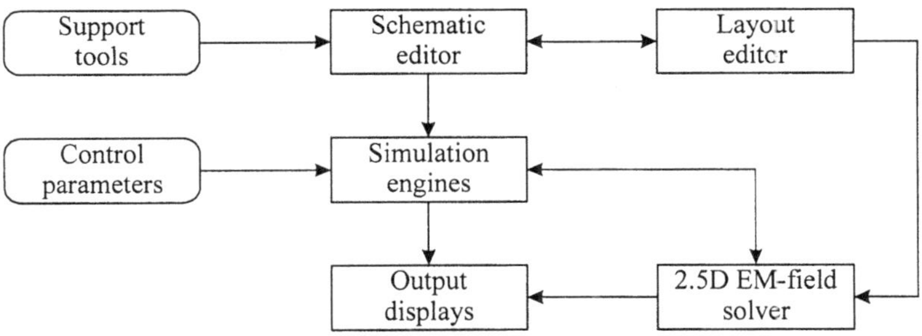

Figure 4.2 Integrated CAD design environment

The second sub-package incorporates all of the simulation engines. The exact number and nature of the engines varies from package to package. Typically, one can expect DC, linear, transient simulator and large-signal steady-state simulators. To invoke a given simulation, the designer must specify a circuit schematic, control parameters and desired outputs. Control parameters include frequency range, swept bias ranges, swept input power ranges, injected waveform shapes and output parameter optimisation controls. Outputs can take the form of network parameters, DC voltages/currents and AC voltages/currents. Depending on the simulator in question, these controls and outputs are specified in special test schematics or directly on the circuit schematic itself.

Having run a simulation, it is necessary to present the results in a format that can be accurately interpreted by the designer. The third sub-package is output data post-processing. Most commercial packages can display rectangular grids, polar grids and Smith charts, onto which a variety of different outputs can be applied. The designer can also inspect tables of collimated data and output them to a file for further processing. If circuit optimisation is required, it is often possible to view the convergence of the optimised outputs as each iteration is performed.

The fourth major sub-package of the integrated design environment is the layout editor. This tool permits the designer to synchronise the circuit schematic with the physical layout at the push of a button. This process can usually work both ways, i.e. the layout is automatically updated by changes in the schematic and conversely, the schematic is automatically updated by changes in the layout (known as back-annotation). Many layout editors include the capability for performing user-defined design rule checks (DRCs), to ensure that layouts meet foundry process requirements.

Finally, some CAD packages contain a 2.5-D EM field solver to simulate the effects of arbitrary single or multilayer planar structures. This sub-package is used when the designer wishes to include the effects of passive structures that do not conform to the standard library components. Since library components are defined in isolation from one another, the EM field solver can also simulate coupling effects when components are placed close to one another. When included as part of the integrated CAD package, the interface between the circuit schematic and the EM field solver is via the layout sub-package. This ensures accurate representation and avoids having to manually draw circuit components. For those CAD packages that do not contain an integrated EM field solver, it is still possible to employ separate packages and to import the results in the form of frequency-dependent *S*-parameter files.

In addition to this general framework of sub-packages, the integrated MMIC design environment has two other major features. Most commercial CAD packages permit considerable customisation. This ranges from simple interface modifications such as programmable toolbars and keyboard commands, to user defined components that are based on independent run-time code. These components can then be added to existing libraries. It is also possible to associate user defined scaleable artwork with particular components, so that the synchronised layout can accurately reflect the specific processing requirements. The second major feature that makes commercial CAD packages attractive is their use of flexible licensing schemes. By licensing the various components of a given package separately, customers can configure their installation to best meet their requirements. For example, many installations will include numerous instances of the schematic license and fewer instances of the more specialised simulator engines. As shall be seen, some simulators go even further, in that they permit third-party vendors to add their own licensed code to the existing framework.

4.3 CAD package features

Present day commercial microwave and RF CAD packages are organised in a very similar fashion to one another. Essentially, the role of the CAD package is to supplement the knowledge and skills of the designer with a suite of tools to make the art of design more efficient and organised. This section will review some of the key components that make this possible.

4.3.1 Schematic capture

With the evolution of computer graphics in the last two decades, from low-resolution monochrome to high resolution colour, circuit CAD has advanced from descriptive netlist files to graphical schematic capture. The modern approach permits the designer to place circuit symbols in a design area and to interconnect them with virtual wires. If all of the available components are represented by appropriate symbols, the resulting circuit description resembles an electrical circuit schematic. The parameters are most commonly displayed by lines of parameter text in the design area below their associated component. An example is given in Figure 4.3.

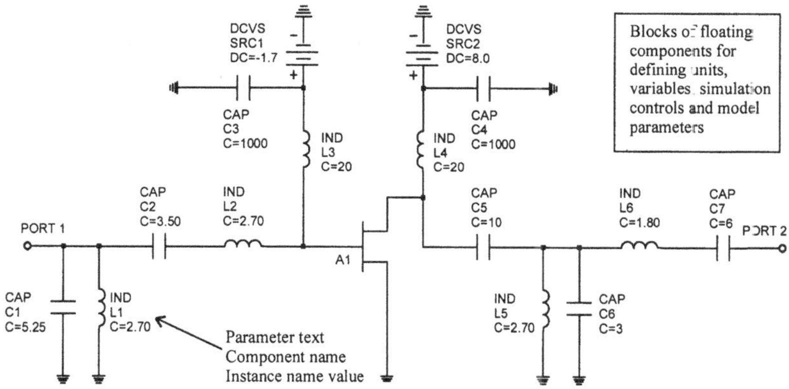

Figure 4.3 Schematic capture for circuit entry

Details on how to choose components and to change their parameters can vary slightly from one CAD package to another. However, they all have libraries with both general and manufacturer specific components organised by sub-libraries. The designer selects the sub-library and then the desired component. With some simulators, the designer will then be prompted to enter parameters directly in a pop-up window. Regardless, once the component is placed in the schematic, its parameters can be edited by changing the parameter text directly or by invoking an editor window. When all of the required components have been placed and edited, a wiring tool allows them to be interconnected. Adding measurement-related items (e.g. ports and/or voltage and current test point indicators) completes the schematic.

The ease of using schematic capture has now virtually eliminated the use of the netlist for circuit entry in CAD packages. It is much faster, avoids confusion, minimises errors and does not require the designer to recall the specific syntax for each and every circuit component. Furthermore, it avoids the need for maintaining additional sketches or drawings of the circuit, since the schematic used by the simulator is exactly the same as a conventional circuit representation.

4.3.2 Support tools

During the preliminary design of MMICs, it is often necessary to refer to standard textbook definitions and techniques for matching networks and synthesising filters or determining physical dimensions of transmission line structures. This can be tedious and time-consuming. To improve design efficiency, most CAD packages provide a collection of tools that can be invoked as required. The transmission line calculator is one such tool that allows the designer to input electrical length, characteristic impedance and dielectric properties to obtain equivalent physical dimensions. Some calculators have more capabilities, and can even be used to generate parameters for coupled-line structures. Another useful tool is the filter synthesiser that can be used in place of filter tables. By inputting the centre frequency, bandwidth, out-of-band rejection and permissible in-band ripple, the synthesiser can generate a corresponding filter with either lumped or distributed components. Smith chart tools are also appearing in CAD packages in response to the popularity of similar stand-alone products. Essentially, these behave like a blank Smith chart, upon which all of the customary pencil, ruler and protractor operations can be applied. The support tools are a valuable addition to the microwave CAD package. Although the results tend to be idealised, they are excellent for obtaining starting values in any design.

4.3.3 Hierarchical design

Hierarchy is a powerful concept that is employed extensively in CAD design, as a way of breaking a complex circuit down into more manageable components. A hierarchical design is developed by defining sub-circuits independently, and importing them into a more complex circuit by symbolic reference only. In circuits with large numbers of individual components, this can avoid the problem of schematics becoming so large that it is necessary to switch between views to inspect different portions. For circuits where groups of components are repeated, importing multiple references of the same sub-circuit saves time and ensures coherence. Hierarchy within the schematic also allows the designer to optimise the response of a repeated sub-circuit, while simultaneously observing the effects on the overall circuit. The balanced amplifier in Figure 4.4 is a good illustration of these principles. Here, the individual single-ended amplifier given in Figure 4.3 and the quadrature coupler are both imported twice to form the complete circuit. Any changes in the single-ended amplifier performance are now automatically updated in the balanced design.

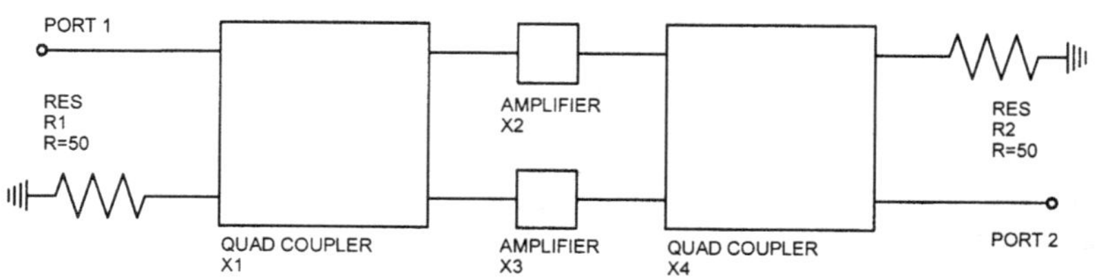

Figure 4.4 Hierarchical schematic entry

Hierarchical concepts are equally prevalent in the layout, with the same intention of avoiding unnecessary duplication and simplifying design. Common sub-circuit layouts can be repeated (or tiled) to avoid having to individually draw them and perform identical edits on multiple instances of the same structure.

4.3.4 Circuit component libraries

Libraries within commercial CAD packages offer a vast array of components for use in circuit design. These can be divided into four broad categories: generic passive components, generic active components, manufacturer-supplied components and user defined components. The generic passive components comprise ideal types (for use in an initial design) and physical types (for use in final implementation). For example, there exists a whole range of ideal lumped element and transmission line components. These can be used to quickly simulate a proof-of-concept or to establish the theoretical performance limits for a circuit topology. For practical MMIC designs above a few GHz, all passive components must be realised with planar equivalents. The most complete planar component libraries available in CAD packages are for microstrip; these include not only transmission lines, but also discontinuities such as bends, junctions, gaps, coupled-line segments and termination effects (opens and shorts). Furthermore, libraries may also contain spiral inductors, interdigital capacitors and MIM capacitors. These components are all termed generic because they are not specific to any given foundry process. They can only be made to model the components of a given process by the appropriate specification of substrate, dielectric and metallisation. Based on these parameters, the simulator applies established models to derive their equivalent electrical performance. Generally speaking, such complete libraries do not exist for other planar transmission line types, such as stripline, coplanar waveguide (CPW), coplanar strip, slotline and finline. Beyond simple transmission line segments, there exist no empirical models for more complicated structures. Nonetheless, as shall be seen later (Section 4.6.4), there exists a package that can integrate a planar EM simulator with the circuit simulator to overcome this limitation for CPW.

The list of generic active components includes standard models for diodes, BJTs, MOSFETs, MESFETs and HEMTs. Like the generic passive components, they are independent of a particular foundry or device. Their behaviour in the simulator is dictated by a lengthy set of parameters, specified by the designer. These parameters can, in some instances, be obtained from the device manufacturer; otherwise they must be extracted from measurements. There are two types of generic active component model: linear and non-linear. Linear models can only be used for small-signal linear simulation, and represent a device biased at one particular level with negligible input power applied. Non-linear components are far more general, their behaviour being a function of both bias and RF input power level. The generic active device models can also be used as building blocks to model more complicated devices. For example, a GaAs HBT is often modelled using a silicon BJT model, standard diode models and a thermal sub-circuit (to account for changes in emitter–base junction temperature with increased collector bias current and RF input power).

A major selling point of most commercial CAD packages is their extensive list of manufacturer-supplied components. These components typically draw on the model definitions of the generic active components and apply their own parameters. However, to protect their intellectual property, the exact parameters are not accessible to the designer. The manufacturer-supplied libraries are of little importance to MMIC designers, because they comprise components for hybrid circuits, such as surface mounted chip capacitors, inductors and resistors. However, most foundries offer add-on simulatable microwave artwork (SMART) libraries for at least one of the major CAD packages. SMART libraries usually include passive and active device models, with intelligent scaleable artwork for layouts.

The final category of library components is user-specified. These can be of three different types: measurement files; symbolically defined models, that are composed of generic components (e.g. the GaAs HBT described previously); and compiled models, based on linked C or Fortran code. If at all possible, compiled models are preferred because they allow unrestricted behavioural representation and are processed much faster than symbolically defined models. In either case, the user can also associate a symbol and layout with each model.

4.3.5 Simulation controls

The simulation controls are parameters given to the CAD package that instruct it how to carry out the simulation. Depending on the particular package, these are specified either in the top-level circuit schematic or in special test areas. There are two types of control parameters: one to set sweep ranges for the circuit stimuli, and another to change the run-time settings of the simulator. The first type basically dictates the characteristics of the input signals: frequency, voltage level, current level, input RF power or waveform shape. In many cases, it is also possible to sweep user-specified parameter values within the circuit itself. An excellent indication of the flexibility of any given CAD package is the number of independent circuit stimuli that can be swept simultaneously. For example, some packages will only permit one user-specified parameter to be swept or limit the number of independently defined sources. Without such restrictions, being able to sweep multiple parameters at the same time can yield considerable insight into circuit behaviour. For example, it is often useful to sweep several component values simultaneously when determining the optimum feedback for a feedback amplifier.

The second type of control parameter is mainly required in large-signal simulations, to specify start- and end-conditions for the numerical solvers. For example, with harmonic balance, the simulator must know how many harmonics of the input tones are permissible in the final solution and how many numerical iterations should be taken. Making appropriate decisions regarding these settings requires experience, and can have a significant impact on the final solution. This emphasises the fact that the simulator is a tool that can be very powerful when used properly. Without an appreciation of how it works, the large number of possible control settings can produce misleading and erroneous results.

4.3.6 Optimisation

One of the most powerful aspects of a CAD package is to allow the simulator to automatically adjust parameter values within a circuit until an optimal result is found. Optimisation is very useful for MMIC design, because the designs include a significant number of components with non-ideal behaviour. This means that it is very difficult to obtain the exact desired response from first principles. For example, a microstrip matching network can be generated using a Smith chart, but this only yields electrical lengths and impedances. As a first guess, the designer might convert these electrical specifications to physical dimensions and enter them on the schematic. However, for good agreement with the fabricated MMIC, additional junction and termination effects need to be added. Furthermore, to reduce the occupied chip area, it may be necessary to meander long transmission lines. This requires accurate models for bends. The unwanted loss and parasitic effects introduced by all these new components will degrade impedance matching. By allowing the optimiser to change line lengths and widths, it is simple to compensate for the unwanted effects and to restore the desired match.

The underlying principle of the optimisation tool is to calculate an error function, which is related to the difference between the target result and the current simulated result. The simulator then uses a search method to minimise this function. Depending on the method selected and the proximity of the initial to the optimal value, the optimiser may converge to a local minimum, the global minimum, or be unable to improve the circuit results. Generally speaking, rapid search methods rely on strong gradients of the error function and will converge to its nearest minimum. Slower search methods, such as the random selection of circuit parameter values, can find the global minimum but do so in a less systematic way. The best approach is to perform successive optimisations using a slower method followed by a faster method.

To perform an optimisation, the designer must first specify which parameter values within the circuit are to be varied and over what range. The designer must then select a search method, specify a range for the swept circuit stimuli over which the optimised performance is sought, and enter the number of iterations to perform. The ability of the optimiser to converge to a desired result requires that the problem being posed should have a solution. If this is the case, the optimiser will only be able to find it if the number of circuit parameters being varied is reasonable and/or the initial seed values give a result close to the minimum of the error function. Once the optimisation is complete and the designer is satisfied with the results, the circuit parameters are updated on the schematic.

Optimisation is a powerful tool that can be of great assistance to the designer in obtaining the best possible results. However, this is only possible when the initial design is based on firm principles. Fitting device models to measured data is a good example of why this is important. Without having prior estimates of the model parameters, the resulting fit is unlikely to be an accurate representation and may even be non-physical. In general, if the equivalent circuit model of a device reflects its physical nature, then the device has a greater chance of being accurately characterised over a broader range of stimuli (e.g. frequency and bias).

4.3.7 Layout

A layout is the equivalent of an engineering blueprint of the circuit. For MMICs, it is a multi-layered scaled drawing that can be interpreted by a foundry to generate photolithographic masks. Most integrated CAD packages now offer synchronisation between electrical schematics and physical layouts. For the MMIC designer to exploit this capability, SMART libraries must be available to ensure correct representation. In so doing, circuit layout becomes a purely automated task and the chances of committing design rule violations are substantially reduced. Although commercial foundries provide SMART libraries, not all of them support the same CAD package or put the same level of effort into preparing the libraries. In these instances it is necessary to generate a layout, even though it can be laborious and prone to error.

Regardless of the approach taken to generate the layout, the end result should be a binary GDSII Stream format file, which is the standard transfer format in the MMIC industry. From this file, the mask manufacturer can separate the individual layers of the MMIC and convert them to Gerber or similar file format. These are then used to control a precision photo-plotter or e-beam writer.

4.4 Circuit simulation engines

Within the integrated CAD package there exists a range of different simulation engines to investigate the response of an MMIC circuit. Table 4.1 presents a brief overview of the circuit simulator engines described in this chapter. For each engine, it shows whether simulation occurs in the time and/or frequency domain, the characteristics of the stimuli, the mathematical representation of the circuit, situations where it might be used and what the generated outputs are. This table is intended only as a rough guide as each one will be explained in further detail. For a comprehensive understanding of the underlying mathematics of circuit simulation engines, the reader is referred to References 1–3.

Although these operate independently of each other, they can all make reference to the same circuit schematic. For many circuits, several engines may be invoked at different stages to test different aspects of the circuit. Consider the design example of a bias-tuned pHEMT oscillator. The first step is to trace the DC I-V curves and to decide on a quiescent bias point. Switching to the linear simulator, source feedback is applied to maximise the negative resistance, and suitable resonator and terminating networks are designed in order to satisfy the conditions for oscillator start-up. Using the linear design as a starting point, the closed-loop transfer function is tuned at the centre frequency, using the harmonic balance simulator, until its magnitude reaches unity and its phase reaches zero degrees. The harmonic balance simulator can also simulate the delivered steady-state output power. To confirm that start-up will occur, a tiny pulse is injected into a DC port and the effects are observed using the transient or convolution simulator. After the start-up transients have died away, the time-domain results give an independent measure of the delivered output power and centre frequency. Other simulation engines can also be used to estimate the circuit's phase noise performance, yield and response to bias tuning.

Table 4.1 Overview of circuit simulator engines

Simulator engine	Domain	Stimuli	Mathematical solution	Applications and examples
DC	Freq.	Multiple DC volt levels	Non-linear algebraic equations	Establish operating levels, trace DC bias curves.
Linear	Freq.	Single small-signal sinusoid	Complex linear algebraic equations	Small-signal steady-state behaviour. Can calculate network parameters, MAG, 2-port simultaneous conjugate matching coefficients, gain circles, K, stability circles.
Harmonic balance	Freq. & time	Multiple large-signal sinusoids	Complex linear algebraic equations, DFT and non-linear ordinary differential equations, ODEs	Large-signal steady-state behaviour of power amplifiers, mixers, oscillators. Can be used to calculate steady-state oscillation conditions, intermodulation distortion, conversion loss, output power.
Volterra series	Time	Multiple large-signal sinusoids	Non-linear ODEs	Accurate large-signal steady-state behaviour of weakly non-linear power amplifiers, mixers, oscillators.
Transient	Time	Multiple arbitrary time varying signals	Non-linear ODEs	Transient response of lumped component networks only. Can be used to perform stability checks on amplifiers, observe effects of non-linear distortion, and confirm oscillator start-up.
Convolution	Freq. & time	Multiple arbitrary time varying signals	Non-linear ODEs	Transient response of networks containing lumped and distributed components. Same uses as the Transient time bench.
Noise	Freq.	Single small/large-signal sinusoids	Cascade of noise correlation matrices	Noise performance of linear or non-linear networks. Can estimate 2-port noise figure, oscillator phase noise.
Envelope	Freq. & time	Multiple arbitrarily modulated large-signal sinusoids	Harmonic balance repeated at intervals of time	Response of non-linear networks to modulated input signals. Particularly useful for simulating non-linear distortion effects.
Mixed mode	Time	Multiple large-signal sinusoids, and/or digital inputs	Digital logic simulator and transient, convolution or envelope simulator	Performance of subsystems including both analogue and digital components.
Yield	Not-applicable or freq.	Single small-signal or multiple large-signal sinusoids	DC, linear or harmonic balance	Statistical estimate of circuit yield from circuit component tolerances.

Before proceeding with the explanation of the simulator engines, it is worth clarifying the definitions of the terms small-signal, large-signal, linear and non-linear analysis. The small-signal approximation assumes that the input RF signal has a negligible voltage and current swing about a quiescent bias point. As a result, it cannot induce changes in the behaviour of the circuit under test. By contrast, a large-signal input is one in which this assumption cannot be made. Linear analysis techniques produce valid results only if they meet one of two conditions: the circuit behaviour does not vary with input RF signal power, or the input RF signal can be classed as small-signal. Non-linear analysis accounts for all cases where neither condition is met, and applies to all circuits whose behaviour does vary as a function of the large-signal input RF power level.

4.4.1 DC simulator

Before any AC simulations are performed, DC analysis is used to establish a circuit's quiescent bias point. To do this, the simulator must replace all circuit components with their DC equivalents. Thus, all capacitors, coupled lines and gaps become open circuits and inductors become short circuits. Transmission lines are replaced by DC resistors, with their values determined by their length, cross-sectional area and conductivity. Linear components are replaced by their DC conductances and non-linear components by their analytic DC functions.

Since the circuit may contain non-linear components, the problem of determining the node voltages and branch currents becomes quite complicated. The simulator must solve a system of N non-linear equations with N unknowns, where N corresponds to the number of circuit nodes. With the exception of trivial examples, such systems of equations can only be solved using numerical methods. The most commonly used is the Newton–Raphson iterative method. This is best described for the case of one non-linear equation, one linear equation and two unknowns (V_d and I_d), as depicted in Figure 4.5. In this approach, the non-linear analytical expression is replaced by a straight line with a slope equal to its tangent at the initial condition. For the simple case shown in Figure 4.5, the result of the first iteration is simply the intersection point between the response of the linear expression and the linearised non-linear expression. Using the intersection point as the basis for the next iteration, the process continues until the difference between successive iterations drops below a specified threshold. For more complicated systems of non-linear equations, the tangents are represented by a Jacobian matrix whose (i,j) entry is the partial derivative of the ith equation with respect to the jth variable. Solving the resulting linear system of equations is fairly straightforward and proceeds using standard methods such as lower-upper triangular (LU) factorisation and back substitution. Once calculated, the solution provides the initial values for the next iteration.

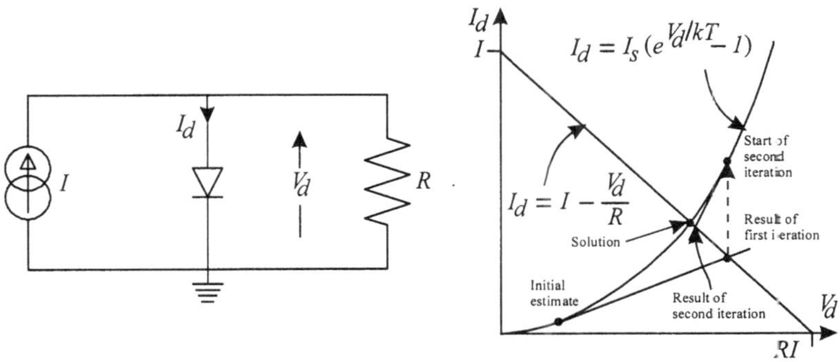

Figure 4.5 Graphical interpretation of Newton–Raphson iteration

When solving non-linear systems of equations, there is a distinct possibility that the analysis procedure may converge on an incorrect solution. Although it may be valid mathematically, it may have no physical significance. To avoid this problem, CAD packages use a variety of techniques to ensure that the correct result is obtained. One such technique is to use source stepping, where the initial node voltages are taken to be zero and the applied sources are dampened to a fraction of their full value. As the analysis proceeds, the damping on the sources is gradually relieved.

4.4.2 Linear circuit simulator

The linear circuit simulator is used to analyse the steady-state response of a circuit whose behaviour does not change with variations in the input RF signal level. It can also be used to analyse a non-linear circuit under small-signal excitation conditions. The simulation procedure starts by converting all circuit components to their phasor representation. For lumped and distributed components, this conversion uses stored analytic equations. Linear models made up of lumped components can be decomposed into their constituent components and similarly converted. If there are any non-linear components, the simulator must first run a DC simulation followed by a harmonic balance simulation. If the input RF signal level is made very small, the resulting power-dependent response of a biased non-linear component closely approximates the small-signal behaviour. This conversion process is completely transparent to the designer.

With all of the components converted, the circuit is now represented by a large web of interconnected nodes. This network can be defined by an equivalent set of N coupled linear algebraic equations, where the N unknowns are the complex phasor representation of the voltages at each node. Compared to DC simulation, which is represented by a non-linear system of equations, the numerical solution of the linear system does not require iterative solution methods. It can be determined directly using standard numerical matrix solution methods such as LU factorisation and back substitution. To improve speed, many linear simulators also use sparse matrix solution methods, in order to exploit the fact that most of the matrix entries are zero. Based on the solution, the node voltages can be related to the branch currents in the network and any of the

S-, *Z*-, *Y*-, *G*- and *H*-parameters can be derived. For 2-port networks, most CAD packages provide post-processing to obtain simultaneous conjugate matching coefficients, gain circles, *K*-factor and stability circles.

4.4.3 Harmonic balance simulator

For the steady-state analysis of general non-linear circuits, stimulated by large-signal input sinusoids, harmonic balance is the preferred simulation engine. It can be used to design power amplifiers, mixers and oscillators, etc. Designing non-linear circuits can be quite difficult since the process being modelled is inherently complicated. As a result, it is normal to begin the design using linear design techniques. Once an acceptable linear design is obtained, harmonic balance can be used to refine the design and to make large-signal predictions.

With harmonic balance, the non-linear network is partitioned into a linear sub-circuit and a non-linear sub-circuit, as shown in Figure 4.6, with *N* ports connecting the two sub-circuits together. In addition, the linear sub-circuit has *M* ports for sources and loads. The simulator first decomposes the non-linear device models, by separating the linear components (e.g. fixed resistors and reactances) from the non-linear lumped-element components (e.g. voltage-dependent capacitances and transconductances). These linear components are then added to the external linear network. At first glance, forming this partition from a complicated circuit topology is difficult to imagine. Nonetheless, it is always possible to rearrange the nodes such that partition can be achieved.

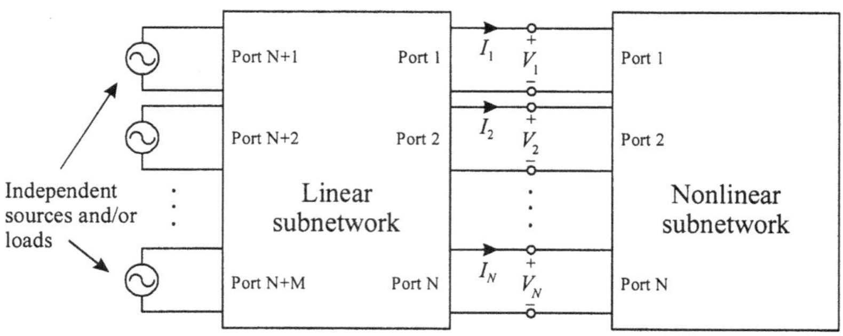

Figure 4.6 Circuit partitioning with harmonic balance

The aim of the harmonic balance method is to determine the response of a non-linear network to sinusoidal input signals. As a result, the output is expected to be a complex Fourier series. The analysis begins by running a DC simulation to establish the quiescent voltages at each node in the network. This is followed by a linear simulation of the entire network to establish an estimate of the AC voltages and currents at the *N* nodes. The *N* node voltage phasors are inverse Fourier transformed (IFT). Using the resulting periodic voltage waveforms, *N* nodal current waveforms are determined. This involves finding the steady-state periodic time-domain solution of a system of non-linear ordinary differential equations (ODEs), where the initial forcing functions are linearly related to the sinusoidal input signals. The solution technique for non-linear ODEs is discussed in Section 4.4.5. The new currents are then reconverted into the frequency domain and used to calculate a new set of nodal voltages. By comparing the new and the old nodal voltages an error value is obtained. The process is repeated with a new estimate that is located somewhere between the old and new values to force a reduction in the error value. Convergence occurs when the error is within a pre-specified tolerance.

One drawback with harmonic balance is the fact that the iterative solution process has no guarantee of convergence. Secondly, as the number of excitation frequencies increases, the mathematical complexity of the problem can require excessive amounts of computer memory and processing power. This latter concern has been addressed in commercial CAD packages, with the advent of the Krylov subspace solver. For a modest loss in accuracy, the Krylov subspace solver substantially reduces memory and processing power requirements. Finally, harmonic balance simulation is not always the most accurate engine for predicting intermodulation distortion. Since intermodulation products tend to be weak signals, they can become obscured by the numerical noise generated by the repeated application of the fast Fourier transform (FFT) algorithm.

As a final note on harmonic balance simulation, many integrated CAD packages contain additional tools to aid in oscillator design. Since an oscillator is a continuous feedback loop, special non-intrusive signal injection ports are provided for the purposes of calculating loop gain. Some CAD packages also provide additional routines to iteratively solve for the oscillation frequency and RF loop power.

4.4.4 Volterra series

Another method for solving non-linear circuit problems is to use Volterra series analysis. This approach is best described by analogy with a simple linear causal system having memory. In this example, the time domain output of the system is described by the convolution of the input signal with the time-varying transfer function. For a non-linear system, this description is insufficient because the system's transfer function changes as a function of the instantaneous applied input RF signal level. A convenient way of accommodating this input dependence is to represent the non-linear transfer function by a Taylor series expansion. When this expansion is combined with the convolution concept, the result is a Volterra functional series. Each term is the Qth-order convolution of the input signal with the Qth-order impulse response. Like any power series expansion, terms can be added *ad infinitum*.

For CAD packages offering Volterra series analysis, the simulation procedure is somewhat reminiscent of the harmonic balance method. The non-linear circuit is first partitioned into linear and non-linear sub-circuits. At the interconnection points between the sub-circuits, the non-linear components are decomposed into a linear term (represented by a resistor) and a set of non-linear current sources, as shown in Figure 4.7. This decomposition represents the Q terms in the Volterra series expansion. Several recursive methods exist for determining the expansion coefficients (i.e. the dependence of the current sources on the input voltage). For small values of Q, the simplest is the harmonic input method, which uses sinusoidal excitation to obtain the coefficients in the frequency domain. An IFT is then applied to obtain the time domain equivalents. Once the sources have been defined, the final solution to the problem can be carried out using transient analysis. The currents are simply treated as excitations at each order.

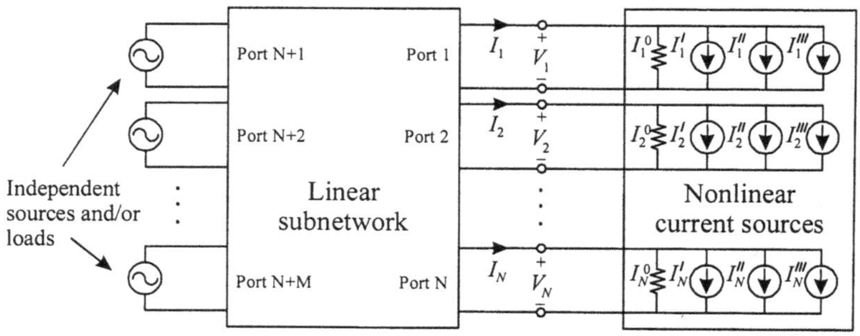

Figure 4.7 Circuit partitioning in Volterra series analysis (Q = 3)

Since the Volterra series simulator avoids repeated Fourier transforms, the numerical precision in the results is considerably higher than with harmonic balance. As a result, many designers prefer using this engine when calculating intermodulation levels. Unfortunately, these attributes are only applicable to weakly non-linear circuits or in cases where the input RF voltage is small. This occurs because most commercial Volterra series packages use the harmonic input method described above to determine the expansion coefficients. Although other pure time-domain procedures exist for calculating much larger numbers of coefficients, they are very complicated. It becomes a self-defeating exercise to add more terms, as the range of input RF signal levels over which the expansion is valid does not increase quickly enough to justify the additional computing effort.

4.4.5 Transient analysis

The transient analysis engine is similar, in many respects, to the well-known SPICE simulator. This type of simulation is customarily used to describe the response of a circuit to an arbitrary excitation in the time-domain. To carry out a time domain simulation of a general non-linear circuit, the problem is represented in terms of an equivalent system of non-linear ODEs. With the exception of a simple linear circuit, containing no more than three time-varying components, no closed-form solution exists. Therefore, numerical methods are required.

Prior to transient analysis, a DC simulation is performed on the circuit to establish the initial conditions. Next, the simulator formulates a set of equations that interrelate all unknown transient node voltages and branch currents. There are a number of ways in which this can be accomplished, although the state-space approach is the most widely employed [3]. If the circuit being analysed only contains linear and non-linear lumped components and linear time-invariant distributed components, then the system can be described entirely by coupled ODEs. If these are then expressed in terms of integral equations, a combined application of numerical integration and Newton–Raphson iteration can be used to find a solution. This is similar in many respects to the techniques used by the DC simulator. Since the integration operations are performed numerically, the solution can only be performed at discrete time points. As a result, the transient simulation engine is referred to as a discrete time simulator.

As mentioned previously, there are limitations imposed on the types of components that can be simulated using this time-domain method. The main restriction applies to distributed element components, because voltages and currents vary along their length. These components can only be described by partial differential equations, which are more complicated to solve than ODEs. Consequently, the only type of distributed component that can be analysed is one where the terminal voltages and currents are linearly related to each other and time invariant. One example is a non-dispersive transmission line with no frequency varying losses. Since MMIC circuits usually contain distributed components that violate these conditions, transient analysis can only generate estimates of circuit performance.

Great care must also be exercised when selecting the time step used for the transient analysis of MMIC circuits. Owing to the limit imposed by the Nyquist sampling rate, it can be no larger than half the period of the highest frequency of the excitation. If the circuit is strongly non-linear, then the time step must be further reduced to accommodate any generated harmonics. If this time step is made too small, relative to the transient responses of any components within the circuit, it may require an unacceptably large number of steps to reach the steady state. This imposes severe restrictions on the use of transient analysis for MMIC applications.

4.4.6 Convolution analysis

Transient analysis is limited to the simulation of circuits that contain only lumped components or distributed components that exhibit no frequency dependent characteristics. For MMIC circuits, this means that all practical distributed components have to be approximated by lumped component equivalents. Convolution analysis is designed to overcome this problem.

For the most part, convolution analysis is identical to transient analysis, since the solution is derived from a set of coupled ODEs expressing the time dependence of the voltages and currents of the circuit. Before the solution of the system can begin, the circuit must be partitioned into simple lumped components and frequency dependent distributed components. These latter components are first represented by their frequency-dependent admittance parameters, which characterise their response to a steady-state sinusoidal excitation. If the IFT of this response is taken, an equivalent impulse response can be obtained. The solution of the combined lumped-distributed network is then solved using conventional transient analysis. However, whenever terms in the system of differential equations refer to distributed components, their response is calculated by numerical convolution of the excitation with their impulse response.

Convolution analysis is indeed a very powerful simulation engine, but it is not without limitations. Firstly, it is very computationally intensive. By introducing convolution calculations within the framework of the iterative differential equation solver, computing time can rapidly become prohibitive. Secondly, all impulse responses must be truncated to some finite time interval. The usual convention is to terminate the impulse response when it drops below a given threshold. Unfortunately, when the impulse response decays very slowly, as with some distributed components, the threshold must be set quite high. This introduces considerable error in the results.

4.4.7 Noise analysis

The noise analysis engine is used to calculate the equivalent noise properties of a circuit. In commercial CAD packages, a general noisy multi-port network is represented as a combination of interconnected noisy 2-port components. If the system is driven by a sinusoidal source, each 2-port can, in turn, be represented by its admittance matrix and four spot noise parameters [4], as shown in Figure 4.8. These noise parameters are the minimum noise figure (F_{min}), noise resistance (R_N), optimum input conductance (G_0) and optimum input susceptance (B_0).

The noise parameters can be derived from measurements and/or physical circuit models (by the techniques outlined in [5]). Lossy passive components are straightforward because they only introduce thermal noise. On the other hand, linear and non-linear active components require more effort. If no measurements or physical models are available, the division into 2-port networks must extend to include all components within the device circuit model. Once fully subdivided, the active device can include components that model thermal noise, shot noise, flicker noise, channel noise and burst noise. The resultant spot noise parameters can then all be described by analytic functions of frequency, bias and temperature.

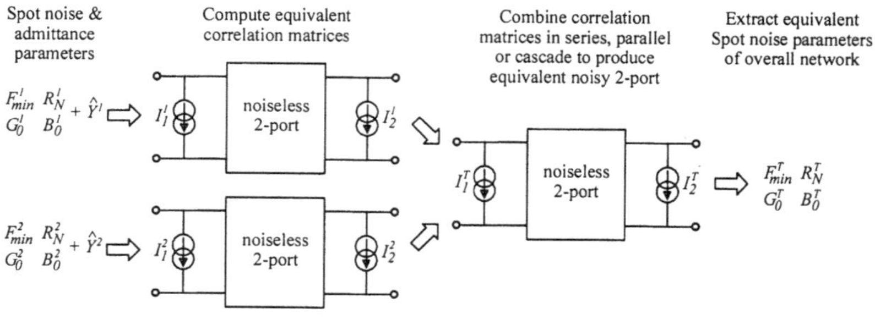

Figure 4.8 Combining equivalent noisy 2-port

The analysis begins by performing a single-point DC analysis for the specified bias levels and operating temperature. This permits the simulator to calculate the spot noise parameters. Based on the spot noise parameters, and the admittance parameters of each individual 2-port, a normalised correlation matrix is calculated. This matrix is a convenient description of the complete noise behaviour of the 2-port, because it can assume three different representations that are suitable for interconnecting 2-port pairs: in parallel, series and cascade (as shown in Figure 4.8). The analysis proceeds by combining all of the 2-ports until the final circuit is represented entirely by a single 2-port correlation matrix. As part of this interconnection process, the simulator must also determine the equivalent admittance parameters of each cascaded pair. Working in reverse, the values within the final correlation matrix are used to express the optimal input admittance, minimum noise figure and noise resistance of the overall noisy 2-port. Since these parameters are only meaningful for 2-port networks, noise analysis is consequently constrained to the class of networks that can be represented with one input and one output port only.

The above procedure only applies to linear steady-state simulations, as the spot noise parameters are only meaningful in the context of a sinusoidal excitation. For large-signal steady-state excitations, the simulation must take into account frequency translation of the noise resulting from non-linearities, and must also consider shifts in bias levels arising from the presence of large-signal RF sources [6]. Noise analysis cannot be performed in conjunction with transient analysis.

4.4.8 Envelope simulation

The envelope simulator uses both frequency- and time-domain methods to calculate the response of a circuit to baseband-modulated input RF signals. This circumvents a major problem with pure time-domain simulations. When circuits are excited by a high frequency sinusoid, with or without modulation, the time step becomes so small that an excessive number is required to reach the steady state. With the envelope simulator, the time step need only satisfy the Nyquist sample rate for the modulation.

The trick behind this technique is to sample the modulating signal and to perform a harmonic balance with an input sinusoid having the amplitude and phase of the sampled point. Thus, every node voltage, including the output, can be represented in time by a discrete spectrum that changes with time. The output modulation is extracted from the spectra by plotting the amplitude and phase of the fundamental spectral component as a function of time. By taking an FFT, this time function can be transformed to an envelope spectrum.

Envelope analysis has widespread applications in circuit analysis. For example, it can be used to compute adjacent channel interference resulting from intermodulation distortion, oscillator start-up and noise performance. At present, it provides the best alternative to transient time-domain analysis, for simulating MMIC performance when excited by modulated input signals.

4.4.9 Mixed mode analysis

In some instances, it is desirable to simulate the performance of a circuit or subsystem in conjunction with baseband DSP components. For example, it may be necessary to simulate the acquisition time or synchronisation time of a phase-locked loop, or the tracking capability of an automatic gain control component. Clearly, it is not possible to simulate the analogue components using a logic simulator, or to simulate the entire digital subsystem at the transistor level using a transient simulator. The logical approach is to simulate the analogue parts using a transient or envelope engine and to simulate the digital parts using a logic simulator. This approach is referred to as cosimulation or mixed mode simulation.

Although these two simulations can be carried out independently of one another, there are important interactions that can have an effect on the overall system performance. For example, the digital circuitry may have an impedance loading effect on the analogue circuitry at the interface between the two domains. It might also be desirable to investigate the transient behaviour of feedback loops, where part of the loop is analogue and the other part is digital (e.g. with digital frequency synthesisers).

The main challenge in combining these disparate simulators is the development of a suitable interface between the two domains. It must be able to convert signals between the two simulators and the two simulators must remain synchronised. In the digital domain, signals are represented by data streams, without reference to an underlying timebase. Furthermore, logic components tend to be unidirectional. By comparison, signals in the analogue domain are represented by time-varying voltages and currents and all nodes are bidirectional.

Of the two challenges, signal conversion is by far the simplest to resolve. The digital inputs can be modelled by a suitably valued capacitor and resistor, both of which may vary in accordance with the current state of the input. The resulting digital signal is derived from the analogue signal as it crosses a specified voltage threshold. By using a complete voltage-dependent model, the loading effect of the digital input on the analogue circuit is taken into consideration. A similar situation exists when converting digital signals to analogue. Although the pure digital domain transitions may be considered instantaneous, this is not so in the analogue domain. It is usually assumed that the transition between the two states has a specified rise- or fall-time. This modified waveform then modulates a non-linear gate model to produce an output with corresponding time-varying resistance and capacitance. The exact relationship between digital input and analogue output is technology dependent. The timebases are also brought into agreement at the signal conversion point, by applying or removing the analogue time step currently in use.

For seamless integration of digital and analogue circuit simulators their results must be synchronised in time. This poses a considerable challenge because the processing requirements of the circuit simulator are orders of magnitude higher than the digital simulator, for equivalent advances in time and numbers of components. This situation arises because the analogue circuit simulator uses much smaller time steps and must solve systems of differential equations. By comparison, the digital simulator must only contend with scheduling a sequence of Boolean logic operations.

There exist several degrees of synchronisation available for cosimulation. At one extreme, the two simulators can be permitted to run as independent programs. Although simplistic, it is clear that the outputs will only be meaningful when the output of one simulator is the input to the other. Another approach is to arrange the simulators in a master–slave mode. In this scheme, the slave simulator behaves like a subroutine of the master and automatically adopts its timebase. This approach is perfectly acceptable provided that the circuit handled by the slave simulator is relatively simple. Otherwise, calls to the slave simulator may dominate CPU time. The final approach is to unify the two simulators and allow them to use their own independent timebases. A synchronisation algorithm monitors the progress of both simulators, and halts the progress of the faster one until the slower one catches up.

4.4.10 Yield analysis

In large-scale MMIC manufacturing, it is often necessary to estimate the number of chips that will be within a required specification, given statistical variations throughout processing. For example, most MMIC processes will quote tolerances in thin-film resistor resistivities, MIM capacitor film thickness, some non-linear model parameters and/or transmission line dimensions. The quoted values are derived from measurements of drop-in test structures, or process control monitors (PCMs), that are inserted alongside circuits during fabrication. The process of yield analysis involves simulating the circuit over a specified number of trials, where the parameters are varied randomly about their nominal values, using probability distribution functions extracted from PCM measurements. The simulator then registers each trial as a pass or a failure, based on the supplied specifications. This method of yield calculation is known as Monte Carlo analysis. It has the property that its accuracy is independent of the number of statistical variables and their probability distributions [7].

The difference between the estimated yield and the actual yield is dependent on the number of trials performed. With a small number of trials, the sample may not be statistically representative of the overall performance. This is expressed by a relatively large error in the estimate that is valid for a small fraction of the trials. This fraction is referred to as the confidence level. To obtain a small error with a high level of confidence, a much larger number must be conducted. The expressions for the exact number of trials required can be obtained from standard references on Monte Carlo trials and confidence levels [8,9].

To improve yield, the circuit can be further optimised by a process known as design centring. The aim is to adjust the nominal circuit component values in order that a greater number of chips meet the design specifications. For each iteration, several individual yield analyses are performed, possibly with varying numbers of trials and confidence levels. Once the desired minimum yield levels are obtained, a final yield analysis should be performed with a user-specified number of trials to obtain the optimised error estimate and confidence level.

4.5 Commercial CAD packages

Over the past five years there have been dramatic changes in the power of commercial microwave CAD software packages. Until relatively recently, there were three main companies supplying integrated microwave CAD simulation packages. These were Compact Software (*Harmonica*™, *Super Compact*™, *Super Spice*™ and *Serenade*™), Hewlett Packard (*MDS*™ and *Impulse*™) and HP-EEsof (*Series IV*™ and *Touchstone*™). Since then, the Compact Software range has been acquired by the Ansoft Corporation and is now referred to as the *Serenade Desktop*™. Hewlett Packard has now changed names to Agilent Technologies and has invested a considerable amount of effort in its new Advanced Design System (*ADS*™). The Agilent EEsof *Series IV*™ and *Touchstone*™ products are still around and continue to be very popular. In 1997, Applied Wave Research Inc. made its debut with the first offering of *Microwave Office*™, an integrated CAD package designed specifically for use on desktop PCs running the *Windows*™ operating system. As a result of these changes, the present microwave CAD software market is now dominated by the Ansoft Corporation, Agilent EEsof EDA (electronic design automation) and Applied Wave Research Inc.

In addition to the major players, there exist several other companies that may be of interest to the MMIC designer. Unlike the companies already mentioned, their products tend to address more specialised applications. For example, Cadence offers its *Analog Artist*™ design environment. This contains the *Spectre RF*™ 'shooting Newton' time-domain simulator, which is excellent for generating designs involving upwards of 5000 devices. The environment also contains a powerful IC layout program and mixed signal simulation capability. Optotek Ltd offers a package called *MMICAD*™, comprising linear, transient and convolution analysis capability as well as a layout tool. Finally, Eagleware's *Genesys*™ is another popular package that contains a linear simulator, noise simulator, yield simulator, layout program and an extensive list of synthesis and support tools.

The remainder of this section will give a brief description of each of the aforementioned packages. A listing of their integral simulation engines is given in Table 4.2. For each of the individual vendors, the capabilities listed in this table are based on the complete software suite. In many cases, it is possible to configure subsets of these suites depending on customer requirements. For completeness, the table also indicates that all of the packages contain MMIC layout tools and that some have integral 2.5-D EM field solvers.

4.5.1 Agilent EEsof EDA Series IV™ (release 6)

Often referred to as *Series IV*™ for short, this package was originally developed by the EEsof company that was subsequently acquired by Hewlett Packard. *Series IV*™ refers to a whole suite of tools for systems design, circuit design, planar EM design and layout. The total package is offered in a large variety of configurations with numerous add-on options. The design environment itself has a very distinct look and feel to it, as it is structurally divided into four major working areas. It has a schematic window for editing systems and circuit designs, a synchronised layout window, a test window for specifying simulation controls and a global defaults window. Figure 4.9 shows examples of schematic, test and output windows. By separating the test controls from the schematic, it is relatively simple to generate standard test templates that can be re-used for different circuits. The package contains an optional filter synthesis tool, *E-Syn*™, and an excellent transmission line calculator, *Linecalc*™. The complete library of components stands at over 90,000. *Series IV*™ is available for both the Unix and PC platform, although the latter does not include the planar EM simulator *Momentum*™ and omits some model libraries.

Table 4.2 Summary of commercial circuit CAD packages

	Agilent EEsof EDA ***Series IV*** 6.0	Agilent EEsof EDA ***ADS*** 1.3	Ansoft Corp. ***Serenade*** 8.5	AWR ***MW office 2000*** 3.20	Cadence ***Analog Artist***	Optotek ***MMICAD suite*** 2.0	Eagleware ***Genesys*** 7.0
Platform	Unix/PC	Unix/PC	PC	PC	Unix	PC	PC
DC	✓	✓	✓	✓	✓	✓	
Linear	✓	✓	✓	✓		✓	✓
Harm. Bal.	✓	✓	✓	✓			
Volt. Ser.				✓			
Transient	✓	✓	✓		✓	✓	
Convolution	✓	✓	✓			✓	
Noise	✓	✓	✓	✓	✓	✓	✓
Envelope		✓	✓				
Mixed-mode		✓	✓	✓	✓		
Yield	✓	✓	✓	✓	✓	✓	✓
Layout	✓	✓	✓	✓	✓	✓	✓
EM simulat.	✓	✓	✓	✓			✓

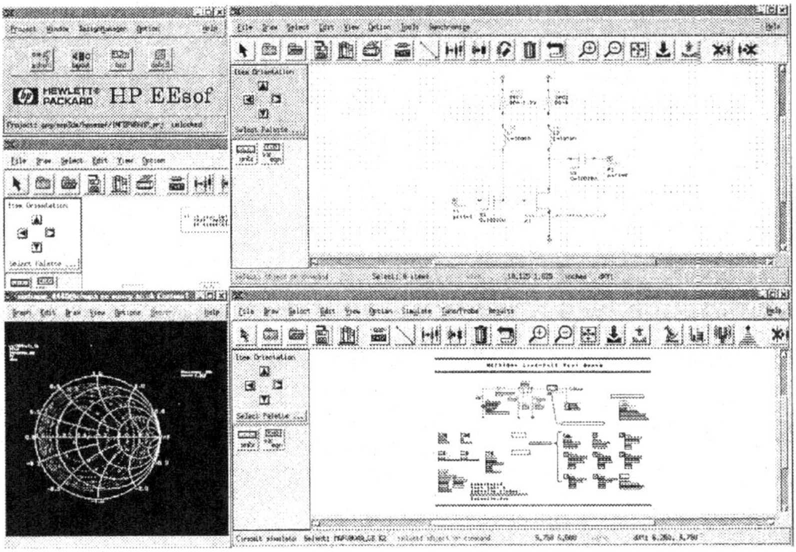

Figure 4.9 Series IV™ screenshot showing circuit, test and output windows

4.5.2 Agilent EEsof EDA ADS™ (release 1.3)

First launched in 1997, Agilent EEsof EDA's Advanced Design System (*ADS*™) is a completely new package [10]. Contrary to what some may believe, *ADS*™ is not a merged version of *Series IV*™ and HP's former Microwave Design System (*MDS*™). This CAD software package offers a complete end-to-end solution to a large number of MMIC and RF systems' problems. In its maximally configured form, *ADS*™ can simulate an entire wireless communications link (including DSP, IF, RF front-end and propagation aspects). Using the *Ptolemy*™ timed synchronous data flow simulator, cosimulation with transient, convolution and envelope analysis is possible. In comparison to *Series IV*™, the layout of this package is somewhat different, as the structural divisions between schematic entry and test control are absent. In *ADS*™, several sets of controls can be placed on a single schematic, thus permitting multiple simultaneous simulations of the same circuit. A data display window offers a neat working space, where a combination of different graph types and simulation results can be placed, along with post-processing equations and explanatory text to form complete design reports. This arrangement is shown in Figure 4.10. *ADS*™ also offers additional 'design guides' for facilitating the design of couplers, filters, power amplifiers, oscillators and PLLs. *Series IV*™ and *ADS*™ are open to considerable user customisation and permit the incorporation of proprietary SMART libraries. They also permit the inclusion of third-party executable code for compiled models and independent simulation code. *ADS*™ shares the same vendor model library as *Series IV*™ and is available in identical formats for both PC and UNIX platforms.

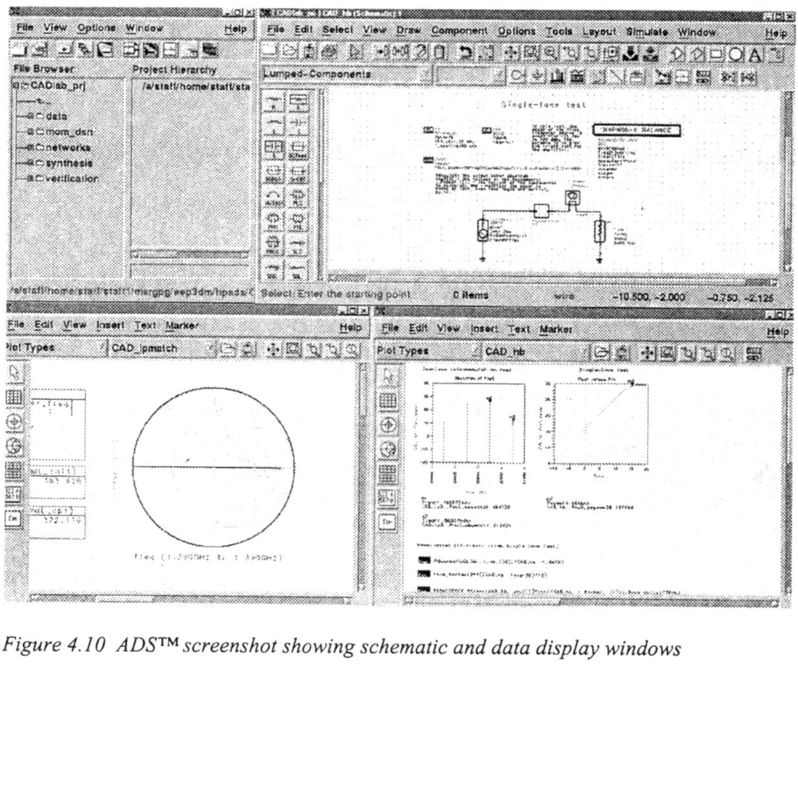

Figure 4.10 ADS™ screenshot showing schematic and data display windows

4.5.3 Ansoft Serenade Desktop™ (version 8.5)

The *Serenade Desktop*™ PC-based high frequency circuit design environment was first released shortly after Ansoft's acquisition of Compact Software in 1997 [11,12]. Previously, Compact Software offered a range of separate packages, covering linear analysis (*Super Compact*™), non-linear analysis (*Harmonica*™), transient analysis (*Super Spice*™), system simulation (*Success*™) and layout (*Serenade*™). Ansoft integrated these packages within *Serenade Desktop*™, which is functionally similar to most other *Windows*™ programs. Structurally, the working space is shown by the screenshot in Figure 4.11. It is divided into a hierarchical project window to relate the components of the design, schematic windows (for circuits and simulation controls) and output windows (for individual simulation results). In addition to the impressive set of simulation engines, *Serenade Desktop*™ integrates seamlessly with Ansoft's 2.5-D and 3-D EM simulators, *Ensemble*™ and *HFSS*™, respectively. This permits combined linear, non-linear and EM analysis. Support tools for *Serenade Desktop*™ include matching and filter synthesisers, transmission line synthesisers and a handy interactive Smith chart. Using the new system simulator, *Symphony*™, cosimulation of mixed functional analogue and digital circuits is possible. Like *Series IV*™ and *ADS*™, *Serenade Desktop*™ permits user-defined compiled components for both linear and non-linear circuit and systems analysis. Finally, the circuit component library offers a total of over 100,000 individual components.

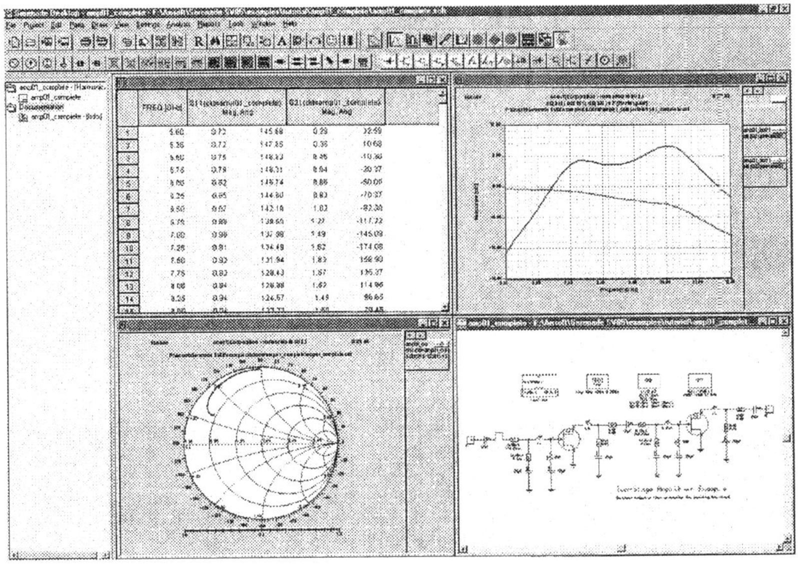

Figure 4.11 Serenade™ screenshot showing project hierarchy, schematics and outputs

4.5.4 AWR Microwave Office 2000™ (version 3.20)

Founded in 1994, Applied Wave Research (AWR) is a relative latecomer to the integrated microwave CAD market. Their flagship product, *Microwave Office 2000™* is the first microwave CAD system that is entirely native to the *Windows™* operating system [13,14]. Its main simulator, *Voltaire XL™*, features the capabilities of both a harmonic balance and a Volterra series simulation engine. As a result, it is one of the fastest and most accurate products available for predicting intermodulation distortion in weakly non-linear circuits. Since the suite has been designed using advanced object-oriented programming methods, exclusively for *Windows™*, it derives considerable processing speed advantages. The simulation engine can also be easily modified in the future to increase its functionality and efficiency, as new computational methods become available. These comments are equally valid for the *Voltaire LS™* linear simulator, whose performance is further augmented by means of its efficient sparse matrix solution method. Other attractive features of *Microwave Office™* include its use of sliders (for real-time tuning of both linear and non-linear circuits), its new integrated 3-D layout visualisation tool, *Artisan™*, and its 2.5-D Galerkin method-of-moments EM simulator, *Emsight™*. The design environment, shown in Figure 4.12, is divided between a project view that displays the design hierarchy, a component browser (containing the libraries), a variable browser for rapidly changing circuit component parameters, a layout browser showing layer definitions, and individual windows for schematics, layouts and data display. The environment is easily navigable to change between browsers and schematics. *Microwave Office 2000™* contains a library with over 40,000 components.

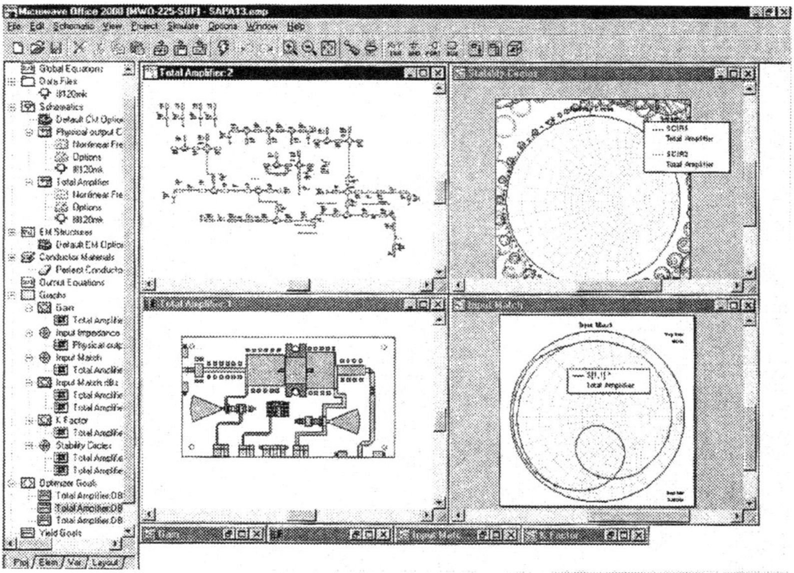

Figure 4.12 Microwave Office 2000™ screenshot showing various windows.

4.5.5 Cadence Analog Artist™

Compared to the previous packages, the Cadence *Analog Artist*™ design environment is primarily for commercial analogue IC development rather than for generalised high frequency RF applications. This is because it is principally an analogue transient circuit simulator, with the *Spectre RF*™ module and *ATS*™ mixed mode simulator offered as additional options. Nevertheless, *Analog Artist*™ is ideally suited to low RF frequency MMIC designs involving large numbers of devices (as many as 5,000) and particularly where mixed mode simulation is required. For this reason, it is now commonly used by many designers of digital wireless products, including cellular telephones, paging systems, mobile computing, wireless local area networks and global positioning systems. At the heart of the *Spectre RF*™ module is a very powerful proprietary Spice-like simulator that uses a set of time-domain solution algorithms. These bypass initial transient behaviour, using the 'shooting Newton' methods. This approach is ideally suited to strongly non-linear circuits such as switching mixers, ring oscillators and sampling and switched-capacitor circuits that can contain thousands of individual transistors. The *Analog Artist*™ environment also has a very versatile layout capability for automatic layout and DRCs of extremely complicated multi-functional ICs.

4.5.6 Optotek MMICAD™ (version 2.0)

The Optotek Ltd MMIC CAD suite, *MMICAD*™, is comprised of a linear simulator, transient/convolution simulator, *WAVEFORM*™, yield simulator, layout tool and filter synthesiser [15–17]. In addition to circuit design, the suite can also incorporate the large- and small-signal FET modelling tool *LASIMO*™ (reviewed in Section 4.6.2). By comparison to the first four CAD packages, *MMICAD*™ is smaller and has fewer simulation options. However, for linear circuit design, filter synthesis and time-domain non-linear design, *MMICAD*™ has ample functionality at relatively low cost.

4.5.7 Eagleware Genesys™ (version 7.0)

Genesys™ is a low-cost CAD suite that offers linear circuit simulation (SuperStar), noise simulation, yield analysis, multilayer EM simulation (EMPOWER), layout (LAYOUT) and an extensive set of synthesis tools for linear oscillator (Oscillator), matching network (MATCH), *R-C* filter (FILTER), microwave filter (M/FILTER), active filter (A/FILTER), group delay equalisation (EQUALIZE), transmission line (T/LINE) and PLL (PLL) synthesis [18,19]. The *Genesys*™ suite presents an impressive array of integrated tools for rapidly designing a host of different circuits. For ease-of-use and efficiency, this package is very useful for linear circuit design.

4.6 Commercial modelling software

For effective MMIC design, models for all circuit components are required. However, in cases where they are not provided by the foundry, it is necessary for the designer to develop their own models for use in subsequent designs. This is greatly facilitated by using one of a number of commercially available modelling packages. This section will review three software packages for generating linear and non-linear device models: Agilent EEsof EDA *IC CAP*™, Optotek Ltd *LASIMO*™ and IMST GmbH *TOPAS*™. It will also discuss IMST GmbH's *COPLAN*™, a supplemental coplanar waveguide model library that integrates seamlessly into Agilent EEsof EDA *Series IV*™/*ADS*™. This is a very useful tool that addresses the absence of accurate CPW models in commercial CAD packages.

4.6.1 Agilent EEsof EDA IC-CAP™ (release 5.0)

The Agilent EEsof EDA Integrated Circuit Characterization and Analysis Program, *IC-CAP*™, is a stand-alone UNIX program used for developing linear and non-linear device models and analysing statistical device performance data [20]. This latter feature applies to process development and will not be discussed here. In terms of modelling, *IC-CAP*™ provides simple extraction modules for a wide range of commonly used non-linear MOS, BJT, MESFET and HEMT device models, as shown in Table 4.3. All of the models are defined by functional equations (except for the Agilent root models), whose parameters are derived from static DC-IV characteristics and small-signal *S*-parameter measurements. This data can be collected automatically, using the extensive list of *IC-CAP*™-supported computer-controlled power supplies and instruments. An intuitive user interface provides sequential extraction routines with plenty of opportunity for parameter tuning and optimisation. *IC-CAP*™ also permits extraction of user-defined models and standard diode models.

Table 4.3 Non-linear models supported by IC-CAP™ (version 5.0)

Device types	Available models
BJT	BCTM VBIC BJT model, Philips MEXTRAM BJT model, Gummel–Poon model, Agilent EEsof HF Gummel–Poon model, EEBJT2 BJT
MOSFET	BSIM3v3.1; BSIM1,2; UCB MOS level 2,3 Agilent EEsof level 3, Agilent EEMOS1 MOSFET, Agilent Root MOSFET model
MESFET	Curtice, Statz, Agilent EEFET3, Agilent Root MESFET
HEMT	Agilent EEHEMT1, Agilent Root HEMT
TFT (thin film transistor)	Si TFT, p-SI TFT

4.6.2 Optotek Ltd LASIMO™ (version 2.1)

The Optotek Ltd large- and small-signal MESFET and HEMT modelling software, *LASIMO*™, is a PC-based extraction tool obtained either as a stand-alone application or as part of its *MMICAD*™ suite [21]. Although restricted to unipolar devices, it is capable of fitting parameters for an impressive list of built-in standard I_{ds}, C_{gs} and C_{ds} models, as shown in Table 4.4. In addition to the standard models, *LASIMO*™ contains reconfigurable models that allow users to modify parameter functional representations or to substitute for their own. *LASIMO*™ also provides extraction routines for small-signal models with up to 16 parameters. The fitting procedure avoids incorrect solutions by using special techniques to ensure convergence to the global minimum. Device characterisation in *LASIMO*™ can be carried out automatically using its data acquisition module.

Table 4.4 Non-linear models supported by LASIMO™

Device types	Available models
I_{ds} (MESFETs)	Curtice, Statz, Materka–Kacprzak, Triquint Own Model (TOM), Advanced Curtice, Curtice–Ettenberg, Lehovic–Zuleeg
C_{gs} and C_{ds} (MESFETs)	Junction Model, Statz, a physically based model
I_{ds} (HEMTs)	Curtice, Advanced Curtice
C_{gs} and C_{ds} (HEMTs)	A physically based model
Reconfigurable	Curtice and basic semiconductor junction, TOM3, Alpha Own Model (AOM)

4.6.3 IMST GmbH TOPAS™ (version 1.1, revision 1)

The TransistOr Parameter Scaleable (*TOPAS*™) model by the Institut für Mobil- und Satellitenfunktechnik GmbH (IMST), is a new generalised non-linear model suitable for both MESFETs and HEMTs [22]. It uses a consistent implementation that ensures excellent agreement between DC, small-signal and large-signal performance, and can simulate the effects of variable gate geometry, temperature dependence and noise. The model is highly adaptable because, unlike many other models, all bias dependent components are represented by spline fits, rather than equations having specific functional forms. At present, this model can only be used in Agilent EEsof EDA's *Series IV*™ & *ADS*™ simulators, as the interpretation of the spline coefficients uses the IMST-specific run-time code, which can only be linked to this platform. As a result, *TOPAS*™ has two parts; a stand-alone extraction tool that accepts bias-dependent *S*-parameter measurements and DC-IV curves, and an add-on run-time module for the simulator. The generalised nature of the model offers excellent agreement between measured and modelled results, and applies equally well to all types of MESFET and HEMT devices.

4.6.4 IMST GmbH COPLAN™ (version 2.3)

Also from the IMST, *COPLAN™* is a third-party module that can be added to the *Series IV™* and *ADS™* simulators to simulate complicated CPW structures. In most commercial CAD packages there exist resident CPW component libraries, but they are usually sparse and contain only simple transmission lines, open circuit stubs and short circuit stubs. *COPLAN™* was specifically designed to address this deficiency because it offers a library of components that puts CPW on a par with microstrip. The reason for this deficiency is rooted in the EM nature of the CPW structures themselves. Unlike microstrip, EM propagation in CPW is not easily confined to a single quasi-TEM mode. To suppress other modes, air-bridges or underpasses need to be included at junctions, bends and in long line segments, to ensure that the potential difference between left and right ground planes remains constant. These structures, however, disturb the EM fields and add shunt capacitance. Moreover, they cannot be modelled by the sort of empirical equations used for modelling microstrip junctions and bends. The situation is further complicated by the fact that CPW structures are defined by more physical dimensions. For accurate representations of these structures, *COPLAN™* invokes its own finite difference method (FDM) engine and fits the results to an empirical model. Some of the models' parameters are scaleable, meaning that it is not always necessary to repeat the EM simulation every time a single dimension is changed. As result, *COPLAN™* is the first package that allows hybrid EM/circuit simulation and parameter optimisation. The *COPLAN™* library seamlessly integrates into *Series IV™* and *ADS™*, allowing CPW components to be manipulated on a schematic, in exactly the same fashion as the components of the resident libraries. The available *COPLAN™* components are divided into distributed and lumped component libraries as shown in Table 4.5. The libraries can be licensed independently of one another. *COPLAN™* also has integrated scaleable artwork that allows for synchronisation between schematic and layout. An extensive set of foundry default options allows the *COPLAN™* elements to accommodate considerable variation in process parameters, such as different layer overlaps and air-bridge geometries.

Table 4.5 CPW components available in COPLAN™ (version 2.3)

Component categories	Available models
Distributed	Transmission line, open, short, gap, step, taper, airbridge, bend, T-junction, cross, inter-metal via, N-coupled lines
Lumped/Coupled line	Interdigital capacitor, rectangular spiral inductor, thin film resistor, MIM-capacitor (series/shunt)

4.7 Electromagnetic simulation tools

Full-wave, 3-dimentional (3-D) electromagnetic (EM) simulation is, arguably, the most demanding area of CAD software, with the hardest task of fulfilling the expectations of its users. This is particularly the case in terms of both its computational intensity and its ability to converge on accurate solutions. Unlike hybrid microwave integrated circuits (HMICs), post-manufacture tuning of

MMICs is generally not possible. This is due to the small physical dimensions of the structures, the use of multi-conductor and multi-dielectric layers, and the presence of the passivation layer. Therefore, because of their physical nature, MMICs are more difficult to design than HMICs. They require an accurate representation of the electrical behaviour of each circuit component. In addition, arbitrary-shaped structures normally associated with precision machining are now being introduced into multi-layer and micromachined MMIC technology. Thus, an EM simulator is an essential tool for achieving right-first-time components and circuits.

4.7.1 Need for EM solvers

Most MMIC designs employ both active and passive components, as well as interconnects and transmission lines. The term 'interconnects' can be used to describe lines that connect different metallisation layers, planar baseband/IF signal lines, control signal lines and bias lines, or it can be used to describe lines that connect chips together. The need for high package density has pushed the trend for multi-layer structures. These may exhibit complex interactions between different parts (and paths) of the circuit. These interactions, in turn, become more and more critical in the circuit performance as the operating frequency increases [23].

At present, the modelling and design of MMICs can be broadly divided into four categories, as summarised in Table 4.6.

Table 4.6 MMIC modelling and design techniques

	Cost	Time	Accuracy	Limitations
Device characterisation	High	Long	High	Extensive
Analytical/empirical modelling	Low	Short	Low	Extensive
Circuit simulations	Medium	Short	Low	Extensive
Full-wave EM analysis	Medium	Long	High	Little

With device characterisation, a great number of test structures have to be realised and measured. The measured data is then used in the modelling of these structures. However, the characterisation of test structures can take a great deal of time and effort and leads to high costs. Although this technique has the advantage of greater accuracy, this is limited to certain dimensions and substrate parameters. In addition, the resulting models are only valid within the measured frequency range.

Another method that is often used in MMIC designs is the use of analytical and empirical modelling. Even though time and cost are reduced in comparison to device characterisation, approximations within the models reduce the accuracy. In addition, circuit designers have a restricted number of models available.

For example, unlike microstrip lines, there are only a small number of such models for CPW structures.

Circuit simulators can be used to overcome the problem of limited frequency range, as in the case of device characterisation, but they are reasonably accurate only with simple structure geometries. No circuit simulator can accommodate the variety of different 3-D structures and layouts being employed in modern MMIC designs, nor can they accurately model all the possible interactions of these structures as package densities increase [42].

An alternative is to use EM simulators, which overcome the problems of cost and small range of validity, and can be used for arbitrary-shaped structures. On the other hand, the use of such tools is complicated, the simulation time is very large and the optimisation of circuit performance is not yet mature. Recent advances in computational speed and power have produced extensive improvements in the scale and complexity of problems that can be addressed by an EM solver (i.e. EM simulator). However, designers would like to see much more use of parallel processing, to allow even larger problems to be tackled and to make optimisation a more viable process [43].

The methods for analysing transmission line structure are based on determining the characteristic impedance and propagation constant (i.e. phase velocity and attenuation constant). These methods can be divided into two groups. In the first, which compromises quasi-static methods (e.g. conformal mapping), the nature of the mode of propagation is considered to be pure-TEM. In this case, the characteristic impedance of the transmission line is related to the phase velocity along the line and the capacitance (per unit length) between the conductors of the line. It is also possible to express the phase velocity in terms of the ratio of the actual capacitance of the line to the capacitance of the same line in air. Therefore, determining the characteristic impedance and phase velocity of the structure essentially reduces to the problem of finding the capacitance of the line. However, this type of analysis is only adequate for designing lower frequency MMICs, where the strip width and substrate thickness are much smaller than the guided wavelength [24,25].

The second group (i.e. full-wave analysis) takes into account the hybrid nature of all the modes of propagation. As a result, an important outcome of this type of analysis is information of the dispersive nature of the line. This gives the frequency variation of the effective permittivity, and therefore the variation in the characteristic impedance and phase velocity [24,25].

In general, traditional quasi-TEM approximations, used in the modelling of passive components within many commercially available CAD software packages, cannot accurately account for frequency dispersion. As the design frequency of the MMIC increases, the effects of the associated parasitics increase, thus changing the electrical characteristics of the components and associated circuits [24,25]. For example, this can have the effect of increasing impedance mismatch losses or reducing coupler directivities.

As previously pointed out, the microstrip configuration is not capable of supporting a pure TEM mode. Therefore, this hybrid mode cannot be fully described in terms of static capacitances and inductances only. As a result, one has to consider time-varying electric and magnetic fields, and solve the wave equation subject to appropriate boundary conditions. In general, EM solvers are better to use when one examines:

1. multiconductor transmission lines
2. cross-talk effects
3. ringing effects
4. timing errors
5. skin effect losses
6. R-C time constant effects
7. effects of discontinuities.

In addition, EM solvers are the only tools to use when one considers interconnect and packaging effects, which will be considered in more detail.

4.7.1.1 Wire-bonding and flip-chip interconnect effects

Bond-wires, used to interconnect MMICs together, often result in serious effects because of their low-pass filter characteristic, as seen in Figure 4.13*a*. In addition, there are associated uncertainties in their length and position(s), which cause uncertainties in the self- and mutual-inductances. Furthermore, MMIC package interconnects require accurate modelling, to avoid unwanted resonances [23].

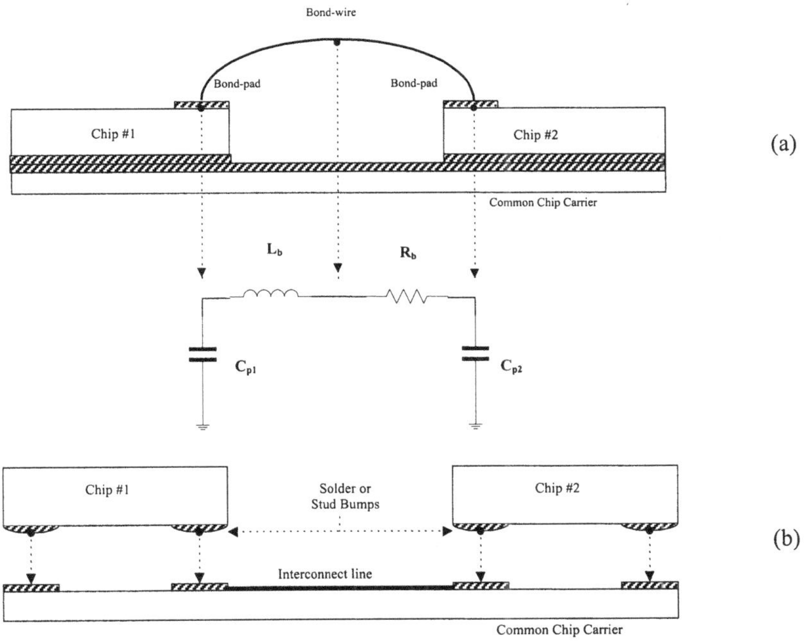

Figure 4.13 Chip-to-chip interconnections: (a) wire-bonding and (b) flip-chip

On the other hand, the main benefit of flip-chip technology, as illustrated in Figure 4.13*b*, is that the 3 dB cut-off frequency, associated with the low-pass nature of the wire-bonded interconnects, can be greatly increased. This is because the bond-wire is completely removed, thus removing the series inductor and greatly reduced shunt capacitances.

4.7.1.2 Limitations of design libraries

Many passive circuit elements cannot be satisfactorily characterised using (semi-) empirical modelling tools or design libraries. This can be due to the elements either departing from the design range or a composite mixture of models becoming unsatisfactory. As an example, consider the case when modelling planar spiral inductors having simple bend discontinuities and parallel-coupled lines. This is illustrated in Figure 4.14. Here, cross-coupling effects due to the bends, the ground plane or non-adjacent lines will be completely ignored.

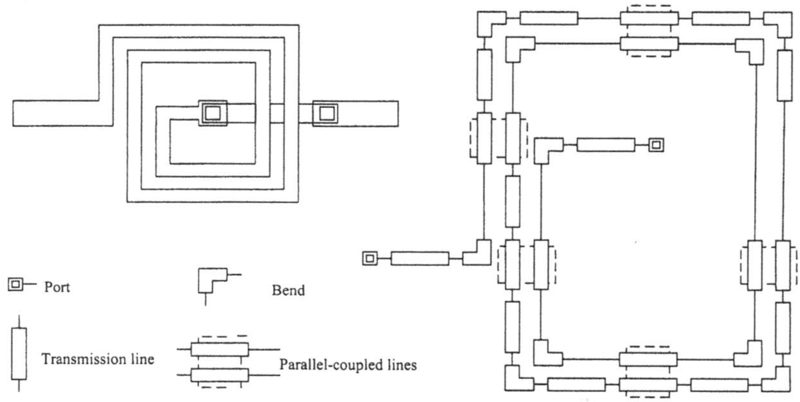

Figure 4.14 A simple spiral inductor model using circuit simulator elements

4.7.1.3 Ground currents

Ground currents can cause major impedance mismatch problems at high frequencies. When connecting MMICs of different thickness or MMICs on different carriers, they may be separated by a deep gap. Often this gap can be effectively modelled as an inductor in series with the substrate's ground, thus changing the transmission line's characteristic impedance [23].

4.7.1.4 RF absorption materials

RF absorption materials might be used to remove cavity modes. Using too little absorber may not remove the resonant fields. Using too much absorber may increase the transmission line losses, resulting in reduced gain, increased noise figure and reduced output power and efficiency [23].

4.7.2 EM software requirements

There are three basic requirements for any EM simulator. These are related to accuracy, computational speed and flexibility. Clearly, accuracy is the most important of these and it usually comes in the form of an error bound (i.e. an error term sufficient to stop the simulation). There are two approaches to achieving a desired accuracy. One is to use closed-form expressions, where for a given structure the error is known to be below a certain level. The other is to use an iterative technique, whereby computational speed is traded-off against increased accuracy [33].

An EM simulator should, preferably, be interactive in order to allow the user to experiment in the software environment. In others words, it is like having an MMIC made in a workshop and a vector network analyser to experiment with, without having to deal with the actual hardware. This requires the turn-around computational speed of such a package to be very high. This speed, in turn, depends on three factors:

1. the chosen numerical technique to solve the EM problem
2. the computer algorithm used to implement this numerical technique
3. the processing speed of the host computer.

EM simulators intended for mounting on mainframe platforms may be prohibitively slow when adapted for personal computers. In this sense, the efficiency of an EM simulator can be said to be hardware-dependent [33]. Finally, the flexibility of an EM simulator is mainly determined by the numerical technique on which it is based. Most techniques require a certain amount of analytical pre-processing, usually based on assumptions about the topology and topography of the structure being studied. This may be a problem when novel structures are investigated [33].

To summarise, designers often require a simulator that is capable of modelling complete structures, having realistic geometries, that can provide information on the electric and magnetic fields (*E*- and *H*-field, respectively) in order to gain a deeper insight into their performance [23]. Table 4.7 summarises some important distinctive features between circuit and EM simulators.

Table 4.7 Circuit and EM simulator comparison

	Circuit simulators	EM simulators
Model parameter range	Restricted	Unrestricted
Model geometry	Restricted to built-in models	Arbitrary
Inter-element coupling	Very limited	Automatically determined
Package effects modelling	Very limited	Automatically determined
Radiation effects	Very limited	Automatically determined
Moding	None	Full
Current distribution, field visualisation	None	Graphical interface

4.7.3 Use and limitations of EM simulators

Full-wave analysis takes into account all modes of propagation. This includes both the TE- and TM-modes that constitute the hybrid mode present in microstrip. Such analysis can account for the effects of dispersion, radiation, etc., to provide the designer with a powerful tool for optimising the performance of an MMIC [23].

The correct use of EM solvers for high frequency designs is the key to realistic circuit performance evaluation. A microwave engineer unfamiliar with EM simulators might attempt to design a complete circuit by trial and error, sometimes even starting with inappropriate structures. This would result in a waste of valuable CPU power and time [34]. Understanding the theoretical operation of the structure/circuit, and conducting an initial simulation using conventional circuit simulators to prove the principle, should be undertaken whenever possible. This is because the EM solver is by no means always the best tool to use in order to investigate all structure/circuits, which may result in valuable time being wasted [34].

As a rule of thumb, EM solvers are best suited to problems in which the physical dimensions are of the order of $0.1\lambda_g$ to $10\lambda_g$. For smaller electrical lengths, quasi-static solvers are the best solution. Designs larger than approximately 100 cubic wavelengths usually impose a very high demand on computing resources. This means that they may require a large memory and several hours of computational time on a typical workstation platform [34].

Most of the numerical methods employed by EM solvers rely on some form of grid, which divides the entire problem into much smaller parts, where the solution is to be calculated. These parts must be small compared to the wavelength in order to have an accurate solution [34].

Given that complex geometries frequently require lengthy simulation times, it is highly recommended that a single frequency pass is first performed, in order to inspect the fields and *S*-parameters. This will show if there are any errors in the set-up, possibly saving time [34].

In the case of complex networks, direct use of EM simulators may be a slow and uncertain process. However, many attempts to alleviate this drawback have been published [35–37], where EM analysis is used to generate equivalent models that can be imported into circuit simulators for further work.

4.7.4 Types of EM simulator

Before starting the discussion on EM simulation techniques, it is useful to briefly examine the different types of EM simulators. An EM solver classification exists in terms of the number of spatial dimensions that they can handle. Lumped elements may be termed as zero-dimensional, since all three spatial dimensions are much smaller than the wavelength of operation. Distributed elements (e.g. microstrip or CPW) with one dimension (i.e. height) that is much smaller than the guided wavelength, but with the planar dimensions (i.e. width and length) comparable to the guided wavelength, may be termed as 2-D or planar components. Finally, rectangular waveguides having cross-sectional dimensions that are comparable with wavelength may be termed as 3-D components [38]. This should not be confused with electrically small 3-D structures found in multi-layer or micromachined MMICs.

A key comparison feature is the computational effort. In principle, the greater the computational effort the more the simulator can perform. Arbitrary 3-D simulators require the highest computational effort. However, they can provide a powerful tool that is able to simulate any arbitrary-shaped structure, taking into consideration all possible EM effects (e.g. dispersion, stray coupling, radiation effects, etc.) [34]. Arbitrary 3-D simulators mainly use finite element methods (like Ansoft's *HFSS*™) to compute 3-D fields and currents associated with 3-D objects. At the other end, requiring less computational effort, 2.5-D simulators mainly use methods of moment (like Ansoft's *Ensemble*™ or Agilent's *Momentum*™) to compute 3-D fields, but only allow 2-D conductors and currents on planar layers. Further classification between 3-D planar 2.5-D solvers exists, with their classification dependant on how they handle vertical currents between different layers. Table 4.8 summarises the basic types of EM simulators [34].

4.7.5 Numerical methods

3-D EM simulators use numerical analysis techniques to solve Maxwell's equations. Real physical structures are defined by three spatial variables (x, y, z) and one time variable. Because the full treatment of such problems require very large memory and very long simulation times, whenever possible simplification to two spatial variables and the time variable should be carried out [39–41].

EM problems are sometimes described by differential equations and sometimes by integral equations. The unknown function is usually continuous and depends on continuous independent variables. There is a vast range of different numerical analysis methods, most of which are variations or combinations of older methods. The most widely used methods for full-wave analysis are summarised in Table 4.9.

Table 4.8 Basic types of EM simulators

EM Simulator Type	Comments
Arbitrary 3-D simulators	Compute 3-D fields and currents on arbitrary 3-D objects.
3-D planar simulators	Compute 3-D fields but only allow 2-D currents on planar layers and planar currents between layers.
2.9-D planar simulators	Allow non-uniform currents (not necessarily orthogonal to interfaces) between layers on vias and surfaces.
2.8-D planar simulators	Allow non-uniform vertical currents (orthogonal to interfaces) between layers on small diameter vias and surfaces.
2.7-D planar simulators	Allow non-uniform vertical currents (orthogonal to interfaces) between layers on small diameter vias.
2.6-D planar simulators	Allow limited uniform vertical currents (orthogonal to interfaces) between layers on small diameter vias.
2.5-D planar simulators	Compute 3-D fields but only allow 2-D conductors and currents on planar layers.

Table 4.9 Comparison of EM simulation methods

Method	Memory requirements	CPU time consumption	Flexibility
Finite-difference	Large	Large	Very good
Finite-elements	Very large	Medium to large	Very good
Method of moments	Medium to large	Medium to large	Very good
Transmission line matrix	Medium to large	Medium to large	Very good
Boundary elements	Medium	Medium	Very good
Mode matching	Medium	Small to medium	Good
Methods of lines	Medium	Small	Good
Integral equation	Small to medium	Small to medium	Good
Transverse resonance	Small to medium	Small to medium	Limited
Spectral domain	Small	Small	Moderate

The finite difference method (FDM) [44–46] is a simple numerical technique used in solving partial differential equations. A finite difference solution to Poisson's or Laplace's equations can be carried out in two basic steps [39–41]:

1. approximate the differential equations and the boundary conditions by a set of linear algebraic equations (referred to as the difference equations), on grid points that lay within the solution region
2. solve this set of algebraic equations.

The solution region is divided into rectangular meshes with nodes (i.e. grid points). A grid point on the boundary of the solution region (at which the potential is specified) is called a fixed node. All interior grid points within the solution region are called free nodes. The objective is to determine the potential of all free nodes [39–41]. This can be seen in Figure 4.15.

Like FDM, the finite element method (FEM) [47–49] is a powerful tool for solving differential equations. While FDM uses an array of grid points, its application becomes difficult with problems having irregularly shaped boundaries. Such problems are best suited for FEM [39–41]. This can be seen in Figure 4.16. This type of numerical analysis can be carried out in four basic steps:

1. discretise the solution region into a finite number of sub-regions or elements
2. derive the equations for a typical element
3. assemble all elements in the solution region
4. solve the system of algebraic equations.

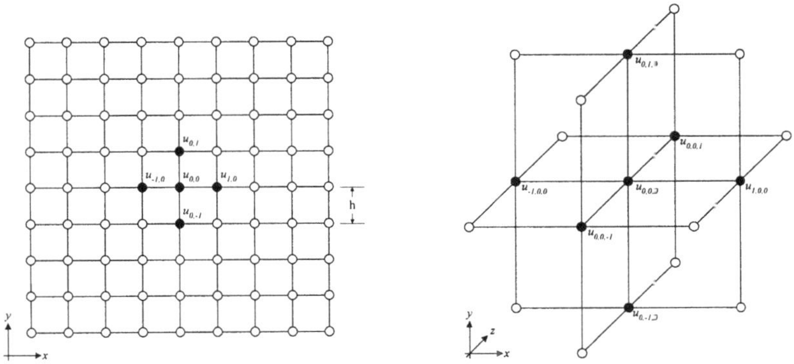

$$\nabla^2 u_{0,0} = \left(\frac{\partial^2 u}{\partial x^2} + \frac{\partial^2 u}{\partial y^2} \right) = \frac{1}{h^2} \left(u_{1,0} + u_{0,1} + u_{-1,0} + u_{0,-1} - 4u_{0,0} \right) + O\left(h^2\right)$$

Figure 4.15 The FDM method (a) in a 2-D cell and (b) in a 3-D cell

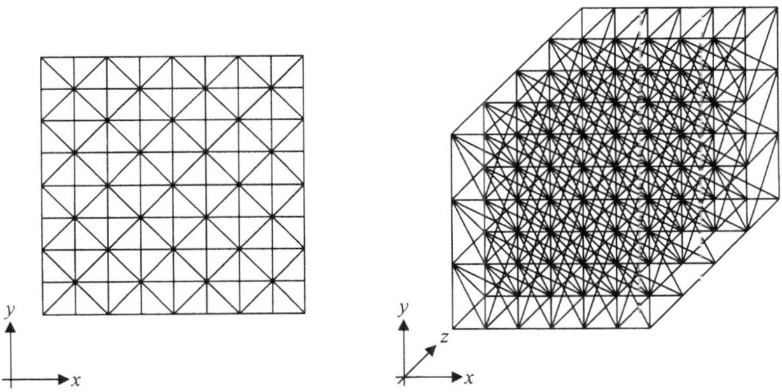

Figure 4.16 The FEM method (a) in a 2-D cell and (b) in a 3-D cell

A major problem with FEM analysis is the relatively large amount of memory required to store all element data, and consequently, the associated high computational time to manipulate the stored data. However, several algorithms have been developed to alleviate the problem to some degree [39–41].

Like FDM, the method of moments (MOM) has the advantage of conceptual simplicity. While FDM is used for solving differential equations, MOM is commonly used to solve integral equations. Here, sophisticated mathematics is used to achieve great accuracy and computational efficiency. Much of the effort with MOM involves the calculation of the Green's function, which is frequently determined as an infinite series [39–41,50].

The method of lines [51,52] is a semi-analytical way for solving the Helmholtz equation, where discretisation is carried out in cross-section (x, y), while the other dimension (z) is treated analytically. This method is particularly suited to irregularly shaped transmission lines.

Since its inception in 1971 [53–55], the transmission line matrix (TLM) method has been developed into a powerful and flexible microwave and millimetre-wave simulator. The TLM models the field components of EM waves in space, by representing equivalencies to voltages and currents flowing along an orthogonal transmission line network. Scattering of the waves on the TLM mesh takes place at each node and at each iteration (or time step). In this way, the impulse propagation is discretised both spatially and temporally. By obtaining the Fourier transform of the impulse response at a specific node, it is possible to obtain the magnitude and phase of the field components at the corresponding location in the model.

The spectral-domain analysis technique [56–58] utilises the Green's function to determine the longitudinal and transverse current distributions. The majority of spectral-domain techniques rely on Galerkin's procedure to determine the unknown coefficients of the expanded current distributions.

4.7.6 Features of EM simulators

In commercial EM simulators there are many different numerical methods used with different types of graphical user interface. However, the underlying principles remain the same. A generalised flowchart that can be applied to almost all types of EM simulator can be seen in Figure 4.17.

An EM simulator operates on a physical layout description of the component/circuit. An attractive feature of this approach is that it is possible to interface the EM simulator with the drafting software used to draw the physical structure and mask layouts. This allows a very close link between the physical design, simulation and the manufactured component, and minimises the risk of error in translating the physical dimensions into the EM simulator.

The EM simulator excites the component/circuit with a voltage potential and solves Maxwell's equations using numerical techniques. The designer specifies the range of frequencies over which the network is to be analysed. By solving for the *E*-field equation at each frequency point, the simulator determines the way current is distributed in the metal surfaces for the given layout, where and how much EM coupling will occur, and what energy will be radiated. After the analysis is complete, designers can extract the component/circuit *S*-parameters or view the field/current distributions in different parts of the structure.

To enter a design into an EM simulator and perform the analysis there are several steps that need to be followed. These basic steps are almost identical with all commercially available EM simulators. In this section they will be briefly examined with the aid of Figure 4.18.

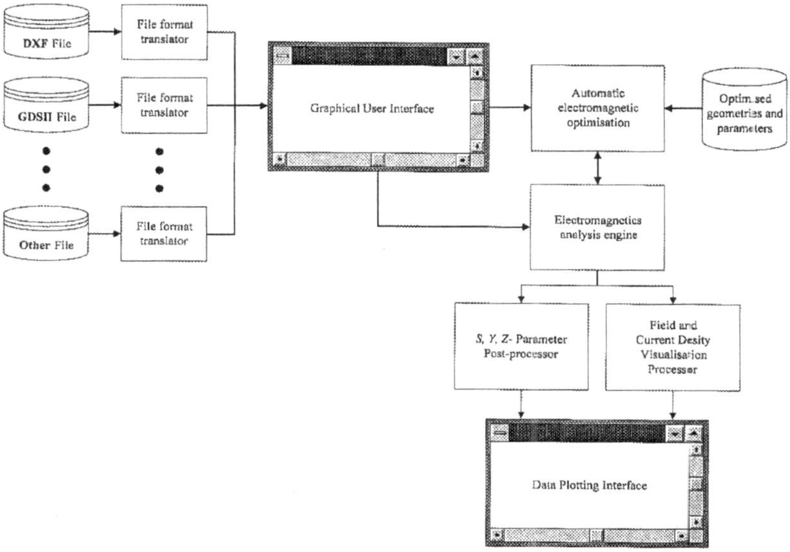

Figure 4.17 The EM simulator flowchart

4.7.6.1 Project set-up

During the course of designing a structure many different files will be generated such as drawing files, solution files, log files, etc. Thus, a project folder is invaluable for organising and navigating around different projects. This is usually the first step when using an EM simulator. This process is similar to other software packages and allows the user to perform 'house-keeping' operations (e.g. to create, review, rename, move, copy or delete files within a specific project folder).

4.7.6.2 Simulation set-up

The general strategy undertaking in creating a structure for EM analysis is to build a collection of 2-D and 3-D objects, treating each different material as a separate object, as can be seen in Figure 4.19.

2-D objects are irregular polygons that lie on a plane. For example, a square is a 2-D object that is bounded by four line segments and a circle is approximated by an *n*-sided polygon. There are several reasons for creating 2-D objects:

1. 2-D objects can be used to represent conductive surfaces (e.g. signal track)
2. 2-D objects can be used as ports, to identify where the signal or wave enters the structure, etc.
3. Creating a 2-D object could be the first step in creating a 3-D object: for example, the user can use polylines and arcs to create an irregularly-shaped object, and then use the Sweep command to extend that polygon through the problem region (e.g. to give finite metallisation thickness).

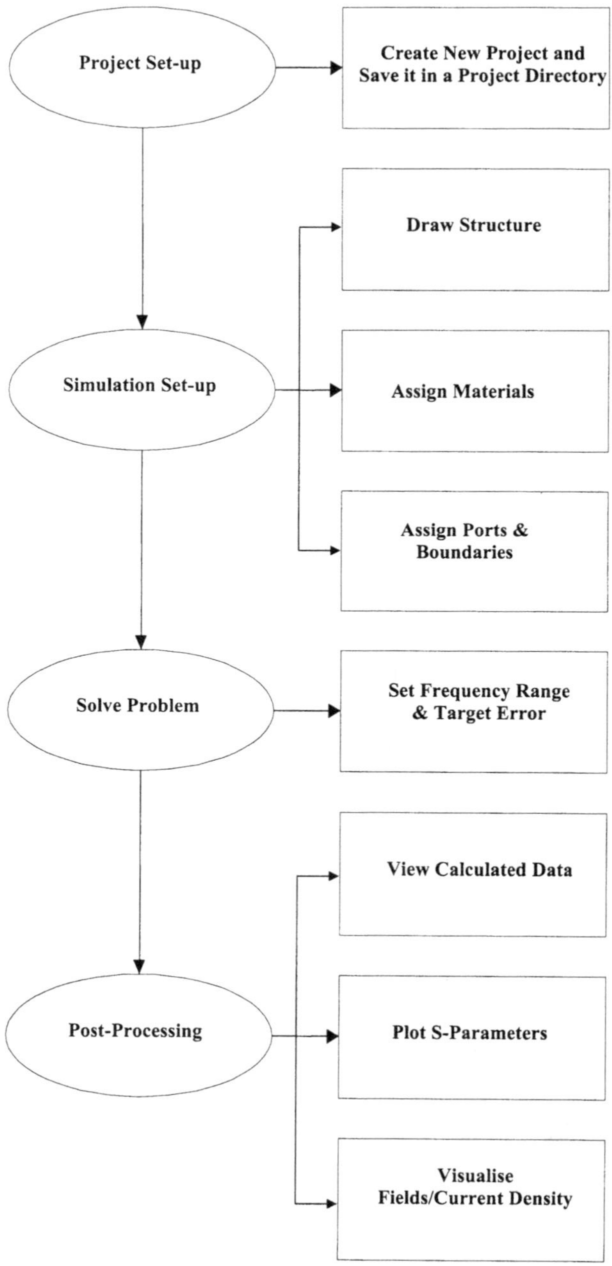

Figure 4.18 The EM simulation procedure

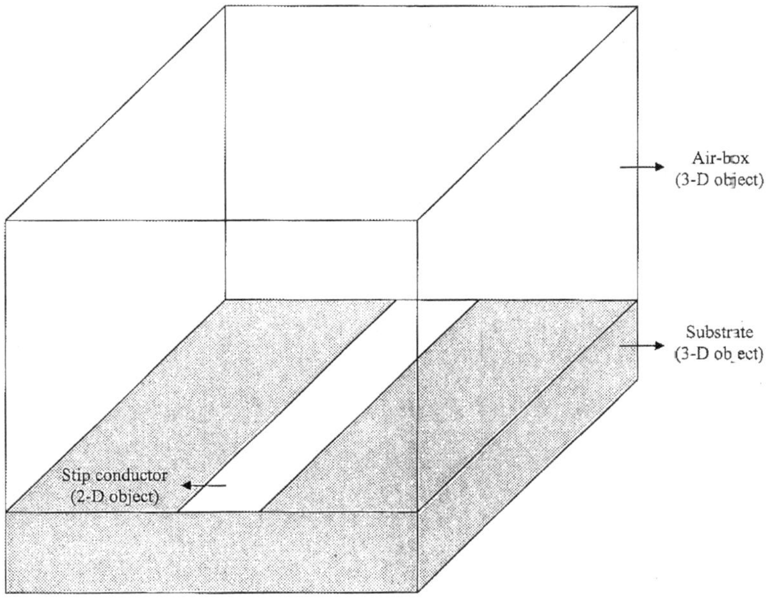

Figure 4.19 Different types of object for EM analysis

The most common commands for 2-D objects are rectangles, polylines, arcs, circles and ellipses. All these objects can be edited (i.e. copied, moved, scaled, rotated, mirrored, etc.), as with traditional drawing packages. In cases where 3-D objects are needed, Boolean operations are available (e.g. unite, subtract and intersect).

After drawing the required structure, the next step is to assign materials to the different sub-components. Most EM simulators allow the user to assign a material to an object, such as perfect metals, lossy metals, semiconductors, isotropic and anisotropic dielectrics, etc. The user can either use a material from a library, which is available within the EM simulator, or create and define custom materials by entering selected values (e.g. conductivity, dielectric constant, dielectric loss tangent, relative permeability, etc.) where applicable.

At this stage, the structure has all the required physical properties, and so the solution boundaries can now be applied. Using these boundaries, the user can specify the location of the ports, conductive surfaces, resistive surfaces, radiation boundaries, symmetry planes, etc. Any plane, object surface or interface between two objects can be usually defined as a boundary, and thus, assigned a boundary condition. The most common boundaries are:

1. **Ports**. This is a type of boundary condition that permits energy to flow into and out of a region of a structure.
2. **Perfect H-boundary**. This represents a perfect magnetic boundary (or wall), where the *H*-field is forced to be normal to the assigned boundary.
3. **Perfect E-boundary**. This represents a perfect electric boundary (or wall), where the *E*-field is forced to be normal to the assigned boundary.
4. **Symmetry planes**. A plane of symmetry exists wherever a structure has geometric and EM symmetry. That is, the fields in one half of the structure are either identical to, or the mirror image of, the fields in the other half of the structure. There are two types of symmetry planes: the symmetric *E*-plane and the symmetric *H*-plane. The symmetric *E*-plane can be used in cases where the symmetry is such as the *E*-field is normal and the *H*-field is tangential to the symmetry plane. The symmetric *H*-plane is the opposite, i.e. it can be used in cases where the *H*-field is normal and the *E*-field is tangential to the plane of symmetry. An example of a microstrip circuit with symmetry can be seen in Figure 4.20.

Symmetry planes dramatically reduce the time required by the EM simulator to solve the given problem. However, it should be noted that when solving for multiple modes, symmetry planes must not be used, since the orientation of *E*- and *H*-fields differ from mode to mode.

In cases where a symmetry plane exists, the computed impedances will not be for the full structure. In this case, impedance multipliers need to be specified to adjust the calculated impedance of the structure. If the structure has a symmetric *E*-plane, then they will have half the voltage differential and half the power flow of the full structure, and thus the resulting impedance needs to be scaled up by a factor of 2. On the other hand, if a symmetric *H*-plane has been defined, then this structure will have the same voltage differential but half the power flow of the full structure, and thus the resulting impedance needs to be scaled down by a factor of 0.5. It will be apparent that structures that have both planes of symmetry need no impedance multiplication.

4.7.6.3 Solve problem

The Solve Problem stage enables the user to set-up the type of solution, to specify the frequencies over which the simulation is to be performed, to specify/refine the mesh seeding, and to set-up the simulation error-bound. Under the same menu, the user should be able to start the simulation, view the status of the simulation job or cancel the job if necessary.

4.7.6.4 Post processing

The post-processor provides a variety of simulation data to be viewed and analysed:

1. *S*-parameter magnitude/phase plots and Smith chart
2. *E*- and *H*-fields (usually in shaded, contour and arrow field plots)
3. current density plots
4. propagation constant and impedance.

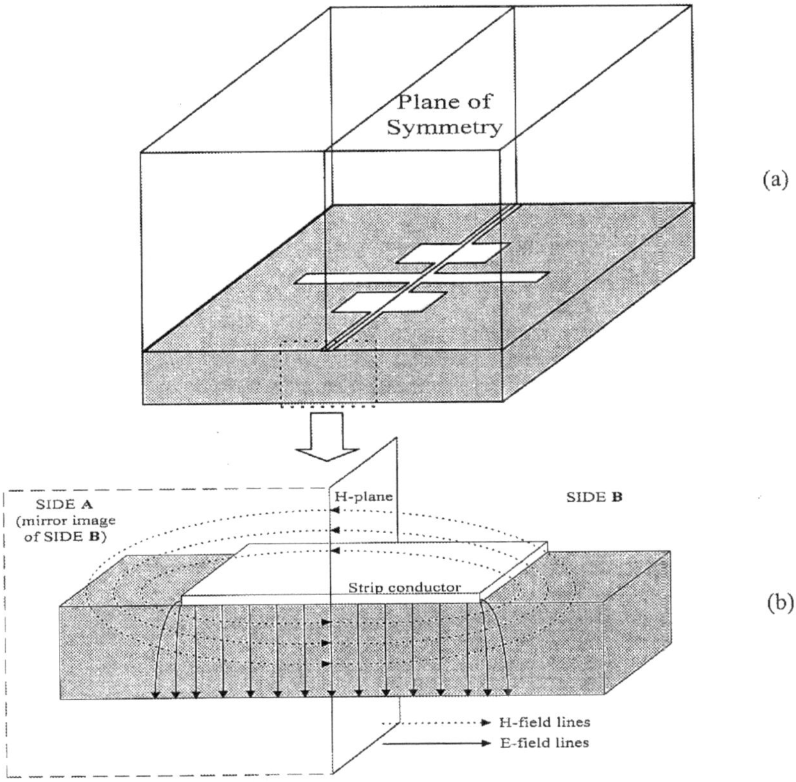

Figure 4.20 Microstrip circuit (a) with symmetry plane (symmetric-H) and (b) close-in view of the fields distribution

The field plot functions usually display a 2-D cross-section of *E*- and *H*-fields within a 3-D outline of the structure. Most EM simulators provide animation functions to illustrate the fields propagating through the structure. Different graphical formats are available, such as arrow plots (where arrows represent the direction, and sometimes the magnitude, of the observed field quantity), contour plots (which displays the equal-value contours of a scalar quantity) and, finally, shaded plots (where the magnitude of a scalar quantity being plotted is represented by a range of different colours). Sample plots can be seen in Figure 4.21.

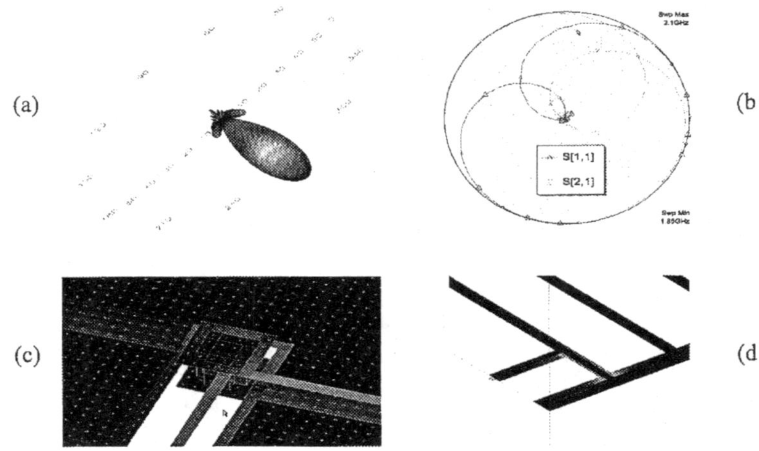

Figure 4.21 Post-processing graphs and plots: (a) antenna radiation pattern, (b) Smith chart, (c) field distribution on a via and (d) field distribution at the input of a filter

4.7.7 Optimising the EM simulator performance

There are simple rules that can be applied to optimise both memory consumption and computational time. The recommended strategies can be divided into four main areas:

1. hierarchical design
2. geometry preconditioning
3. problem size reduction
4. mesh refinement techniques (if available).

A hierarchical design process can save a great deal of computational time. As an example, consider the case where one needs to design a microstrip phased antenna array. It would be wise to start looking at individual elements first (i.e. single patches with the corresponding feeding networks) and then connect them up as blocks. Therefore, simulate and optimise their performances, instead of simulating the whole structure in one go. Due to long run times, an iterative design process can be difficult to justify [36].

Most 3-D simulators have the ability to use geometries imported from mechanical drawing software (such as *AutoCAD*™, a well known package). In such cases, it is essential to remove any unnecessary details. For example, a circular patch antenna might be represented with as many as 100 individual facets to provide reasonable visual rendering. If this were copied directly into an EM simulator, the initial mesh would far exceed current computational capability. In this case, 10 to 20 facets on a circle are more than adequate for most applications. The same applies to the number of segments on surfaces created using curves. Moreover, drawings for EM simulation need to consider the air and other media in and around the structure, which would not normally be required in a mechanical drawing. Thus, the relationship between mechanical and EM design differs [36].

Finite metallisation thickness should be used when edge coupling is important. If it is not important, simple 2-D objects will do the job. Narrow gaps between parallel faces of objects need to be avoided. Otherwise, this will result in an exponential growth in the number of meshing elements. In addition, large aspect ratios must be avoided. Some 3-D simulators have the option to use dummy elements (objects), to reduce the meshing element aspect ratio [36].

Use symmetry planes wherever possible. This will considerably reduce the circuit size, and consequently, the computation time. Therefore, it is a good idea to avoid making geometries larger than necessary. A great deal of attention should be paid when designing the air-boxes around a structure. Keep this boundary far away from the structure, so that it does not significantly influence the fields in the vicinity of the structure. However, avoid having the boundary too far away, since this will result in a large volume of space that will have very low field values [36].

Use mesh refinement techniques where possible. Some simulators give the ability to create constraints that help in the meshing process. Meshing should be much smaller than the wavelength. It is suggested that in some scenarios (e.g. at capacitively-loaded features) aggressive meshing may be needed [36]

4.7.8 Commercial EM software packages

Nowadays, there are many vendors of EM simulation software. They usually come in two forms: full 3-D EM simulators and 3-D planar simulators. The remainder of this section will give brief descriptions of some of the more popular EM simulators. In addition, Table 4.10 summarises the features of each.

Table 4.10 EM simulator summary

Package name	EM simulation method	Stand-alone	Platform
Ansoft's HFSS™	Finite element method	yes	PC/Unix
FullWave™	Finite element method	yes	PC/Unix
Momentum™	Method of moments	no	PC/Unix
Ensemble™	Method of moments	yes	PC/Unix
EM Suite™	Method of moments	yes	PC/Unix
IE3D™	Method of moments	yes	PC/Unix
Fidelity™	Finite difference time-domain	yes	PC/Unix
Empire™	Finite difference time-domain	yes	PC/Unix
Concerto™	Finite difference time-domain	yes	PC/Unix
XFDTD™	Finite difference time-domain	yes	PC/Unix
EMA3D™	Finite difference time-domain	yes	Unix
Microwave Studio™	Perfect boundary approximation	yes	PC
EMSight™	Spectral-domain moment-method	yes	PC
EMpower™	Method of lines	no	PC
MEFiSTo-2D™	Transmission line method	yes	PC
Mafia™	Finite integration method	yes	PC/Unix

4.7.8.1 Agilent Momentum ™

Momentum ™ is a planar EM simulator that enables RF and microwave designers to expand the range and accuracy of their passive circuits. *Momentum ™* takes into account real-world design geometries to simulate coupling and parasitic effects, and can analyse multi-layer arbitrary shapes. A key advantage of *Momentum ™* over other planar EM simulators is its complete integration into *ADS ™*. This allows existing users of Agilent EEsof EDA circuit simulators to access EM analysis quickly, and allows new users to learn a common interface for both EM and circuit analysis. Some of its major features are:

1. side wall coupling
2. box resonance calculation
3. DC and low frequency calculation
4. adaptive frequency sampling
5. edge meshing.

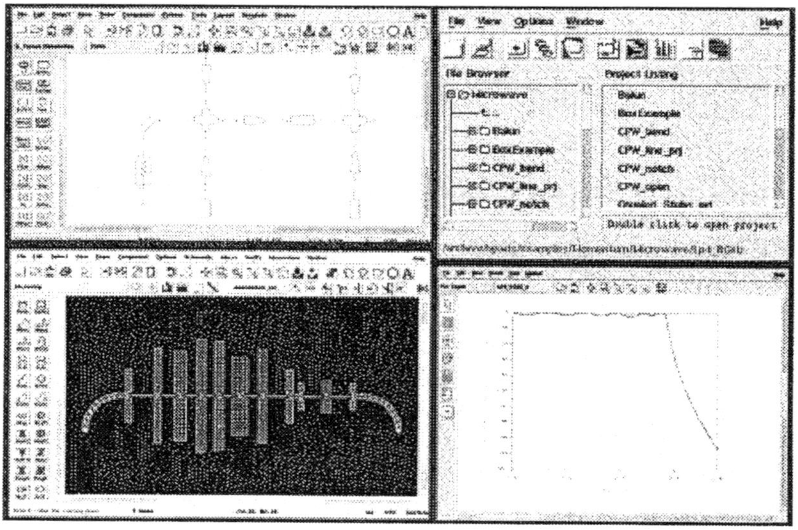

Figure 4.22 Screenshot of Agilent's Momentum ™

4.7.8.2 Ansoft HFSS ™ and Ensemble ™

Ansoft's *HFSS ™* is a complete solution for EM design, simulation and optimisation of arbitrarily-shaped, passive 3-D structures. It has a powerful FEM engine that solves and optimises for a wide array of applications with high accuracy and speed. It has a friendly and intuitive user interface to simplify design entry, fast solution times, efficient memory usage, quick design entry and powerful display capabilities. Some of its major features are:

1. calculates *S*-parameters for single or multi-port structures
2. unrestricted geometries with unlimited number of dielectrics and conductors
3. analysis includes conductor and dielectric loss
4. analysis based solely on Maxwell's equations with dispersion included
5. complete solution of *E*- and *H*- fields, energy densities
6. powerful drawing and visualisation capabilities, including field animation
7. fully automated EM design optimisation through geometric parameterisation, and a sophisticated set of optimisation tools.
8. wideband fast frequency sweep
9. powerful features for antenna design
10. parametric analysis
11. powerful field post-processor.

HFSS ™ brings the power of FEM with automatic adaptive mesh generation and refinement, tangential vector finite elements and Adaptive Lanczos Pade Sweep (ALPS). *HFSS* ™ automatically computes multiple adaptive solutions until a user-defined convergence criterion is met. Field solutions, calculated from first principles, accurately predict all high frequency behaviour such as dispersion, mode conversion and losses due to materials and radiation.

Ansoft also markets *Ensemble* ™, a 2.5-D planar EM field simulator that can be stand-alone software or integrated within *Serenade Desktop* ™.

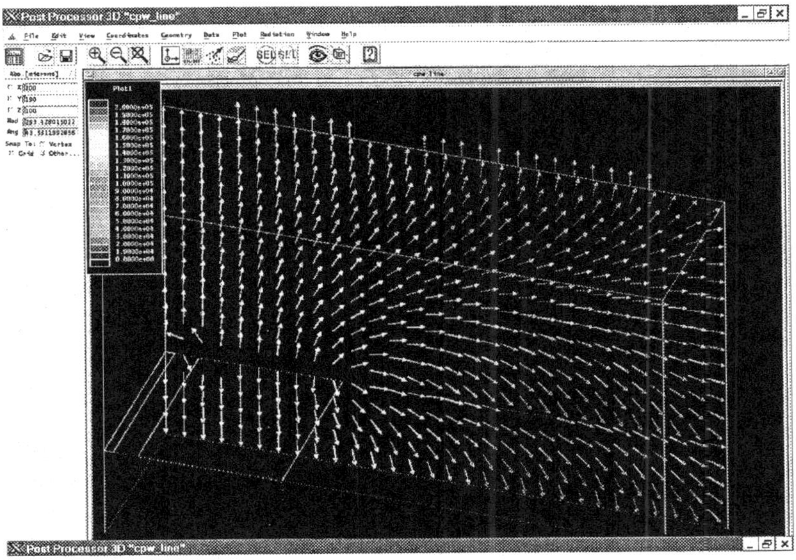

Figure 4.23 Screenshot of Ansoft's HFSS ™

4.7.8.3 Sonnet Software EM Suite™

Sonnet Software's suite of high frequency EM software is for 3-D planar circuits such as stripline, microstrip, CPW and printed circuit boards (single and multiple layers). It incorporates vias, vertical metal sheets and any number of layers of metal embedded in stratified dielectric material. EM analysis is performed inside a six-sided metal box and includes the effects of package resonances. Changing the boundary conditions enables analysis under open environment conditions. Some of its major features are:

1. multi-frequency caching reduces matrix fill time for frequency sweeps
2. chirp-Z algorithms speed non-binary box size analysis
3. parameter-driven element layout generation in *xgeom™*, with a palette of standard geometries
4. direct SPICE model extraction from arbitrary *S*-parameter data.

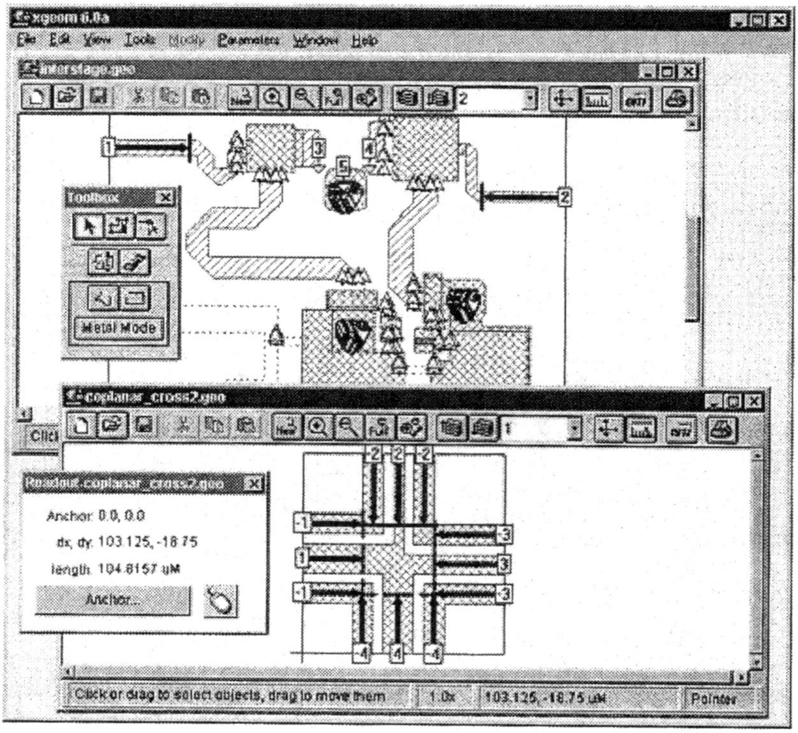

Figure 4.24 Screenshot of Sonnet's EM Suite™

4.7.8.4 *Sonnet Software and CST Microwave Studio™*

Sonnet Software markets a powerful full 3-D high frequency EM analysis package, called *Microwave Studio™*, a joint product with Computer Simulation Technology (CST). This is in addition to Sonnet's more widely known *EM Suite™* for 3-D planar structure analysis. *Microwave Studio™* has an intuitive, yet powerful, interface and well designed graphics that quickly gives the user an insight into the EM behaviour for high frequency designs. Some of its major features are:

1. fully automated meshing for analysis, so no manual meshing required
2. user-defined structures are easily incorporated through macros with a powerful integrated macro language (VBA), allowing the user to design the structure once, then instantly modify the design by changing a simple variable
3. transient solver for time-domain analysis based on finite integral technique
4. eigenmode solver for efficient simulation of high-Q structures
5. model stimulation methods
6. time- and frequency-domain sense 'probe' points to detect and display time- and frequency-domain fields at any number of locations in a structure
7. automated parametric optimisation of structures, including user-defined optimisation functions
8. parameter sweep analysis to investigate structure response changes with respect to controlled geometry changes
9. SPICE model extraction for 3-D structures.

4.7.8.5 *Zeland Software IE3D™ and Fidelity™*

Zeland Software is marketing *IE3D™*, a full-wave 3-D and planar EM simulation and optimisation package, and also *Fidelity™*, a full-wave 3-D EM simulator. *IE3D™* is a full-wave MOM simulator, solving the current distribution on 3-D and multi-layered structures of arbitrary shape. It has been widely used in the design of microwave circuits, microwave and millimetre-wave MMICs, RF printed circuit-boards, high temperature superconductor circuits, filters, microstrip patch antennas, wire antennas, various kinds of wireless antennas, IC transmission lines and IC packaging. *IE3D™* has the following major features:

1. simplified and re-usable simulation and optimisation set-up
2. simplified and template based geometry importing
3. sophisticated symbolic EM optimisation for S-, Y-, Z-parameters
4. implementation of random optimiser and improving of existing optimisation schemes.

Fidelity™ is a time-domain full 3-D simulator for high speed and high frequency circuits. In addition, it is used for wireless and RF antennas, EMC and EMI applications. The following lists the features of this software:

1. PC-based non-uniform FDTD simulation engine
2. accepts planar and 3-D metallic and dielectric structures
3. accepts an unlimited number of materials (isotropic and non-isotropic)
4. provides multiple 2-D and 3-D views of the simulated structure.

4.7.8.6 Applied Wave Research EMSight™

EMSight™ is a *Windows™* based EM simulator for analysing the behaviour of high frequency, RF and microwave circuits and antennas. *EMSight™* can analyse arbitrary planar circuits, including interconnecting vias, on an unlimited number of dielectric layers, with an unlimited number of ports. *EMSight™* uses the full-wave spectral-domain Galerkin MOM. The circuit is analysed inside a multi-layered rectangular enclosure.

A gridded variable cell size mesh is automatically generated. This places smaller cells in areas that have high variations in current density and larger cells in areas with more uniform current variations. The user can control the mesh by increasing the meshing density of specific polygons. The generated mesh can be viewed while the geometry is being edited, so that the effect of changing the meshing density can be seen instantly.

4.7.8.7 Eagleware EMpower™

EMpower™ is an EM simulator for the analysis of planar-type circuits, such as stripline, microstrip, CPW, slotline, finline and suspended/inverted substrates. *EMpower™* can analyse circuits with lumped elements (including transistors), extending the proven method of lines technique that was first commercialised in *EMpower™*. Since *EMpower™* is integrated into *Genesys™*, software interface issues are eliminated.

4.7.8.8 Vector Fields Concerto™

Concerto™ is a fully 3-D high frequency modelling package using the conformal FDTD method. An optimiser and a module for modelling high-Q systems are included. FDTD gives a much faster analysis than FEM, as it is not limited to single frequency driving functions.

Concerto™ has an enormous advantage over traditional finite difference methods in that the usual block or 'staircase' problems, normally associated with modelling curved surfaces, have been overcome. A conformal meshing technique, researched and developed over many years, is implemented to achieve this. As *Concerto™* is a time-domain package, transient excitation functions can be used and the frequency response or S-parameters of a system are easily displayed. Moreover, the software allows dynamic viewing of results, so that evolution of field distributions or S-parameter plots can be monitored.

4.7.8.9 Faustus Scientific MEFiSTo-2D™

MEFiSTo-2D™ is a multi-purpose EM simulation tool for *Windows™*. It allows the user to solve and visualise 3-D fields and waves that are functions of two planar dimensions and time. An extensive user interface and powerful processing capabilities give it the functionality of:

1. a time-domain reflectometer
2. a microwave/millimetre-wave vector network analyser
3. a spectrum analyser
4. a signal processor.

MEFiSTo-2D™ solves Maxwell's equations in the time-domain; the user can simulate both transient and time harmonic fields. *MEFiSTo-2D™* features numerical convolution in time, enabling the design of arbitrary, frequency-dispersive boundary behaviour and partitioning of large structures in the time domain. *MEFiSTo-2D*™ allows the designer to move boundaries and observe phenomena such as the Doppler effect in the time domain.

4.7.8.10 Infolytica FullWave™

FullWave™ generates full-wave solutions of arbitrary-shaped 3-D high frequency structures, including field animations. Resonant modes, *S*-parameters and port solutions are all available in *FullWave*™. *FullWave*™ incorporates parameterisation to simulate and optimise design variations quickly.

FullWave™ models 3-D high frequency structures with access ports through which energy enters and leaves. A different arbitrary-shaped waveguide or transmission line is easily connected at each port of the model. *FullWave*™ can compute the scattering matrix or resonant modes of the structure or only port modal solutions. The results can be exported to circuit simulation packages that accept *Touchstone*™, *Citifile*™ and *Supercompact*™ file formats. In computing a solution, *FullWave*™ first does a 2-D modal analysis at the ports; the user may choose to do this independently of the 3-D solution. There is also a choice of definitions for characteristic impedance.

4.7.8.11 CST Mafia™

Mafia™ is a multi-purpose CAD system designed to solve all kinds of EM problems. This package has been in use worldwide in industry and research laboratories for over 20 years. It uses the finite integration technique, a theoretical basis for solving Maxwell's equations in integral form. The analytical equations are discretised onto two grids, which are orthogonal to each other. This yields a set of matrix equations, each of which is the discrete analogue of one of the original Maxwell equations.

4.7.8.12 Remcom XFDTD™

Remcom offers a suite of high performance, low cost FDTD method EM solvers called *XFDTD*™. Although the *XFDTD*™ packages are complete, there are a series of options available for specific applications. The multiple processor modules allow efficient use of PCs and workstations with more than one processor.

4.7.8.13 IMST Empire™

Empire™ is a versatile 3-D EM field simulator developed at the IMST. It is based on the powerful 3-D FDTD, which has become a new standard for high frequency simulations. Its applicability ranges from analysing planar structures, interconnects and even multiport packages to waveguide, antennas and EMC problems. Thus, it covers the whole area of 3-D field simulations for RF designers. *S*-parameters, radiation patterns, field plots, etc., are generated for a user-defined frequency range within only one simulation process. A smart 3-D editor, which is part of the basic *Empire*™ package, can be used for structure definition or *AutoCAD*™ can be used as the graphical user interface. Monitoring and animation can give physical insight into the EM wave phenomena, while accurate results are obtained with little effort.

4.7.8.14 Jansen Microwave SFPMIC™

SFPMIC™ is a 2.5-D full-wave EM simulator supporting irregular gridding. As part of *LINMIC+/N™* comes an advanced CAD package for the linear and non-linear design of hybrid and monolithic microwave integrated circuits. *SFPMIC™* is capable of dealing with arbitrary conductor shapes and via-holes between different metallisation layers. In addition, it is the world's first commercial tool to feature the enhanced spectral-domain diakoptics technique for fast block-wise circuit computation, including inter-block coupling. Some of the *SFPMIC™* features are:

1. irregular grid discretisation, using rectangles and triangles, for design flexibility and efficiency
2. accurately simulates user-defined structures with via holes between metallisation layers
3. fast per frequency analysis using the spectral operator expansion technique, which is particularly powerful for broadband simulation
4. accommodates complex surface resistivities.

4.8 References

1. LITOVSKI, V., and ZWOLINSKI, M.: 'VLSI: Circuit simulation and optimization' (Chapman & Hall, London, 1997)
2. MAAS, S.: 'Nonlinear microwave circuits' (Artech House, USA, 1988)
3. RODRIGUES, P. J. C.: 'Computer-aided analysis of nonlinear microwave circuits' (Artech House, USA, 1998)
4. HAUS, H. A. *et al.*: 'Representation of noise in linear two-ports', *Proceedings of the IRE*, 1960, **48**, pp. 69–74
5. HILLBRAND, H., and RUSSER, P. H.: 'An efficient method for computer aided noise analysis of linear amplifier networks', *IEEE Transactions on Circuits and Systems*, 1976, **CAS-23** (4), pp. 235–238
6. HEINEN, S., KUNISCH, J., and WOLFF, I.: 'A unified framework for computer-aided noise analysis of linear and nonlinear microwave circuits', *IEEE Transactions on Microwave Theory and Techniques*, 1991, **MTT-39** (12), pp. 2170–2175
7. BALABAN, P. A., and GOLEMBSKI, J. J.: 'Statistical analysis for practical circuit design', *IEEE Transactions on Circuits and Systems*, 1975, **CAS-22** (2), pp. 100–108
8. SHRIED, Y. A.: 'Method of statistical testing: Monte Carlo method' (Elsevier, London, 1964)
9. DALY, F., HAND, D. J., JONES, M. C., LUNN, A. D., and McCONWAY, K. J.: 'Elements of statistics' (Addison-Wesley, Harlow, England, 1999)
10. McCLEARNON, D.: 'Company's rebirth includes major EDA software release', *Microwaves & RF*, 1999, **38** (11)
11. WILLIAMS, L., ALBRECHT, G., SAFFIAN, M., and NEWKIRK, D.: 'Integrated software suite enables PC-based high-frequency circuit design', *Applied Microwave & Wireless*, 1997, **9** (12), pp. 82–88

12. VYE, D., WILLIAMS, L., and O'ROURKE: 'Ansoft serenade design environment: end-to-end software solution for high-frequency design', *Microwave Product Digest*, 1999
13. APPLIED WAVE RESEARCH INC.: 'A new suite of RF/microwave design tools', *Microwave Journal*, 1998, **6** (6), pp. 122–124
14. APPLIED WAVE RESEARCH INC.: 'An advanced integrated layout editor', *Microwave Journal*, 1999, **42** (7), p. 166
15. OPTOTEK LTD: 'Software suite improves design-cycle efficiency', *Microwaves & RF*, 1996, **35** (9)
16. OPTOTEK LTD: 'Nonlinear waveform prediction in fast microwave circuits', *Microwave Journal*, 1999, **42** (11), pp. 168–175
17. OPTOTEK LTD: 'A major schematic capture/layout package upgrade', *Microwave Journal*, 1999, **42** (5), pp. 374–375
18. EAGLEWARE CORP.: 'Integrated software for RF/microwave design', *Microwave Journal*, 1996, **39** (1), pp. 188–189
19. EAGLEWARE CORP.: 'Low-cost CAE suite adds power and speed', *Microwaves & RF*, 1999, **38** (8), pp. 141–145
20. HP EESOF DIVISION, HEWLETT PACKARD CO: 'An IC characterization and analysis program update', *Microwave Journal*, 1997, **4**, pp. 146–154
21. OPTOTEK LTD: 'Advanced large-signal models for transistor parameter extraction software', *Microwave Journal*, 1999, **42**, pp. 112–113
22. FOLLMANN, R., TEMPEL, R., SPORKMANN, T., and WOLFF, I.: 'A new spline based FET model for MESFETs and HEMTs', *Proceedings of the 27th European Microwave Conference, Jerusalem*, 1997, pp 1360–1366
23. 'Engineering applications of electromagnetic field solvers', *IEEE MTT-S Workshop Notes*, San Diego, 1994
24. MONGIA, R., BAHL, I., and BHARTIA, P.: 'RF and microwave coupled-line circuits' (Artech House, Norwood, 1999)
25. GUPTA, K. C., GARG, R., BAHL, I., and BHARTIA, P.: 'Microstrip lines and slotlines' (Artech House, Norwood, 1996)
26. HELSZAJN, J.: 'Microwave planar passive circuits and filters (John Wiley & Sons, Chichester, 1994)
27. HELSZAJN, J.: 'Synthesis of lumped element, distributed and planar filters' (McGraw-Hill, Berkshire, 1990)
28. GOEL, A. K.: 'High-speed VLSI interconnect' (John Wiley & Sons, 1994)
29. WHEELER, H. A.: 'Formulas for the skin effect', *Proceedings of the IRE*, 1942, pp. 412–424
30. HORTON, R., EASTER, B., and GOPINATH, A.: 'Variation of microstrip losses with thickness of strip', *Electronics Letters*, 1971, **7**, pp. 490–491
31. GARG, R., GUPTA, K. C., and SHARAN, R.: 'Optimum thickness of metal in waveguiding structures, ground planes and reflectors', *International Journal of Electronics*, 1975, **39**, pp. 525–527
32. 'Microwave engineer's handbook' (Artech House)
33. McKENZIE, W., BARRY, D. M., SNOWDEN, C. M., and HOWES, M. J.: 'Computer-aided design of MMIC structures', *IEE Proceedings*, 1986, pp. 405–410
34. ANSOFT HFSS: 'Training course manual', 1999

35. LEVY, R.: 'Derivation of equivalent circuits of microwave structures using numerical techniques', *IEEE Transactions on Microwave Theory and Techniques*, 1999, **MTT-47**, pp. 1688–1695
36. BANDLER, J. W., BIERNACHI, R. M., CHEN, S. H., GROBENLY, P. A., and HEMMERS, R. H.: 'Space mapping technique for electromagnetic optimisation', *IEEE Transactions on Microwave Theory and Techniques*, 1994, pp. 2536–2544
37. BANDLER, J. W., BIERNACHI, R. M., CHEN, HEMMARS, R. H., and MADSEN, K.: 'Electromagnetic optimisation exploiting aggressive space mapping', *IEEE Transactions on Microwave Theory and Techniques*, 1995, pp. 2874–2882
38. 'State-of-the-art in microwave CAD: Part I', *International Journal of Microwave and Millimeter-wave CAD*, special issue, 1991
39. SADIKU, M. N. O.: 'Elements of electromagnetics' (Saunders College Publishing, 1994)
40. BOOTON, R. C.: 'Computational methods for electromagnetics and microwaves' (John Wiley & Sons, 1992)
41. PAUL, C. R., NASAR, S. A.: 'Introduction to electromagnetic fields' (McGraw-Hill, 1987)
42. 'Advanced CAD for microwave filters and passive devises', *Proceedings of ESA workshop*, 1995, Noordwijk.
43. 'The current state of CAD- a user's perspective', *Microwave Engineering Europe*, 2000, pp. 15–20
44. YEE, K. S.: 'Numerical solution of initial boundary value problems involving Maxwell's equations in isotropic media', *IEEE Transactions on Antenna and Propagation*, 1966, **AP-14**, pp. 302–307
45. ZHANG, X., FANG, J., MEI, K. K., and LIU, Y.: 'Calculation of the dispersion characteristics of microstrips by the time-domain finite-difference method', *IEEE Transactions of Microwave Theory and Techniques*, 1988, **MTT-36**, pp. 263–267
46. LIANG, G., LIU, Y., and MEI, K. K.: 'Full-wave analysis of coplanar waveguide and slotline using the time-domain finite-difference method', *IEEE Transactions on Microwave Theory and Techniques*, 1989, **MTT-37**, pp. 1949–1957
47. SILVESTER, P. P., and FERRARI, R. L.: 'Finite elements for electrical engineers' (Cambridge University Press, New-York, 1983)
48. SVEDIN, J. A. M.: 'A numerically efficient finite-element formulation for the general waveguide problem without spurious modes', *IEEE Transactions on Microwave Theory and Techniques*, 1989, **MTT-37**, pp. 1708–1715
49. JACKSON, R. W.: 'Full-wave, finite-element analysis of irregular microstrip discontinuities', *IEEE Transactions on Microwave Theory and Techniques*, 1989, pp. 81–89
50. CHIU, C. W., and WU, R. B.: 'A moment method analysis for Coplanar waveguide discontinuity inductance', *IEEE Transactions on Microwave Theory and Techniques*, 1993, **MTT-41**, pp. 1511–1514
51. CHEN, Z., and GAO, B.: 'Deterministic approach to full-wave analysis of discontinuities in MIC's using the method of line', *IEEE Transactions on Microwave Theory and Techniques*, 1989, **MTT-37**, pp. 606–611

52. NAM, S., LING, H., and ITOH, T.: 'Characterisation of uniform microstrip line and its discontinuities using the time-domain method of lines' *IEEE Transactions on Microwave Theory and Techniques*, 1989, **MTT-37**, pp. 2051–2057
53. JOHNS, P. B., and BEURLE, R. L.: 'Numerical solution of 2-dimensional scattering problems using a transmission-line matrix', *Proceedings of the IEE*, 1971, **118**, pp. 1203–1208
54. HOEFFER, W. J. R.: 'The transmission-line matrix method – theory and applications', *IEEE Transactions on Microwave Theory and Techniques*, 1985, pp. 882–893
55. JOHNS, P. B.: 'A symmetrical condensed node for the TLM method', *IEEE Transactions on Microwave Theory and Techniques*, 1987, **MTT-35**, pp. 370–377
56. ITOH, T., and MITTRA, R.: 'Spectral-domain approach for calculating the dispersion characteristics of microstriplines', *IEEE Transactions on Microwave Theory and Techniques*, 1973, **MTT-21**, pp. 496–499
57. ITOH, T.: 'Spectral-domain immitance approach for dispersion characteristics of generalised printed transmission lines', *IEEE Transactions on Microwave Theory and Techniques*, 1980, **MTT-28**, pp. 733–736
58. JANSEN, R. H.: 'The spectral-domain approach for microwave integrated circuits', *IEEE Transactions on Microwave Theory and Techniques*, 1983, **33**, pp. 1043–1056
59. GREEN, H. E.: 'The numerical solution to some transmission line problems', *IEEE Transactions on Microwave Theory and Techniques*, 1987, **MTT-35**, pp. 882–893

Chapter 5

Amplifiers

I. D. Robertson and M. W. Geen

5.1 Introduction

This chapter describes the most important techniques used for MMIC amplifier design. The classical theory and analysis of *S*-parameters, two-port stability and unilateral transducer gain are briefly reviewed. The key impedance matching techniques and DC biasing circuits are described before introducing the five major MMIC amplifier topologies: the reactively matched amplifier, the lossy match amplifier, the feedback amplifier, the distributed amplifier and various forms of actively matched amplifier. Finally, the design of low noise and high power amplifiers is considered. For further information on the mathematical treatment of microwave amplifier design, the reader is referred to one of the many excellent texts on this subject [1–6]. The justification for not covering this topic in great detail is that MMIC amplifier design techniques are rather different from conventional microstrip MIC amplifier design. This is partly because the size limitations of MMICs have forced designers to develop new topologies, and partly because the low component parasitics on MMICs have enabled more complicated circuit topologies to be used. In addition, the extensive use of CAD has removed the need for many of the original graphical and analytical methods of matching network synthesis and amplifier design (e.g. [7,8]), although a good understanding of the principles of amplifier design is still vital.

Compared with hybrid-MIC amplifiers, MMIC amplifiers have the major advantages of low cost and excellent reproducibility when manufactured in large quantities. However, MMIC amplifier performance can suffer from their small size because this increases the matching network losses and restricts the choice of topology. Furthermore, because of the need for high yield, the active devices on MMICs often have inferior performance compared with their discrete counterparts. This is particularly true of high power devices, where the monolithic device performance suffers from a number of constraints. Generally, therefore, MMIC amplifiers do not achieve the state-of-the-art noise figure and output power performance. However, the removal of bond-pad and bond-wire parasitics in monolithic amplifiers means that very large bandwidths can be achieved, especially with the distributed amplifier. Also, millimetre-wave amplifiers can be designed very easily compared with hybrid MIC techniques, with which the required precision of mechanical assembly is almost impossible to achieve.

The performance of MMIC amplifiers has improved dramatically as high performance MESFET, HEMT and HBT devices have been developed. MMIC amplifiers using HEMTs have been demonstrated at over 100 GHz, and at the time of writing, HBT amplifiers have been demonstrated at 90 GHz. A general comparison of the different MMIC technologies and devices can be found in Chapters 1 and 2. For amplifier designs, the MESFET finds applications at frequencies up to approximately 30 GHz, and has quite good noise figure performance up to 18 GHz. The HEMT has the best noise figure performance and is currently the best device for low noise and millimetre-wave applications. The GaAs HBT has demonstrated excellent output power capability, with over 12 W of CW power having been achieved at 8.5 GHz from a monolithic amplifier [9]. The silicon bipolar has continued to improve, and amplifiers operating at over 10 GHz have been reported. Although the silicon bipolar device does have a relatively poor noise figure performance, it is extremely economical to produce in large volumes and can be integrated with CMOS circuitry in BiCMOS technology. SiGe HBT technology is developing very rapidly, and at the time of writing an f_{max} of >160 GHz has been reported, and a device noise figure of less than 1 dB has been achieved at 10 GHz [10].

It is certain that device performance will improve further, but the amplifier design techniques presented in this chapter will continue to apply. However, as the gain available from the devices increases, and as different circuit functions become more closely integrated (rather than being individual 50 Ω stages), the need for the classical matching techniques is reduced. As a result, designs at low microwave frequencies will use active matching and DC-coupled techniques more and more, because of their high packing density. There will be some exceptions to this trend, in the cases where low noise figure, high output power or low DC power consumption is the overriding concern. In millimetre-wave applications, where transistor gain is limited, traditional matching techniques such as stub matching and impedance transformers will continue to be used for many years.

5.2 Classical stability and gain analysis

The gain provided by a transistor is very dependent on the source and load impedances presented to its input and output terminals. It is the job of the matching networks to ensure that the desired performance (e.g. minimum noise figure, maximum gain or maximum output power) is achieved over the required frequency range, and the design of the matching networks is the major part of the amplifier design task. In addition, real devices have significant reverse transmission coefficients, which means that the problem of stability must also be considered in amplifier design. In order to analyse the stability and gain of a small-signal amplifier, the transistor can be represented by its *S*-parameters, and Figure 5.1 shows the flowgraph representation of a transistor with the impedance presented to its output being represented as a load reflection coefficient Γ_L.

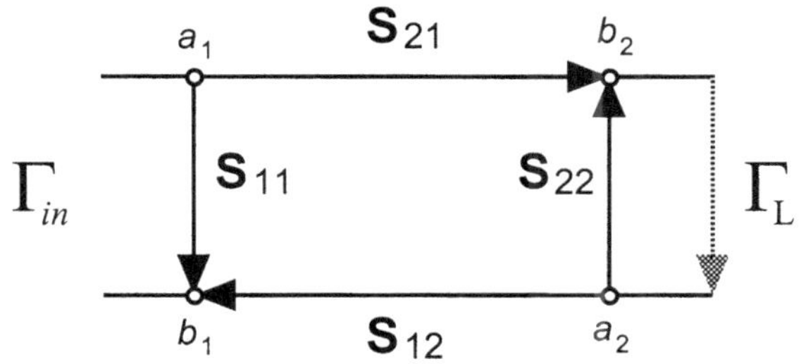

Figure 5.1 Flowgraph representation of device with output load

The first point to note is that the presence of the reverse transmission coefficient, S_{12}, means that the load reflection coefficient, Γ_L, will affect the input impedance of the transistor. Using Mason's non-touching loop rule it can easily be shown that

$$\Gamma_{in} = S_{11} + \frac{S_{21}S_{12}\Gamma_L}{1 - S_{22}\Gamma_L} \tag{5.1}$$

Similarly, for the output reflection coefficient in terms of the source reflection coefficient, Γ_s:

$$\Gamma_{out} = S_{22} + \frac{S_{21}S_{12}\Gamma_S}{1 - S_{11}\Gamma_S} \tag{5.2}$$

There are two important consequences of eqns. (5.1) and (5.2). First, in general, the input and output impedances to which one must match are only given by S_{11} and S_{22} when the transistor has S_{12}=0, which is not the case in practice. Second, there are certain values of Γ_S and Γ_L which can result in Γ_{in} and Γ_{out} becoming greater than unity, and the transistor can have negative input/output resistance. This means that the transistor can become unstable for certain source and load impedances. If we consider the input side, then by setting Γ_{in} =1, we can establish the boundary condition for stability:

$$\left| S_{11} + \frac{S_{21}S_{12}\Gamma_L}{1 - S_{22}\Gamma_L} \right| = 1 \tag{5.3}$$

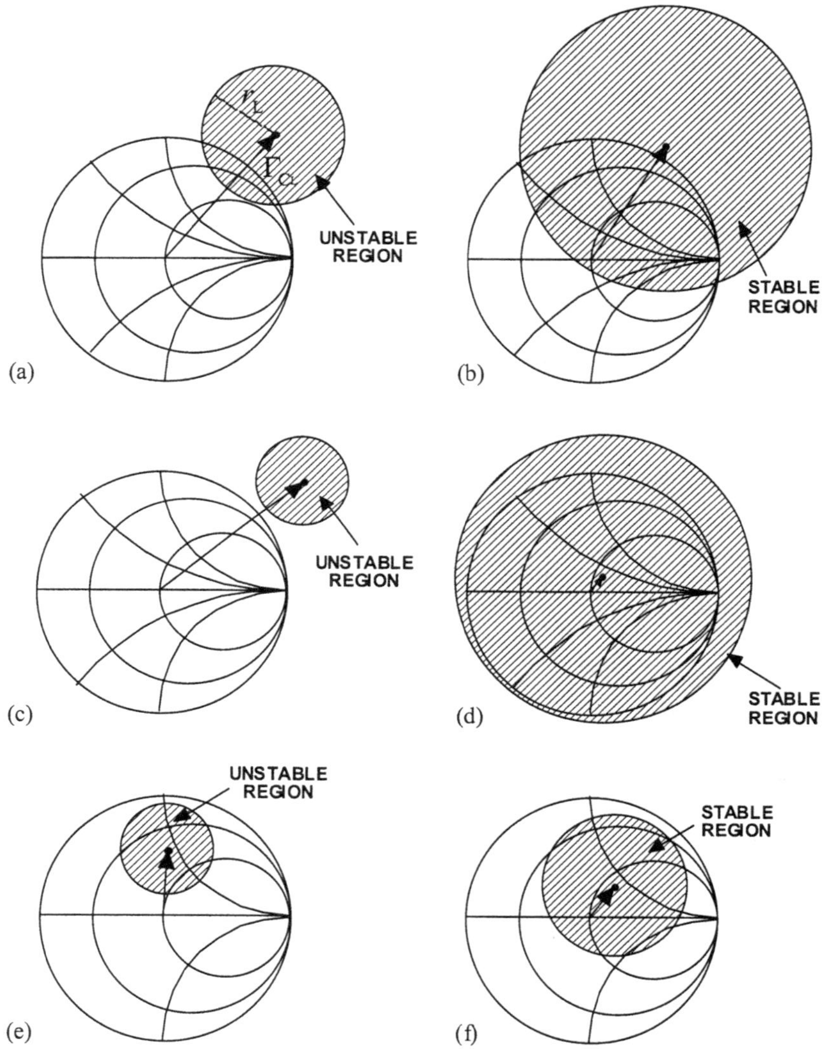

Figure 5.2 Stability circles on the Smith chart: (a) stability circle partially inside the Smith chart, (b) partially inside and encompassing the 50 Ω point, (c) completely outside, (d) completely encompassing the Smith chart, (e) completely inside but not encompassing the 50 Ω point and (f) completely inside and encompassing the 50 Ω point

It can be shown that this equation can be rearranged to give the equation of a circle, whose radius is given by:

$$r_L = \frac{|S_{12}S_{21}|}{\left||\Delta|^2 - |S_{22}|^2\right|} \tag{5.4}$$

and whose centre is given by:

$$\Gamma_{C_L} = \frac{(S_{22} - \Delta S_{11}^*)^*}{|S_{22}|^2 - |\Delta|^2} \tag{5.5}$$

where $\Delta = S_{11}S_{22} - S_{12}S_{21}$

This circle represents a region in the Γ_L plane (on the Smith chart), inside or outside of which all load impedances will make the transistor input reflection coefficient greater than unity, leading to instability. The circle is known as the load stability circle, and it can be partially inside the Smith chart, completely inside the Smith chart, or completely outside the Smith chart. Figure 5.2 shows the possible cases graphically. Whether the stable region is inside or outside the circle can be determined by inspecting where the origin (the 50 Ω point) lies, since this point is within the stable region (assuming that $|S_{11}|$ and $|S_{22}| < 1$). Similar equations can be found to define the source stability circle: source impedances lying within the unstable region will lead to an output reflection coefficient greater than unity. Since the *S*-parameters vary with frequency, the centre and radius of the stability circles are different at different frequencies.

It is evident from Figures 5.2*c* and *d* that there are cases when there are no source or load reflection coefficients within the Smith chart which lead to instability. This is known as unconditional stability. Unconditional stability means that no combination of passive source and load terminations can lead to instability. However, it should be stressed that in multi-stage amplifiers one transistor might present negative resistance to another, and so the stability problem is more severe.

It is very important for the designer to know straight away whether the transistor is unconditionally stable or not, Rollett's stability factor [11] gives an immediate indication of stability and is given by:

$$K = \frac{1 - |S_{11}|^2 - |S_{22}|^2 + |\Delta|^2}{2|S_{12}S_{21}|} \tag{5.6}$$

If $K > 1$ then the transistor is unconditionally stable, and if $K < 1$ the stability depends on the position of the source and load impedances relative to the stability circles.

5.2.1 Constant gain circles

The transducer power gain of the transistor, G_T, is defined as the ratio of the power delivered to the load to the power available from the source, and is given by:

$$G_T = \frac{|S_{21}|^2(1-|\Gamma_S|^2)(1-|\Gamma_L|^2)}{|1-S_{11}\Gamma_S - S_{22}\Gamma_L + \Delta\Gamma_S\Gamma_L|^2} \tag{5.7}$$

This equation determines how the source and load impedances affect the gain. However, in order to visualise this clearly it is necessary to simplify the equation by making the assumption that the transistor has $S_{12} = 0$; i.e. that the transistor is unilateral and has no reverse transmission. This leads to an expression for the unilateral transducer gain, G_{TU}, which is as follows:

$$G_{TU} = \frac{|S_{21}|^2(1-|\Gamma_S|^2)(1-|\Gamma_L|^2)}{|(1-S_{11}\Gamma_S)(1-S_{22}\Gamma_L)|^2} \tag{5.8}$$

This can be separated into three factors:

$$G_{TU} = G_S . G_0 . G_L \tag{5.9}$$

where

$$G_S = \frac{1-|\Gamma_S|^2}{|1-S_{11}\Gamma_S|^2}$$

$$G_0 = |S_{21}|^2$$

and

$$G_L = \frac{1-|\Gamma_L|^2}{|1-S_{22}\Gamma_L|^2}$$

With the unilateral assumption, and with this separation of the three gain factors, we now find that circles of constant unilateral transducer gain exist in both the Γ_S and Γ_L planes. Figure 5.3 shows the constant gain circles (in the Γ_L plane) for a typical 300 μm MESFET at 8 GHz. These gain circles can be used to develop a matching strategy for broadband amplifier design. They must, however, be used in conjunction with the stability circles, because the higher gain circles will lie either partially or wholly inside the unstable region when the *K*-factor is less than unity.

When the *K*-factor is less than unity, the point of maximum gain will be inside the unstable region. In this case we have to accept that the maximum transducer gain is not achievable due to instability. When *K* is less than unity, the maximum gain that can be safely achieved is called the maximum stable gain (MSG) and is given by:

$$\mathrm{MSG} = \left|\frac{S_{21}}{S_{12}}\right| \tag{5.10}$$

When K is greater than unity (and the magnitudes of S_{11} and S_{22} are both less than unity [5]), the device is unconditionally stable and the maximum gain that can be achieved is called the maximum available gain (MAG), given by:

$$\mathrm{MAG} = \left|\frac{S_{21}}{S_{12}}\right|\left(K - \sqrt{K^2 - 1}\right) \tag{5.11}$$

Typically, a transistor is conditionally stable ($K < 1$) at low frequencies, and the maximum stable gain rolls off at 3 dB per octave. At a certain frequency $K = 1$, and beyond that the device is unconditionally stable ($K > 1$) and the maximum available gain rolls off at 6 dB per octave.

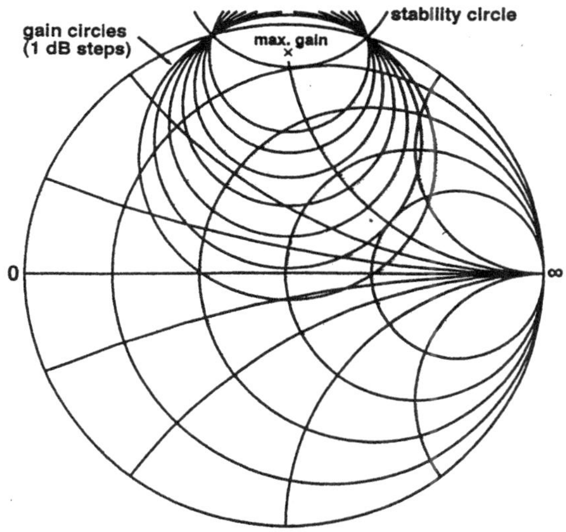

Figure 5.3 Gain circles (in the Γ_L plane) at 8 GHz for a typical 300 μm MESFET

The maximum gain response for typical 0.5 μm MESFETs of different gate-widths is shown in Figure 5.4. Note that the number of fingers is fixed, and the unit gate-width is varied: apart from some differences due to distributed effects, a 4 × 100μm FET has the same MSG/MAG frequency response as a 2 × 100μm or 6 × 100μm FET, because identical device subsections are being placed directly in parallel. However, at low frequencies a small device has very high impedance and so a large gate-width device is easier to match to 50 Ω. Conversely, at high frequencies a large device has very low impedance, and so a small gate-width device is easier to match to 50 Ω. The unit gate-width has a marked effect on the MSG/MAG transition frequency: a small unit gate-width leads to a lower gate resistance and a lower source inductance (in most cases). These changes in the

FET parasitic elements are the reason why the 4 × 150μm, 4 × 100μm and 4 × 50μm have different MSG/MAG responses in Figure 5.4. These important subtleties of maximum gain, stability, and device impedance should be closely studied early in the design process for the design frequency and devices of interest in order to find the optimum choice of device geometry.

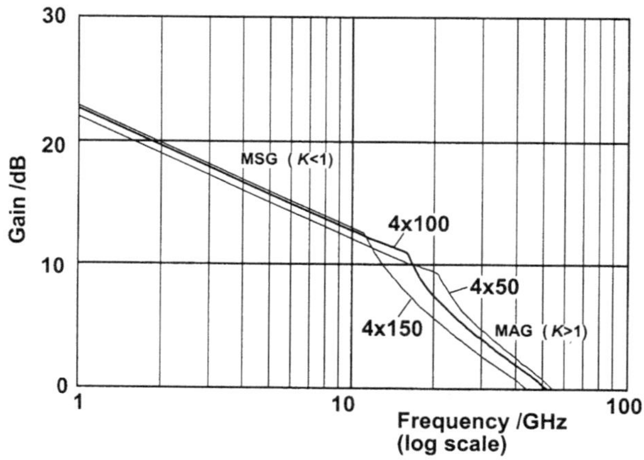

Figure 5.4 Maximum stable/available gain for typical 0.5 μm MESFETs of different geometries

5.2.2 The practical implications of the theory

The classical analysis has been summarised here in order to illustrate that it is wrong to think that amplifier design is simply a matter of matching to the transistor's S_{11} and S_{22}. In fact, although it might provide a useful starting point in the design, strictly speaking the designer should *never* match to S_{11} and S_{22}, and the cases of conditional and unconditional stability must be handled differently.

When the transistor is unconditionally stable, the designer is able to conjugately match the input and output, in order to achieve maximum gain and good input and output matching. However, because of the transistor's S_{12} the reflection coefficients that must be matched to are those given by eqns. (5.1) and (5.2), and not S_{11} or S_{22}. These reflection coefficients, which are only defined when $K > 1$, are known as the simultaneous conjugate match reflection coefficients. It is the transistor's input reflection coefficient, when its output is perfectly matched, which must be plotted on the Smith chart and then manoeuvred into the 50 Ω point with suitable matching elements. The values are most readily obtained from CAD programs. However, care must be taken to establish whether the CAD program is displaying the transistor's input reflection coefficient under conjugate matching conditions or the required matching network reflection coefficient that must be presented to the transistor's input to achieve conjugate matching. In *Libra*™ the parameters are called the simultaneous match reflection coefficients (*GM1* and *GM2*), and these values are those that the matching networks must present to the transistor. Hence, for the matching methods described later in this chapter, the conjugate of the *Libra*™ values must be plotted on the Smith chart. *Libra*™ and other CAD programs can also display stability and gain circles.

When the device is only conditionally stable it cannot be conjugately matched at the input and output because this would lead to oscillation. It is possible to design a working amplifier by deliberately mismatching the input and output and staying a safe distance from the instability regions. However, such an amplifier would not be a very useful component as the mismatch would lead to gain ripples, and the amplifier might still oscillate if presented with non-50 Ω source and load impedances. Hence, it is necessary to take steps to stabilise the device.

5.2.3 Amplifier design under conditional stability conditions

5.2.3.1 Resistive loading

A transistor can be stabilised by adding small series resistors or large shunt resistors to its input and/or output, as shown in Figures 5.5*a* and *b*. These lossy elements ensure that the transistor cannot be presented with impedances inside the instability regions, irrespective of what source and load impedances are connected. However, this technique cannot be used for low noise amplifiers because the resistors will degrade the noise figure. The resistor values can be tuned on the computer until the *K*-factor is just greater than unity. It is important to note that purely reactive elements cannot change the *K*-factor.

5.2.3.2 Parallel feedback

The introduction of negative feedback by adding a resistor network from the output to the input, as shown in Figure 5.5*c*, has a very beneficial effect on the transistor stability. In addition, the effect of feedback is to make the input and output impedances more convenient for matching. Feedback amplifiers using FETs are described in Section 5.7. Resistive feedback will, however, degrade the noise figure.

5.2.3.3 Series feedback

Series feedback entails inserting a resistor or inductor into the common-lead of the device. Most commonly an inductor is inserted into the source of an FET, as shown in Figure 5.5*d*, in order to make the device stable at a lower frequency. With an inductor the noise figure may actually improve, and the noise matching impedance may be brought closer to the power matching point. The special case of the series feedback FET LNA is discussed in Section 5.11.

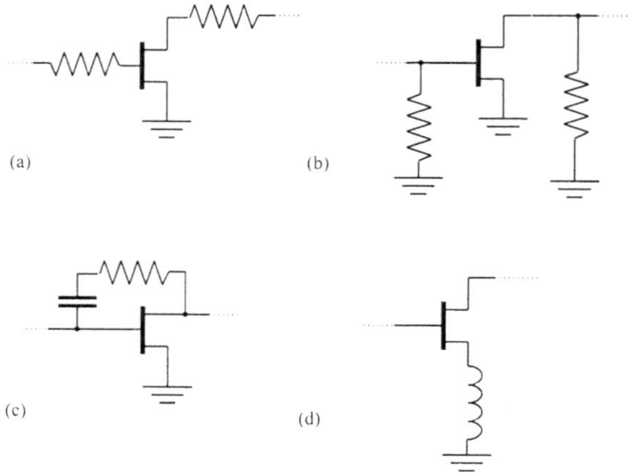

Figure 5.5 Stabilisation methods: (a) series resistors, (b) shunt resistors, (c) parallel feedback and (d) series feedback

5.2.3.4 The balanced amplifier

The balanced amplifier is used extensively in MIC low noise and power amplifiers. In LNAs the usual problem is that matching for low noise figure does not give a good 50 Ω input match. The balanced amplifier solves this problem by terminating the reflected signals in a matched load. In a similar way, if a conditionally stable device is deliberately mis-matched to avoid instability, then the balanced amplifier can be used as a means of terminating the unwanted reflected signals. Figure 5.6 shows the schematic diagram of a balanced amplifier. Two identical amplifiers are placed between a pair of quadrature couplers (e.g. Lange couplers). The result of this arrangement is that the amplifiers are fed 90° out-of-phase. The reflected signals at the input become 180° out-of-phase after passing back through the coupler, and thus cancel out. The reflected signals at the coupler termination are absorbed. The result is that the balanced amplifier has excellent input and output matches, and the designer is free to optimise the amplifier for stability, flat gain, noise figure, etc.

Compared with the individual single-ended amplifier, the balanced amplifier has the following features:

Noise figure	Single amplifier's noise figure + the coupler loss
Gain	Single amplifier's gain – 2 coupler loss
Output power	Single amplifier's + 3 dB – the coupler loss
DC power	Doubled
Reliability	Some redundancy: gain drops by 6 dB if one amplifier fails
Port matches	Excellent, easily cascaded without ripple
Stability	Excellent
Intermodulation	Third-order IMPs are 6 dB lower for a given input power

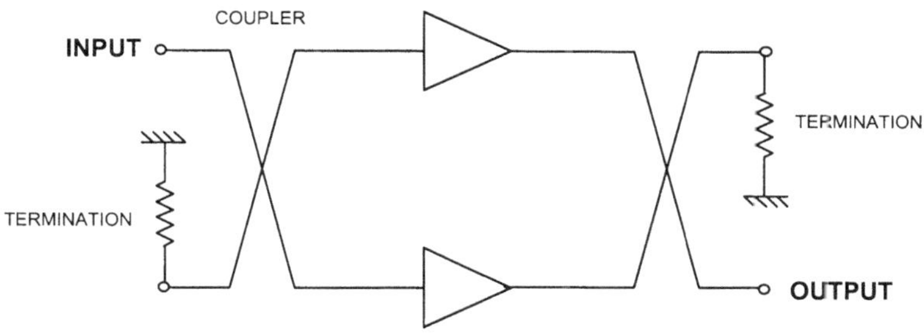

Figure 5.6 Block diagram of a balanced amplifier

5.3 Matching techniques

5.3.1 Lumped element matching

Lumped-element matching networks are attractive because of their small size and their smooth frequency characteristic. MMIC lumped elements are extensively described in Chapter 3. Here, we first show how an arbitrary impedance (such as a transistor input impedance, for example) can be matched to 50 Ω by using two lumped elements. Then we consider how two complex impedances can be matched, and finally how the *Q*-factor can be controlled by the use of three-element matching networks.

5.3.1.1 L-networks

In principle, any impedance can be matched to 50 Ω by using just two reactive lumped elements. On the Smith chart, a series inductance will move the load clockwise along the constant resistance contour. A series capacitance will move the load anti-clockwise along the same constant resistance contour. A shunt inductance will move the load anti-clockwise along the constant conductance circle, and a shunt capacitance will move the load clockwise along the constant conductance circle. As the constant resistance and constant conductance contours are essentially orthogonal, a suitable choice of series and parallel components can move any load impedance into the centre of the chart.

In total there are eight series/shunt inductor/capacitor combinations, and at least one of these will be able to match any specific impedance to 50 Ω. Figure 5.7 shows the eight combinations that can be used and summarises the required Smith chart operations for series and shunt inductors and capacitors. These networks are often called L-networks, and are the most basic type of lumped element matching network. Often, more than one of the L-networks can be used to match a given impedance, and a choice must be made by considering factors such as the value of the components and the convenience of applying DC bias. For example, if it were suitable, network 4 would be convenient for a transistor matching network, because the grounded end of the inductor can be used to apply DC bias and the series capacitor doubles up as the DC block.

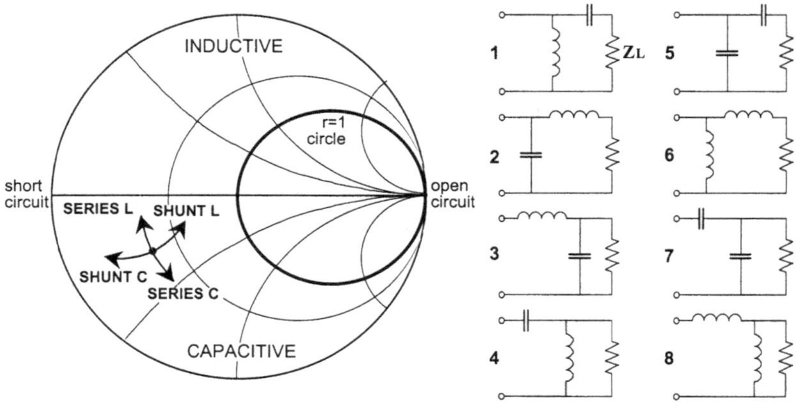

Figure 5.7 Lumped element impedance matching with L-networks

5.3.1.2 Matching two complex impedances with L-networks

There are cases where two complex impedances are required to be matched, and this can be achieved with a simple extension of the L-network technique. The reactance/susceptance of one of the impedances can either be absorbed into the matching network or resonated out by introducing an additional component. An example of the former case, shown in Figure 5.8*a*, is that the shunt output capacitance of an FET might be absorbed into the first shunt capacitor of the matching network. An example of the latter case, shown in Figure 5.8*b*, is that the series input capacitance of an FET might be resonated out with a series inductor (which might in turn be absorbed into the matching network).

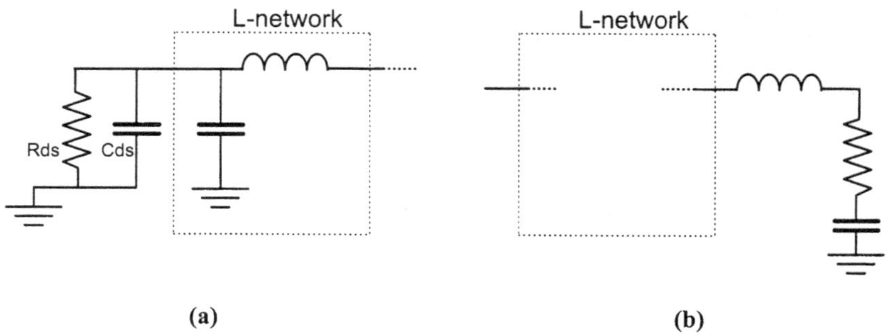

Figure 5.8 Matching complex impedances with L-networks: (a) absorb the reactance and (b) resonate out the reactance

5.3.1.3 T- and π-networks

The major disadvantage of L-networks is that the *Q*-factor of the circuit is entirely fixed by the source and load resistances. In other words, the designer has no control over the bandwidth of the amplifier with this matching strategy. The use of T- or π-networks overcomes this limitation by using a two stage matching process. The load impedance is first transformed to an intermediate resistance, which can be chosen in order to control the amplifier bandwidth. The intermediate resistance is then matched to the source resistance. The T- or π-network can be visualised as a cascade of two L-networks, where the two components in the centre are the same type (series/shunt inductor/capacitor) and can be combined into a single component, resulting in a three-component matching network. The principle of a T-network, for example, is illustrated in Figure 5.9. The first L-network, connected to the load, consists of a series reactance followed by a parallel reactance. This transforms the load impedance to the chosen intermediate resistance. The second L-network consists of a parallel reactance followed by a series reactance. This matches the intermediate resistance to the source resistance. The two shunt reactances are combined to form a single component, and the intermediate resistance stage doesn't exist at any real node.

Design equations can be derived in order to calculate the component values for the various T- and π-networks [5]. The source and load impedances and the operating Q are specified, and the equations yield the required component reactances. On the Smith chart, T- and π-matching networks can be designed by superimposing curves of constant Q onto the chart. The extremes of the Q-curves are the real axis ($Q = 0$; no reactance at all) and the outside of the chart ($Q = \infty$; no resistance). It is important to note that the overall operating Q of the circuit is determined by the highest Q of all the nodes. The L-networks will give lower Q-factor and more bandwidth than these T- and π-networks. Any move towards the edge of the chart will increase the Q-factor. This is an important point to remember for matching network design: a topology which involves moving near the edge of the chart will have narrow bandwidth, require extreme component values, and be more sensitive to process parameter variations.

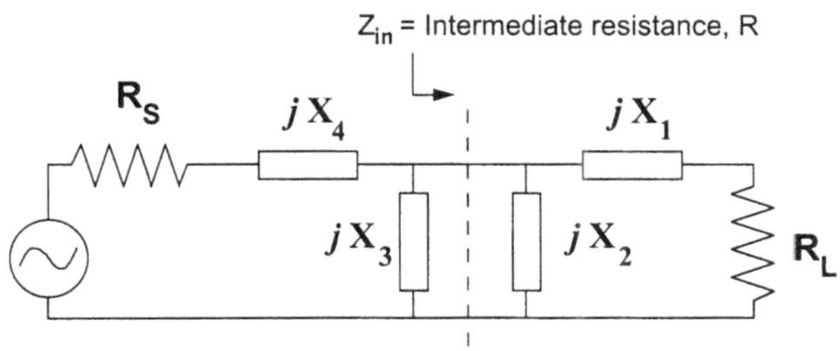

Figure 5.9 The principle of matching with a T-network

5.3.2 Distributed matching networks

At high frequencies the stray capacitance, spurious resonances and distributed effects of spiral inductors mean that transmission-line matching elements are preferred to lumped-elements. The three methods that will be described here are the single stub matching technique, the quarter-wave transformer and the short transformer. There are, of course, many other matching techniques, but the limited space available on MMICs makes many of these unattractive. When using distributed matching elements, it should be remembered that these have a cyclical frequency response, and the amplifier's stability should be checked at the harmonics and sub-harmonics of the design frequency.

5.3.2.1 Series line and parallel stub

In the microstrip medium, series stubs cannot be used because the ground plane cannot be accessed readily. Parallel (shunt) open circuit or short circuit stubs are used instead. This necessitates the use of the admittance chart for some of the operations, and some familiarity with the Smith chart is required to avoid mistakes. In the lossless case, the open circuit stub has an input admittance given by:

$$Y_{in_{OC}} = jY_o \tan \beta l \tag{5.12}$$

and the short circuit stub has an input admittance given by:

$$Y_{in_{SC}} = -jY_o \cot \beta l \tag{5.13}$$

where β is the propagation constant of the line and is equal to $2\pi/\lambda_g$, λ_g being the guided wavelength. βl is the electrical length of the stub, normally expressed in terms of degrees or fractions of a wavelength, and Y_0 is the characteristic admittance of the stub. The input admittance against electrical length for an open circuited stub is shown in Figure 5.10.

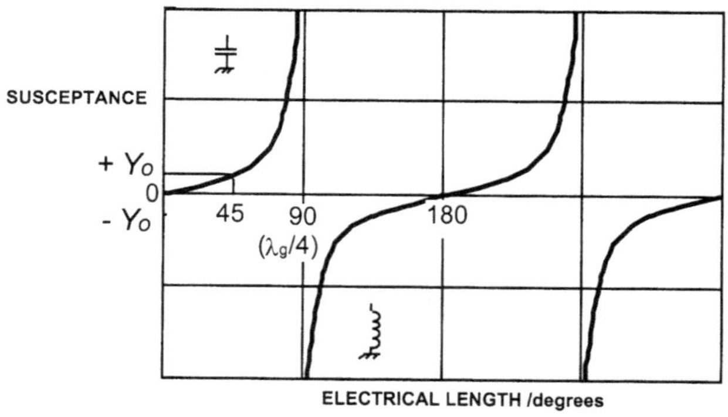

Figure 5.10 Input admittance against electrical length for an open circuited stub

Note that in principle any value of susceptance can be achieved. This enables the stub to be used as a matching element for any impedance on the chart. As the shunt stub can only change the imaginary part of a load's admittance, an additional transmission-line element is required to adjust the real part. The most common method is to use a series line before the stub, as shown in Figure 5.11*a*. The first step is to choose the characteristic impedances of the series line and stub, Zo_{ser} and Zo_{stub}: 50Ω lines are rarely used in MMICs due to their excessive width. In this example, 70Ω has been chosen for both. The design procedure, referring to the Smith chart of Figure 5.11*b*, can be summarised with an example load impedance of 25–*j*100 Ω as follows:

1. Plot the load admittance on the admittance chart, normalised to the chosen characteristic admittance of the series line (i.e. plot $Y_L \times Z_{O ser}$ using constant conductance and susceptance contours); point A.

2. Rotate the load admittance around the centre of the chart until it meets the $g = (Z_{O ser} \div 50)$ circle (point B). The electrical length of the series line, βl_{ser}, is half this angle of rotation, giving a series line length of 41° for this example.

3. The required stub susceptance can now be read by following the constant susceptance contour to the edge of the chart (point C). As the stub has to cancel the susceptance at point B, the opposite sign is used, remembering that in admittance the bottom half of the chart represents positive susceptance.

4. The value should be de-normalised (divided by $Z_{O ser}$): in the example, the required stub susceptance is $b_{stub} = -3$, so $B_{stub} = (-3 \div 70)\ \Omega^{-1}$.

5. From eqn. (5.12) or (5.13) the required stub electrical length can be calculated for the desired stub characteristic impedance and type (short or open circuit). In the example, for an open circuit stub:

$$Y_{o_{stub}} \tan \beta l_{stub} = B_{stub}$$

$$\therefore \text{ stub electrical length, } \beta l_{stub} = \tan^{-1}\left(B_{stub} \times Z_{o_{stub}}\right) = \tan^{-1}(-0.0428 \times 70)$$

This gives a stub length of 108.5°, after adding 180° to the negative result from the calculator. The design can then be converted to physical dimensions, and effects such as T-junctions and bends incorporated into the simulation. It is important to use the Smith chart first because the matching problem can be solved with many other combinations of stub length and position, which means that a CAD optimiser will have many local error function minima to contend with. Furthermore, some of the alternative solutions will involve longer lines, narrower bandwidth and excessive sensitivity.

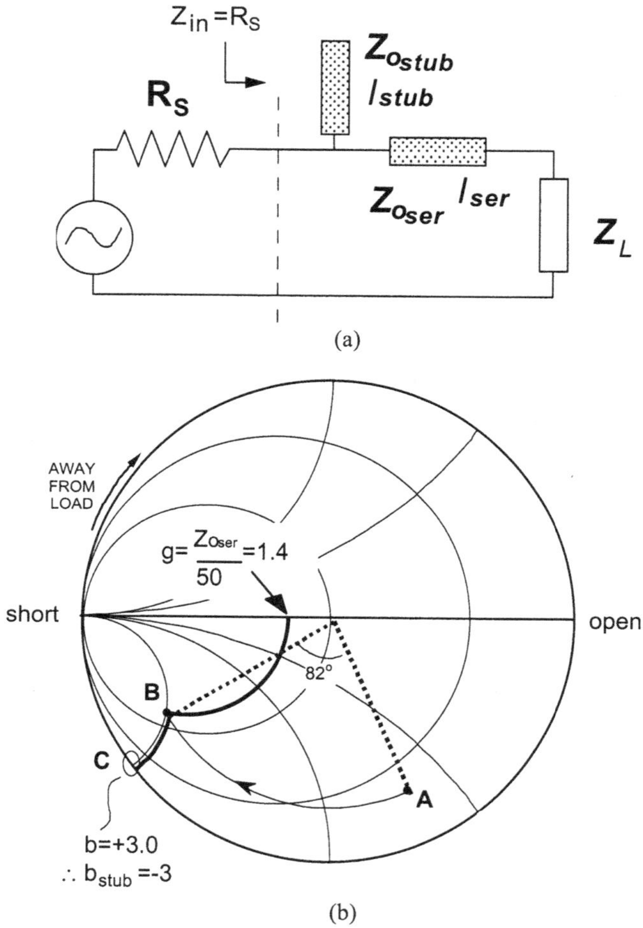

Figure 5.11 Single-stub impedance matching: (a) network and (b) Smith chart operations

5.3.2.2 Quarter-wave transformer

This well known technique can match a load resistance, R, to Z_0 using a series transmission line one quarter-wavelength (90°) long, with a characteristic impedance of $Z_T = \sqrt{Z_0 \times R}$. Since a transistor's impedance is rarely purely real, first a reactive element must be used to resonate out the imaginary part. This can be achieved with either a series or shunt inductor/capacitor or with a transmission-line stub.

5.3.2.3 Short transformer

The short impedance transformer technique can be used to directly match a load impedance $R + jX$ to Z_0 by selecting the transformer characteristic impedance, Z_T, and electrical length, θ, such that:

$$Z_T = \sqrt{Z_0 R - \frac{X^2 Z_0}{Z_0 - R}} \tag{5.14}$$

and

$$\theta = \tan^{-1}\left(\frac{Z_T (Z_0 - R)}{X - Z_0} \right) \tag{5.15}$$

provided that $X^2 < R(Z_0 - R)$ and $R \neq Z_0$.

This matching technique is more compact than the quarter-wave transformer, but often requires unrealistic values of Z_T. A useful solution to the length and width problems encountered with the quarter-wave and short transformer is to convert the transmission line into an equivalent lumped or lumped-distributed π-network, as described in Chapter 3.

5.4 DC bias injection

The transistor must have DC bias applied to it, in order to set the operating point on its I-V characteristic about which the AC microwave signal swings. Figure 5.12 shows schematically the I-V curves of a generic FET and indicates the four most common operating points used for amplifiers. V_{ds} is the drain-source voltage, V_{gs} is the gate-source voltage and I_d is the drain current. I_{dss}, the saturated drain current, is used as a reference current, and is the drain current obtained when the FET is biased at its specified drain voltage and with $V_{gs} = 0$ V. For the generic FET characteristic shown in Figure 5.12, point I represents the low noise operating point; with the FET biased at $V_{ds} = 3$ V (or lower for many HEMTs) and 10 to 20% of I_{dss}, the best noise figure performance is obtained, along with low DC power consumption. However, because the gate is biased near to pinch-off, the gate voltage swing, and therefore the power handling, is limited. At point II the FET is operated with V_{gs} close to 0 V and I_d is close to I_{dss}. At this bias point the FET gives its maximum small-signal gain. However, the high DC power consumption and increased noise figure make this operating point unattractive. Point III is placed almost exactly in the middle of the I-V curves, and this operating point allows the maximum linear output power to be achieved for class A operation. For higher efficiency, but degraded linearity, operating point IV can be used for class AB operation of the amplifier.

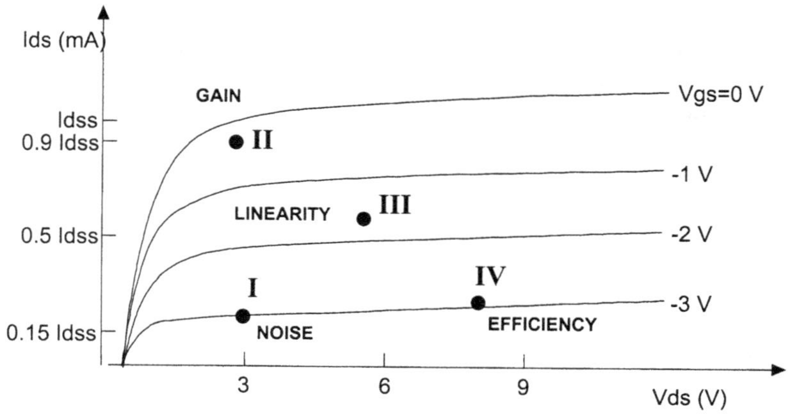

Figure 5.12 Typical FET I-V curves and operating bias points

The gate and drain DC bias voltages can be applied to a FET in a number of ways, as illustrated in Figure 5.13. In method (a), inductors act as bias chokes, with DC blocking capacitors used at the input and output to isolate the bias from other circuits, and with decoupling capacitors to prevent the leakage of RF signals into the power supplies. In this bias arrangement, the drain voltage is required to be positive (e.g. +3 V) and the gate voltage is required to be negative (e.g. −1 V). If inductors are introduced as stand-alone bias chokes, considerable chip area will be wasted. Hence, when designing the matching networks, it is very convenient to find a topology that has shunt inductors, so that bias can be applied through them.

In method (b), high value resistors are used to apply the bias voltages. With the exception of low-noise and high-power amplifiers, this is a very suitable technique for the gate bias, since the gate draws no current and the resistor can be used to stabilise the device. However, on the drain side, it is likely that the drain current will be too high to allow resistor biasing unless the device has a very small gate-width. For example, even a modest 20 mA drain current would lead to a 6 V drop and 120 mW dissipated in a 300 Ω drain bias resistor. The use of smaller bias resistors will start to lower the amplifier gain considerably.

In method (c), short circuit microstrip stubs are used. The short circuit end is grounded with a decoupling capacitor so that DC bias can be applied. As stand-alone elements these bias lines would have to be a quarter wavelength long, and this would be too large in most cases. However, if the stubs are part of the input and output matching networks then the technique is quite attractive. The decoupling at the end of these matching stubs is very critical, and care must be taken to avoid oscillations due to the resonant nature of the stub and its cyclical frequency response.

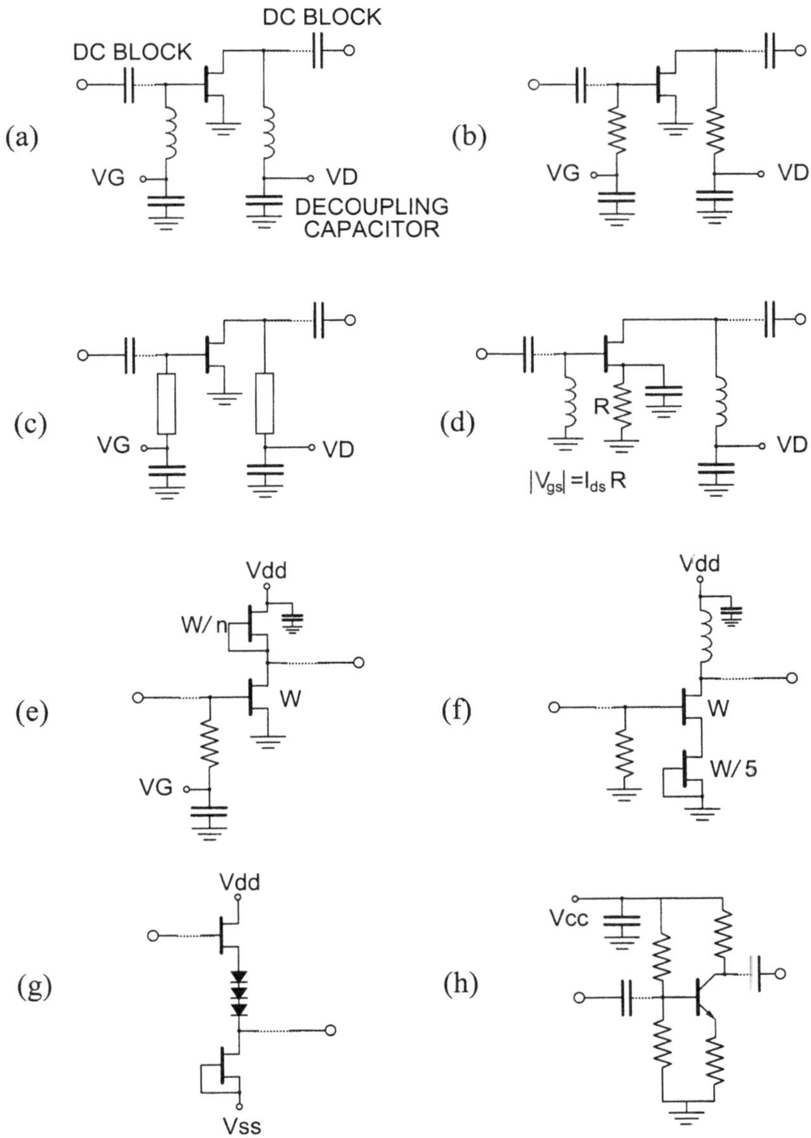

Figure 5.13 DC bias networks: (a) inductors as bias chokes, (b) high value resistors, (c) microstrip stubs, (d) self-biasing, (e) active load, (f) constant-current source self-biasing, (g) DC coupling and (h) bipolar transistor biasing

Method (d) is known as the self-biasing technique and enables the FET to be biased from a single power supply rail. The gate is grounded at DC through either an inductor or high value resistor. The source is raised to a positive DC potential, equal in magnitude to the desired gate-source voltage, by inserting a small resistor in the source (whose value is equal to the required V_{gs} divided by the desired drain current). In order to prevent a loss of RF gain, the source is grounded with a large decoupling capacitor. The advantage of the single power supply is considerable, particularly for battery operation, but there is a penalty because of the slightly increased DC power consumption. One drawback of the technique is that it removes one useful post-fabrication tuning mechanism from the amplifier, because the bias is fixed. Also, the amplifier is likely to be more sensitive to process variations, such as changes in pinch-off voltage. Some foundries have facilities for laser trimming of resistors, and this would solve these problems (at a cost). Alternatively, a number of different resistors can be connected to the source and the optimum one selected during testing. Figure 5.14 shows a microphotograph of a single-stage feedback amplifier using the self-bias technique; a very large decoupling capacitor can be seen at the FET source, this being necessary to maintain good low frequency performance.

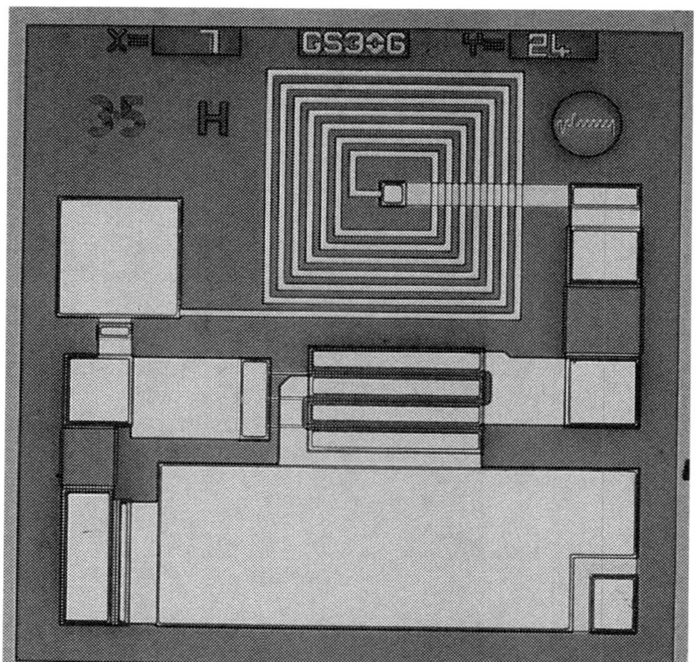

Figure 5.14 Microphotograph of a single-stage amplifier employing the self-bias technique

Method (e) is the well known active load technique, in which the active load has V_{gs} = 0 and hence operates at its saturated drain current. Therefore, if the active load FET has a gate-width of $W_g/2$, the amplifying FET with gate-width W_g runs at a drain current of $I_{dss}/2$. The active load presents a large resistance in parallel with a small capacitance. It does cause some reduction in gain as a result, but is very compact and suitable for high packing density and DC coupled amplifiers. It is particularly useful for IF amplifiers, where bias choke inductors would be too large. However, with FETs the standard active load has been shown to be very sensitive to the gate-source voltage, and an improved active load has been proposed for MMIC applications [12].

Method (f) is the constant-current source self-bias technique: the gate of the amplifying FET is at 0 V, but its source is raised to a positive DC voltage by the drain-source voltage of the current source FET. Thus, the gate-source voltage is negative and the amplifying device runs at the I_{dss} of the current source FET. Method (g) employs level shifting diodes to drop the DC output voltage of one stage down to the input voltage of the following stage. This technique is used for DC-coupled amplifiers and is discussed later.

The bipolar transistor can be controlled via its base–emitter voltage with many of the same techniques described here for FETs. However, the base–emitter junction is forward biased, and so a positive supply is required (for an *n-p-n* transistor). Also, the base current must be carefully controlled, with at the very least a current limiting resistor. Generally, the standard bipolar bias configuration shown in (h) is used. An emitter stabilisation resistor can be added if desired.

5.4.1 Stacked bias

There are many applications where low DC current is more important than the ability of the device to operate at low drain voltages. Such applications might include battery operated communications terminals, in which the supply voltage is fixed but where ultra-low DC power consumption is required. For this type of application, the stacked bias technique is very attractive. This is best illustrated by studying the three-stage amplifier circuit diagram shown in Figure 5.15. This amplifier uses both self-biasing (evident on the source of the first stage's FET) and stacked biasing. The drain current from the third stage's FET passes from its source into the drain of the second stage FET, and then from the second FET's source into the drain of the first FET. This is achieved by having the sources grounded at AC with decoupling capacitors, and by using inductors to connect successive sources and drains at DC. Hence, if each FET is required to operate at 3 V drain voltage with 10 mA drain current ($V_{gs} \approx$ -1 V) then the complete amplifier requires a single 10 V supply and will draw only 10 mA. The first stage FET has ≈ +1 V on its source, and 4 V on its drain. The second FET has 4 V on its source and 7 V on its drain. The third FET has 7 V on its source and 10 V on its drain. The amplifier only requires a single supply rail and draws a minimum of current. The operating points of the transistors can be adjusted during RFOW testing by control of the first FET's gate bias voltage and by the inclusion of select-on-test self-bias resistors. Figure 5.16 shows a 1.5 to 2.3 GHz two-stage low-noise amplifier, which employs this biasing technique.

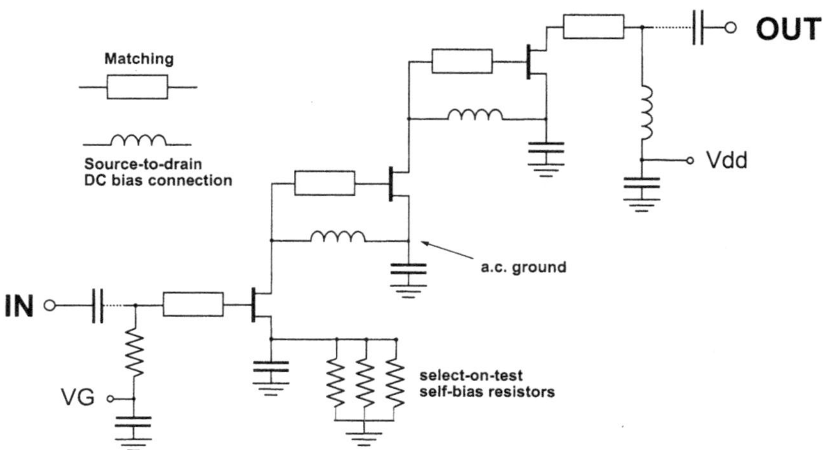

Figure 5.15 The stacked bias technique used in a three-stage amplifier

5.4.2 Off-chip components

On-chip bias decoupling capacitors are limited in value to approximately 20 pF because of their large area. However, this is rarely enough to ensure that the amplifier is unconditionally stable at frequencies below 1 GHz, and so off-chip decoupling capacitors and other bias circuitry usually have to be included. It is very important that this is considered early in the design procedure and included in the simulations. For example, it is often found that an off-chip decoupling capacitor has to be placed directly next to the chip and connected with very little bond-wire inductance, and this may have an influence on the layout of the DC bias bond pads. Alternatively, it may be the case that off-chip components cannot solve the stability problem and that on-chip resistive loading is the only solution. As a result, it is very unwise to send a chip for fabrication thinking: 'I can worry about the off-chip bias circuits later'.

When an off-chip decoupling capacitor is wire-bonded to the MMIC, you can be assured that its capacitance will resonate with the bond-wire inductance at some frequency. Such resonances must be kept out of band and closely scrutinised using CAD for signs of possible instability. Figure 5.17 shows the typical arrangement of an MMIC amplifier mounted in a package with off-chip bias components.

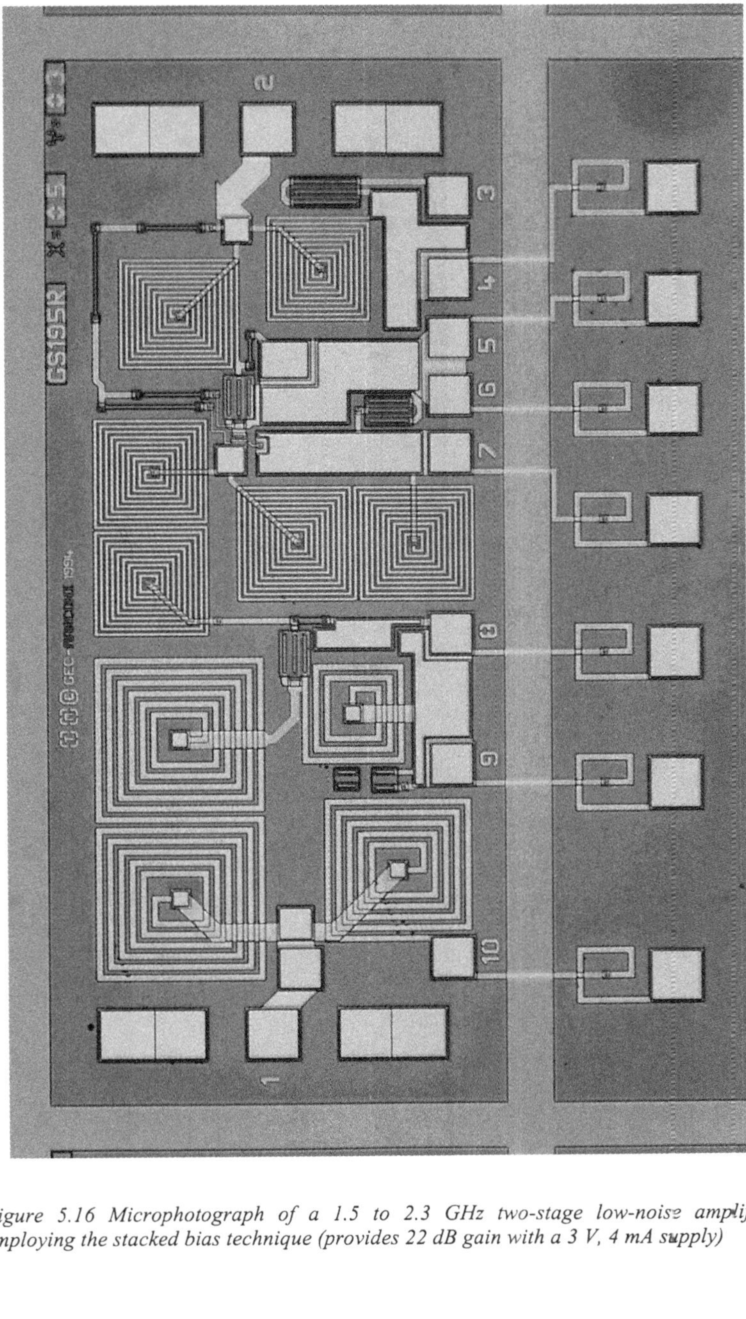

Figure 5.16 Microphotograph of a 1.5 to 2.3 GHz two-stage low-noise amplifier employing the stacked bias technique (provides 22 dB gain with a 3 V, 4 mA supply)

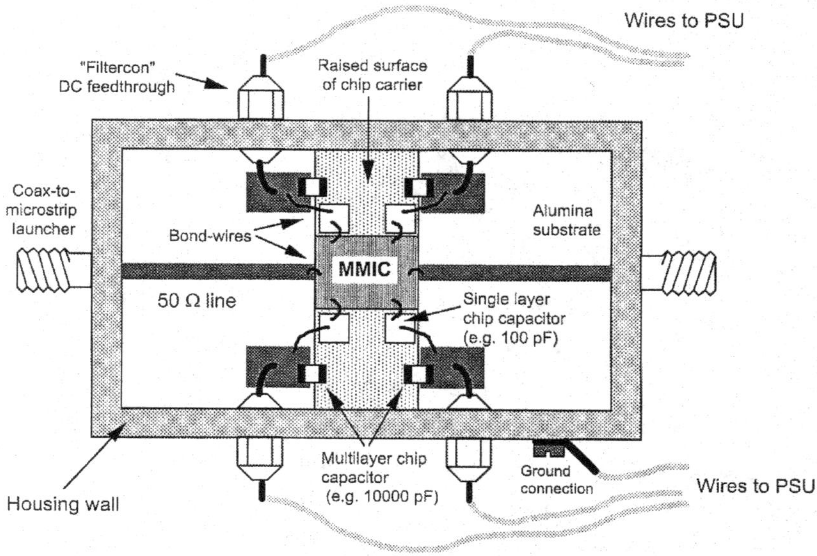

Figure 5.17 Packaged MMIC amplifier with off-chip bias components

5.4.3 RFOW testing considerations

For RFOW testing of an amplifier it is not essential that it is unconditionally stable at all frequencies, because the amplifier will, ideally, be terminated with the 50 Ω impedances of the network analyser test ports. However, it is of course essential that the amplifier does not go into oscillation during measurement! The inductance of DC probe needles is very high, however, and so it is very difficult to add effective extra decoupling capacitance to the supply lines. The amplifier must be modelled to take the RFOW scenario into account to make sure that low frequency oscillations will not result from this probe needle inductance.

5.5 Reactively matched amplifier design

The reactively matched amplifier uses lossless matching networks, with either lumped- or distributed-elements. Since the matching networks are lossless, the reactively matched amplifier can be designed for optimum gain, noise figure or output power. For a single-stage narrowband small-signal amplifier using a transistor which is unconditionally stable, the design is a matter of matching the input and output simultaneous match reflection coefficients to 50 Ω, using the techniques described in Section 5.3, and designing suitable bias networks. However, for broadband amplifiers, the inherent gain roll-off requires that the input and output matches have to be traded off in order to maintain flat gain.

Furthermore, when the device is only conditionally stable, the input and output cannot be matched to 50 Ω because of instability. Hence, the disadvantages of the reactively matched amplifier are that it is difficult to achieve good input/output matches, flat gain and good stability. This makes the reactively matched amplifier difficult to cascade, unless isolators or a balanced amplifier topology are employed. In hybrid-MIC design, isolators or balanced stages can readily be used for multi-stage amplifiers. However, for MMIC amplifiers it is preferable to use an amplifier topology that is more amenable to cascading; more suitable techniques are lossy matching, negative feedback and the distributed amplifier.

5.5.1 Multi-stage design

In multi-stage reactively matched amplifiers the two main problems are stability and gain ripple. In a multi-stage design, the inter-stage matching networks can be designed to give a positive gain slope, which counteracts the gain roll-off of the transistors. Figure 5.18 shows the block diagram of a generic two-stage amplifier. If the devices are unconditionally stable, the input and output matching networks are synthesised to give a good 50 Ω match over the desired frequency range: maximum power is thus transferred throughout the operating frequency range, and the input and output matching networks have a flat frequency response. However, the transistors themselves, shown as FETs, will have a 6 dB/octave gain roll-off. Hence, in order to achieve flat overall gain, the inter-stage matching network must introduce a 12 dB/octave positive gain slope. For a reactively matched amplifier the matching networks are lossless, and so this gain slope must be achieved by introducing frequency-dependent mismatch between the output of the first stage and the input of the second. If the transistor output and input impedances are represented as simple resistor–capacitor networks, the required inter-stage network can be synthesised from first principles by considering the poles and zeros needed to give the desired insertion loss response [2]. Nowadays, many CAD tools exist which can perform filter and matching network synthesis. However, great care must be exercised in defining the desired frequency response parameters: these types of synthesis tools will readily provide solutions, which are overcomplicated and require extreme element values.

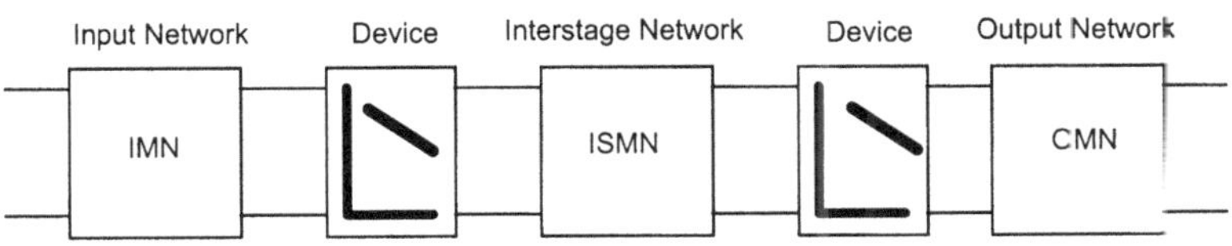

Figure 5.18 A generalised two-stage amplifier

Stability is a major problem in reactively matched multi-stage amplifiers. The overall K-factor of an amplifier is not a sufficient indication of the amplifier's stability. For example, an intermediate stage may have $K < 1$, and the stage before it (which is not necessarily matched to 50 Ω) may present it with an impedance that is in an unstable region. Inter-stage oscillations can thus exist which cannot be diagnosed from the overall two-port S-parameters.

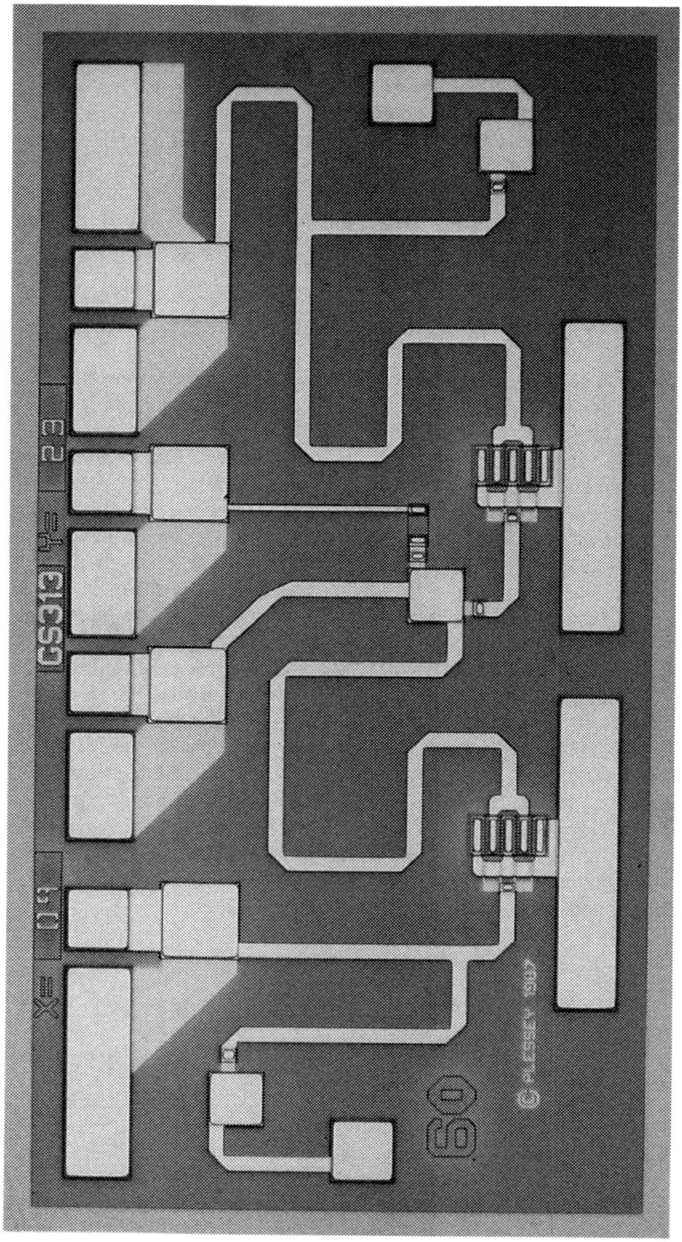

Figure 5.19 Microphotograph of an X-band two-stage reactively matched amplifier (with an inter-stage stabilising resistor)

Ensuring stability is particularly challenging at low frequencies, where the device has high gain and $K < 1$. The stability of each amplifier stage should be investigated individually. Furthermore, as discussed in Chapter 3, unwanted feedback paths can result from improper grounding and poor decoupling of the DC power supplies. Figure 5.19 shows a microphotograph of a two-stage X-band amplifier, which employs reactive matching with a single inter-stage stabilising resistor. Note that this design features many of the good grounding practices described in Chapter 3.

5.6 Lossy matching

The lossy match amplifier uses resistors within its matching networks to enable flat gain to be achieved over a broad bandwidth. The most typical topology is to employ resistors in series with high impedance stubs on both the input and output, as shown in Figure 5.20*a*. At low frequencies the stubs have little reactance, and the resistors load the transistor and lower its gain. At high frequencies the stubs have high reactance (going to infinity when they are a quarter-wavelength long), and the resistors have little effect on the transistor. Hence, the matching networks can introduce a positive gain slope to compensate the transistor's gain roll-off without resorting to mismatching. The lossy match amplifier has quite high gain, which is very flat, and has good input and output matches. In addition, the resistors can greatly ease the stability problem at low frequencies. The disadvantages of this approach, compared to the reactively matched amplifier, are that it has lower gain, lower output power and higher noise figure.

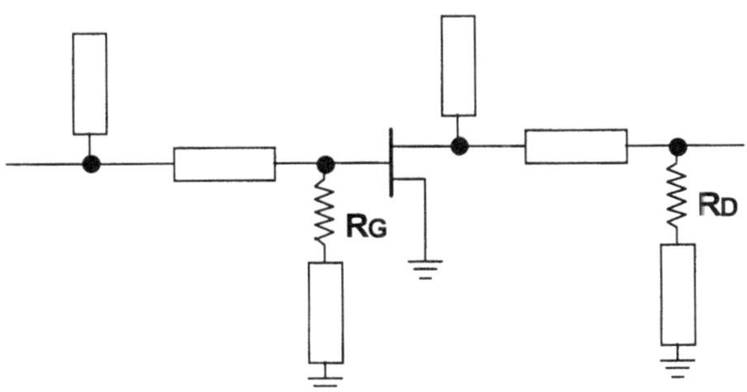

(a)

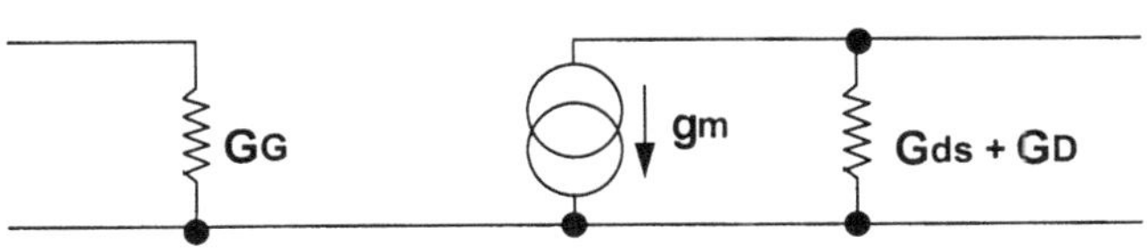

(b)

Figure 5.20 Lossy matching: (a) basic topology and (b) low frequency model

From the low frequency model of a lossy match FET amplifier, shown in Figure 5.20*b*, it can be shown that [13]:

$$S_{11} \approx \frac{1 - G_G Z_0}{1 + G_G Z_0}, \tag{5.16}$$

$$S_{22} \approx \frac{1 - (G_{ds} + G_D)Z_0}{1 + (G_{ds} + G_D)Z_0} \tag{5.17}$$

and

$$\text{Gain} \approx \left[\frac{g_m Z_0}{2}(1 + S_{11})(1 + S_{22}) \right]^2 \tag{5.18}$$

where G_{ds} is the drain-source conductance ($1/R_{ds}$) and G_D is the lossy match drain loading conductance. From eqn. (5.18), it can be seen that if $G_G = (G_{ds}+G_D) = 1/Z_o$ (i.e. the gate loading resistor is 50 Ω and the drain loading resistor in parallel with R_{ds} gives 50 Ω, then $S_{11} = 0$, $S_{22} = 0$ and the gain (in dB) is equal to $20\log(g_m Z_0/2)$, which gives typically 8 dB for an 800 μm MESFET. This low frequency model clearly shows how the gate-width of the FET determines the low frequency gain of the lossy match amplifier in the ideally matched case. Figure 5.21 shows the microphotograph of a PHEMT MMIC lossy match amplifier operating from 20 to 40 GHz.

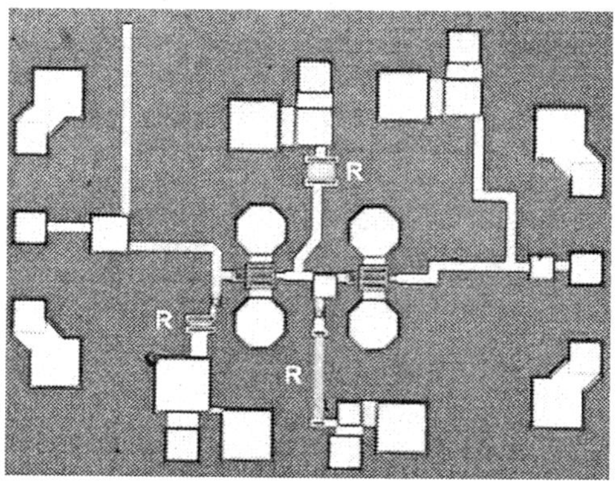

Figure 5.21 20–40GHz PHEMT MMIC amplifier using lossy matching

5.7 FET feedback amplifier

The FET feedback amplifier is a very common solution for wide-band MMIC amplifiers. The technique gives very flat gain with good input and output matches and can achieve moderate power levels (hundreds of mW). It is useful as a general purpose gain block and as a wideband IF amplifier for millimetre-wave systems. The basis of the technique is that negative feedback is applied to the FET by connecting a resistance (of the order of hundreds of Ohms) from the drain to the gate. This has the effect of stabilising the device and can make the input and output impedances much closer to the desired 50 Ω. This is very beneficial because FETs can have very low input resistance and high output resistance at low frequencies, and matching these to 50 Ω is rather difficult because of the limited range of component values provided by MMIC lumped-elements. Through the use of feedback the matching networks can be kept simple [14]. Furthermore, the amplifier performance becomes less sensitive to process-related variations of the FET parameters.

For the best response, a number of additional components are required in the amplifier. Figure 5.22 shows the schematic of a matched feedback amplifier [15] suitable for use up to approximately 10 GHz. Each component has a particular role to play in achieving maximum performance. R_{FB} is the key feedback element, and its value determines the basic gain and bandwidth, which have to be traded off against one another. L_{FB} introduces a degree of frequency dependence into the feedback loop: at the lowest frequencies it has no effect and R_{FB} controls the gain level, but at high frequencies the reactance of L_{FB} increases, which reduces the amount of negative feedback. Hence, the effect of L_{FB} is to maintain flat gain and give operation up to a higher frequency. L_D is chosen to compensate for the C_{ds} of the FET; the overall effect is to provide gain peaking at the upper edge of the frequency response, and this provides extended bandwidth. Although the feedback network achieves input and output impedances quite close to 50 Ω, there is still a need for additional matching elements to give good input and output return loss. These matching elements are L_{in}, C_{in} and C_{out}. Finally, C_{FB} is a DC block, which is required to isolate the positive drain bias from the negative gate bias. This completes the basic matched feedback amplifier. In practice, the matching networks may have to be slightly more complicated than this, and, after the basic design has been established, the whole circuit should be optimised using CAD. Bandwidths as high as 1 to 10 GHz and 6 to 18 GHz have been achieved [16,17]. Figure 5.23 shows a microphotograph of a basic 1 to 6 GHz feedback amplifier.

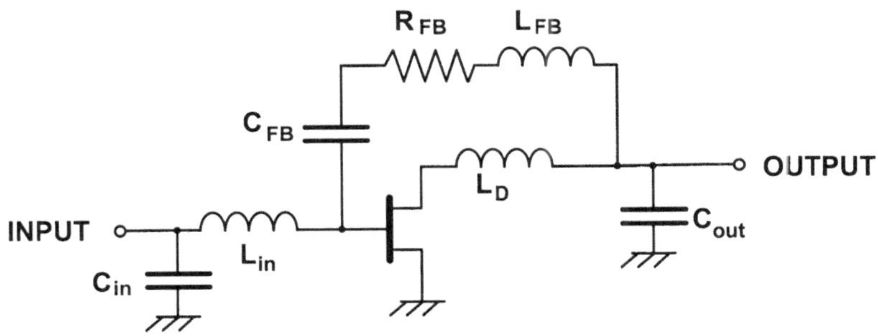

Figure 5.22 The matched FET feedback amplifier

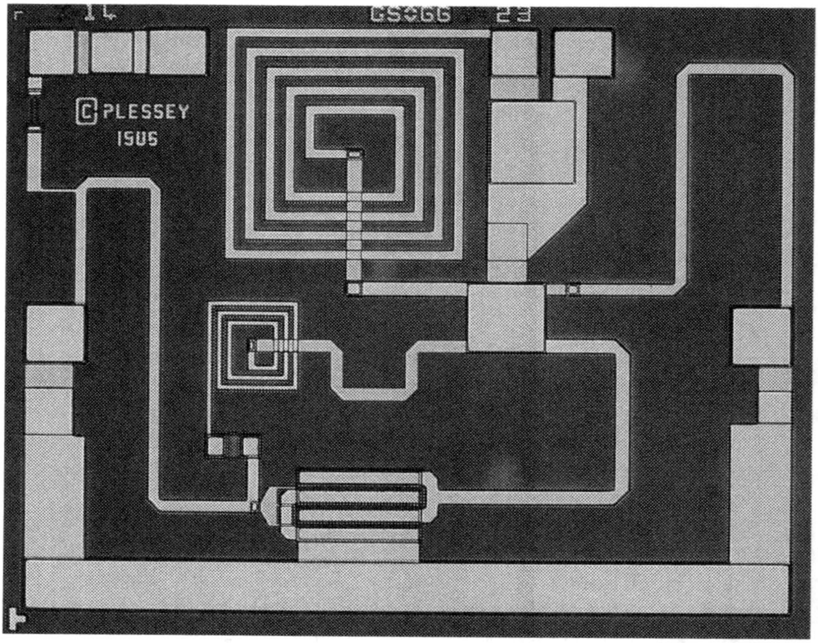

Figure 5.23 Microphotograph of a basic 1 to 6 GHz feedback amplifier

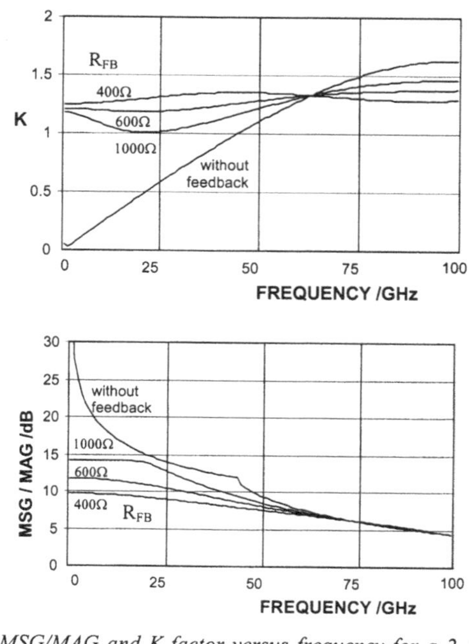

Figure 5.24 MSG/MAG and K-factor versus frequency for a 2 × 60 μm PHEMT, with and without parallel feedback

For a design operating at higher frequencies, the device no longer has an S_{21} phase near 180° and the resistive feedback can, instead, be used to limit the low frequency gain, thus aiding stability, whilst having little effect on the high frequency gain. This is illustrated in Figure 5.24, where the MSG/MAG of a 2 × 60 μm PHEMT is plotted against frequency with the value of feedback resistance as a parameter (the feedback resistor is assumed to be ideal). Figure 5.25 shows a microphotograph of a three-stage 60 GHz MMIC amplifier employing this approach for ensuring stability, in conjunction with stub matching using high impedance microstrip lines.

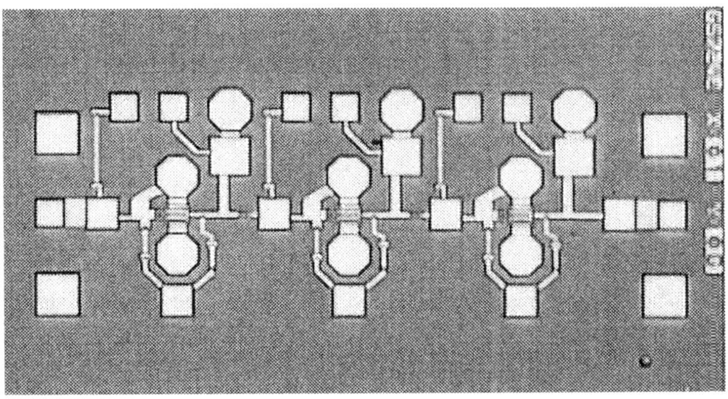

Figure 5.25 Microphotograph of a 60 GHz PHEMT amplifier using parallel feedback

5.8 Distributed amplifiers

In the distributed (or travelling-wave) amplifier, the problem of achieving a broadband match to the transistor input and output impedances is overcome by incorporating the input and output capacitances of a number of transistors into artificial transmission-line structures. The amplifier thus consists of an input line, incorporating the input capacitances of the transistors, and an output line incorporating the output capacitances. The signal on the input line is amplified and fed into the output line. The result of this arrangement is that the amplifier can operate over a very broad frequency range, from very low frequencies up to the cut-off frequency of the artificial transmission lines. The technique was originally patented by Percival [18] as a technique for realising broadband amplifiers with thermionic valve devices. In the early 1980s the technique suddenly became very popular with the realisation that GaAs MESFETs could be used in a monolithic distributed amplifier to achieve decade bandwidth microwave amplification [19,20]. Its real strength is that with MMIC technology the distributed amplifier can achieve extraordinarily wide bandwidths with a simple circuit topology that is insensitive to process variations.

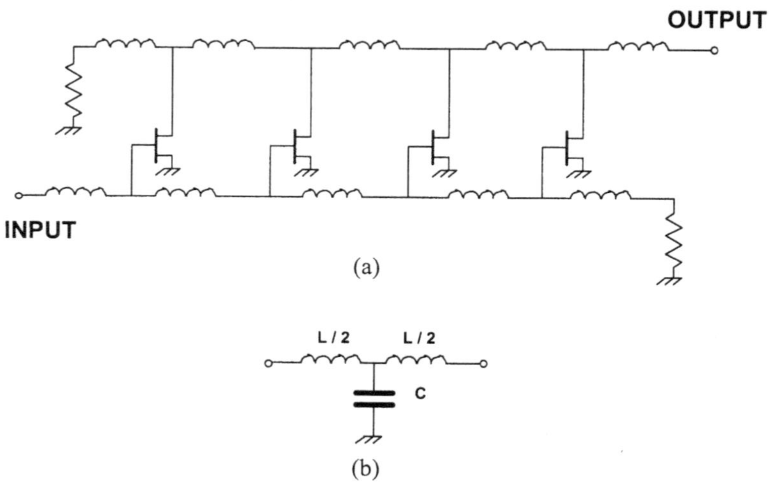

Figure 5.26 The distributed amplifier: (a) circuit diagram and (b) a single section of constant-K ladder

In the most common configuration, the capacitances become the shunt capacitances in a constant-*K* low-pass filter ladder network. In the case of the FET distributed amplifier there is a gate-line and a drain-line, as shown in Figure 5.26. The input signal travels down the gate-line, exciting each FET in turn, before being absorbed by a terminating resistor. The transconductance of the FETs amplifies the signal and feeds it into the drain-line. If the phase velocities in the gate-line and drain-line are roughly equal, then the signals from each FET will add constructively at the output port. At high frequencies the signals will largely cancel out at the reverse end of the drain-line, but this is not the case at low frequencies. A drain-line termination absorbs any undesired signals present at the reverse end.

The basic design of a FET distributed amplifier can be carried out very easily by considering the characteristics of the constant-*K* ladder. The characteristic impedance of the artificial transmission line is given by:

$$Z_0 = \sqrt{\frac{L}{C}} \tag{5.19}$$

and its cut-off frequency is given by:

$$f_c = \frac{1}{\pi\sqrt{LC}} \tag{5.20}$$

It is evident that for a given Z_0 (normally 50 Ω) and a given transistor capacitance, the inductance, *L*, and cut-off frequency, f_c, are both fixed. Also, since the gate-source capacitance is considerably larger than the drain-source capacitance, this will determine the cut-off frequency and thus the maximum frequency of operation. As an example, an amplifier employing 240 μm gate-width FETs with C_{gs} = 0.3 pF requires an inductance of 0.75 nH, and the cut-off frequency would be 21.2 GHz. A higher cut-off frequency could be achieved by using smaller gate-width FETs with a lower C_{gs}, but the transconductance, g_m, would also be correspondingly lower, giving less gain. Likewise, a larger device has higher g_m and gives more gain, but the higher C_{gs} lowers the cut-off frequency, as illustrated in Figure 5.27.

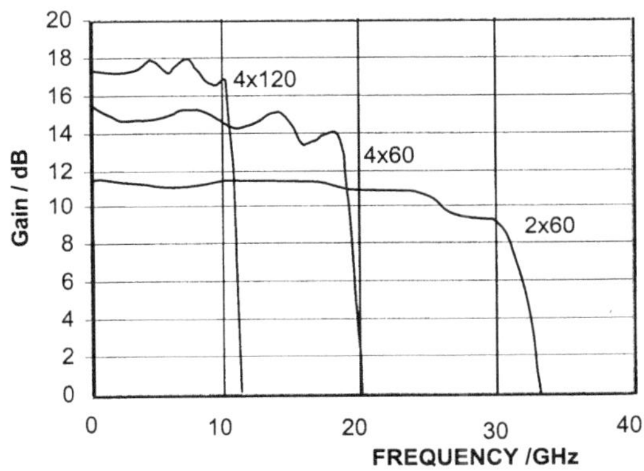

Figure 5.27 Four-section distributed amplifier simulated responses (0.25 μm PHEMTs)

The only way to achieve both a low C_{gs} and a high g_m is to use a short gate-length device with a high f_t (since $f_t = g_m/2\pi C_{gs}$). With ideal FETs, the distributed amplifier can achieve more gain by employing more sections. However, in reality the parasitics of the FET mean that there is a limit to the number of sections that can usefully be employed.

5.8.1 Gate and drain-line losses

In the case of ideal lossless constant-*K* lines, it has been shown [20] that a distributed amplifier with *n* sections has a power gain given by:

$$G = \frac{n^2 g_m{}^2 Z_0{}^2}{4} \tag{5.21}$$

In the ideal case, the gain can be increased by adding more sections. However, in practice the number of sections that can be used is limited by gate and drain-line losses.

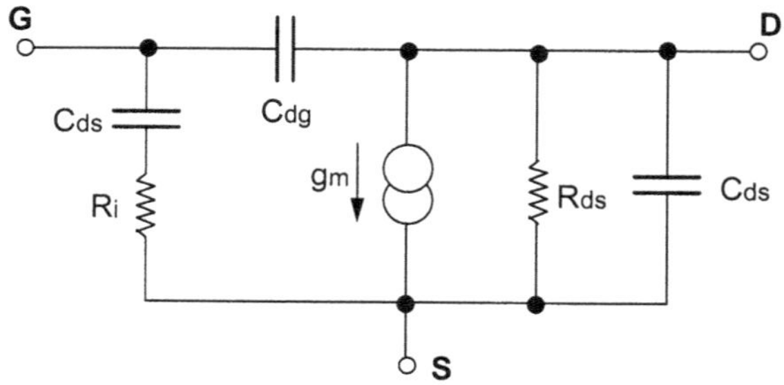

Figure 5.28 A simplified FET equivalent circuit

Referring to the simplified MESFET equivalent circuit in Figure 5.28, the resistance R_i of each FET will absorb some of the signal travelling down the gate-line, and the resistance R_{ds} will absorb some of the signal travelling up the drain-line. It is found that the drain-line loss is essentially constant with frequency, whereas the gate-line loss increases rapidly with frequency and is the more dominant limitation in most cases. The result of gate-line loss is that successive FETs along the gate-line receive a steadily decreasing level of input signal. The effect is so pronounced that after a certain number of sections the input signal becomes so weak that there is no benefit in adding more sections: there is found to be an optimum number of sections for given levels of gate and drain-line loss. Additional stages beyond this optimum number add virtually no signal to the drain-line because their input signal has been attenuated so much. Furthermore, these additional sections add more and more drain-line loss, so they are actually detrimental to the performance.

The effects of gate and drain-line losses have received considerable attention in the literature. Conventionally, the analysis is presented in terms of the gate and drain-line angular cut-off frequency, ω_c, and in terms of the FET's intrinsic cut-off frequencies ω_g and ω_d. The gate-line loss is given by [21]:

$$A_g = \frac{(\omega_c/\omega_g)(\omega/\omega_c)^2}{\sqrt{1-\left[1-(\omega_c/\omega_g)^2\right](\omega/\omega_c)^2}} \tag{5.22}$$

and the drain-line loss is given by:

$$A_d = \frac{\omega_d/\omega_c}{\sqrt{1-(\omega/\omega_c)^2}} \tag{5.23}$$

where $\omega_g = 1/R_i C_{gs}$, $\omega_d = 1/R_{ds}C_{ds}$, and $\omega_c = 2/\sqrt{L_g C_{gs}} = 2/\sqrt{L_d C_{ds}}$.

Gate- and drain-line losses are plotted against normalised frequency in Figure 5.29, for different relative values of ω_g and ω_d. The gate-line losses are generally found to limit the bandwidth and cause high frequency gain roll-off, whereas the drain line losses affect mainly the low frequency gain level. When these losses, A_g and A_d, are incorporated into the analysis, the gain of the distributed amplifier is found to be [20]:

$$G = \frac{g_m{}^2 Z_o{}^2}{4} \frac{\left[\exp(-A_g n) - \exp(-A_d n)\right]^2}{(A_g - A_d)^2} \tag{5.24}$$

and the optimum number of sections has been shown to be [21]:

$$N_{opt} = \frac{\ln(A_g / A_d)}{A_g - A_d} \tag{5.25}$$

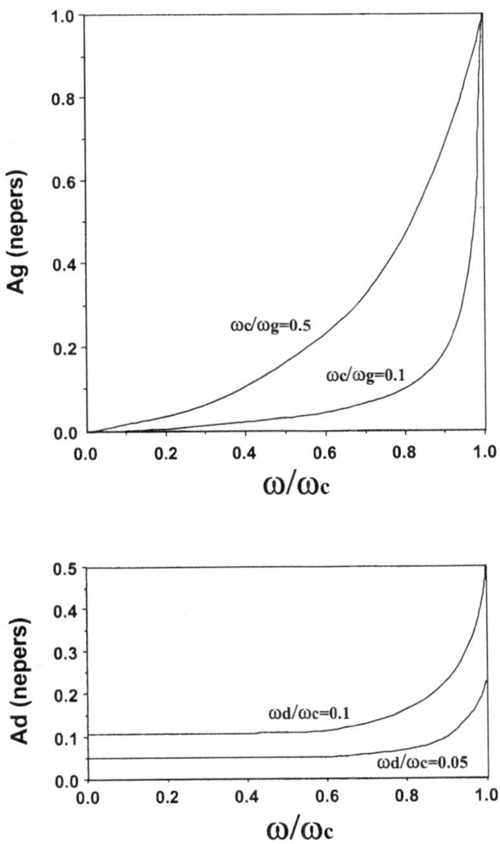

Figure 5.29 Gate- and drain-line losses against normalised frequency [21]

In practice, the optimum number of sections has been found to be typically four or five, although as many as seven sections have been used to good effect. Figure 5.30 shows a microphotograph of a 2 to 20 GHz distributed amplifier with four sections. A frequency range of 0.5 to 20 GHz is typical of that which can be achieved with a 0.5μm MESFET MMIC process. The low frequency coverage is determined by the design of the bias networks, which would certainly require off-chip components to get down to 100 MHz, say. Realising 'simple' DC blocks over such multi-decade bandwidths is no small task, as the large capacitors required for low frequency operation will have very large parasitic elements which will destroy the high frequency performance. One of the most popular solutions to this problem is to mount the chip capacitors in a suspended-microstrip environment so that their shunt capacitance to ground is reduced.

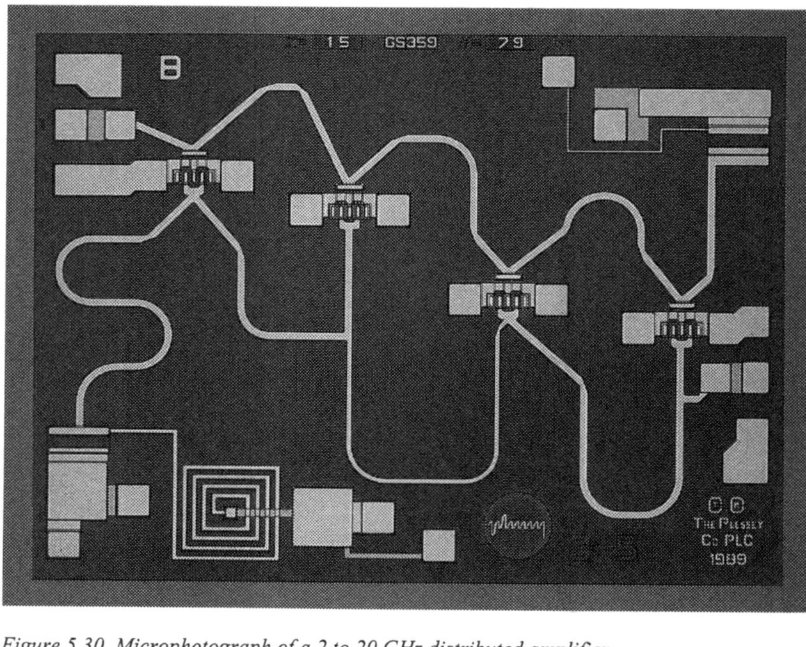

Figure 5.30 Microphotograph of a 2 to 20 GHz distributed amplifier

5.8.2 Equalisation of gate- and drain-line phase velocities

In the above description, it has been assumed that the gate- and drain-lines have equal characteristic impedance and equal phase velocities. For the constant-K topology this can only be achieved if C_{gs} and C_{ds} are equal. However, in practice C_{ds} is much smaller than C_{gs}, and so steps must be taken to equalise the gate and drain-line phase velocities. One method is to add an additional shunt capacitance in parallel with C_{ds}, in order to make the overall capacitance equal to C_{gs}. However, this brute force approach is far from optimum and considerably higher performance can be achieved by employing m-derived sections on the drain line. In this topology, shown in Figure 5.31, a series inductance αL_d is connected at the drain of the FET. This results in equal gate and drain-line phase velocities and improved performance.

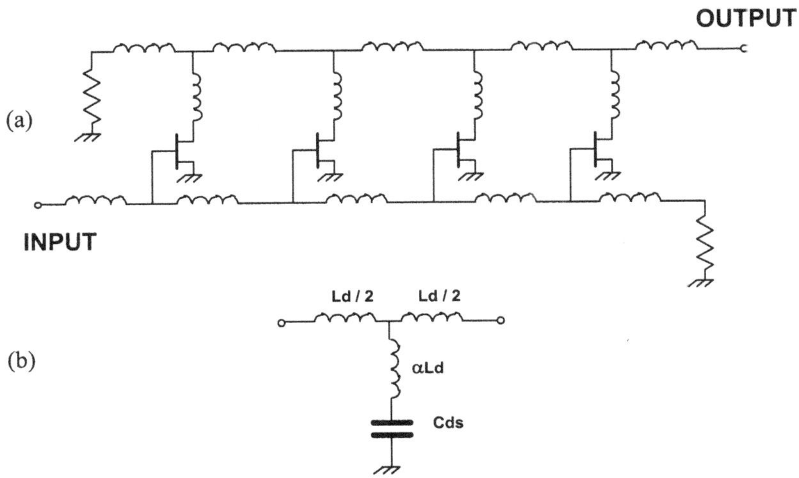

Figure 5.31 Gate and drain-line velocity compensation using an m-derived section

5.8.3 Distributed amplifier with constant-R networks

The constant-K network is the most obvious choice for the artificial transmission lines of the distributed amplifier, but it does not give the best performance. Chase and Kennan [22] demonstrated that constant-R networks could be used instead. The constant-R network requires a bridged-T element using a mutual inductor, bridging capacitor and the FET capacitance, as shown in Figure 5.32. The cut-off frequency for the constant-R network is given by:

$$f_c = \frac{\sqrt{2}}{\pi R_L C} \tag{5.26}$$

The advantage of this topology is that the cut-off frequency is increased by a factor of two, which yields a significant improvement in bandwidth. Furthermore, the constant-R network presents a conveniently constant resistance to the transistors, whereas the constant-K presents a reactive impedance which changes rapidly near the cut-off frequency.

The real beauty of the circuit presented by Chase and Kennan is that the mutual inductor and bridging capacitor required were implemented with a single spiral inductor structure: the coupling between turns provided the mutual inductance, and the normally unwanted crossover capacitance provided the bridging capacitance. In addition, the four-section amplifier used cascode FET pairs in order to achieve higher performance and used the capacitive coupling technique (described later in this chapter) to achieve more power. Figure 5.33 shows a microphotograph of a distributed amplifier employing constant-*R* networks.

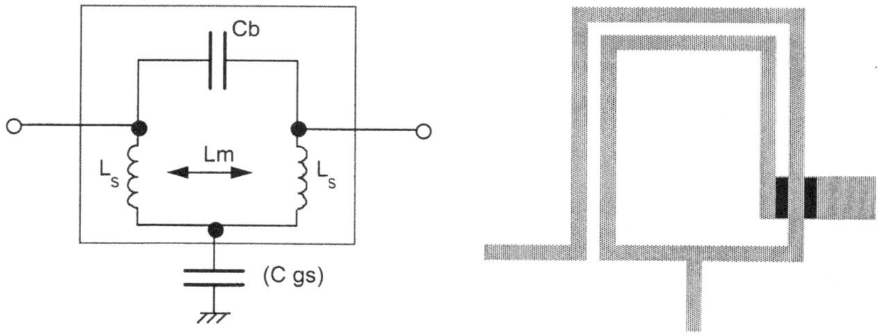

Figure 5.32 Constant-R network using a bridged-T element

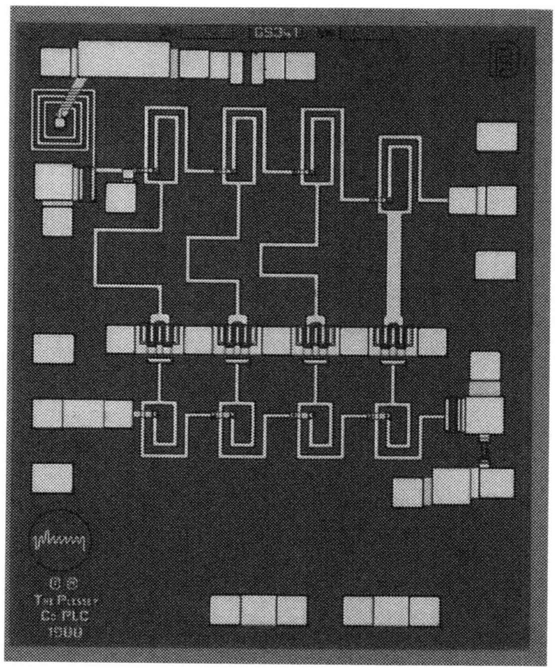

Figure 5.33 Microphotograph of a four-section 2 to 18 GHz distributed amplifier employing constant-R networks

5.8.4 Cascode distributed amplifier

The cascode arrangement of two FETs provides high gain, high output resistance and high reverse isolation. In addition, the maximum output voltage swing is higher than for a single FET because the common-gate FET's gate is grounded at AC. As a result of these advantages, the cascode FET is often used in distributed amplifiers, as shown in Figure 5.34, and this yields a significant improvement in performance with very little additional circuit area. Furthermore, the common-gate FET's gate bias may be used as an effective means of achieving gain control; Figure 5.35 compares the gain control range of a conventional distributed amplifier with that achievable from a cascode distributed amplifier. A small series inductance between the two FETs can provide a useful gain peaking mechanism to extend the bandwidth. Often, dual-gate FETs are used to implement the cascode FETs rather than two separate devices. This means that the increase in circuit area becomes negligible [23]. Dual-gate FETs in the cascode arrangement are routinely used as gain elements by many designers. However, care should be exercised regarding the stability of this device. In particular, grounding inductance in the common-gate FET can cause serious problems, and distributed amplifiers have been known to oscillate at frequencies above the cut-off frequency of the amplifier.

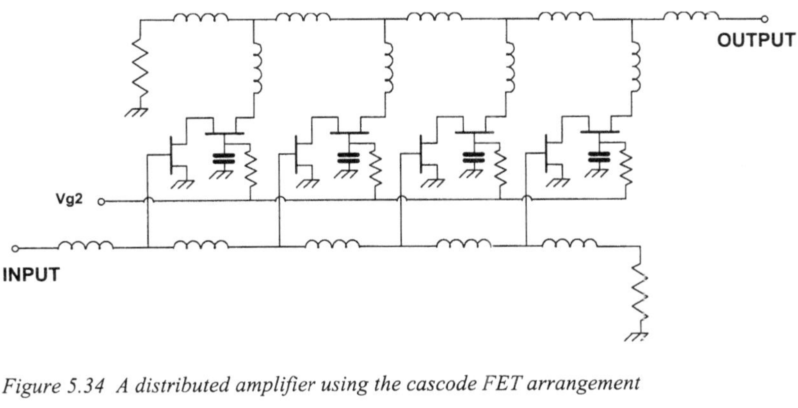

Figure 5.34 A distributed amplifier using the cascode FET arrangement

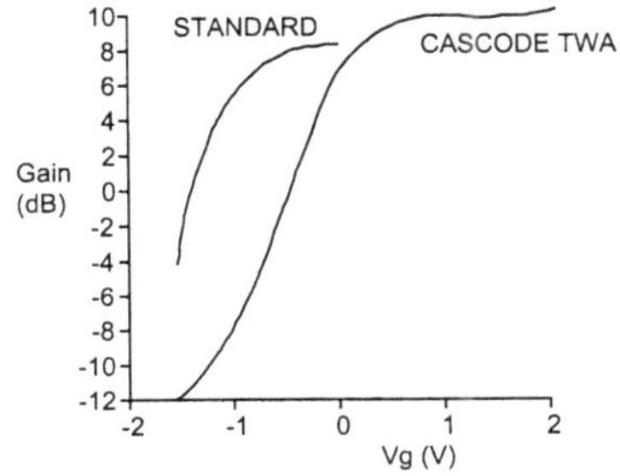

Figure 5.35 Comparison of the gain control range of a conventional and cascode distributed amplifier

5.8.5 Single-section distributed amplifier

The travelling-wave matching technique has been extended to the design of compact, broadband single-stage amplifiers by Minnis [24]. Whilst the distributed amplifier is attractive because of its excellent bandwidth and good port matches, a single-stage conventional distributed amplifier would have very low gain because of the n^2 dependence of the gain on the number of section; see eqn. (5.21). The first step towards increasing the gain is to use drain-line tapering (see Section 5.10.4.3) and, in fact, completely remove the drain-line termination so that all the drain current passes into the load. In addition, by increasing the size of the transistor, the larger g_m can compensate for the low gain. However, the increased capacitance will then lower the cut-off frequency of the conventional low-pass input and output networks. Minnis showed that this could be overcome by converting the low-pass networks to band-pass ones, using classical transformations for lumped-element filters. The result is an amplifier topology, which could be likened to the lossy matching technique. However, the reactive and lossy elements are treated as a single network, so that exact synthesis methods can be used.

5.8.6 Matrix distributed amplifier

The matrix distributed amplifier [25,26] consists of a number of 'tiers' of FETs connected into artificial transmission lines, as shown in Figure 5.36. The distinctive feature of this topology is that the signal is amplified from one tier to the next before the final output signal leaves the uppermost drain-line output. This means that both multiplicative gain (one transistor's output feeding the next's input) and additive gain (the summing of transistor outputs on the same drain-line) are provided within the single module. A cascade of conventional distributed amplifiers also provides additive gain and multiplicative gain, but multiple amplifiers are required.

If there are m tiers, and n sections in each tier, then there is an $m \times n$ matrix of active devices, and hence the name matrix amplifier. In practice, a two-tier matrix amplifier with four sections has very similar performance to a cascade of two conventional four-section distributed amplifiers. In fact, the cascade of conventional amplifiers can achieve slightly more bandwidth, possibly because the central artificial transmission lines of the matrix amplifier suffer from both gate and drain losses. However, the matrix topology is highly suitable for monolithic integration and a number of compact, high-performance realisations have been reported [27–30]. A further advantage of the matrix amplifier is that it can easily employ the stacked biasing technique to lower the current consumption.

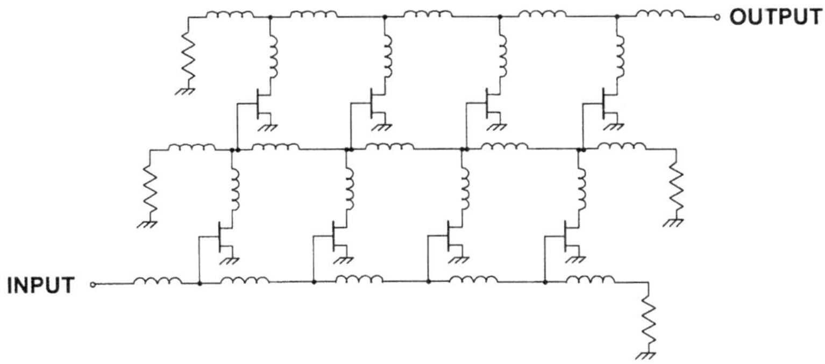

Figure 5.36 The circuit diagram of a matrix distributed amplifier with four sections and two tiers (bias circuitry not shown)

5.8.7 Practical design guidelines

The most important ingredient of a successful distributed amplifier design is a high performance transistor. The MMIC designer has complete freedom over the transistor geometry, the number of sections and the layout of the interconnecting lines, but the extent to which the basic transistor performance determines the maximum gain/bandwidth that can be achieved is remarkable. The transistors must be grounded effectively, and through-substrate via-holes are essential for a 20 GHz microstrip design. For operation into the millimetre-wave band, the inductance of these via-holes may become too high, and so CPW is an attractive alternative.

The inductors required for the artificial transmission lines are normally realised as narrow microstrip tracks (ribbon inductors). Spiral inductors tend to have too high a parasitic capacitance to be used, except for designs that are only required to operate up to 10 GHz or so. Indeed, even ribbon inductors have some capacitance to ground, and this will lead to a reduction in bandwidth compared with a simulation employing ideal inductors. This degradation can be reduced by making the tracks as narrow as possible. The minimum width is determined by the DC current handling requirement (which may be quite high for the drain-line), the fabrication limits, and the high series resistance encountered with very narrow tracks. The track lengths required usually necessitate the use of meandering. However, the effects of coupling between lines and the effects of bend discontinuities will both tend to add more capacitance into the circuit, further restricting the bandwidth. It is not uncommon for a design using straight tracks to start with a promising simulated performance of, say, 6 dB gain from 1 to 20 GHz, only for this to drop disappointingly to 5 dB gain up to 15 GHz once the tracks have been meandered and the resulting layout modelled properly. An important lesson is to not waste time optimising a circuit before these unavoidable discontinuities and coupling have been introduced into the simulation.

DC bias can be applied to the FET gates through the gate-line terminating resistor. However, the drain-line termination cannot normally be used for the drain biasing because the total current required by the FETs is rather high. Instead, an inductor bias choke network needs to be used. Because of the very wide frequency range of the distributed amplifier, it is difficult to find a single inductor that has both high inductance (for the low frequency end) and high self-resonant frequency (for the high frequency end). In order to extract the full bandwidth potential of the distributed amplifier, the drain bias network may have to be a fairly sophisticated low-pass filter structure, and a number of off-chip components may be required.

As an alternative, active loads have been demonstrated as a means of biasing distributed amplifiers [12]. This technique offers very wideband operation with negligible off-chip circuitry. However, it does increase the power consumption, lower the gain and increase the noise figure. Whichever biasing method is used, it is important to realise that the DC current in the drain-line is highest at one end, but steadily drops as it flows into successive transistors. Hence, although the total DC current may necessitate the use of a inconveniently wide line at one end of the drain-line, this width does not need to be maintained throughout the drain-line, and the tracks can be made narrower later on down the drain-line in order to improve performance.

5.8.8 Other circuit functions employing distributed amplifier techniques

The beauty of the distributed (or travelling-wave) amplifier technique is that impedance matching can easily be achieved over very wide bandwidths, such as 1 to 40 GHz. As a result, the technique has found applications in a wide range of other circuit functions such as mixers, baluns, frequency doublers, power splitters and combiners, switches, isolators and quasi-circulators, and impedance transformers. Distributed mixers [31–33], baluns [34–36], splitters [37] and combiners [38,39] are described in more detail in Chapter 7. The balanced frequency doubler [40] shown in Figure 5.37*a* employs common-source and common-gate FET pairs in order to achieve inherent cancellation of the input fundamental signal. The travelling-wave switch [41] has been demonstrated with DC to 20 GHz operation by incorporating multiple switching FETs in artificial transmission lines, as shown in Figure 5.37*b*.

Another application of the distributed amplifier structure is in the area of active isolators and unilateral circuits. These can make use of the reverse drain-line termination isolation properties [42], or employ a balanced circuit approach [43,44]. Figure 5.37*c* shows the topology of a wideband quasi-circulator using a distributed balun and combiner pair. Lastly, distributed amplifier impedance transformers using a number of topologies have been demonstrated [45]. It is feasible that different circuit functions can be integrated into a matrix structure, and Figure 5.37*d* illustrates a matrix distributed mixer which incorporates RF amplification, mixing and IF amplification functions in individual tiers [33].

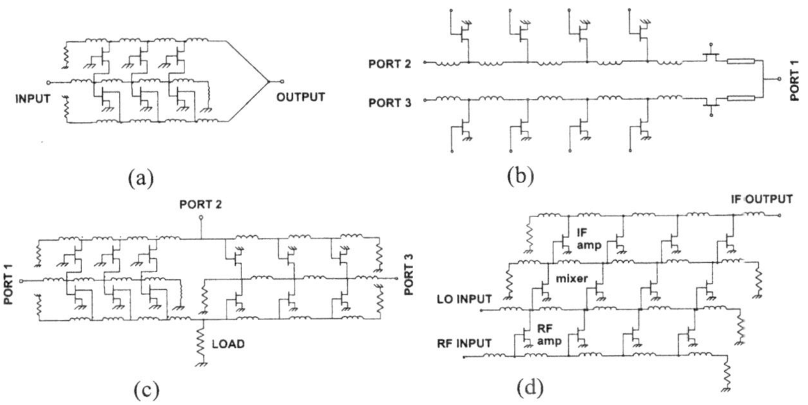

Figure 5.37 Distributed circuits: (a) balanced frequency doubler, (b) switch, (c) quasi-circulator and (d) matrix mixer

5.9 Active matching

This section describes a number of techniques in which the impedance matching problems are eased or overcome by using more than just a single common-source FET or common-emitter BJT.

5.9.1 Common-gate/common-source/common-drain amplifier

At low frequency, the input admittance of a common-gate FET can be approximated to:

$$Y_{in} = g_m + j\omega C_{gs} \tag{5.27}$$

For a transistor with $g_m = 20$ mS and a low value of C_{gs}, the input impedance of the common-gate configuration is, therefore, very close to 50 Ω at low frequencies. In a similar manner, the common-drain (or source follower) configuration can have an output impedance very close to 50 Ω at low frequencies. As a result, it is possible to construct a broadband amplifier with good input and output matches using a minimal number of passive matching elements. This amplifier topology, illustrated in Figure 5.38, uses a common-gate input stage followed by a common-source stage and a common-drain output stage. Resistors are employed both for DC biasing and for stabilisation: both the common-gate and common-drain stages are significantly more prone to oscillation than the common-source FET. The use of small inductances between stages provides a degree of gain-peaking to extend the high frequency response.

The advantage of this topology is its small size, but it does have quite high DC power consumption, and the noise figure of the common-gate configuration is significantly worse than the common-source FET. This type of amplifier can achieve operation from approximately 0.5 to 5 GHz with typical 0.5 μm MESFETs. Since the common-gate FET is non-inverting with an input impedance close to 50 Ω, whereas the common-source FET is inverting, with a high input impedance, common-gate and common-source FET pairs have found use in active baluns, and these are described in Chapter 7.

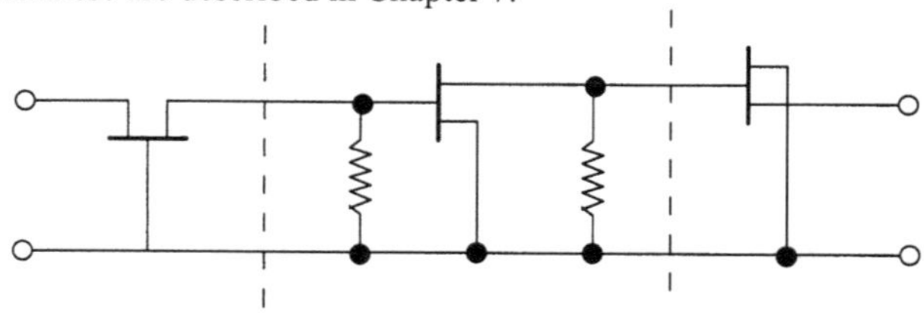

Figure 5.38 Actively matched amplifier using common-gate and common-drain FETs

5.9.2 Darlington pair

The Darlington pair has become a very familiar amplifier configuration through its use in the ubiquitous silicon bipolar modular MMIC amplifier (e.g. the MODAMP™, which stands for monolithic Darlington amplifier). The Darlington pair has a current gain equal to the product of the individual transistor current gains and gives an increased input impedance and larger S_{21} bandwidth. The circuit diagram of a typical Darlington pair amplifier is shown in Figure 5.39. The Darlington pair ensures a high gain-bandwidth product, and the resistor feedback ensures a smooth broadband response with good input and output matches. Emitter resistors are employed for stabilisation of the operating point and for series feedback. External decoupling capacitors and a bias inductor or resistor are often used to enable DC coupling or operation down to low frequency. With either silicon- or GaAs-based heterojunction bipolar transistors, this same amplifier topology is capable of achieving at least DC-to-20 GHz bandwidth [46]. The Darlington pair has also been used for broadband transimpedance amplifier design, for application in fibre-optic communications [47].

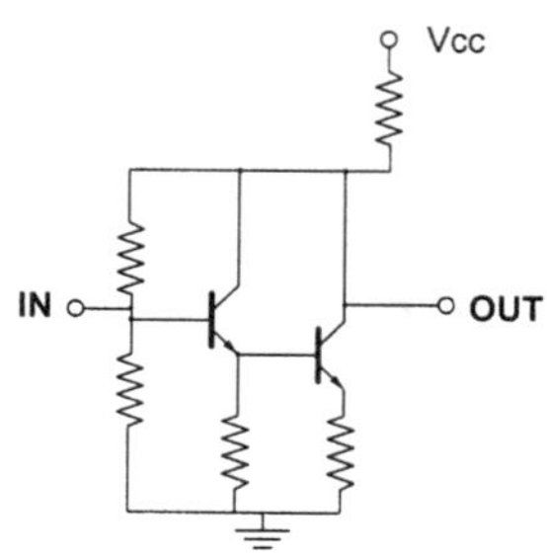

Figure 5.39 Darlington pair amplifier

5.9.3 DC-coupled amplifiers

Standard microwave amplifier design techniques require large numbers of passive components for matching and DC biasing. In contrast, 'low frequency' IC amplifier designs use transistors exclusively, in order to facilitate the use of a simple process technology and realise high packing density designs. Passive components such as capacitors, resistors and inductors are not used and are not part of the IC technology. Instead, DC-coupled amplifiers are designed using active loads for biasing, rather than inductors or resistors as bias chokes, and using level-shifting diode chains for inter-stage coupling, rather than using capacitors. The problem of impedance matching is avoided by the judicious use of small periphery devices and by employing common-gate or common-base input stages and source- or emitter-followers as buffer amplifiers. This type of amplifier design can be classed as precision analogue IC design. This is a huge field in its own right, and it is impossible to do this topic justice in the confines of this book. Nevertheless, the early work of Hornbuckle and Van Tuyl [48,49] and Estreich [50] should be highlighted, since it serves to demonstrate the great improvement in packing density that these techniques achieve compared to traditional microwave amplifier design approaches. Figure 5.40 shows the circuit diagram of one of Hornbuckle's single-stage amplifiers (with feedback). These techniques are limited to fairly low microwave frequencies because of the impedance matching limitations, but the high packing density means that this design approach can lead to very small (and therefore cheap) circuits. Hence, this type of design should be seriously considered for low cost applications such as mobile radio communications in the low-GHz frequency range. However, for situations where performance is critical, such as low noise and high power amplifiers, the traditional lumped-element matching approach is superior.

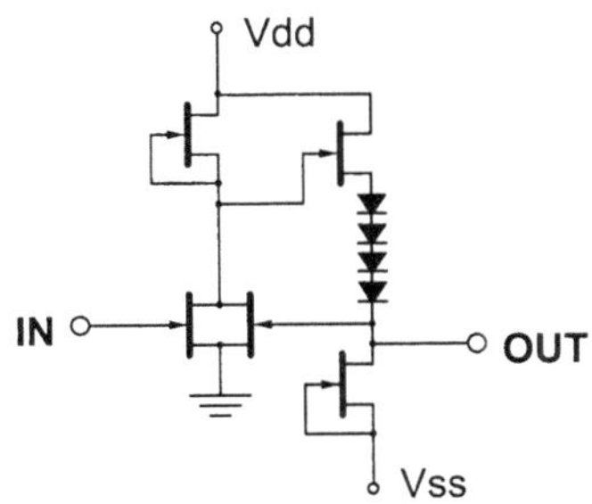

Figure 5.40 DC-coupled feedback amplifier stage used by Hornbuckle

5.10 Power amplifiers

In power amplifier design, the overriding concern is to ensure that the load impedance presented to the transistor gives the optimum load-line on the I–V characteristic, in order to maximise the voltage and current swings. This load-line, for a FET operating in class A, is illustrated in Figure 5.41. At the input, the gate voltage swing is limited by the pinch-off voltage and the forward barrier potential.

If the transistor has low gain it is this maximum swing that determines the output power. Otherwise, the drain voltage swing becomes the limitation, since this is limited by the knee voltage and the gate-drain breakdown voltage. Similar limits occur in the bipolar transistor. If the transistor structure and material parameters are optimised for maximum power, then it is the breakdown voltages, which ultimately limit the output power. In order to overcome the limited input and output voltage swing, a power transistor is a large periphery device which can handle a large current. For FETs, this means using the largest gate-width possible, and for bipolar transistors it means using the largest device area possible. In a reactively matched power amplifier, the transistor operates in a low impedance environment (less than a few ohms), and matching transformers are required to bring this up to the desired 50 Ω level. However, there are four major limitations to this:

1. the thermal problem of extracting heat
2. the problem of matching to the resulting low impedances
3. the yield problem
4. distributed effects; non-uniform power distribution within the device.

The thermal problem is compounded in any MMIC because a thick substrate is required for low loss passive elements, whereas a thin substrate is required for good heat dissipation. GaAs MMICs are even worse off because of the low thermal conductivity of GaAs. Perversely, because the HBT has achieved extraordinary power density, it suffers more from thermal limitations, and many quoted results are for pulsed operation. The matching problem is caused by the size limitation of MMICs: a hybrid MIC power amplifier can readily use (say) a 10 Ω microstrip transformer, even though it may be 8 mm wide. However, this is clearly not possible on a MMIC. The yield problem is encountered when a large power device is integrated onto a multi-function chip. There are a number of methods for easing the thermal and impedance matching limitations:

1. bath-tub vias (selective thinning of the wafer with a plated heatsink)
2. flip-chip mounting
3. power dividing/combining
4. cluster matching
5. off-chip components integrated into the package or MCM.

The bath-tub via process is important for multi-function MMICs because it means that the substrate is thick, so low loss passive components are available throughout the chip, and the substrate is only thinned underneath the power devices. Flip-chip mounting is common with discrete power FETs, and the technique has been extended to MMICs, although the passive elements would first need to be characterised when mounted in this way. Power combining and cluster matching are ways of distributing the signal among a number of smaller devices in order to overcome the problems encountered with a single large device. With power combining and cluster matching the signal is shared amongst a number of smaller devices, so that the heat is spread over a wider area and many via-holes can be used to improve heat dissipation.

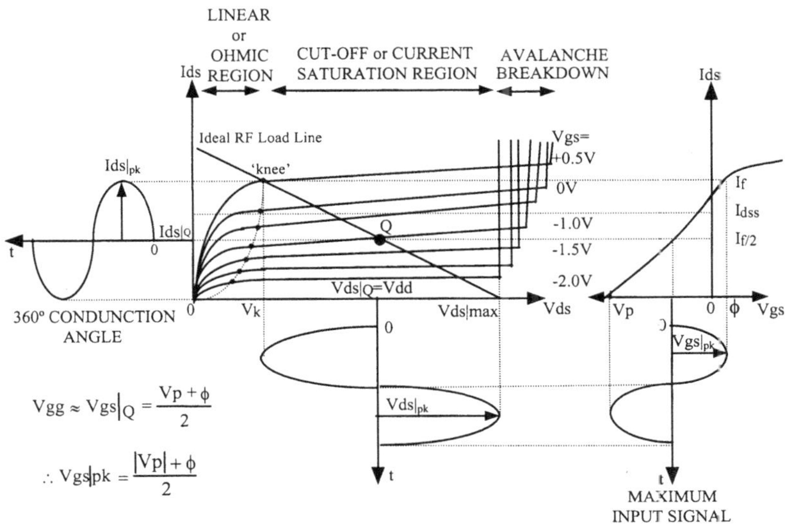

Figure 5.41 Class A load-line for an FET

5.10.1 Device characterisation

The most important device parameter is the output load impedance required for delivering maximum output power. This can be measured using the load-pull technique with either manual or automatic tuners. With manual tuners, the device is tuned for maximum output power, and the tuners are then removed and individually measured on a network analyser to determine the optimum source and load impedance. However, with manual tuners the optimum impedance may take some time to find, and there is no guarantee that the optimum point has been found. With automatic tuners under computer control, the contours of constant output power can be plotted on a Smith chart, and this has the advantage that the optimum point can be found with accuracy and a high level of confidence. The source impedance is also important because it will be quite different to the small-signal value. Once these optimum source and load impedances have been determined, the matching networks can be designed and the amplifier gain and stability checked using the small-signal *S*-parameters. This is the basic design procedure for a narrowband power amplifier.

Load-pull measurements are very time consuming and do not normally take into account the effects of the terminating impedances at the signal harmonics. Hence, it may be better to use a large-signal device model if one is available. Using computer simulations, the emulated load pull 'measurements' can be conducted to determine the optimum source and load impedances. In addition, the harmonic frequency terminations can be optimised to improve efficiency and linearity. However, it has to be said that large-signal models often lack sufficient accuracy to make a purely CAD-based power amplifier design wise. CAD is essential for broadband power amplifier design, but load-pull data should ideally still be used in order to verify the model accuracy.

5.10.2 Power combining and cluster matching

There are many well-known techniques for power dividing/combining [51], but many of these are unsuitable for MMIC power amplifiers: *N*-way in-phase Wilkinson power dividers/combiners are not entirely suitable for planar constructions. Fork dividers/combiners are difficult to realise with equal path-lengths. The serial power combining method requires couplers with a logarithmically varying sequence of coupling levels. This leaves three main methods:

1. corporate (or binary) power combining, with Wilkinson dividers/combiners (Figure 5.42*a*)
2. travelling-wave power combining, with Wilkinson dividers/combiners (Figure 5.42*b*)
3. dual-balanced amplifiers with quadrature couplers (Figure 5.42*c*).

As an example, Figure 5.43 shows a microphotograph of an X-band transmitter chip consisting of a VCO feeding a two-stage power amplifier. The output stage consists of two 1.5 mm MESFETs, power combined using lumped-element Wilkinson splitters and combiners.

Other methods include bus-bar combining [52,53], cluster matching and the distributed amplifier. However, with true power combining the individual amplifiers are isolated from one another due to the port isolations of the divider/combiners or couplers. This is not the case with device-level combining or with the distributed amplifier. This lack of isolation between devices leads to stability problems associated with power combining that must be thoroughly investigated. In general, an amplifier consisting of n devices connected in parallel, and even a single device with n fingers, is able to support n possible modes of oscillation. This problem has been addressed in detail by Freitag [54], using Z-parameter analysis with eigenvectors and eigenvalues, and by Ohtomo [55], using feedback loop analysis and Nyquist plots. The problem is akin to the analysis of multiple coupled lines. In the case of two devices, there is an even mode and an odd mode (also referred to as the push-pull or differential mode). The even mode is the desired one, in which the input and output voltages and currents of each transistor are in-phase. In the odd mode, however, anti-phase operation occurs and an internal feedback loop exists which may cause oscillation. The stability parameters calculated by standard CAD packages (K and B1 factor) only relate to the even mode, and do not take this internal feedback loop into account. Therefore, even when $K > 1$ and B1 is positive, the amplifier can still oscillate. This problem can be alleviated by placing bridging resistors across the inputs and/or outputs of the transistors; these have no effect on the even mode, since the terminal voltages are all equal and in-phase, but the odd-mode oscillation is strongly damped.

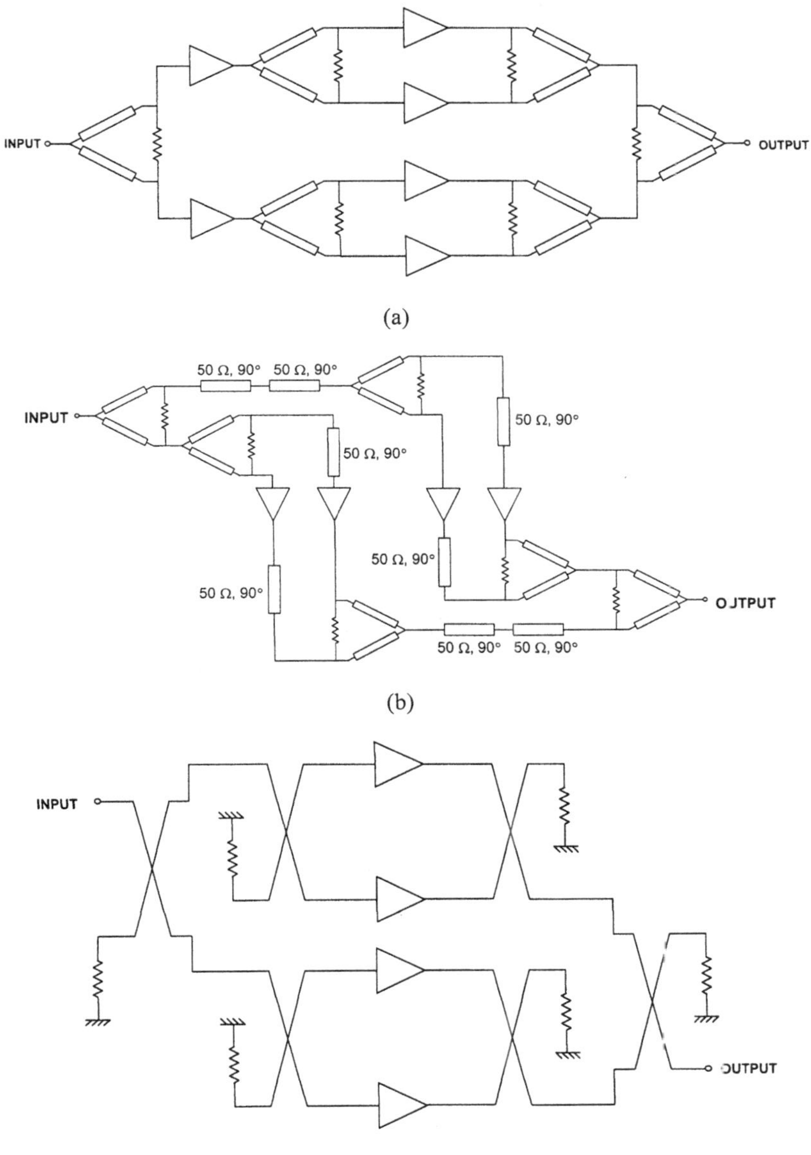

Figure 5.42 Power combining techniques: (a) corporate combining, with Wilkinson dividers and combiners, (b) travelling wave combining, with Wilkinson dividers and combiners and (c) dual-balanced amplifiers

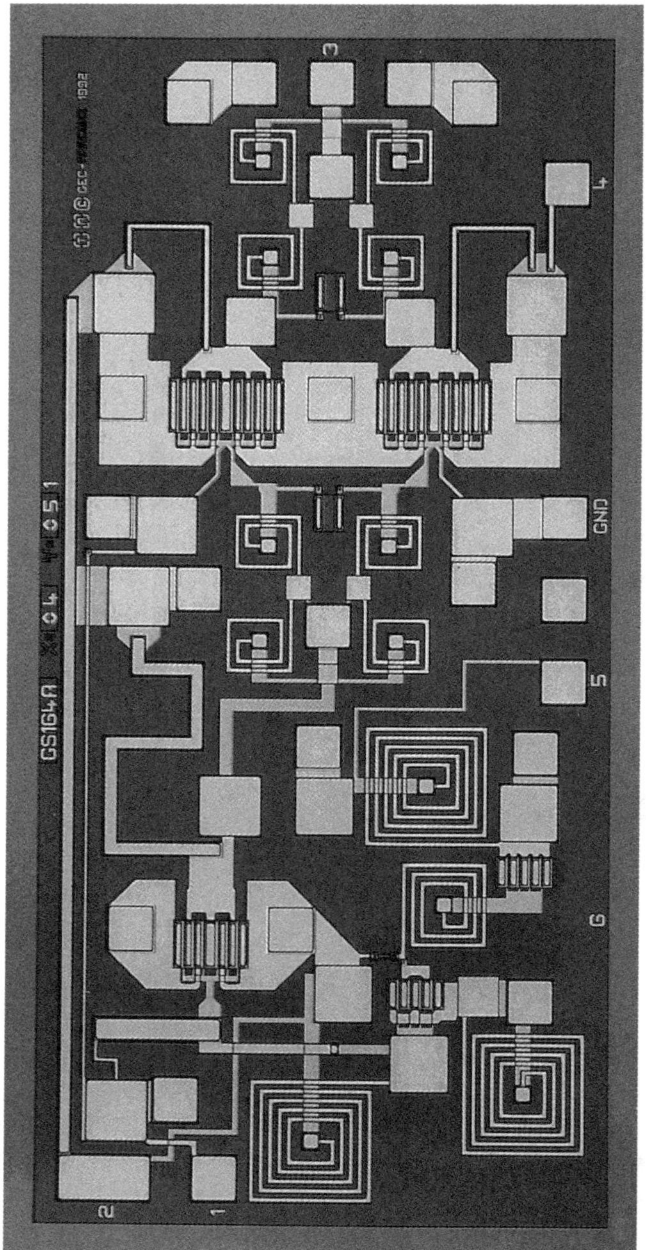

Figure 5.43 Microphotograph of an X-band transmitter chip

In the cluster matching technique [56,57] the amplifier is split-up into cells, and the cells are connected with input and output manifolds which incorporate impedance matching. In this way, the matching operation is carried out on clusters of small periphery devices, and extremes of impedance are avoided. In the enhanced method demonstrated by Pavlidis [58], the input manifold is replaced by an active splitting network. Each cell has one input and two (or more outputs) feeding the next stage cells. Once the required number of paralleled devices has been reached, the outputs are combined together again using an output manifold. Impedance matching takes place within the individual cells, so that low impedances are avoided. A schematic of this topology is shown in Figure 5.44. The advantage of this approach for MMIC power amplifiers is that the devices are spaced well apart for improved thermal dissipation, the power splitting/combining does not required large couplers, the impedance matching is simplified and the problem of unequal phasing is overcome. The use of active splitting means that the transistor inputs are isolated, typically by 20 dB, and this helps to prevent the problem of instability caused by internal feedback loops.

The relatively high loss of microstrip on MMICs means that the overall power efficiency of a power combined amplifier can be reduced considerably. For example, if four 2.5 W devices with individual efficiencies of 40% are combined using couplers with 0.5 dB loss each, then only 7.9 W is obtained. The other 2.1 W is lost as heat in the combining network and the efficiency drops to 32%. Not surprisingly, therefore, it is very common for the final output matching and combining to be performed off-chip with conventional microstrip lines. This has the additional benefit of overcoming the current handling limitation of monolithic microstrip tracks.

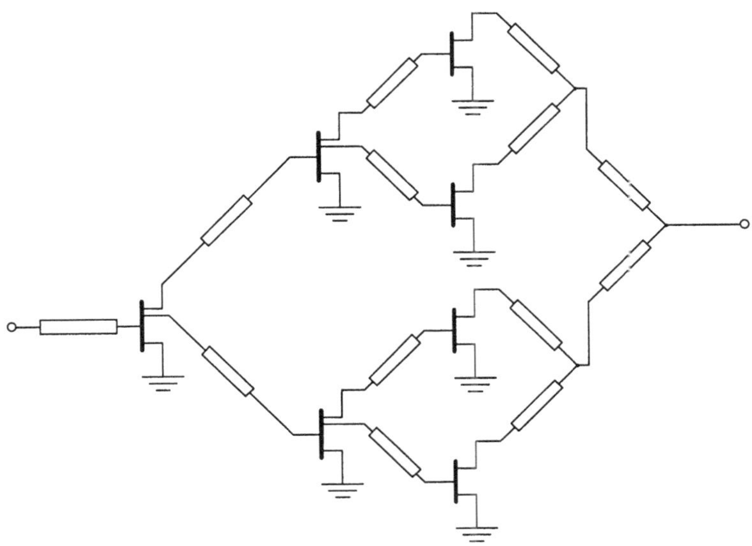

Figure 5.44 The enhanced cluster matching technique

5.10.3 Class B operation

So far, class A operation has been assumed, and this gives the best linearity. However, class A operation has a maximum theoretical efficiency of 50%, and 30% power-added efficiency is typical of most MMIC GaAs FET amplifiers. With the increasing importance of high efficiency power amplifiers for mobile communications and phased-array antennas, there has been a corresponding increase in the study of high efficiency amplifiers. Class C amplifiers, while giving better efficiency, are the most non-linear, and are not well suited for digital communication systems using varying-envelope signals. In class B, transistors are operated with almost zero quiescent current, which means that the efficiency is high, with a theoretical maximum efficiency of 78.5%.

In a push-pull amplifier, two transistors are employed in tandem: one amplifies the positive half cycles, and the other amplifies the negative half-cycles. At microwave frequencies it is not easy to employ complementary transistor pairs (*n-p-n*/*p-n-p* BJTs or *n*-channel/*p*-channel FETs) because of the large difference in the electron and hole mobilities. Hence, microwave class B amplifiers have to employ two identical transistors fed by complimentary signals generated by transformers or baluns. Figure 5.45 shows the schematic of a GaAs FET push-pull amplifier using planar spiral transformers. The major problem with the true class B amplifier is that 6 dB of the transistor available gain is lost because only half of the voltage waveform is amplified.

A second problem encountered with GaAs FET push-pull amplifiers is that the g_m of a standard device drops rapidly towards pinch-off. Hence, the device can either be operated with a small quiescent current (10 or 20% of I_{dss}), which is actually class AB operation, or the channel doping profile has to be redesigned specifically for class B operation. The doping density has to increase with depth so that g_m is kept high towards pinch-off. Unfortunately, this is the opposite requirement to that of low-noise FETs, where high doping at the surface is required for low contact resistance. Some excellent results have been achieved with class B and push-pull power amplifiers [59,60].

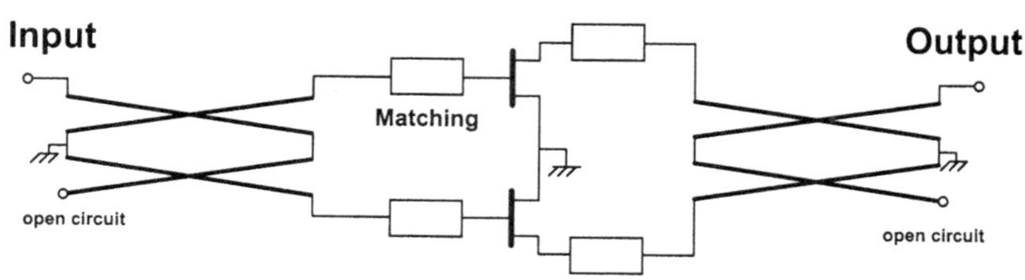

Figure 5.45 A push-pull FET amplifier using planar baluns

5.10.4 High power distributed amplifiers

The distributed amplifier suffers from a fundamental power limitation that is not encountered with reactively matched power amplifiers. In a reactively matched power amplifier, the transistor can operate in a low impedance environment in order to limit the voltage swing. However, in the distributed amplifier the minimum impedance of the gate- and drain-lines is determined by other factors, such as the gate- and drain-line loss. In the standard distributed amplifier, the first FET on the gate-line sees the 50 Ω source impedance almost directly, and the last FET on the drain-line sees the 50 Ω load impedance almost directly. As a result, with a typical FET breakdown voltage of 18 V, one can calculate that the output power cannot be more than about 400 mW, unless some enhancements are made to the topology. Unfortunately, whilst a number of circuit techniques can be used to improve the output power, fundamentally a distributed amplifier with n devices cannot achieve n times the power of a single device.

5.10.4.1 The capacitively-coupled distributed amplifier

A further limit to the achievable output power is the fact that large gate-width transistors, suitable for high power, have a high C_{gs}, which leads to a reduction in bandwidth. In addition, gate-line loss means that the first FET on the gate-line will receive a considerably greater input voltage swing than the others further down the gate-line. Hence, this FET will go into gain compression too early. An extremely neat solution to these limitations is to capacitively couple the FETs to the gate-line [61], as shown in Figure 5.46. This technique allows larger FETs to be employed without lowering the cut-off frequency of the gate-line. In addition, if the coupling capacitors are varied, the effect of gate-line loss can be accommodated so that each FET has the same gate voltage swing applied. In other words, the first FET has a relatively small coupling capacitor because the input signal is large, whereas the last FET on the gate-line is almost directly coupled to the gate-line with a large coupling capacitor. The penalty, however, with the capacitive coupling is that the gain is reduced because of the potential divider action across C_{gs} and the coupling capacitor. The coupling capacitors are shunted with high value resistors in order to allow the DC gate bias to reach the gates. This can impose a restriction on the high power performance, however, because the gates need to be fed from a low impedance DC supply for maximum saturated power.

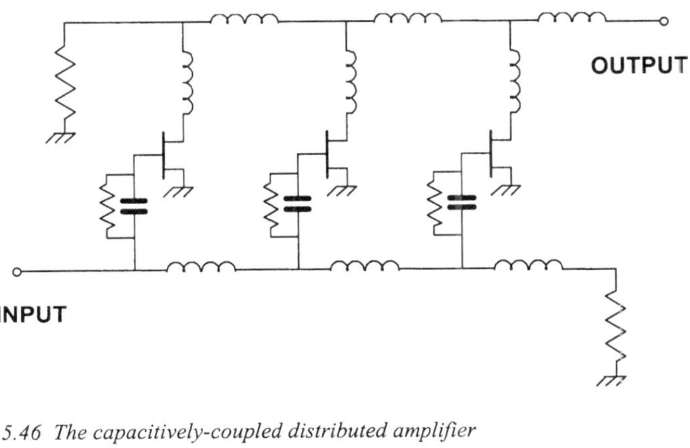

Figure 5.46 The capacitively-coupled distributed amplifier

5.10.4.2 The split gate-line distributed amplifier

It has already been stated that C_{gs} is much larger than C_{ds} and that a high gate-line cut-off frequency is hard to maintain with large gate-width FETs required for high power. A novel solution to this limitation is the split gate-line amplifier [62]. As shown in Figure 5.47, the input line splits into two separate gate-lines, but the FETs are all connected to the same drain-line. The individual gate-line cut-off frequencies can be kept high, since each FET has half the C_{gs} of the standard amplifier. The cut-off frequency of the drain-line is still high because C_{ds} is much smaller than C_{gs}. Hence, the technique allows a greater total gate-width to be used. Ideally, a matched splitter should be used to split the signal; clearly, if the two gate-lines are increased to a 100 Ω impedance level (in order to maintain an overall 50 Ω input) then some of the power advantage is lost because the voltage swing is fixed by the gate pinch-off voltage and the built-in barrier voltage.

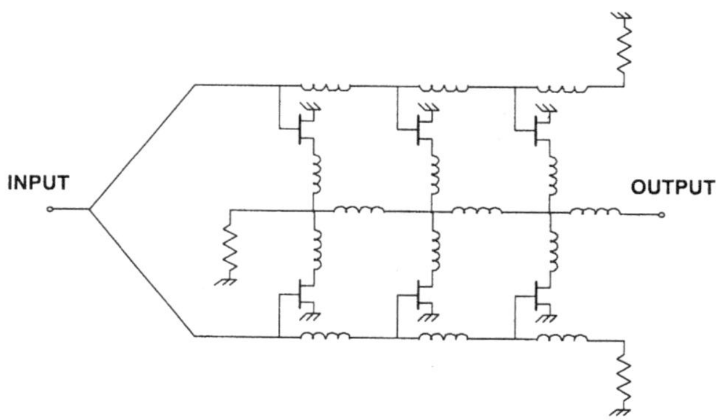

Figure 5.47 The split gate-line distributed amplifier

5.10.4.3 The tapered drain-line technique

This topology, shown in Figure 5.48, attempts to force the total FET output current at the drain-line reverse termination to zero and ensure that all the current from the FETs is fed in the forward direction only. This is accomplished by tapering the gate-widths of the FETs [63]. In the ideal case, with a properly designed tapered distributed amplifier, the first FET feeds a section of line with impedance Z_0, the second feeds a section of line with impedance $Z_0/2$, the third $Z_0/3$, and so on [1]. This arrangement means that the proportion of current produced by section $n-1$ that is reflected from the impedance change at section n will cancel the reverse current produced by the FET in section n. Since the drain-line impedance is decreasing rapidly, there is a limit to the number of stages that can be employed. For a 50 Ω system, the output impedance would need to be matched with an off-chip microstrip taper. However, an advantage of the low load impedance presented to the final FET in the drain-line is that it alleviates the problem of the limited voltage swing.

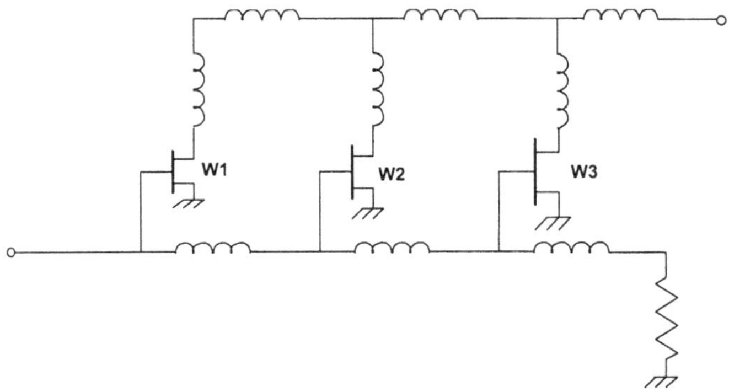

Figure 5.48 A distributed amplifier with a tapered drain-line

5.11 Low noise amplifiers

The noise factor, F, of an amplifier chain or receiver can be calculated by Friis' well-known formula:

$$F = F_1 + \frac{F_2 - 1}{G_1} + \frac{F_3 - 1}{G_1 G_2} + \frac{F_4 - 1}{G_1 G_2 G_3} + \ldots \tag{5.28}$$

where F_n is the noise factor of component n in the chain, and G_n is the power gain (not in dB) of component n in the chain. This formula shows that the noise figure (which is given by $10 \log F$) of the first stage is most critical, because the following stages' noise contributions are reduced by the gain of the preceding stages. (An important exception to this is a millimetre-wave receiver with no LNA: in this case, the overall noise figure is approximately equal to the mixer noise figure plus the IF amplifier noise figure.) The noise measure, M, of an amplifier gives the noise factor that would result from an infinite cascade of identical amplifiers, and is calculated from:

$$M = \frac{F - 1}{1 - 1/G} \tag{5.29}$$

The noise factor is given by:

$$F = 1 + M \tag{5.30}$$

This is, therefore, the lower limit on the overall noise factor. Generally, the first amplifier stage should be designed for optimum noise measure, rather than optimum noise figure, because the gain must be kept sufficiently high to minimise the noise contributions from the following stages. In a multi-stage LNA, the second and third stages can be designed for higher gain, with their noise figure becoming steadily less important. If the gain of a single stage is quite low (significantly less than 10 dB), which might well be the case for millimetre-wave amplifiers, then all three (or more) stages may need to be designed for optimum noise measure. The first step in LNA design is to calculate the number of stages needed and the noise figure and gain that is required from each particular stage.

5.11.1 Noise matching

Conjugate matching between the source and the input impedance of a transistor achieves maximum power transfer, and hence maximum gain, but this does not generally give minimum noise figure. Instead, there is an optimum impedance, Z_{opt} (or reflection coefficient, Γ_{opt}), which needs to be presented to the transistor input to achieve the minimum noise factor, F_{min}. It can be shown that as one moves away from this optimum noise matching point on the Smith chart, circles of constant noise figure are formed. The noise factor, F, with an arbitrary source impedance Γ_S can be expressed as:

$$F = F_{\min} + 4R_n \frac{\left|\Gamma_s - \Gamma_{opt}\right|^2}{\left(1 - \left|\Gamma_s\right|^2\right)\left|1 + \Gamma_{opt}\right|^2} \tag{5.31}$$

where R_n is known as the equivalent noise resistance of the device. It can be seen that R_n determines how rapidly the noise figure degrades as Γ_S moves away from Γ_{opt}. A small R_n means that the noise figure is less sensitive to the source impedance, making the circuit easier to design.

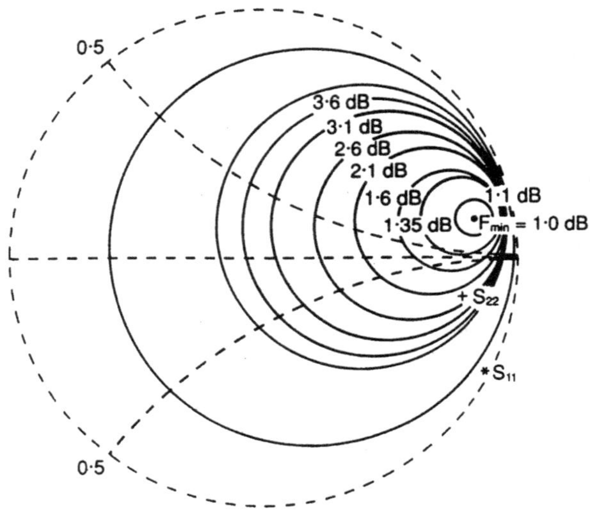

Figure 5.49 Noise figure circles for a 300μm MESFET at 1.6 GHz

Figure 5.49 shows the noise circles at 1.6 GHz for a 300 μm gate-width MESFET with 1 μm gate length. Note that Γ_{opt} is considerably different from S_{11}* (and hence from the simultaneous match input reflection coefficient). This implies that an amplifier designed for minimum noise figure must be mismatched to some extent from the maximum power transfer condition, unless the device has S_{11}* and Γ_{opt} very close or they are brought close together by using feedback. This poor input match problem has traditionally been overcome by using a balanced amplifier topology, and this has been described earlier.

5.11.2 Simultaneous match LNA

For an MMIC low noise amplifier, the balanced amplifier solution is rather unattractive because it more than doubles the chip area. A technique was proposed many years ago for using common lead inductance to improve the signal-to-noise ratio of triode valve circuits [64]. Later, Engberg [65] showed that a combination of series and shunt feedback, with suitable output loading, could be used to make the power matching and noise matching reflection coefficients converge. He reported that the noise measure was unaffected by these lossless feedback elements. In 1985, Lehmann and Heston [66,67] first demonstrated that this series feedback technique could be applied to monolithic low noise amplifier design.

When applied to a FET, the principle of the technique is that a carefully controlled value of source inductance can be used to move S_{11}* and Γ_{opt} closer together, resulting in an LNA with simultaneous noise and power match. With reference to Figure 5.50, it can be shown that:

$$Z_{in} = (R_{ch} + R_{ds} + \frac{g_m L_S}{C_{gs}}) + j\omega L_s + \frac{g_m R_s}{j\omega C_{gs}} + \frac{1}{j\omega C_{gs}} \tag{5.32}$$

It is found that S_{11}* varies with both the output loading and the series feedback, whereas Z_{opt} is only affected by the series feedback. As a result, it is possible to find a combination of feedback inductance and output loading which results in S_{11}* and Γ_{opt} being equal. It is actually possible that the noise figure might improve at the expense of the gain. In a practical design, the gate-width of the device is an important parameter that controls the proximity of S_{11}* and Γ_{opt}. For each different frequency of application, a different gate-width is optimum. In any LNA design, the use of resistors for biasing or stabilisation must be avoided because of their thermal noise contribution. Furthermore, in MMIC LNA design it is often found that the resistance of spiral inductors can introduce a significant noise penalty. Unfortunately, at low frequencies the simultaneous match LNA may require a spiral inductor of one or two turns for the series feedback, and the resistance of this inductor can cause a significant degradation in noise figure.

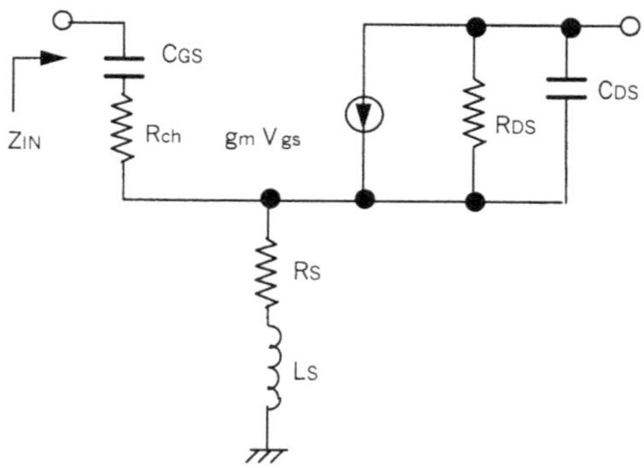

Figure 5.50 Equivalent circuit of the series feedback FET LNA

5.12 Summary

The first part of this chapter reviewed the theory and analysis of microwave amplifiers in terms of the device's two-port *S*-parameters. This provides the foundations for the understanding of the problems of impedance matching and stability, and demonstrates that amplifier design is far more involved than simply matching to S_{11} and S_{22}. The practical implications of the theoretical analysis have been identified, and a wide range of matching techniques and DC biasing circuits have been described. The key MMIC amplifier topologies have been studied, and the design of high-power and low-noise amplifiers has been discussed. The remaining references [68–126] provide specific details of a number of MMIC amplifier designs, as well as some further general information on amplifiers.

To conclude the chapter, Table 5.1 presents a comparison of the features of each of the major MMIC amplifier topologies: the general advantages and disadvantages of each amplifier type are listed, along with an indication of the typical frequency range that they can cover. This summary table should assist in the first and most important step in MMIC amplifier design, which is choosing the most suitable topology. Figure 5.51 shows the microphotograph of a 6–18 GHz 5 W power amplifier MMIC, designed at the Defence and Evaluation Research Agency (DERA), UK. This chip, which produced true state-of-the-art results when first reported [127], uses a combination of many of the techniques described in this chapter in order to achieve this level of performance: firstly a high technology 0.25 μm PHEMT process (Texas Instruments) was chosen in order to provide high power and high gain-per-stage. Secondly, the amplifier was subdivided into a pair of 2.5 W channels, in order to improve the chances of a first pass design success and enable lower loss external power combining to be used. The first stage is a three-section distributed amplifier, chosen because it comfortably achieves 6 to 18 GHz bandwidth with a good input match, and designed to provide a positive gain slope to compensate for the gain roll-off of the PHEMTs over the design bandwidth. The second stage uses a pair of 1.2 mm periphery devices, and the final stage uses four 1.2 mm devices. Corporate splittter/combiner networks are employed, and resistors are strapped across the gates and drains in order to prevent odd-mode oscillations.

The inter-stage matching network uses the lossy matching technique in order to provide flat gain. The output network was designed for maximum power matching conditions using experimental load-pull data. The complete design was simulated using both linear and non-linear CAD, and electromagnetic simulation was employed for the matching networks. The complete amplifier achieves 20 dB gain and 3.8 W power (CW) across the whole 6-18 GHz band.

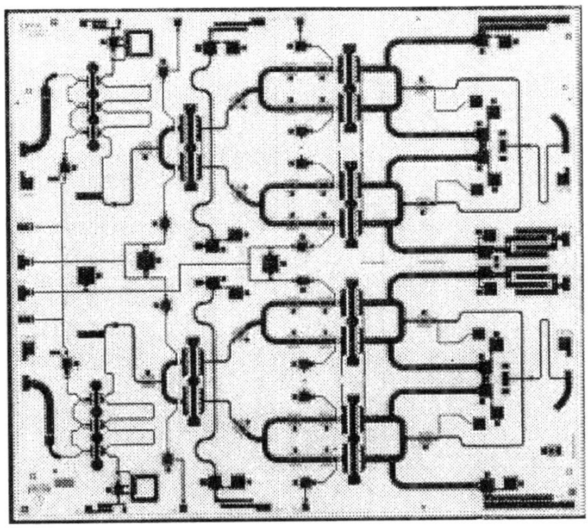

Figure 5.51 Microphotograph of a 6–18 GHz 5 W power amplifier MMIC (Courtesy of QinetiQ: Crown copyright)

Table 5.1 Comparison of the key MMIC amplifier toplogies

Amplifier type	Advantages	Disadvantages	Typical bandwidth
Reactively matched	Best gain per stage, noise figure and output power	Difficult to control the input and output VSWR: Poor gain flatness when cascaded. Hard to achieve unconditional stability. Very sensitive to process variations	Typically 20 % bandwidths at any centre frequency from 1 GHz to over 100 GHz (device permitting). Octave bandwidths just feasible
Lossy matched	Good VSWRs and gain flatness: Broadband design easier, and readily cascadable	Reduced gain per stage compared with the reactive match. Some degradation in noise figure and output power	Octave bandwidths readily achievable. Frequency range again determined by device performance
Feedback	Excellent gain flatness over a wide frequency range, good stability, and less sensitive to transistor spreads	Noise figure is mediocre and gain is reduced significantly. Tends to need large gate-widths, leading to high DC power	0.1 to 6 GHz (MESFET) 6 to 18 GHz (MESFET) 20 to 50 GHz (pHEMT)
Distributed	Ultra-wide bandwidths achievable (more than one decade). Good VSWRs, easy to cascade	Low gain, mediocre noise figure and limited output power. High DC power. Large total gate-width required	1 to 20 GHz (MESFET) 1 to >50 GHz (pHEMT) 1 to > 100 GHz (LM-HEMT)
Darlington pair	Very simple topology with small chip size. Broadband, with quite high gain per stage and good VSWRs. Can be DC coupled	Mediocre noise figure and fairly high DC power	DC to >5 GHz (Si bipolar) DC to >20 GHz (HBT)
Actively matched	A simple topology for broadband amplifiers at lower frequencies. Small chip size	Poor noise figure and high DC power requirement	1 to 6 GHz (MESFET)
DC coupled (OP-AMP style)	The highest packing density: No passive elements required. Ideally suited to BiCMOS technology. Highly competitive for mobile radio applications	Limited noise figure and power handling performance compared with traditional techniques	DC to >5 GHz This technique is generally best suited to bipolar transistors

5.13 References

1 VENDELIN, G. D., PAVIO, A. M., and ROHDE, U. L.: 'Microwave circuit design using linear and nonlinear techniques' (John Wiley and Sons, 1990)

2 PENGELLY, R. S.: 'Microwave field effect transistors – Theory, design and applications' (Research Studies Press, 1986)

3 LIAO, S.: 'Microwave circuit analysis and amplifier design' (Prentice Hall, 1987)

4 SOARES, R. (Ed.): 'GaAs MESFET Circuit Design' (Artech House, 1988)

5 YIP, P. C. L: 'High-frequency circuit design and measurements' (Chapman and Hall, 1990)

6 POZAR, D. M: 'Microwave Engineering' (Addison Wesley, 1990)

7 YOUNG, G. P., and O' SCANLAN, S.: 'Matching network design studies for microwave transistor amplifiers', *IEEE Transactions on Microwave Theory and Techniques*, 1981, pp. 1027–1034

8 VILLAR, J. C., and PEREZ, F.: 'Graphic design of matching and interstage lossy networks for microwave transistor amplifier, *IEEE Transactions on Microwave Theory and Techniques*, 1985, pp. 210–215

9 KHATIBZADEH, A. *et al.*: 'High-efficiency X-band HBT power amplifier', *Proceedings of IEEE microwave and millimeter-wave monolithic circuits symposium*, 1994, pp. 117–120

10 SCHUMACHER, H., ERBEN, U., and GRUHLE, A.: 'Low-noise performance of SiGe heterojunction bipolar transistors', *IEEE Microwave and Millimeter-wave Monolithic Circuits Symposium*, 1994, pp. 213–216

11 ROLLETT, J.: 'Stability and power gain invariants of linear two-ports', *IRE Transactions on Circuit Theory*, 1962, **CT-9**, pp. 29–32

12 ROBERTSON, I. D., and AGHVAMI, A. H.: 'Ultrawideband biasing of MMIC distributed amplifiers using an improved active load', *Electronics Lettters*, 1991, **27** (21), pp. 1907–1908

13 NICLAS, K. B.: 'On the design and performance of lossy match GaAs MESFET amplifiers', *IEEE Transactions on Microwave Theory and Techniques*, 1982, pp. 1900–1906

14 ULRICH, E.: 'Use negative feedback to slash wideband VSWR', *Microwave Journal*, 1978, pp. 66–70

15 NICLAS, K. B., WILSER, W. T., GOLD, R. B., and HITCHENS, W. R.: 'The matched feedback amplifier: ultra-wideband microwave amplification with GaAs MESFETs', *IEEE Transactions on Microwave Theory and Techniques*, 1980, pp. 285–294

16 RIGBY, P. N., SUFFOLK, J. R., and PENGELLY, R. S.: 'Broadband monolithic low-noise feedback amplifiers', *IEEE Microwave and Millimeter-wave Monolithic Circuits Symposium*, 1983, pp. 71–75

17 BEALL, J. M., NELSON, S. R., and WILLIAM, R. E.: 'Design and process sensitivity of a two-stage 6-18 GHz monolithic feedback amplifier', *IEEE Transactions on Microwave Theory and Techniques*, 1985, **MTT-33**, pp. 1567–1571

18 PERCIVAL, W.S.: 'Thermionic valve circuits'. British patent 460562, 1936

19 STRID, E. W., and GLEESON, K. R.: 'A DC-12 GHz monolithic GaAs FET distributed amplifier', *IEEE Transactions on Microwave Theory and Techniques*, 1982, **MTT-30** (7), pp. 969–975

20 AYASLI, Y., MOZZI, R. L., VORHAUS, J. L., REYNOLDS, L. D., and PUCEL, R. A.: 'A monolithic GaAs 1-13 GHz traveling-wave amplifier', *IEEE Transactions on Microwave Theory and Techniques*, 1982, **MTT-30**, pp. 976–980

21 BEYER, J. B., PRASAD, S. N., BECKER, R. C., NORDMAN, J. E., and HOHENWARTER, G. K.: 'MESFET distributed amplifier design guidelines', *IEEE Transactions on Microwave Theory and Techniques*, 1984, **MTT-32**, pp. 268–275

22 CHASE, E. M., and KENNAN, W.: 'A power distributed amplifier using constant-R networks, *IEEE MTT-S International Microwave Symposium Digest*, 1986, pp. 811–815

23 KENNAN, W., ANDRADE, T., and HUANG, C. C.: 'A 2-18 GHz monolithic distributed amplifier using dual-gate GaAs FETs', *IEEE Transactions on Microwave Theory and Techniques*, 1984, pp. 1693–1697

24 MINNIS, B. J.: 'The traveling wave matching technique for cascadable MMIC amplifiers', *IEEE Transactions on Microwave Theory and Techniques*, 1994, **MTT-42**, pp. 690–692

25 NICLAS, K. B., and PEREIRA, R. R.: 'The matrix amplifier: A high-gain module for multioctave frequency bands', *IEEE Transactions on Microwave Theory and Techniques*, 1987, **MTT-35**, pp. 296–306

26 NICLAS, K. B., PEREIRA, R. R., and CHANG, A. P.: 'A 2-18 GHz low-noise/high-gain amplifier module', *IEEE Transactions on Microwave Theory and Techniques*, 1989, **MTT-37**, pp. 198–207

27 CHANG, A. P., NICLAS, K. B., CANTOS, B. D., and STRIFLER, W. A.: 'Design and performance of a 2-18 GHz monolithic matrix amplifier', *IEEE MTT-S International Microwave Symposium Digest*, 1989, pp. 287–289

28 CHANG, A. P., NICLAS, K. B., and CANTOS, B. D.: 'Monolithic 2-18 GHz matrix amplifier', *IEEE Transactions on Microwave Theory and Techniques*, 1989, **MTT-37**, pp. 2159–2162

29 CHU, S. L., TAJIMA, Y., COLE, J. B., PLATZKER, A., and SCHINDLER, M. J.: 'A novel 4-18 GHz monolithic matrix distributed amplifier', *IEEE MTT-S International Microwave Symposium Digest*, 1989, pp. 291–295

30 SIMON, K. M., CHU, S. L. G. and WOLHERT, R.: 'A millimeter-wave monolithic matrix distributed amplifier', *Proceedings of IEEE Microwave and Millimeter-wave Monolithic Circuits Symposium*, 1992, pp. 73–75

31 TANG, O. S. A., and AITCHISON, C. S.: 'A practical microwave travelling wave MESFET Gate Mixer', *IEEE MTT-S International Microwave Symposium Digest*, 1985 , pp. 605–608

32 HOWARD, T. S., and PAVIO, A. M.: 'A dual-gate 2-18 GHz monolithic FET distributed mixer', *IEEE Microwave and Millimeter-wave Monolithic Circuits Symposium Digest*, 1987, pp. 27–30

33 ROBERTSON, I. D., and AGHVAMI, A. H: 'A novel 1 to 15 GHz matrix distributed mixer', *Proceedings of 21st European Microwave Conference, Stuttgart*, pp. 489–494

34 PAVIO, A. M., *et al.*: 'Double balanced mixers using active and passive techniques', *IEEE Transactions on Microwave Theory and Techniques*, 1988, **MTT-36**, pp. 1948–1957

35 ROBERTSON, I. D., and AGHVAMI, A. H: 'A novel wideband MMIC active balun', *Proceedings of 20th European Microwave Conference, Budapest, Hungary*, 1990

36 ROBERTSON, I. D., READER, H. C., and AGHVAMI, A. H: 'Operating modes in wideband monolithic distributed FET mixers', *Proceedings of IEEE Asia-Pacific Microwave Conference, Adelaide*, 1992, pp. 747–750

37 BARTA, G. *et al.*: 'Surface-mounted GaAs active splitter and attenuator MMICs used in a 1-10 GHz leveling loop', *IEEE Transactions on Microwave Theory and Techniques*, 1986, **MTT-34** (12), pp. 1569–1575

38 GERARD, R. E. J.: 'Multisignal amplification', U.S. Patent no. 4423386, Marconi Company Limited, Chelmsford, England

39 ROBERTSON, I. D., and AGHVAMI, A. H.: 'A practical distributed FET mixer for MMIC applications', *IEEE MTT-S International Microwave Symposium Digest*, 1989, pp. 1031–1032

40 PAVIO, A. M., *et al.*: 'A distributed broadband monolithic frequency multiplier', *IEEE MTT-S International Microwave Symposium Digest*, 1988, pp. 503–504

41 TAJIMA, Y. *et al.*: 'Broadband GaAs FET 2 × 1 switches', *IEEE GaAs IC Symposium Digest*, 1984, pp. 81–84

42 LEISTEN, O. P., COLLIER, R. J., and BATES, R. N.: 'Distributed amplifiers as duplexer/low crosstalk bidirectional elements in S-band', *Electronics Letters*, 1988, **24** (5), pp. 264–265

43 ROBERTSON, I. D., and AGHVAMI, A. H: 'Novel monolithic ultra-wideband unilateral 4-port junction using distributed amplification techniques', *IEEE MTT-S International Microwave Symposium Digest*, 1992, pp. 1051–1054

44 ROBERTSON, I. D., and AGHVAMI, A. H: 'A novel reflectometer using a wideband monolithic active unilateral 4-port junction', *Proceedings of 23rd European Microwave Conference*, Madrid, 1993, pp. 296–298

45 CIOFFI, K. R.: 'Active broadband impedance transformations using distributed techniques', *IEEE MTT-S International Microwave Symposium Digest*, 1989, pp. 1043–1046.

46 KOBAYASHI, K. W. *et al.*: 'GaAs HBT MMIC broadband amplifiers from DC to 20 GHz', *Proceedings of IEEE Microwave and Millimeter-wave Monolithic Circuits Symposium*, 1990, pp. 19–22

47 NAGANO, N. *et al.*: 'Monolithic ultra-broadband transimpedance amplifiers using AlGaAs/GaAs heterojunction bipolar transistors', *IEEE Transactions on Microwave Theory and Techniques*, 1994, **MTT-42**, pp. 2–9

48 HORNBUCKLE, D. P.: 'GaAs ct-coupled amplifiers', *Proceedings of IEEE International Microwave Symposium*, 1980, pp. 387–389

49 HORNBUCKLE, D. P., and VAN TUYL, R. L.: 'Monolithic GaAs direct-coupled amplifiers', *IEEE Transactions on Electron Devices*, 1981, pp. 175–182

50 ESTREICH, D. B.: 'A monolithic wide-band GaAs IC amplifier', *IEEE Journal of Solid-State Circuits*, 1982, **SC-17** (6), pp. 1166–1173

51 WALKER, J. L. B.: 'High-power GaAs FET amplifiers' (Artech House, 1993)

52 MARSH, S. P., LAU, D.K.Y., SLOAN, R., and DAVIS, L.E.: 'Design and analysis of an X-band MMIC bus-bar power combiner', *1999 EDMO Symposium Digest (Cat. No. 99TH8401)*, pp. 164–169

53 MARSH, S. P.: 'MMIC power splitting and combining techniques', *IEE Tutorial Colloquium on MMIC Design*, London, 1997, pp. 6/1–7

54 FREITAG, R. G.: 'A unified analysis of MMIC power amplifier stability', *IEEE MTT-S International Microwave Symposium Digest*, 1992, pp. 297–300

55 OHTOMO, M.: 'Stability analysis and numerical simulation of multidevice amplifiers', IEEE *Transactions on Microwave Theory and Techniques*, 1993, **MTT-41** (6/7), pp. 983–991

56 SCHELLENBERG, J. M., and YAMASAKI, H.: 'A new approach to FET power amplifiers', *Microwave Journ*al, 1982, pp. 51–66

57 DEGENFORD, J. E., FREITAG, R. G., BOIRE, D. C., and COHN, M.: 'Broadband monolithic MIC power amplifier development', *Microwave Journal*, 1982, pp. 89–97

58 PAVLIDIS, D. *et al.*: 'A new specifically monolithic approach to microwave power amplifiers', *Proceedings of IEEE Microwave and Millimeter-wave Monolithic Circuits Symposium*, 1983, pp. 54–58

59 LANE, J. R. *et al.*: 'High efficiency 1-, 2-, and 4-W class-B FET power amplifiers', *IEEE Transactions on Microwave Theory and Techniques*, 1986, **MTT-34** (12), pp. 1318–1326

60 HENRY, H. G. *et al.*: 'A compact 3W X-band GaAs MMIC amplifier based on a novel multi-push-pull circuit concept', *IEEE GaAs IC Symposium Digest*, 1991, pp. 327–330

61 AYASLI, Y., MILLER, S. W., MOZZI, R. L., and HANES, L. K.: 'Capacitively coupled traveling-wave power amplifier', *IEEE Transactions on Microwave Theory and Techniques*, 1984, pp. 1704–1709

62 AYASLI, Y., REYNOLDS, L. D., MOZZI, R. L., and HANES, L. K.: 2-20 GHz GaAs traveling-wave power amplifier', *IEEE Transactions on Microwave Theory and Techniques*, 1984, **MTT-32** (3), pp. 290-294

63 JONES, K. E., BARTA, G. S., and HERRICK, G. C.: 'A 1 to 10 GHz tapered distributed amplifier in a hermetic surface mount package', *IEEE MTT-S International Microwave Symposium Digest*, 1985, pp. 137–140

64 STRUTT, M. J. O., and VAN DER ZIEL, A.: 'Suppression of spontaneous fluctuations in amplifiers and receivers for electrical communication and for measuring devices', *Physica*, 1942, **IX** (6), pp. 513–538

65 ENGBERG, J.: 'Simultaneous input power match and noise optimization using feedback', *Proceedings of European Microwave Conference*, 1974, pp. 385–389

66 LEHMAN, R. E., and HESTON, D. D.: 'X-band monolithic series feedback LNA', *IEEE MTT-S International Microwave Symposium Digest*, 1985, pp. 51–54

67 LEHMAN, R. E., and HESTON, D. D.: 'X-band monolithic series feedback LNA', *IEEE Transactions on Microwave Theory and Techniques*, 1985, **MTT-33**, pp. 1560–1566

68 NICLAS, K. B.: 'Reflective match, lossy match, feedback and distributed amplifiers: A comparison of multi-octave performance characteristics', *IEEE MTT-S International Microwave Symposium Digest*, 1984, pp. 215–217

69 NICLAS, K. B.: 'Multi-octave performance of single-ended microwave solid-state amplifiers', *IEEE Transactions on Microwave Theory and Techniques*, 1984, **MTT-32**, pp. 896–908

70 NICLAS, K. B.: 'Noise in broad-band GaAs MESFET amplifiers with parallel feedback', *IEEE Transactions on Microwave Theory and Techniques*, 1982, pp. 63–70

71 NICLAS, K. B.: 'The exact noise figure of amplifiers with parallel feedback and lossy matching circuits', *IEEE Transactions on Microwave Theory and Techniques*, 1982, **MTT-30**, pp. 832–834

72 NICLAS, K. B., WINSER, W. T., KRITZER, T. R., and PEREIRA, R. R.: 'On theory and performance of solid-state microwave distributed amplifiers', *IEEE Transactions on Microwave Theory and Techniques*, 1983, **MTT-31**, pp. 447–456

73 KURDOGHLIAN, A. *et al.*: 'High-efficiency InP-based HEMT MMIC power amplifier', *IEEE GaAs IC Symposium Digest*, 1993, pp. 375–377

74 FUNABASHI, M., and HOSOYA, K.: 'High gain V-band heterojunction FET MMIC power amplifiers', *IEEE GaAs IC Symposium Digest*, 1993, pp. 379–382

75 ZDEBEL, P. J.: 'Current status of high performance silicon bipolar technology', *IEEE GaAs IC Symposium Digest*, 1992, pp. 15–18

76 TAN, K. L. *et al.*: 'A manufacturable high performance 0.1-μm pseudomorphic AlGaAs/InGaAs HEMT process for W-band MMICs', *IEEE GaAs IC Symposium Digest*, 1992, pp. 251–253

77 WU, C. S. *et al.*: 'High efficiency X-band power HBTs', *IEEE GaAs IC Symposium Digest*, 1992, pp. 259–262

78 HO, W. J., WANG, N. L., and CHANG, M. F.: 'Producibility and performance of the microwave power HBT', *IEEE GaAs IC Symposium Digest*, 1992, pp. 263–266

79 LE, H. M. *et al.*: ' Ku-band high efficiency ion-implanted amplifier', *IEEE GaAs IC Symposium Digest*, 1991, pp. 335–338

80 MOGHE, S. B. *et al.*: 'A monolithic direct-coupled GaAs IC amplifier with 12 GHz bandwidth', *IEEE Transactions on Microwave Theory and Techniques*, 1984, pp. 1698–1703

81 PLATZKER, A.: 'Monolithic broadband power amplifier at X-band', *Proceedings of IEEE Microwave and Millimeter-wave Monolithic Circuits Symposium*, 1983, pp. 59–61

82 NICLAS, K. B., and TUCKER, B. A.: 'On noise in distributed amplifiers at microwave frequencies', *IEEE Transactions on Microwave Theory and Techniques*, 1983, pp. 661–668

83 LIU, L. C., MAKI, D. W., and FENG, M.: 'Single and dual stage monolithic low noise amplifier', *IEEE GaAs IC Symposium Digest*, 1982, pp. 94–97

84 HO, W. J., and WANG, N. L.: 'Producibility and performance of the microwave power HBT', *IEEE Transactions on Microwave Theory and Techniques*, 1982, **MTT-30**, pp. 263–266

85 TSERNG, H. Q., MACKSEY, H. M., and NELSON, S. R.: 'Design, fabrication and characterization of monolithic microwave GaAs power FET amplifiers', *IEEE Transactions on Electron Devices*, 1981, pp. 183–190

86 SECHI, F. N.: 'Design procedure for high-efficiency linear microwave power amplifiers', *IEEE Transactions on Microwave Theory and Techniques*, 1980, pp. 1157–1163

87 MARTINES, G., and SANNINO, M.: 'The determination of the noise, gain and scattering parameters of microwave transistors (HEMT's) using only an automatic noise figure test-set', *IEEE Transactions on Microwave Theory and Techniques*, 1994, **MTT-42**, pp. 1105–1113

88 CROUCH, M. A., and HILTON, K. P.: 'An introduction to GaAs/GaAlAs HBTs for power applications', *Microelectronics Journal*, 1993, **24**, pp. 779–794

89 WINSLOW, T. A., and TREW, R. J.: 'Principles of large-signal MESFET operation', *IEEE Transactions on Microwave Theory and Techniques*, 1994, **MTT-42**, pp. 935–942

90 ALI, F.: 'Design considerations for high efficiency GaAs HBT MMIC power amplifiers', *Proceedings of European Microwave Conference*, 1994, pp. 156–176

91 KASODY, R. E., and DOW, G. S.: 'A high efficiency V-band monolithic HEMT power amplifier', *IEEE Microwave and Guided Wave Letters*, 1994, **4**, pp. 303–304

92 LIU, W. *et al.*: 'First demonstration of high-power GaInP/GaAs HBT MMIC power amplifier with 9.9 W output power at X-band', *IEEE Microwave and Guided Wave Letters*, 1994, **4**, pp. 293–295

93 ASBECK, P. M. *et al.*: 'GaAs-based heterojunction bipolar transistors for very high performance electronic circuits', *Proceedings of the IEEE*, 1993, **81**, 1709–1725

94 BAYRAKTAROGLU, B.: 'GaAs HBT's for microwave integrated circuits', *Proceedings of the IEEE*, 1993, **81**, pp. 1762–1785

95 PLATZKER, A., HETZLER, K., and COLE, J. B.: 'High density dual-channel C-X-Ku and 6-18 GHz MMIC power amplifiers', *IEEE GaAs IC Symposium Digest*, 1991, pp. 339–342

96 YUEN, C. *et al.*: '5-60 GHz high-gain distributed amplifier utilizing InP cascode HEMTs', *IEEE GaAs IC Symposium Digest*, 1991, pp. 319–322

97 HIGGINS, J. A.: 'GaAs heterojunction bipolar transistors: A second generation microwave power amplifier transistor', *Microwave Journal*, 1991, pp. 176–194

98 HUANG, J. C. *et al.*: 'An AlGaAs/InGaAs pseudomorphic high electron mobility transistor (PHEMT) for X-and Ku-band power applications', *IEEE MTT-S International Microwave Symposium Digest*, 1991, pp. 713–716

99 MATLOUBIAN, M. *et al.*: 'High power and high efficiency AlInAs/GaInAs on InP HEMTs', *IEEE MTT-S International Microwave Symposium Digest*, 1991, pp. 721–724

100 GAT, M. *et al.*: 'A 3.0 Watt high efficiency C-band power MMIC', *IEEE GaAs IC Symposium Digest*, 1991, pp. 331–334

101 OHMURO, K., FUJISHIRO, H. I., and ITOH, M.: 'Enhanced-mode pseudomorphic inverted HEMT for low noise amplifier', 1991, *IEEE MTT-S International Microwave Symposium Digest*, 1991, pp. 709–712

102 D'AGOSTINO, S. *et al.*: 'A 0.5-12 GHz hybrid matrix distributed amplifier using commercially available FETs', *IEEE MTT-S International Microwave Symposium Digest*, 1991, pp. 289–292

103 YUEN, C. *et al.*: 'Monolithic InP cascode HEMT distributed amplifier from 5 to 40 GHz', *Electronics Letters*, 1990, **26**, pp. 1411–1412

104 MAJIDI-AHY, R. *et al.*: '94 GHz InP MMIC five-section distributed amplifier', *Electronics Letters*, 1990, **26**, pp. 91–92

105 HO, W. J. *et al.*: 'A multifunctional HBT technology', *IEEE GaAs IC Symposium Digest*, 1990, pp. 67–70
106 RIAZIAT, M. *et al.*: 'HEMT millimetre wave monolithic amplifier on InP', *Electronics Letters*, 1989, **25**, pp. 1328–1329
107 PERDOMO *et al.*: 'A monolithic 0.5 to 50 GHz MODFET distributed amplifier with 6dB gain', *IEEE GaAs IC Symposium Digest*, 1989, pp. 91–94
108 BAYRAKTAROGLU, B., KHATIBZADEH, M. A., and HUDGENS, R. D.: 'Monolithic X-band heterojunction bipolar transistor power amplifiers', *IEEE GaAs IC Symposium Digest*, 1989, pp. 271–274
109 AUST, M. *et al.*: 'A family of InGaAs/AlGaAs V-band monolithic HEMT LNA's', *IEEE GaAs IC Symposium Digest*, 1989, pp. 95–98
110 AVASARALA, *et al.*: 'A 1.6-Watt high efficiency X-band power MMIC', *IEEE GaAs IC Symposium Digest*, 1989, pp. 263–266
111 AYAKI, N. *et al.*: 'A 12 GHz-band monolithic HEMT low-noise amplifier', *IEEE GaAs IC Symposium Digest*, 1988, pp. 101–104
112 YUEN, C. *et al.*: 'A monolithic 3 to 40 GHz HEMT distributed amplifier', *IEEE GaAs IC Symposium Digest*, 1988, pp. 105–108
113 GAMAND, P. *et al.*: '2 to 42 GHz flat gain monolithic HEMT distributed amplifiers', *IEEE GaAs IC Symposium Digest*, 1988, pp. 109–111
114 KOMIAK, J. J.: 'S-band eight Watt power amplifier MMICs', *IEEE GaAs IC Symposium Digest*, 1988, pp. 45–48
115 LIU, C. S., WANG, K. G., and CHANG, C. D.: 'A 6-18 GHz monolithic low noise amplifier', *IEEE GaAs IC Symposium Digest*, 1987, pp. 211–214
116 ERON, M., TAYLOR, G., and MENNA, R.: 'X-band MMIC amplifier on GaAs/Si', *IEEE GaAs IC Symposium Digest*, 1987, pp. 171–173
117 WANG, S. K. *et al.*: 'Producibility of GaAs monolithic microwave integrated circuits', *Microwave Journal*, 1986, pp. 121–133
118 BINGHAM, S. D., MCCARTER, S. D., and PAVIO, A. M.: 'A 6.5-16 GHz monolithic power amplifier module', *IEEE Transactions on Microwave Theory and Techniques*, 1985, **MTT-33**, pp. 1555–1558
119 SHIBATA, K., ABE, B., and KAWASAKI, H.: 'Broadband HEMT and GaAs FET amplifiers for 18-26.5 GHz', *IEEE MTT-S International Microwave Symposium Digest*, 1985, pp. 547–550
120 WATKINS, E. T., SCHELLENBERG, J. M., and YAMASAKI, H.: 'A 30 GHz low noise FET amplifier', *IEEE MTT-S International Microwave Symposium Digest*, 1985, pp. 321–323
121 AITCHISON, C. S.: 'The intrinsic noise figure of the MESFET distributed amplifier', *IEEE Transactions on Microwave Theory and Techniques*, 1985, **MTT-33**, pp. 460–466
122 BERENZ, J. J., NAKANO, K., and WELLER, P.: 'Low noise high electron mobility transistors', *IEEE MTT-S International Microwave Symposium Digest*, 1984, pp. 98–101
123 YODER, M. N.: 'Distributed, traveling wave monolithic amplifiers', *IEEE GaAs IC Symposium Digest*, 1984, pp. 73–76
124 SCHELLENBERG, J. M., YAMASAKI, H., and ASHER, P. G.: '2 to 30 GHz monolithic distributed amplifier', *IEEE GaAs IC Symposium Digest*, 1984, pp. 77–79

125 CALANDRA, E., MARTINES, G., and SANNINO, M.: 'Characterization of GaAs FET's in terms of noise, gain, and scattering parameters through a noise parameter test set', *IEEE Transactions on Microwave Theory and Techniques*, 1984, **MTT-32**, pp. 231–242
126 TAJIMA, Y., and MILLER, P. D.: 'Design of broad-band power GaAs FET amplifiers', *IEEE Transactions on Microwave Theory and Techniques*, 1984, **MTT-32**, pp. 261–267
127 BARNES, A. R., MOORE, M. T., and ALLENSON, M. B.: 'A 6 to 18 GHz broadband high power MMIC for EW applications', *IEEE MTT-S International Microwave Symposium Digest*, 1997, **3**, pp. 1429–1432

Chapter 6

Oscillators

K. K. M. Cheng

6.1 Introduction

The oscillator is an essential component for microwave systems such as communications, radars and instrumentation. An ideal oscillator produces a pure sinusoidal carrier with fixed amplitude, frequency and phase. Practical oscillators, however, produce carrier waveforms with parameters (oscillation frequency, output power) that may vary in time, due to temperature changes and component ageing. Oscillator frequency instabilities are characterised primarily by frequency drift (long term variations) and random noise (short term fluctuations). The latter appears as phase and amplitude fluctuations at the oscillator output and will always be of major concern in modulated systems. Phase noise generated by any local oscillator in the transmit/receive chain can significantly affect the performance of digital communication systems.

The early development of microwave solid-state oscillators was mainly based on Gunn and IMPATT diodes. Since the 1970s, the advent of GaAs MESFETs at microwave and millimetre-wave frequencies has given greater freedom to engineers in designing microwave oscillators. GaAs MESFET devices, compared to Gunn and IMPATT diodes, offer several advantages such as suitability for monolithic integration, higher power efficiency and lower phase noise, and they have no threshold current requirements. Free-running oscillators employing GaAs MESFETs, or similar devices, have been reported [1–9] for frequencies up to W-band. Broadband tuneable oscillators using YIG resonators and varactor diodes have also been published [2,8,9] in the literature. Recent investigations have been directed towards very low phase-noise designs in MMIC form, especially those using HEMT and HBT devices. Low phase-noise oscillators based on high temperature superconductive materials operating at cryogenic temperature have also been demonstrated in the past few years. A summary of some research reports on the development of solid-state microwave oscillators is depicted in Table 6.1. In recent years, more and more oscillator circuits have been implemented using monolithic technology, because the size and cost can be significantly reduced.

Table 6.1 Summary of some scientific reports on microwave oscillator designs

Year	Active device	Types of resonator	Operating frequency (GHz)	Output power/ efficiency	Phase noise level (dBc/Hz)	Remarks
1975 [1]	GaAs MESFET	Microstrip resonator	8.8	12.1 mW / 7.8 %		MIC
1977 [2]	GaAs MESFET	YIG	5.9 - 12.5	8 - 22 mW max. 8 %		MIC
1980 [3]	GaAs MESFET	Dielectric resonator	11.85	70 mW 20 %		MIC
1982 [4]	GaAs MESFET	On-chip varactors	11.1 - 14.4 16.0 - 18.7			MMIC
1984 [5]	GaAs MESFET	On-chip varactors	11.6 - 20.0	5.5 - 15.8 dBm		MMIC
1987 [6]	GaAs MESFET	Dielectric resonator	10.74	16 dBm	- 80 (10 kHz)	Quasi - MMIC
1988 [8]	GaAs MESFET	On-chip varactors	5.9 - 12.6	14-19 dBm	- 100 (1 MHz)	MMIC
1988 [9]	GaAs MESFET	External varactors	2.5 - 6.0	17.5 dBm 9.3 dBm	- 92 (100 kHz)	Quasi - MMIC
1988 [10]	Silicon Bipolar	DR - VCO	1.69 0.5% range	8-12 dBm / 3 %		MIC
1994 [11]	HBT	Dielectric resonator	46.3	2.6 dBm 6%	- 132 (1 MHz)	Quasi - MMIC
1997 [12]	HEMT	Distributed elements	29 (700 MHz)	20dBm (buffered)	- 95 (1 MHz)	0.18 μm process
1997 [12]	AlGaAs HBT	Distributed elements	40.5 (200 MHz)	0dBm 5%	- 62 (100 kHz)	60 GHz-f_T process
1998 [13]	BiCMOS	LC (bondwire)	1.55 – 2.05	0.25 mW 1%	- 86 (100 kHz)	0.8 μm BiCMOS
2000 [14]	SiGe HBT	LC (spiral inductor)	4.63 – 4.90	- 13 dBm	- 100 (100 kHz)	50 GHz-f_T SiGe process
2000 [15]	MOSFET	LC (bondwire)	1.1 – 1.45	0.5 mW 0.4%	- 119 (600 kHz)	0.35 μm digital CMOS

In the quasi-MMIC construction, the negative resistance circuit, along with all the necessary bias and de-coupling components, is produced on a single chip and the only external element is the resonator. For this combination, the ultimate noise performance is retained by using off-chip high-Q resonators, but the fabrication procedure is greatly simplified and the reliability of the circuits is also improved.

The design of oscillators based upon linear circuit theory is rather straightforward, but there are many considerations when tight specifications must be met. Modern microwave systems require stable, low phase-noise and wide-band tuneable oscillators with sufficient output power and efficiency. There are two main design issues in MMIC oscillator design: (1) to achieve the lowest phase noise with the minimal power consumption and (2) to enhance the tuning range to encompass process variations. Phase noise depends on the carrier-to-noise ratio in the oscillator and on the Q-factor of the resonant circuit. The on-chip spiral inductor inevitably limits the Q-factor of an integrated resonator.

6.2 Design principles

Figure 6.1 shows a conventional oscillator configuration where the external embedding circuits are denoted as Z_g, Z_f and Z_L. Basically, there are two different approaches in describing the operating principle of the oscillator circuit: the feedback and negative-resistance methods. They differ from each other mainly in the design methodology.

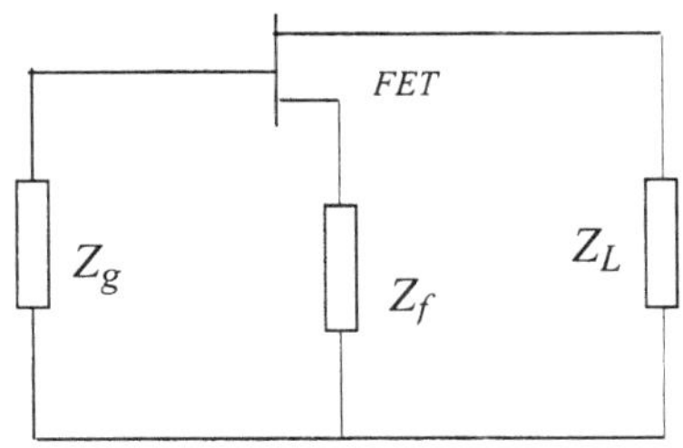

Figure 6.1 Generic FET oscillator topology

6.2.1 Feedback approach

The equivalent circuit shown in Figure 6.2 represents the feedback model of the conventional oscillator. This configuration can be viewed as a combination of a transconductance amplifier and a frequency selective circuit. The feedback network is intended to be used for providing the proper phase shift and signal level for the input of the FET. By applying linear circuit analysis, an expression for the oscillation condition may be derived as follows:

$$v_{ds} = -i_{ds} \times \frac{Z_{ds}(Z_i + Z_L)}{Z_{ds} + Z_i + Z_L}$$

$$v_{gs} = v_{ds} \times \frac{Z_i}{Z_i + Z_L}\alpha$$

$$\textit{where} \qquad \alpha = \frac{1}{1 + j\omega C_{gs}(R_i + Z_g)}$$

Combining the above two equations, we obtain

$$Z_{ds} + Z_i + Z_L = -g_m \alpha Z_{ds} Z_i \tag{6.1}$$

where Z_{ds}, C_{gs} and R_i are intrinsic parameters of the FET. The formulation of eqn. (6.1) is based on the assumption that the gain of the amplifier must be sufficiently large to compensate for the loss in the resonator, and the electrical phase shift round the loop must be an integral multiple of 360°. In practical oscillators, the oscillation frequency and close-to-carrier noise performance are mainly determined by the selectivity of the resonator, and hence the effective quality factor of the frequency selective network. All oscillators are essentially Q-multipliers, which use positive feedback to enhance many times the natural Q-factor of the resonant element. The Q-multiplier amplifies and filters the inherent circuit noise to give the output spectrum of the oscillator. Note that the amplifier gain has to be made higher than the minimum value to avoid start-up problems. The final gain will drop as the oscillation amplitude increases towards its steady-state level, due to the device's nonlinearity.

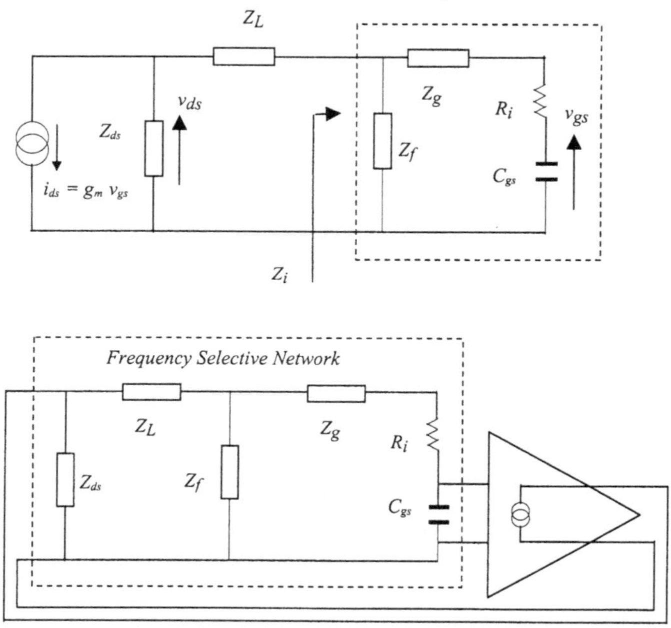

Figure 6.2 Feedback model of an oscillator circuit

6.2.2 Negative resistance approach

Consider the circuit shown in Figure 6.3, where Z_S and Z_L are the input impedance of the left and right parts of the circuit, cut at an arbitrary plane AB, respectively (a most popular choice is to use a plane between the resonator and the transistor). Based on the simplified circuit model, the derivation of impedance Z_S is therefore given by:

$$(i_s - i_{ds}) Z_{ds} + i_s Z_i = v_s$$

$$\alpha i_s Z_i = -v_{gs}$$

$$\Rightarrow \quad (Z_i + Z_{ds}) i_s + g_m \alpha Z_{ds} Z_i i_s = v_s$$

$$\begin{aligned} \Rightarrow \quad Z_S &= \frac{v_S}{i_S} \\ &= Z_i + Z_{ds} + g_m \alpha Z_{ds} Z_i \\ &= -R_S + jX_S \end{aligned} \tag{6.2}$$

For oscillation to occur, we have

$$Z_S + Z_L = 0 \tag{6.3}$$

According to eqn. (6.3), the negative resistance and reactance of the active block (Z_S) must be equal in magnitude and opposite in sign to the resistance and reactance of Z_L. For oscillations to begin, the magnitude of the negative resistance has to be bigger than the value determined from the small-signal condition. Due to the excess negative resistance in the circuit, the oscillation will grow continually until the negative resistance is reduced in magnitude, by non-linear effects, to its steady-state value. By inserting eqn. (6.2) into eqn. (6.3), we get:

$$Z_i + Z_{ds} + Z_L + g_m \alpha Z_{ds} Z_i = 0 \tag{6.4}$$

It should be noted that the two expressions as stated in eqns. (6.1) and (6.4) are identical. In theory, both the feedback and negative-resistance approaches can be employed for the design and analysis of oscillator circuits. In applying the negative resistance formulation, the variations of the real and imaginary parts of the input impedance versus frequency are usually obtained by a linear circuit simulator. Adjustment of circuit parameters (e.g. capacitor and inductor values) is then carried out in a systematic manner until the balanced condition in eqn. (6.3) is attained, for a desired value of oscillation frequency. However, in practice, eqn. (6.3) may fail to work if Z_L appears to behave like a parallel-tuned circuit over the frequency band of interest. Under these circumstances, the problem may be solved either by using an admittance approach in formulating the oscillation condition, or by placing the cutting plane at a different location. Furthermore, similar expressions may also be developed using the concept of power waves if the *S*-parameters of the components are available.

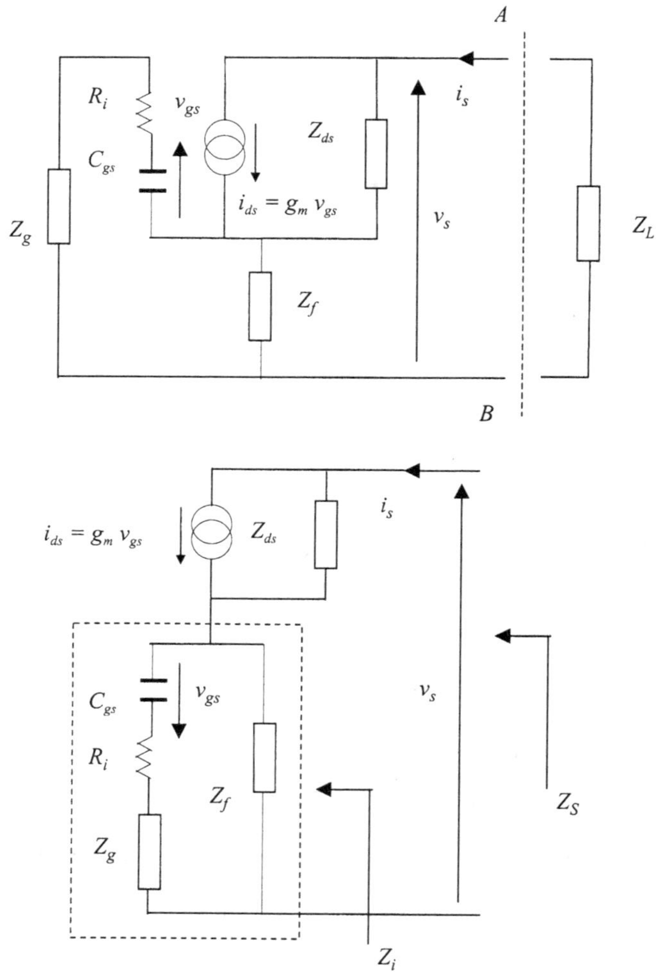

Figure 6.3 Negative resistance model of an oscillator circuit

The two most common FET circuit configurations that have been used to produce negative resistance, namely the common-gate inductive feedback and the common-source capacitive feedback, are shown in Figure 6.4. The values of L_f and C_f are adjusted in such a way that, for a given transistor type, the negative resistance is maximised within the tuning range. In many cases, enhanced tuning bandwidth and negative resistance can be achieved by adding a suitable matching circuit between the transistor's output and the load.

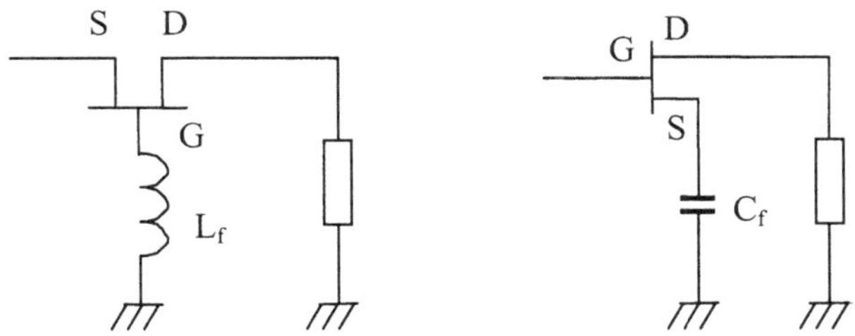

Figure 6.4 Basic circuit topologies for generating negative resistance

6.3 Active devices

The performance of oscillators is largely dependent on the choice of a suitable type of active device. Most microwave oscillators nowadays use either silicon bipolar transistors or GaAs devices. It is very important that the active device is capable of creating sufficient negative resistance, or forward gain, over the band of interest. Other factors that must also be considered for different applications and requirements include flicker noise level, output power, etc.

6.3.1 GaAs MESFET versus silicon bipolar transistors

In most cases the choice of transistors is initially between a silicon bipolar transistor and a GaAs MESFET [16]. It should be noted that the GaAs MESFET has a higher frequency of oscillation, higher gain, higher output power and higher efficiency. The higher output power of the GaAs MESFET is a direct consequence of the higher critical field and higher saturated drift velocity. The higher gain is due to higher mobility of the electrons as compared to silicon. However, silicon bipolar transistors usually have much lower flicker noise level than GaAs devices. A difference of 20 to 30 dB in close-to-carrier noise performance can easily be found in oscillators based on the two types of device. In the past, bipolar transistors were the only practical devices to be used for a low phase-noise microwave oscillator. However, new types of device such as HEMT and HBT are also candidates for low noise applications, especially at microwave and millimetre-wave frequencies.

6.3.2 Heterojunction bipolar transistors

Research into GaAs and other III-V compounds, has led to the widespread application of heterojunction bipolar transistors (HBTs) [17]. These devices offer the prospect of obtaining performance features similar to those of silicon bipolar transistors translated to substantially higher frequency. They have shown excellent gain and noise performance in microwave and millimetre-wave circuits. The vertical structure of the HBT eliminates the surface-state problem associated with the GaAs MESFET, thus resulting in superior phase-noise characteristics. The low phase-noise characteristics make them very attractive for microwave source applications. Broadband tuneable oscillators are also one of the key application areas for HBTs, because these devices are capable of generating a wideband negative resistance due to their high transconductance. In addition, the high breakdown voltage and current handling capability of HBTs could lead to higher output power in oscillators. A drawback of HBTs is the need to access several different layers of the vertical structure, which can lead to non-planar structures in some circumstances. This non-planarity complicates device processing, although it does not preclude monolithic integration of HBTs. A MMIC voltage-controlled oscillator (VCO) that uses an AlGaAs/GaAs HBT has been reported, having a very wide tuning range of 7 to 15 GHz and minimum output power of 9 dBm. This VCO circuit consists of a common-base oscillator with a capacitively-loaded emitter to generate the wideband negative resistance. This circuit requires two external varactors to form a complete VCO.

InP-based HBTs have also been developed by commercial companies for applications in fibre-optic telecommunications. The InP-based technologies take advantage of the intrinsic material properties such as higher electron mobility, higher thermal conductivity and lower band-gap energy for lower voltage operation in order to gain an advantage over GaAs, silicon and SiGe based semiconductors.

6.3.3 SiGe technology

Bipolar technology has matured, and today provides low-parasitic, self-aligned device structures with abrupt doping profiles that enable cut-off frequencies of more than 30 GHz. This figure was further improved by introducing SiGe alloys to the base region of the transistor, which has pushed the figure to over 60 GHz. SiGe bipolar technology [18,19] is an attractive candidate for wireless communications applications since it has the potential to exceed the performance of conventional silicon technology to a level where it can equal GaAs, yet provide the cost advantage and manufacturing volume capabilities of silicon.

SiGe bipolar transistors have shown excellent performance for a wide variety of RF applications. The SiGe HBT has demonstrated high PAE and high linearity performance at wireless frequencies. Current generation SiGe technologies have demonstrated a good combination of low noise figure and associated gain, and thus, have generated a great deal of interest to date for applications in the microwave frequency range. On-wafer F_{min} and G_A values at 2 GHz of 1 dB and 14 dB, respectively, have been demonstrated in a commercial 0.5 μm technology at collector currents of less than 5 mA.

In order to take full advantage of the high frequency solution offered by SiGe technology, it is important to integrate the microwave passive elements onto the chip. This poses many challenges since the silicon substrate is inherently lossy at microwave frequencies when compared to the more expensive GaAs. Basically, there are two main problems in constructing high-Q passive components on standard silicon substrates used for SiGe processes: (1) high series resistance in the relatively thin aluminium conductors and (2) high losses in the silicon substrate. In a GaAs circuit, the use of a semi-insulating substrate allows the construction of low loss transmission lines and spiral inductors. High resistivity silicon substrates have also been proposed, but they are not compatible with the processing temperatures required for SiGe BiCMOS. There are other design tricks, such as special ground plane geometries, the use of thick oxides, thick metal processes, micromachining and multi-level techniques that can improve inductor Q.

6.3.4 RF CMOS technology

The majority of RFICs are implemented using GaAs or silicon bipolar technologies. However, silicon CMOS technology has a major cost advantage over GaAs, due to its high volume throughput capability. It also offers higher levels of integration, being able to combine mixed signal, power management and RF functions on a single chip. This technology provides high gain and noise figures below 1 dB at 2 GHz without the need for sub-micron lithography, and hence avoiding narrow emitter effects occurring on downscaled BJTs. Furthermore, as the linewidths continue to scale downwards, CMOS technology will start challenging GaAs and bipolar technologies for RF markets, even in the higher frequency bands. However, when considering other important performance parameters such as noise and linearity, CMOS still suffers from significant impairments. To alleviate these limitations, improvements are required in all areas of integrated design. This includes device optimisation and modelling [20], circuit topology [21], as well as system level considerations [22]. The limited RF performance of CMOS technology demands greater creativity and innovation in circuit design. Due to the low substrate resistivity obtained with silicon, significant coupling through parallel conducting paths may occur. Also, circuits operating under large-signal conditions, as well as noise contributions from digital circuitry, may cause the substrate potential to 'bounce'. Possible solutions include the use of guard rings and fully balanced designs. One penalty of using balanced designs is the inevitable increase in power consumption, resulting from running identical circuits in parallel. Currently, the LO and PA are considered to be the main circuit challenges in terms of enabling a single-chip CMOS transceiver. The LO design suffers from the lack of high-Q inductors. This makes it difficult to meet phase noise requirements while simultaneously complying with switching-time requirements. The highest levels of performance have been reported for designs using bond-wire inductors, which generally suffer from poor accuracy and repeatability.

6.3.5 Biasing and spurious response suppression

The DC bias condition of transistors is usually established independently of the RF design. Power efficiency, stability and ease of use are the main concerns when selecting a biasing configuration. For the design of a low phase-noise oscillator, the biasing circuit should be properly regulated and filtered to avoid any unwanted signal modulation or noise injection. Variations on the supply voltages or currents may also cause undesirable output power fluctuations and frequency drift. Two commonly used biasing configurations are shown in Figure 6.5. These circuits use either one or two power supplies.

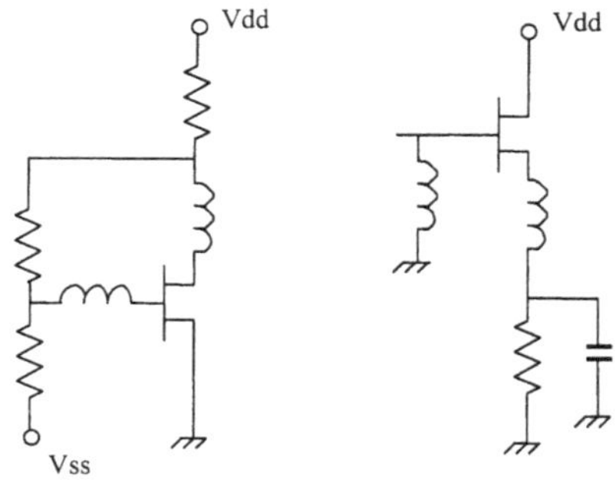

Figure 6.5 Passive biasing circuits for MESFET devices

All oscillators can produce spurious oscillations at one or more undesired frequencies. They can occur at any frequency where eqn. (6.1) or (6.3) is satisfied by the microwave and bias circuits. Prevention of these spurious oscillations requires great care. The rule is to prevent the magnitude and phase angle conditions, required for oscillation, from occurring simultaneously. In practice, the amplifier gain provided by the transistor must be made as narrow band as possible, but still operate over the designated frequency range. Furthermore, the bias lead inductance, de-coupling capacitors and all other off-chip components should be included in the design.

6.4 CAD techniques for large-signal oscillator design

Although an oscillator is in reality a non-linear component, small-signal considerations are usually sufficient to ensure that oscillation conditions are met and to establish the desired frequency. However, the linear theory is incapable of predicting output power precisely and determining the voltage waveforms across critical components in the circuit, such as a tuning varactor. The latter may be very valuable in maximising power and efficiency or minimising noise in VCOs. The traditional methods of designing oscillators are mainly based on empirical methods, quasi-linear models [23] or large-signal *S*-parameters. The accuracy of these methods, however, is questionable, and they do not give information on harmonic content. Today, high performance PCs and workstations are so commonplace that the application of full numerical methods to these non-linear problems is becoming routine. Current research activities have been concentrating on developing accurate device models (both noise and non-linear models) and efficient analysis methods. Analysis techniques generally fall into three categories: time-domain method, harmonic balance algorithm and Volterra series.

6.4.1 Time-domain method

Time-domain methods enjoy a wide range of applications [24], including the steady-state and transient solutions of both analogue and digital circuits. Circuit differential equations are solved in the time domain by numerical integration, requiring the solution of a set of non-linear algebraic equations at each time step. The starting point is usually chosen as the result of a DC analysis. Methods that work entirely in the time domain represent the most natural and straightforward approach to the simulation problem. However, there are several factors that limit the applicability of these methods to microwave circuit analysis. Firstly, the actual circuits may contain dispersive transmission lines that are difficult to analyse in the time domain. Secondly, the circuit may have time constants that are large compared to the inverse of the fundamental excitation frequency. If the steady-state response is of major interest, a large number of time steps may be required, thereby consuming much CPU time. Considerable effort [25] has been spent on devising time-domain methods which allow the transient calculation to be at least partially bypassed and the steady-state solution to be reached more quickly. Long computer time and dynamic range limitations are the major drawbacks of this method.

6.4.2 Harmonic balance algorithm

In this method, variables are represented by their Fourier coefficients rather than as functions of time. The transients are naturally avoided, and therefore, the computational complexity depends only on the size of the circuit and not on the actual frequencies or the time constants present in the circuit. In this method, the linear elements are analysed in the frequency domain, whereas the non-linear elements are simulated in the time domain. The interfacing between the two domains is accomplished by fast Fourier transformation (FFT). The Fourier coefficients of the unknown variables are then optimised to give the approximated solution. Harmonic balance techniques have been used extensively to analyse virtually all kinds of non-linear microwave subsystems [26–28]. However, some problems arise when applying this technique to analyse oscillators. Firstly, in an oscillator, the frequency can change as the oscillation develops, and therefore, can only be known approximately. If the frequency used in the harmonic balance analysis is not equal to the large-signal oscillation frequency, the results may be erroneous. A way to overcome this is by treating the unknown oscillation frequency as an extra optimising variable. The second problem is due to the fact that non-physical, spurious oscillations might exist. An algorithm is therefore required to find the physical solution and separate it from the spurious ones. Newton's method is known to offer fast convergence, but it is generally difficult to find suitable initial values, especially for an oscillator. Hence, the chance of obtaining spurious solutions is high. Furthermore, the harmonic balance method requires the FFT and an optimisation algorithm, which could use up considerable CPU time.

6.4.3 Volterra series analysis

Volterra series theory was first introduced in 1930 and was further developed by Weiner in the 1950s, for the expansion of functionals in terms of orthogonal polynomial series. Weiner's functional expansions, now known as Volterra non-linear transfer functions, can handle frequency-dependent systems with single valued input–output characteristics. For a linear system, the output $y(t)$ can be related to its input $x(t)$ as,

$$y(t) = \int_{-\infty}^{\infty} h_1(u_1)\, x(t - u_1)\, du_1 \tag{6.5}$$

where $h_1(u_1)$ is the impulse response of the system. When the system becomes non-linear, the output can be related to its input by the Volterra functional series [29]:

$$y(t) = \int_{-\infty}^{\infty} h_1(u_1)\, x(t - u_1)\, du_1 + \int_{-\infty}^{\infty}\int_{-\infty}^{\infty} h_2(u_1, u_2)\, x(t - u_1)\, x(t - u_2)\, du_1\, du_2 + \ldots \tag{6.6}$$

where $h_n(u_1, \ldots, u_n)$ is known as the '*n*th order impulse response' of the system. Volterra series analysis in a non-linear system can be viewed as the generalisation of convolution integrals used in linear systems. In the frequency domain,

the output spectrum $Y(f)$ can be expressed as a function of the input spectrum $X(f)$, given by:

$$Y(f) = \sum_{n=1}^{\infty} \int_{-\infty}^{\infty} \ldots \int_{-\infty}^{\infty} H_n(f_1, \ldots, f_n) \cdot \delta(f - f_1 - \ldots - f_n) \prod_{1}^{n} X(f_i)\, df_i \tag{6.7}$$

where $H_n(f_1, \ldots, f_n)$ is known as the non-linear transfer function, which is the Fourier transform of the impulse response $h_n(u_1, \ldots, u_n)$. For a system with mild non-linearity, it has been suggested that the first few terms of the expansion may be sufficient to represent $y(t)$. An efficient method of solving the response of a non-linear system with power series type non-linearity is the current method described by Bussgang *et al.* [29]. Volterra series analysis is very attractive because computations are performed in the frequency domain and neither Fourier transformation nor iteration is required. Volterra series analysis is, in general, restricted to weakly non-linear systems because of the algebraic complexity of determining Volterra non-linear transfer functions of high order (as required with more strongly non-linear systems). The application of the Volterra series formulation to the analysis of microwave oscillators was proposed and experimentally demonstrated by Cheng and Everard in 1990 [30].

6.5 Phase noise in oscillators

6.5.1 Spectral characteristics of phase noise

There are two main sources of phase noise in an oscillator, namely thermal and flicker noise. As shown in Figure 6.6, thermal noise of the active device has a flat spectrum and is responsible for the $1/f_m^2$ spectral characteristic of an oscillator. Flicker or $1/f$ noise is primarily a baseband noise, being upconverted by non-linearities in the active devices to the carrier frequency. The transposition gain is dependent upon the device and the operating conditions. The loop transfer characteristic then filters the output voltage, producing a $1/f_m^2$ characteristic [32].

The value of the offset frequency at which the transition between $1/f_m^3$ and $1/f_m^2$ laws occurs is the transition frequency f_c, and is often is different from the baseband flicker noise corner frequency. The transition frequency is usually less than a few kHz for a silicon bipolar transistor and over hundreds of kHz for a GaAs transistor. Although the $1/f$ baseband characteristic will still be converted to a $1/f_m^3$ characteristic in an oscillator, modulation noise actually behaves very differently as compared to thermal noise. For example, increasing the oscillator power level does not result in a direct reduction in oscillator phase noise as it does for thermal noise. Another consequence of direct modulation noise is that if the amplifier is the dominant source of noise, increasing Q_L will reduce the oscillator flicker noise level. A theoretical analysis which is capable of predicting the noise performance of oscillators, as well as VCOs, is obviously an important subject.

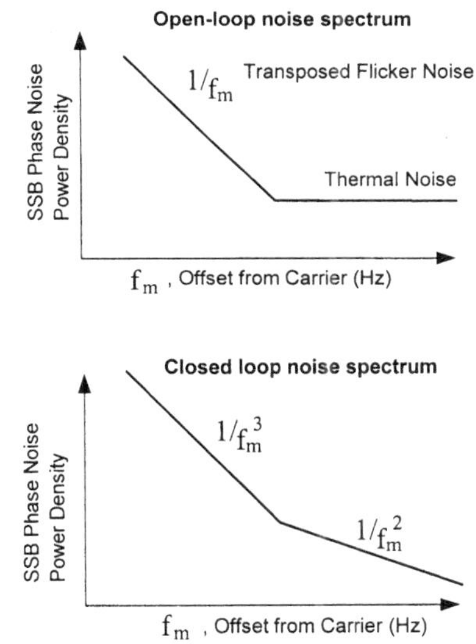

Figure 6.6 Open- and closed-loop noise spectrum in oscillators

6.5.2 Noise analysis of negative resistance oscillators

A negative resistance oscillator is considered to be stable if any perturbation (such as noise) in the RF voltage or current of the circuit at any instant decays, bringing the oscillator back to its point of equilibrium. The general 1st-order treatment was given by Kurokawa [33], who analysed the oscillator quasi-statically. This work indicates that the angle of interaction between the circuit locus and the device line is a measure of the oscillator's stability and noise performance at that particular operating point. Low noise is associated with an intersection angle of 90°, which means that no frequency fluctuations will amplitude modulate the oscillation signal, and, more importantly, that no amplitude fluctuations will modulate the phase of the resonator, and hence the frequency of oscillation. Therefore, for phase-noise minimisation, the resonator phase shift with frequency must be maximised, which implies maximising Q. Kurokawa's method was later generalised by Eddison and Howes [34], to account for the frequency dependency of the device conductance. As a result, the graphical interpretation of the device admittance should then be considered as a surface instead of a line. Esdale and Howes [35] extended Kurokawa's method by converting it to a reflection coefficient analysis, which uses readily available device and circuit reflection coefficient information to predict oscillation conditions. In theory, it should be possible, for any given device, to design a passive circuit to give an oscillator structure with a given stability and noise performance. However, it has been pointed out [34] that the orthogonality condition is not easily satisfied with practical component values. Consequently, the intersection angle is likely to be small, resulting in a large noise output for microwave oscillators, particularly the phase noise, which is by far the largest noise component.

A number of investigations into the upconversion of baseband noise in fixed-frequency oscillators have been published in the literature. Debney and Joshi [36] presented the first detailed version of Kurokawa's analysis, and this involved the transformation of a baseband noise source through a non-linear, 2-port device model to an equivalent RF noise source in the 1-port representation of the oscillator. The interaction of this source with the device and the load impedance was analysed to produce the amplitude and phase-noise spectra. A more general model was considered by Siweris and Schiek [37] in their analysis of a fixed frequency oscillator using the quasi-static approach. They included the gate-source capacitance in the intrinsic non-linear device model, an element neglected by Debney and Joshi, and showed this to be the important element in the upconversion of baseband noise to close-to-carrier phase noise. Siweris and Schiek have also considered the inclusion of a non-linear, baseband noise source to give improved agreement between measured and simulated results for fixed frequency oscillators. A more general solution of the frequency conversion process, by injecting a small RF deterministic signal into a non-linear circuit supporting a periodic steady-state solution, has also been reported [38]. When the RF source is replaced by a set of random noise generators as the perturbing mechanism, it is quite reasonable to expect that the same arguments will lead to a noise analysis of the steady state. Of course, in this case the problem is much more complicated, since the free sources can only be described in a statistical sense. If several noise generators exist, they may be statistically dependent, and their correlation must be accounted for in evaluating the noise power. Further correlation is established among the noise sidebands, because of the intermodulation of noise waveforms with the periodic steady state. All such effects need to be included in the noise analysis if a general and accurate prediction of noise level in oscillators is desired.

6.6 MMIC voltage-controlled oscillator design

MMIC technology is the key to making microwave and millimetre-wave transceivers affordable for widespread use in applications such as wireless data communications and cellular mobile communications. Oscillators form an essential part of these transceivers and for many applications a phase-locked loop (PLL) employing a VCO is desirable. The design of varactor-tuned MESFET VCOs has received widespread attention in the literature and convenient design methods have been identified [39]. This section describes the modelling and design of a MMIC VCO using MESFET devices.

6.6.1 Varactor diode modelling

Voltage-dependent capacitors (varactors) are required for the design of the integrated voltage-controlled oscillator for on-chip carrier frequency generation. Varactor diodes are usually realised on MMICs by using the standard MESFET structure, with source and drain connected together to form the cathode, and the gate forming the anode. In this way, a varactor with acceptable performance can be realised without additional processing steps. An accurate model has been developed [40] for these interdigitated planar Schottky varactor diodes (IPSVDs). In the model shown in Figure 6.7, the significant extrinsic elements are combined with the reverse and forward bias intrinsic elements of the low frequency equivalent circuit. The model includes the voltage-dependent capacitor C_j and the shunt resistance R_D representing the junction leakage. The extrinsic parameters are based on the physical structure of the device. The combined series resistance associated with the gate, drain and source fingers are include in R_S, and the combined series inductance of the gate, drain and source fingers are represented by L_S. Here, the drain/source capacitance to ground is represented by C_1; the much smaller gate capacitance to ground is represented by C_2. In addition, frequency dispersion and delay models are included. The latter is required since the lumped-element model does not take into account the physical length of the gate fingers. The result is a bias-dependant model that is accurate well into the millimetre-wave frequency range, with either reverse or forward bias applied. Small-signal modelling using this bias-dependant varactor diode model and FET *S*-parameters, or an FET model, can be used for most of the design procedure. A bias-dependant varactor diode model enables the tuning range of the VCO to be simulated and optimised. However, a full large-signal model is required in order to optimise the output power of the oscillator, which is best carried out using a harmonic-balance simulator, and to check the oscillator's start-up performance with a time-domain simulator.

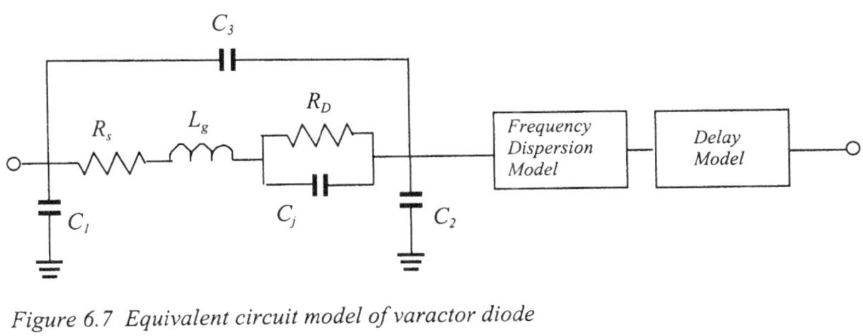

Figure 6.7 Equivalent circuit model of varactor diode

6.6.2 VCO design example

The basic circuit diagram of the VCO is shown in Figure 6.8, and the four major parts are:

1. FET with a feedback capacitance, C_f, at the source
2. resonator on the gate terminal, consisting of L_1 and the varactor
3. output matching on the drain
4. DC bias networks.

The feedback capacitor on the source is required in order that the FET presents a negative resistance at the gate terminal. In the ideal case this negative resistance is given by:

$$R = -\frac{g_m}{\omega^2 C_{gs} C_f} \tag{6.8}$$

where C_{gs} is the gate-source capacitance of the FET, g_m is its transconductance and ω is the angular frequency. The resonator at the gate terminal consists of L_1 connected in parallel with the tuning varactor diode. At resonance these allow positive feedback to occur. The output matching is accomplished by L_4, along with L_3, which is also the drain bias choke. For the proper biasing of the FET, the gate is connected to a negative power supply through resistor R_g, whereas the source is shorted to ground through a bias choke, L_2.

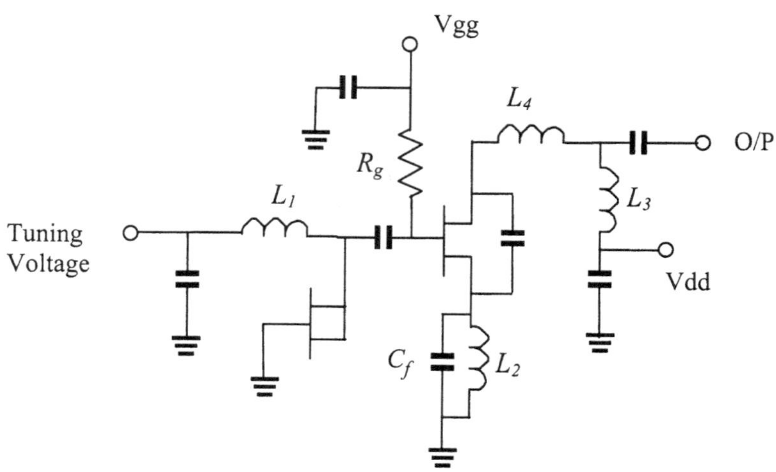

Figure 6.8 VCO circuit diagram

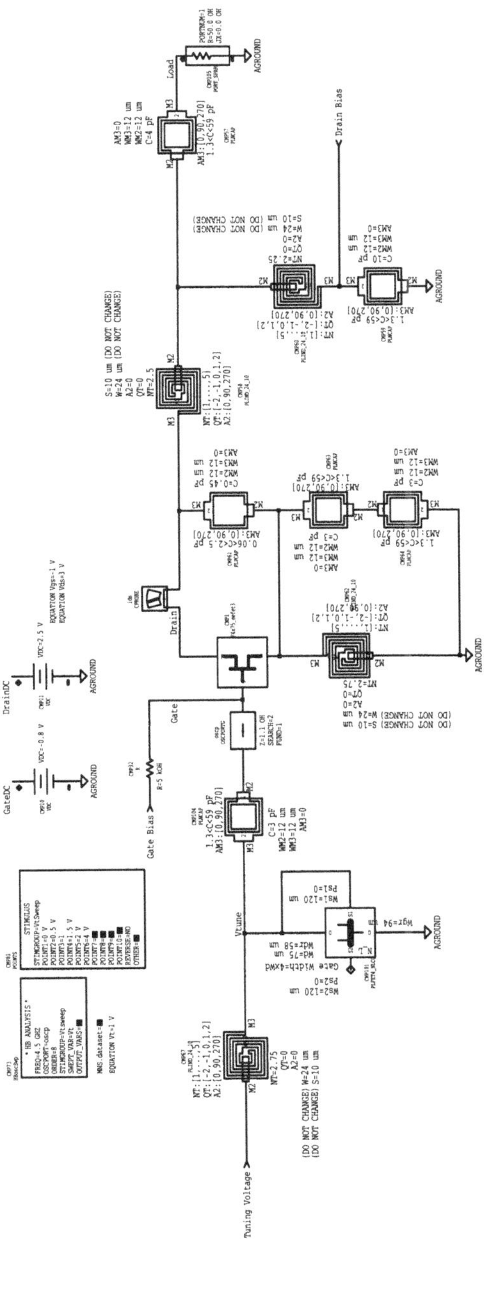

Figure 6.9 MDS™ schematic diagram of VCO design example

The major part of the design is to use small-signal analysis to design the feedback and the resonator elements, in order to satisfy the oscillation conditions over the required tuning bandwidth. The circuit is analysed as a 2-port network by breaking the connection between the tuning varactor, L_1, and ground. The components are optimised to meet the oscillation conditions; at resonance the S_{21} response should show a large gain peak, and the input impedance response should show a negative resistance of more than 10 Ω, with zero reactance. The tuning range of the oscillator can be ascertained by varying the bias in the bias-dependant varactor model and observing the frequency range over which the required level of negative resistance is maintained. For this C-band design, both the FET and varactor diode were chosen to be a 4 × 75 μm device. Having carried out the initial small-signal optimisation, the Foundry models were substituted for all the lumped elements, as shown in Figure 6.9. In addition, after the majority of the design was complete, Microwave Spice™ was used with the Foundry's large-signal MESFET model in order to verify the correct operation of the VCO from start-up. Some manual adjustment of the circuit was also carried out using Microwave Spice™ in order to maximise the output power.

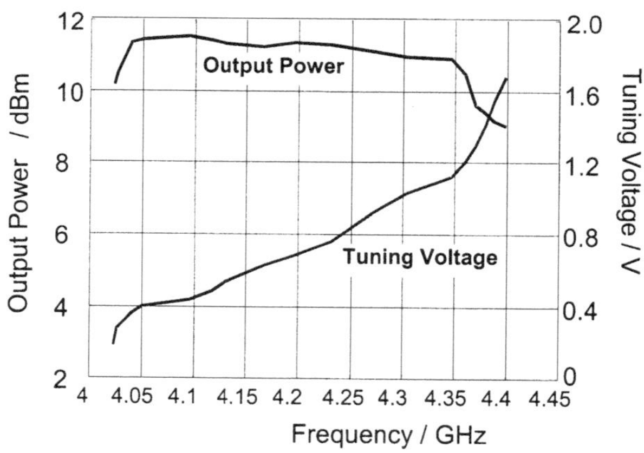

Figure 6.10 Measured VCO output power and tuning characteristics

This VCO design was fabricated using the standard Marconi Caswell F20 process. The chip was measured using a Cascade prober and HP8562A spectrum analyser. The drain voltage and current were nominally 3 V and 15 mA, respectively, and the varactor bias voltage varies from -0.4 to -1.2 V. The frequency versus voltage tuning characteristic is plotted in Figure 6.10, for the design with a fixed feedback capacitance. The output power, corrected for probe and cable losses, is 11.5 ±0.3 dBm over a tuning range of 300 MHz. A microphotograph of the VCO chip, which uses a single varactor diode as the tuning element, is shown in Figure 6.11. The actual chip size is less than 1.5 × 1.5 mm^2.

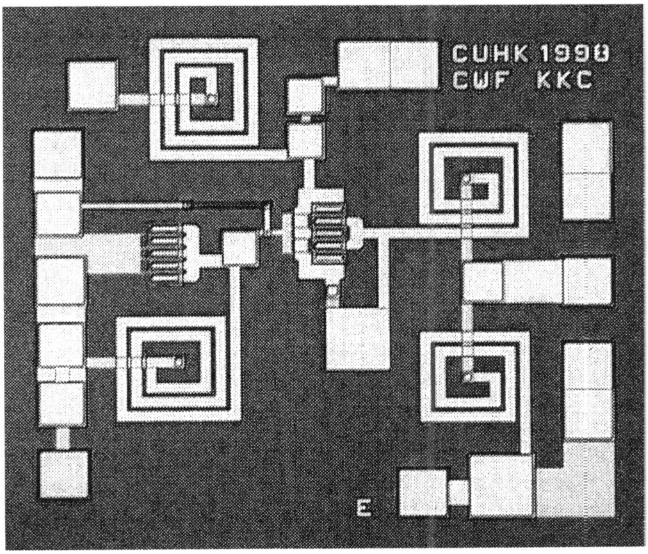

Figure 6.11 Microphotograph of the MMIC VCO chip

These results are very satisfactory considering that no buffer amplifier is used. A buffer amplifier is usually inserted between the VCO output and the mixer input for isolation and to reduce pulling effects. The tuning range of the oscillator can be increased considerably by the use of a varactor diode as the feedback capacitance C_f. This enables the negative resistance value to be varied, and if the two varactors are controlled with independent tuning voltages then a four-fold increase in tuning range can easily be achieved. Naturally, however, due to the inherent phase noise limitation of MESFET oscillators, for very demanding applications it would be advantageous to consider alternative devices such as the HBT. In the case of a bipolar transistor VCO, the device can be used in a common-base configuration with an inductor as the feedback element. Apart from minor changes to the bias network, the VCO design principle is the same.

6.7 MMIC injection-locked oscillator design

Frequency- and phase-locked devices are essential components for many microwave systems such as phase-locked loops (PLLs), frequency synthesisers, frequency modulation subsystems and adaptive phase arrays. These devices include injection-locked oscillators and analogue frequency dividers. Injection locking [41] is an interesting phenomenon in oscillator design, and many applications have evolved from its special characteristics. Injection locking describes the phenomenon where the oscillator frequency shifts and locks to the frequency of an externally applied signal.

An injection-locked oscillator (ILO) is most easily obtained by applying an external RF signal to the input of a free-running oscillator. This signal is referred to as the locking signal or injection signal. Naturally, a free running oscillator is not able to lock to the injection signal if the power and frequency of the injection signal does not fulfill certain criteria. The bandwidth over which the oscillator is able to lock to the injection signal is known as the locking bandwidth or locking range. Locking gain may be realised if the power of the locking signal is lower than that of the oscillator output, where locking gain is defined as:

$$\textit{locking gain} \;=\; 10\log\left[\frac{\textit{output power}}{\textit{injection signal power}}\right] \qquad dB$$

In practice, the oscillator's locking bandwidth can be widened by increasing the power level of the injected signal. However, the locking gain is reduced accordingly. ILOs may be designed using either two-terminal or three-terminal devices. For two-terminal devices, a circulator is usually used to separate the injection signal and the output signal. Three-terminal devices such as transistors have distinct input and output ports, and the injection signal can be applied to form a transmission- or reflection-type ILO. When the frequency of the output signal is equal to the harmonic frequency of the injection signal, a sub-harmonic ILO is obtained [42]. In cases where the frequency of the injected signal is equal to the harmonic frequency of the output signal, then a harmonic ILO is obtained. The regenerative analogue frequency divider falls into this latter type of application [43].

6.7.1 Unified formulation of injection locking

A number of reports have been published on the analysis of fundamental ILOs, including time-domain methods [44], frequency-domain methods [45] and harmonic balance algorithms [46]. This section presents a general formulation [47] for the analysis of ILOs, based upon the Volterra series theory. Unlike the time-domain technique and harmonic-balance algorithms, both long CPU time and problems of convergence are avoided.

ILOs (fundamental-, harmonic- or sub-harmonic type) can generally be represented by the circuit configuration shown Figure 6.12*a*. It consists of a linear (Z_L) and a non-linear block (Z_{NL}), driven by an external current source with amplitude I_j and angular frequency ω_j. Furthermore, when the oscillator is locked, the amplitude, angular frequency and phase angle of the fundamental component of the current flowing through Z_L are denoted as I_0 , ω_0 and ϕ, respectively. Besides, ω_0 and ω_j are harmonically related by:

$$\omega_j = \begin{cases} \omega_0 & \text{fundamental - mode} \\ \mathrm{N}\times\omega_0 & \text{harmonic - mode} \\ \dfrac{1}{\mathrm{N}}\times\omega_0 & \text{subharmonic - mode} \end{cases}$$
$$\textit{where } N = 2, 3, \ldots$$

In addition, the two-element oscillator may also be transformed into the equivalent circuit shown in Figure 6.12*b*, where $Z_S(\omega)$ and $Z_P(\omega)$ are band-stop and band-pass filters defined by:

$$Z_S(\omega) = \begin{cases} \infty & \omega = \omega_0 \\ Z_L(\omega) & \omega \neq \omega_0 \end{cases}$$

$$Z_P(\omega) = \begin{cases} Z_L(\omega_0) & \omega = \omega_0 \\ \infty & \omega \neq \omega_0 \end{cases}$$

Moreover, it is possible to decompose the oscillator into two sub-circuits as depicted in Figure 6.12*c*. It is clear that the voltages across the two circuits are identical, and hence we have:

$$Z_L(\omega_0)\frac{I_0}{2}e^{j\omega_0 t+\phi} + V_{NL}(\omega_0) = 0 \tag{6.9}$$

where $V_{NL}(\omega_0)$ is the frequency component of the voltage across the input of the non-linear subcircuit when driven by the two current excitations. Note that this voltage is a function of four variables, namely ω_j, I_0, I_j and ϕ, and it can be evaluated very efficiently by the nonlinear current method described in [29]. Upon examining eqn. (6.9), it is clear that the injection-locking problem has now been reduced to solving two non-linear algebraic equations (real and imaginary parts of the determining equation) with only two unknowns (I_0 and ϕ), when both ω_j and I_j values are pre-specified.

The accuracy of the new formulation has been verified [47] by comparing the locking bandwidths evaluated by the proposed method and a time-domain simulator, for a MESFET analogue 1/3 frequency divider circuit.

6.7.2 Analogue frequency divider design example

Frequency division is the subject of increasing interest in the field of microwave systems, especially in PLLs. Analogue regenerative frequency dividers are a good choice to simultaneously reach very high operating frequencies, low phase noise and low power consumption. MMIC analogue frequency dividers have been reported [48,49].

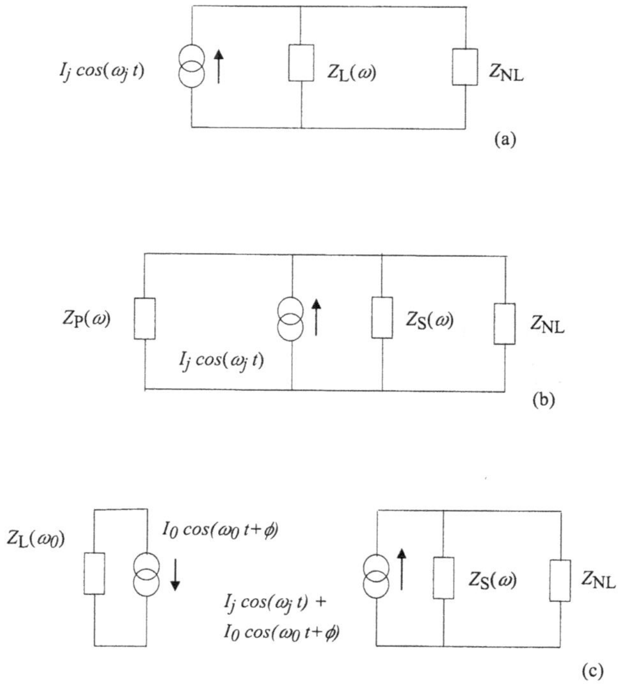

Figure 6.12 Injection locking oscillator analysis: (a) general configuration, (b) equivalent model and (c) open-loop model

Based on the previous analysis, the determining equation for an analogue 1/2 frequency divider may be derived as

$$Z_L\left(\frac{\omega_j}{2}\right)\frac{I_0}{2} + H_1\left(\frac{\omega_j}{2}\right)\frac{I_0}{2} + H_2\left(\omega_j, -\frac{\omega_j}{2}\right)\frac{I_0 I_j}{2}e^{-2j\phi} + H_3\left(\frac{\omega_j}{2}, \frac{\omega_j}{2}, -\frac{\omega_j}{2}\right)\frac{3I_0^{\,3}}{8} = 0 \tag{6.10}$$

where $H_n(\ )$ is the Volterra kernal of nth order [29]. With a low injection current assumption, a simplified expression for I_j may be obtained from eqn. (6.10), by neglecting all Volterra kernals higher than the 3rd order:

$$I_j = \frac{\left| Z_L\left(\frac{\omega_j}{2}\right) + H_1\left(\frac{\omega_j}{2}\right) + H_3\left(\frac{\omega_j}{2}, \frac{\omega_j}{2}, -\frac{\omega_j}{2}\right)\frac{3I_0^{\,2}}{4} \right|}{I_0 \left| H_2\left(\omega_j, -\frac{\omega_j}{2}\right)\right|} \tag{6.11}$$

From eqn. (6.11), a few guidelines for the design of an analogue 1/2 frequency divider can be derived as follows:

1. Note that the numerator in the above expression is identical to the determining equation developed for a free-running oscillator [30]. It is clear that the magnitude of the numerator should be kept small, within the locking range, in order to reduce the amount of injection power (small I_j) required. This condition is equivalent to making the circuit marginally stable (non-oscillating) or unstable (self-oscillating).

2. Using Volterra series analysis, the term $H_2(\ )$ can be shown to be proportional to the 1st-order voltage component (ω_j) appearing across the gate-to-source capacitance of the MESFET. Consequently, a matching network should be placed between the gate terminal of the device and the signal source, in order to maximise $H_2(\ldots)$, and thus reducing the value of I_j required.

3. The MESFET can be modelled by a simple equivalent circuit, in which the controlled drain current source is the only non-linear element. The drain-to-source current is usually represented by:

$$i_{ds} = g_{m1} v_{gs} + g_{m2} v_{gs}^{2} + g_{m3} v_{gs}^{3}$$

   where g_{m1}, g_{m2} and g_{m3} are bias-dependent coefficients [50]. It might be possible to further improve the locking characteristics of the divider circuit by optimising the bias condition of the active device.

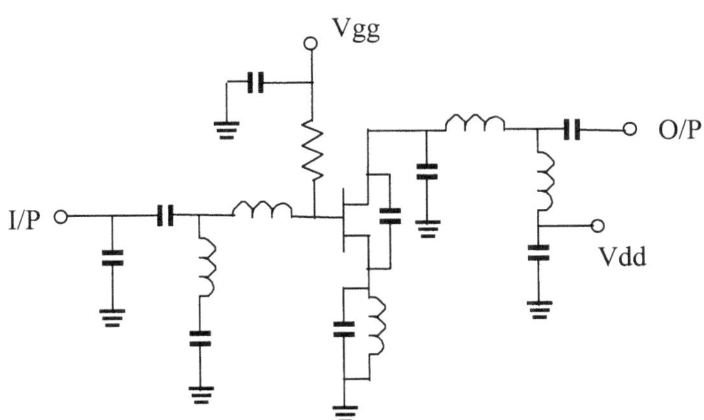

Figure 6.13 Analogue frequency divider circuit diagram

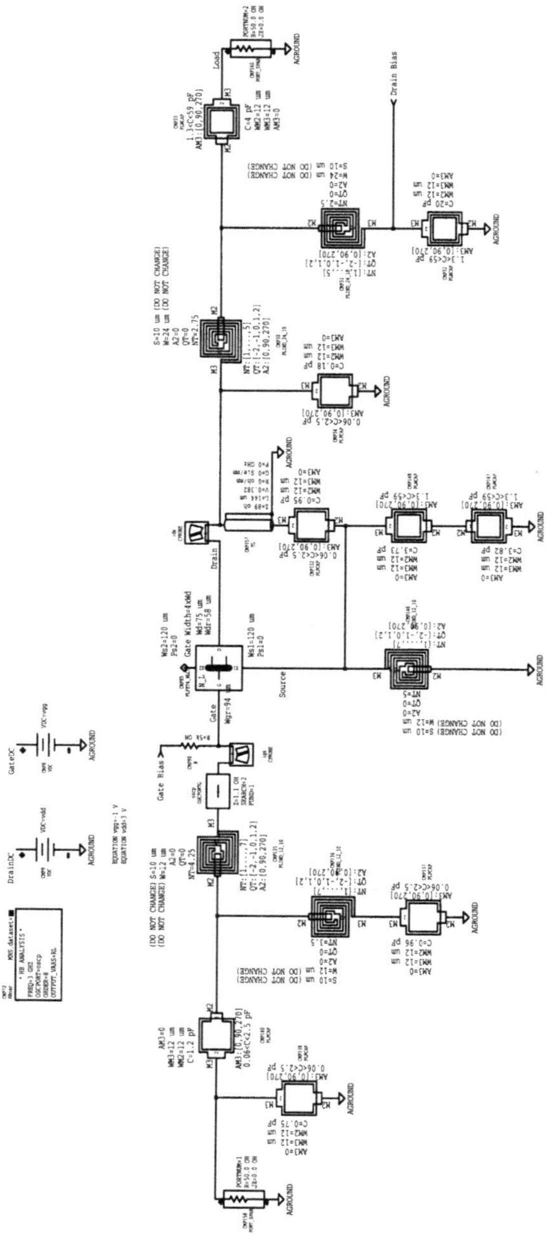

Figure 6.14 MDS™ schematic diagram of the analogue frequency divider design example

The first part of the design is to use small-signal analysis to develop a free-running oscillator that satisfies the oscillation condition over the required locking range. The circuit was analysed as a 2-port network, by breaking the connection between the input port of the oscillator and the driving source. The circuit was initially designed to have a free-running frequency at approximately 3.0 GHz. For the purpose of analysis, design information such as locking range and threshold input power level were determined by solving eqn. (6.9) numerically. Some circuit parameters were also optimised, particularly the series feedback and the matching networks, in order to maximise the locking bandwidth of the divider. Next, the Foundry models were substituted for all the lumped elements, as shown in Figure 6.14. In addition, Microwave Spice™ was used with the Foundry's large-signal FET model in order to verify the correct operation of the frequency divider.

This frequency divider circuit was fabricated using a standard 0.5 μm ion-implanted MMIC process. A microphotograph of the frequency divider chip is given in Figure 6.15. The actual size of the MMIC is less than 1.75 × 1.75 mm^2. The drain bias voltage was nominally 3 V. The drain current was set to 10 mA and 3 mA for the self-oscillating and non-oscillating mode of operation, respectively. Figures 6.16 and 6.17 show the measured locking performances of the divider circuit at different injected power levels. It is clear from the diagram that the locking characteristics of the two different modes are very similar. However, for the non-oscillating case, the circuit actually consumes much less DC power at the expense of some locking gain. For example, at an input power of 6 dBm, the self-oscillating circuit exhibited an output power of approximately 8 ± 0.5 dBm, over the operation frequency range from 5.4 to 6.2 GHz. A wide frequency bandwidth of almost 15% was achieved. The locking gain of the circuit was in the range of 2 to 15 dB, with an overall DC power consumption of about 30 mW.

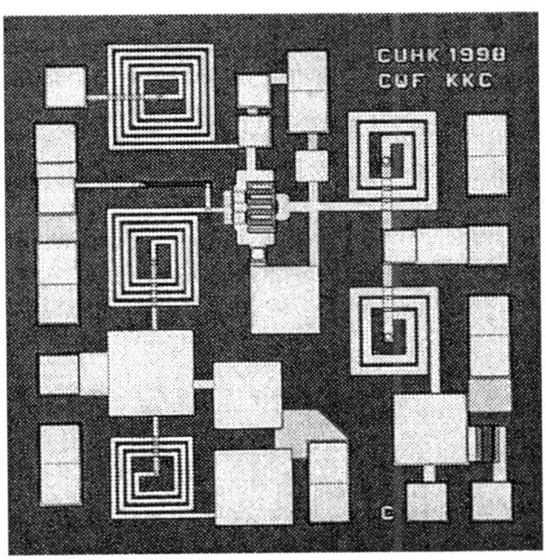

Figure 6.15 Microphotograph of the MMIC analogue 1/2 frequency divider

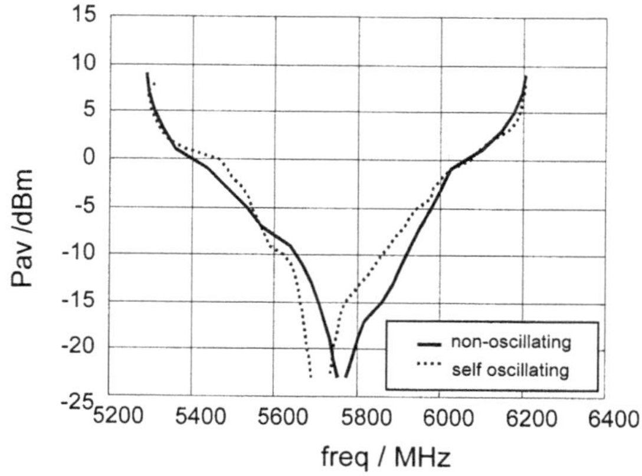

Figure 6.16 Locking characteristics of the analogue 1/2 frequency divider

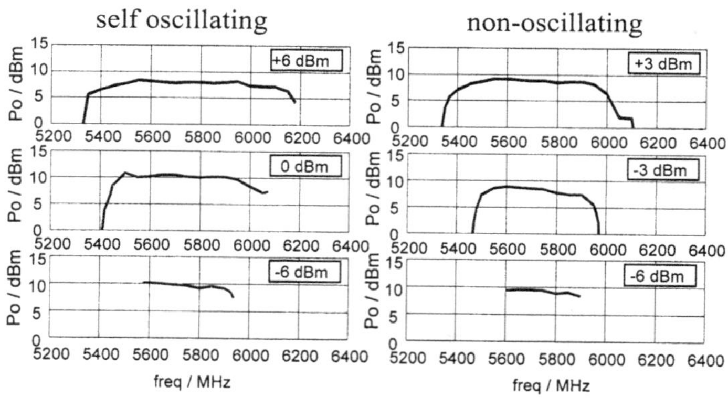

Figure 6.17 Power performance of the analogue 1/2 frequency divider

6.8 References

1 PUCEL, R. A. *et al.*: 'Experiments on integrated gallium-arsenide FET Oscillators at X Band', *Electronics Letters*, 1975, **11** (10), pp. 219–220

2 TREW, R. J.: 'Octave-band GaAs FET YIG-tuned oscillators', *Electronics Letters*, 1977, **13** (21), pp. 629–630

3 ISHIHARA, O. *et al.*: 'A highly stabilized GaAs FET oscillator using a dielectric resonator feedback circuit in 9-14 GHz', *IEEE Transactions on Microwave Theory and Techniques*, 1980, **MTT-28** (8), pp. 817–824

4 SCOTT, B. N., and BREHM, G. E.: 'Monolithic voltage controlled oscillator for X- and Ku-bands', *IEEE Transactions on Microwave Theory and Techniques*, 1982, **MTT-30** (12), pp. 2172–2177

5 SCOTT, B. N. *et al.*: 'A family of four monolithic VCO MIC's covering 2-18 GHz', *IEEE Microwave and Millimeter-wave Monolithic Circuits Symposium Digest*, 1984, pp. 58–61

6 MOGHE, S. B., and HOLDEN, T. J.: 'High-performance GaAs MMIC oscillators', *IEEE Transactions on Microwave Theory and Techniques*, 1987, **MTT-35** (12), pp. 1283–1287

7 ASBECK, P. M. *et al.*: 'Heterojunction bipolar transistors for microwave and millimeter-wave integrated circuits', *IEEE Transactions on Microwave Theory and Techniques*, 1987, **MTT-35** (12), pp. 1462–1468

8 REESE, E., and BEALL, J. M.: 'Optimized X & Ku-band GaAs MMIC varactor tuned FET oscillators', *IEEE MTT-S International Microwave Symposium Digest*, 1988, pp. 487–490

9 ANDREWS, J. E. *et al.*: '2.5 - 6.0 GHz broadband GaAs MMIC VCO', *IEEE MTT-S International Microwave Symposium Digest*, 1988, pp. 491–494

10 CHAN, O. Y., and KAZEMINEJAD, S.: 'Voltage-controlled oscillator using dielectric resonator', *Electronics Letters*, 1988, **24** (13), pp. 776–777

11 CHEN, S. *et al.*: 'U-band MMIC HBT DRO', *IEEE Microwave and Guided Wave Letters*, 1994, **4** (2), pp. 50–52

12 HIEN, D. *et al.*: 'Ka-band MMIC voltage-controlled oscillator', *1997 Asia Pacific Microwave Conference Digest*, Hong Kong, 1997, pp. 545–548

13 CHEN, W. *et al.*: 'A 2-V 2-GHz BJT variable frequency oscillators', *IEEE Journal of Solid-State Circuits*, 1998, **33** (9), pp. 1406–1410

14 JACOBSSON, H. *et al.*: 'Low phase noise, low power IC VCOs for 5-8GHz wireless applications', *IEEE MTT-S International Microwave Symposium Digest*, Boston, 2000, pp. 723–730

15 SVELTO, F. *et al.*: 'A 1.3 GHz low-phase noise fully tunable CMOS LC VCO', *IEEE Journal of Solid-State Circuits*, 2000, **35** (3), pp. 356–361

16 BURNS, L. *et al.*: 'Applications for GaAs and silicon integrated circuits in next generation wireless communication systems', *IEEE Journal of Solid-State Circuits*, 1995, **30** (10), pp. 1088–1095

17 KROEMER, H.: 'Heterostructure bipolar transistors and integrated circuits', *Proceedings of the IEEE*, 1982, **70**, pp. 13–25

18 BURGHARTZ, J. *et al.*: 'Integrated RF components in a SiGe bipolar technology', *IEEE Journal of Solid-State Circuits*, 1997, **32** (9), pp. 1440–1445

19 GOTZFRIED, R. *et al.*: 'Design of RF integrated circuits using SiGe Bipolar Technology', *IEEE Journal of Solid-State Circuits*, 1998, **33** (9) pp. 1417–1422

20 MANKU, T. *et al.*: 'Microwave CMOS — device physics and design', *IEEE Journal of Solid-State Circuits*, 1999, **34** (3), pp. 277–285,

21 ROFOUGARAN, A. *et al.*: 'A 1 GHz CMOS RF front-end IC for a direct-conversion wireless receiver', *IEEE Journal on Solid State Circuits*, 1996, **31** (7), pp. 880–889

22 LARSON, L. *et al.*: 'Integrated circuit technology options for RFIC's — present status and future directions', *IEEE Journal of Solid-State Circuits*, 1998, **33**(3), pp. 387–399

23 ABE, H.: 'A GaAs MESFET oscillator quasi-linear design method', *IEEE Transactions on Microwave Theory and Techniques*, 1986, **34** (1), pp. 19–25

24 ROBERTSON, I. D., CHENG, K. K. M., and AGHVAMI, A. H.: 'L-band 20 Watts GaAs FET amplifier for on-board digital mobile satellite systems', *Proceedings of IEEE International Conference on Communications Digest*, 1988, pp. 532–536

25 APRILLE, T. J., and TRICK, T. N.: 'A computer algorithm to determine the steady-state response of nonlinear oscillators', *IEEE Transactions on Circuit Theory*, 1972, **CT-19** (4), pp. 354–359,

26 RIZZOLI, V., and LIPPARINI, A.: 'A computer-aided approach to the nonlinear design of microwave transistor oscillators', *IEEE International Microwave Symposium Digest*, 1982, pp. 453–455

27 CHENG, K. K. M., and EVERARD, J. K. A. 'Nonlinear circuit analysis using the Newton-SOR continuation method', *Electronics Letters*, 1990, **26** (25), pp. 2120–2121

28 CURTICE, W.: 'Nonlinear analysis of GaAs MESFET amplifiers, mixers, and distributed amplifiers using the harmonic balance technique', *IEEE Transactions on Microwave Theory and Techniques*, 1987, **MTT-35** (4), pp. 441–447

29 BUSSGANG, J. J. *et al.*: 'Analysis of nonlinear systems with multple inputs', *Proceedings of the IEEE*, 1974, **62**, pp. 1088–1119

30 CHENG, K. K. M., and EVERARD, J. K. A.: 'A new and efficient approach to the analysis and design of GaAs MESFET microwave oscillators', *IEEE MTT-S International Microwave Symposium Digest*, 1990. pp.1283–1286

31 CURTICE, W. R.: 'GaAs MESFET modeling and nonlinear CAD', *IEEE Transactions on Microwave Theory and Techniques*, 1988, **36** (2), pp. 220-230

32 LEESON, D. B.: 'A simple model of feedback oscillator noise spectrum', *Proceedings of the IEEE*, 1966, **54** (2), pp. 329–330

33 KUROKAWA, K.: 'Some basic characteristics of broadband negative resistance oscillator circuits', *Bell System Technical Journal*, 1969, pp. 1937–1955

34 EDDISON, I., and HOWES, M. J.: 'Circuit aspects of the noise performance of microwave oscillator modules', *IEE Journal on Microwaves, Optics and Acoustics*, 1977, **1** (3), pp. 103–109

35 ESDALE, D. J., and HOWES, M. J.: 'A reflection coefficient approach to the design of one-port negative impedance oscillators', *IEEE Transactions on Microwave Theory and Techniques*, 1981, **MTT-29** (8), pp. 770–776

36 DEBNEY, B. T., and JOSHI, J. S.: 'A theory of noise in GaAs FET microwave oscillators and its experimental verification', *IEEE Transactions on Electron Devices*, 1983, **Ed-30** (7), pp. 769–775

37 SIWERIS, H. J., and SCHIEK, B.: 'Analysis of noise upconversion in microwave FET oscillators', *IEEE Transactions on Microwave Theory and Techniques*, 1985, **33** (3), pp. 233–241

38 ANZILL, W., and RUSSER, P.: 'A general method to simulate noise in oscillators based on frequency domain techniques', *IEEE Transactions on Microwave Theory and Techniques*, 1993, **MTT-41** (12), pp. 2256–2263

39 KARACAOGLU, U., LUCYSZYN, S., and ROBERTSON, I. D.: 'Modeling and design of X-band MMIC varactor-tuned VCOs', *Proceedings of IEE Colloquium on Modeling, Design, and Application of MMICs*, London, 1994, pp. 13/1–5

40 LUCYSZYN, S., LUCK, J., GREEN, G., and ROBERTSON, I. D., 'Enhanced modeling of interdigitated planar Schottky varactor diodes', *IEEE Asia-Pacific Microwave Conference Digest*, Adelaide, 1992, pp. 273–278

41 KUROKAWA, K.: 'Injection locking of microwave solid-state oscillators', *Proceedings of the IEEE*, 1973, **61**, pp. 1386–1410

42 GUSTAFSSON, L. *el al.*: 'Basic properties of subharmonic injection locking', *IEEE Transactions on Microwave Theory and Techniques*, 1973, **21** (1), pp. 28–34

43 RAUSCHER, C.: 'Regenerative frequency division with a GaAs FET', *IEEE Transactions on Microwave Theory and Techniques*, 1984, **32** (11), pp. 1461–1468

44 DIXON, J. *et al.*: 'Nonlinear time-domain analysis of injection-locked microwave MESFET oscillators', *IEEE Transactions on Microwave Theory and Techniques*, 1997, **45** (7), pp. 1050–1057

45 HUANG, C. C., and CHU, T. H.: 'Analysis of MESFET injection-locked oscillators in fundamental mode of operation', *IEEE Transactions on Microwave Theory and Techniques*, 1994, **42** (10), pp. 1851–1857

46 QUERE, R. *et al.*: 'Large signal design of broadband monolithic microwave frequency dividers and phase-locked oscillators', *IEEE Transactions on Microwave Theory and Techniques*, 1993, **41** (11), pp. 1928–1937

47 CHENG, K. K. M., and FAN, C. W.: 'A novel approach to the analysis of microwave regenerative analog frequency dividers', *IEEE Microwave and Guided Wave Letters*, 1998, **8** (7), pp. 266–267

48 HONJO, K., and MADIHIAN, M.: 'Novel design approach for X-band GaAs monolithic analog ¼ frequency divider', *IEEE Transactions on Microwave Theory and Techniques*, 1986, **34** (4), pp. 436–441

49 DESGREZ, S. *et al.*: 'Wide-bandwidth Ku-band monolithic analog frequency divider', *IEEE Microwave and Guided Wave Letters*, 1998, **8** (2), pp. 84–86

50 FAN, C. W., and CHENG, K. K. M.: 'A new method in characterizing the nonlinear current model of MESFETs using single-tone excitation', *IEEE MTT-S International Microwave Symposium Digest*, 2000, pp. 449–452

Chapter 7

Mixers

K. S. Ang and S. J. Nightingale

7.1 Introduction

Mixers are key elements in nearly all RF and microwave communication receivers, transmitters and signal generators. They have the prime function of converting signals from one frequency to another, where other circuit functions like amplification, filtering, detection and transmission can be performed more effectively. Mathematically, this frequency translation can be performed by simply multiplying the input RF signal, at frequency ω_{RF}, by another signal, at frequency ω_{LO}, referred to as the local oscillator (LO):

$$\cos(\omega_{RF}t) \,.\, \cos(\omega_{Lo}) = 0.5\cos(\omega_{RF}t - \omega_{LO}t) + 0.5\cos(\omega_{RF}t + \omega_{LO}t) \qquad (7.1)$$

This is shown symbolically in Figure 7.1. The mixer output consists of the sum and difference frequencies of the input signal and the LO. In the case of down-conversion, the input is a high frequency RF signal and the desired output is the lower frequency IF signal, $|\omega_{RF} - \omega_{LO}|$. The reverse applies in the case of up-conversion, where the input is a low frequency IF signal and the desired output is the higher frequency RF signal, $|\omega_{RF} + \omega_{LO}|$. In each case, the desired signal is usually obtained by filtering.

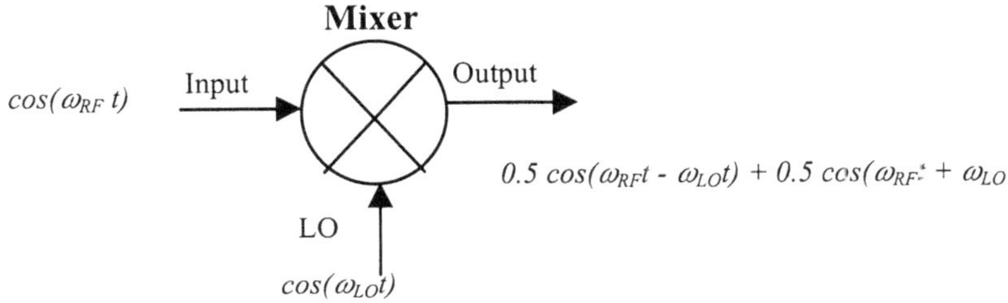

Figure 7.1 Frequency translation by multiplication of two sinusoids

Practically, eqn. (7.1) can represent a circuit equation in which a current waveform $i(t)$ is generated by the product of the signal voltage, $v_{RF}(t)$, and a conductance, $g(t)$:

$$i(t) = v_{RF}(t) \cdot g(t) \tag{7.2}$$

where the conductance has a time-varying behaviour which varies at the rate of the LO frequency. This can be generated by a voltage-controlled conductance which varies directly with the LO voltage waveform:

$$g[v_{LO}(t)] \propto v_{LO}(t) \tag{7.3}$$

Thus, when the LO voltage is applied, a conductance waveform which varies linearly with LO voltage is generated. Applying the RF signal to this time-varying conductance waveform will generate the desired sum and difference frequencies of the LO and RF. In general, these frequencies can be generated by any linear voltage- or current-controlled circuit element having the transfer function:

$$g[V] = a + bV, \qquad \text{where } a, b \text{ are constants} \tag{7.4}$$

Unfortunately, most of the available voltage- or current-controlled circuit elements are non-linear in nature. Their non-linear transfer function can be expressed as a power series:

$$g[V] = a + bV + c\,V^2 + \mathrm{d}\,\mathrm{V}^3 \ldots, \text{ where } a, b, c, d \ldots \text{ are constants} \tag{7.5}$$

Consequently, the time-varying waveform generated by the LO voltage contains a large number of LO harmonics. Applying the RF voltage to this time-varying waveform will result in mixing products between the RF and all harmonics of the LO. The frequencies of these mixing products have the general form:

$$\omega_n = \pm\omega_{RF} \pm n\ \omega_{LO} \qquad n \in [0, 1, 2, 3..] \tag{7.6}$$

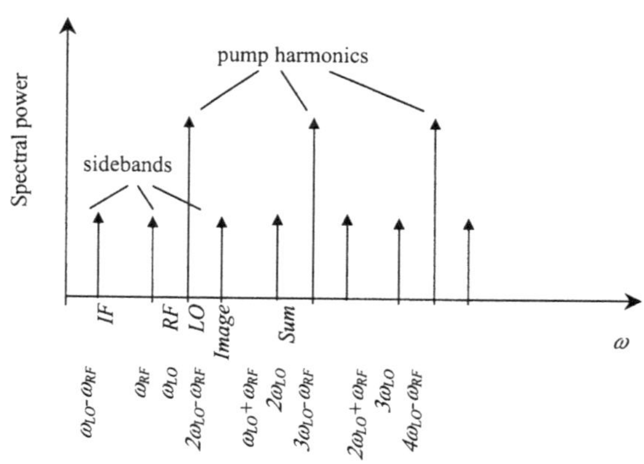

Figure 7.2 Spectrum of frequencies of the form $\pm\omega_{RF} \pm \mathrm{n}\ \omega_{LO}$

Figure 7.2 shows the spectrum of mixing frequencies. In addition to the desired $\pm\omega_{RF} \pm\omega_{LO}$ signals, a large number of frequency components where $n \neq 1$ are also generated. These are referred to as spurious signals or spurious responses.

So far, we have assumed that the voltage- or current-controlled circuit element is only a function of the LO voltage or current. This is normally the case when the LO signal is much larger than the RF signal, as its sole function is to generate the time-varying circuit element waveform. When the RF signal amplitude increases, however, these circuit elements can be a function of both the LO and RF signal amplitudes. In this case, the time-varying waveform will contain harmonics of both the LO and RF. Consequently, the mixer output spectrum will contain mixing products of all harmonics of the LO and RF. These frequencies can be expressed in the general form:

$$\omega_{mn} = \pm\, m\ \omega_{RF} \pm n\ \omega_{LO} \quad \text{where } m, n \in [0, 1, 2, 3..] \tag{7.7}$$

Therefore, a major drawback of employing non-linear devices as mixing elements is the generation of a large number of spurious mixing products.

7.2 Mixer analysis

7.2.1 General analysis

In the general analysis, no assumptions are made about the relative magnitudes of the LO or input signal(s). To analyse this general case, a full non-linear analysis must be performed where both signals are present. A theoretical implementation is difficult, and such analyses are computationally intensive compared with the restricted case. It requires a robust convergence algorithm using time-domain or harmonic-balance methods, which makes use of a special type of Fourier transform for multiple non-harmonically related signals. Such a general analysis can be simulated with a range of different CAD software packages described in Chapter 4. Mixer products are produced with the form in eqn. (7.7).

7.2.2 Restricted analysis

In this analysis, it is assumed that the LO is much larger than the RF signal and that it is solely responsible for the time-varying circuit elements. The analysis can then be made in two parts: first, a large-signal analysis is performed using only the LO, and second, a linear analysis is performed using the time-varying circuit elements derived from the large-signal analysis. The former is performed using time-domain or harmonic-balance techniques. The latter is usually performed in the frequency domain using the Fourier components of the time-varying circuit elements to set up a conversion matrix. The mixing products generated have the form in eqn. (7.6), and there is a linear relationship between the RF and IF signal amplitudes. This is commonly referred to as linear mixing.

7.3 Background reading

There have been numerous publications over the past 40 years relating to the theoretical and practical aspects of mixer design. One of the most useful books, in the authors' opinion, is that by Tucker [1]. This book treats mixers, modulators and parametric converters and amplifiers purely in circuit terms, and lays down the basic principles in a clear and concise way. Although the examples given are for two terminal devices (diodes), the analytical techniques can also be applied to three (single-gate FET) or four (dual-gate FET) terminal devices.

In 1971, Saleh [2] wrote a research monograph on mixers using non-linear resistors (ideal diodes). This work was a milestone in classifying circuits into different types, such as the Z-, Y-, G- and H-mixers, and increased the general understanding of mixer operation. His classification was based on the parameters and matrix representation required to perform conversion loss analysis. The book by Maas [3], published in 1986, presents an up-to-date treatment of mixers covering both diode and FET-based circuits. Another book [4] by the same author, published in 1988, covers the main techniques in use today for solving for the large-signal performance of mixers and power amplifiers. The book covers solid-state device modelling, harmonic balance and large-signal/small-signal analysis, Volterra-series and power-series analysis, together with applications to diode and FET circuits such as mixers, multipliers and power amplifiers.

7.4 Analysis of mixer circuits

In order to provide some understanding of mixer operation, a simple single loop circuit will now be analysed.

7.4.1 Analysis of a simple single-loop circuit

Consider the simple single-loop circuit shown in Figure 7.3. This represents a single-ended mixer design, where all mixed products and intermodulation products are terminated in the source and load resistances, i.e. a wideband mixer design. This circuit will now be examined using both general non-linear analysis and a method of linear circuits with time-varying parameters. The latter method is also referred to as the conversion loss matrix method.

7.4.2 General non-linear analysis

We begin by combining the source and load resistances with the non-linear device to form a new non-linear equation:

$$I = aV + bV^2 + cV^3 + dV^4 + \dots \tag{7.8}$$

where a, b, c, d are constants and V is the sum of all applied voltages. In the present analysis, we will restrict the power series to the first two terms. After substituting for V and some trivial manipulation, we have:

$$I = aV + bV^2$$

$$= aV_b + b(V_b^2 + \frac{1}{2}(V_{LO}^2 + V_{RF}^2)) \quad \text{(DC)}$$

$$+ (a + 2\,bV_b)\,V_{RF}\cos\omega_{RF}t \quad \text{(Signal)}$$

$$+ \frac{1}{2}\,bV_{RF}^2\cos 2\omega_{RF}t \quad \text{(Signal second harmonic)}$$

$$+ (a + 2\,bV_b)V_{LO}\cos\,\omega_{LO}t \quad \text{(Pump)}$$

$$+ \frac{1}{2}\,bV_{LO}^2\cos 2\omega_{LO}t \quad \text{(Pump second harmonic)}$$

$$+ bV_{LO}V_{RF}\cos(\,\omega_{LO} + \omega_{RF})t \quad \text{(Sum frequency)}$$

$$+ bV_{LO}V_{RF}\cos(\,\omega_{LO} - \omega_{RF})t \quad \text{(Difference or intermediate frequency)} \qquad (7.9)$$

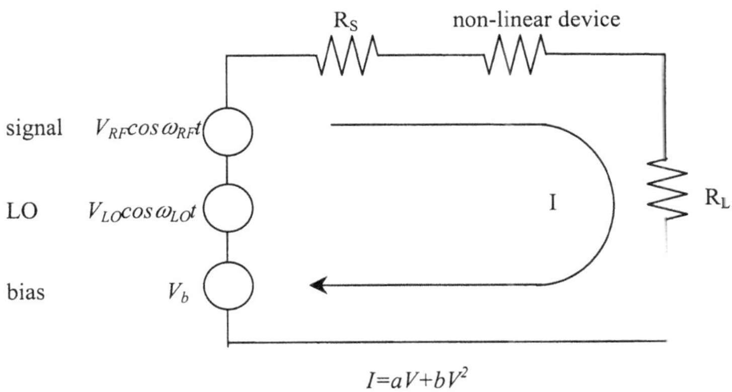

Figure 7.3 A simple single-ended mixer

7.4.2.1 Conversion loss

The power available from the signal, P_{av}, is given by:

$$P_{av} = \frac{V^2}{8R_S} \qquad (V = V_{LO} \text{ or } V_{RF}) \qquad (7.10)$$

Power into the load at any frequency is given by:

$$P_{out} = \frac{I^2 R_L}{2} \tag{7.11}$$

Conversion loss (*CL*) is therefore given by:

$$CL = \frac{V^2}{4I^2 R_S R_L}, \qquad \text{where } V = V_{LO} \text{ or } V_{RF} \tag{7.12}$$

For ω_{RF} to $\omega_{LO} - \omega_{RF}$ and $\omega_{LO} - \omega RF$:

$$CL = \frac{1}{4R_S R_L (bV_{LO})^2} \tag{7.13}$$

For ω_{RF} to $2\omega_{RF}$ and ω_{LO} to $2\omega_{LO}$:

$$CL = \frac{1}{2\, R_S R_L b^2} \tag{7.14}$$

7.4.2.2 Input and output impedance

The input or output impedances of the circuit at a particular frequency are obtained by exciting the circuit with a voltage generator at this frequency and finding the current which flows at this frequency. In any practical measurement of this parameter it must be possible to separate the required current from others that are flowing in the circuit.

$$Z_{in} = \frac{V \text{ at excitation frequency}}{I \text{ at excitation frequency}} - R_S \tag{7.15}$$

$$Z_{out} = \frac{V \text{ at excitation frequency}}{I \text{ at excitation frequency}} - R_L \tag{7.16}$$

For ω_{RF} and ω_{LO}:

$$Z_{in} = \frac{1}{(a + 2bV_b)} - R_S \tag{7.17}$$

For $\omega_{LO} - \omega_{RF}$ to $\omega_{LO} + \omega_{RF}$:

$$Z_{out} = \frac{1}{(a + 2bV_b)} - R_L \tag{7.18}$$

7.4.2.3 Isolation

The isolation (IS) between ports is given by the ratio of the power available from the source to the power dissipated in the load at the same frequency. For ω_{RF} and ω_{LO}:

$$IS = \frac{1}{4R_S R_L(a + 2bV_b)^2} \tag{7.19}$$

7.4.3 Conversion loss matrix method

In this approach, the LO is assumed to be solely responsible for mixing. Under these conditions, the mixer can be analysed in two stages:

1. a non-linear analysis to determine the steady-state time-varying conductance waveform
2. a linear analysis to determine the small-signal performance.

The current waveform is given by:

$$I = aV + bV^2 \tag{7.20}$$

where V is the sum of LO and bias voltages.

The conductance waveform is given by

$$G = \frac{dI}{dV} = a + 2bV \tag{7.21}$$

and therefore:

$$G = a + 2b(V_{LO} \cos \omega_{LO} t + V_b) \tag{7.22}$$

The currents flowing in this time-varying conductance waveform are now given by:

$$I(t) = V(t)G(t) \tag{7.23}$$

$$I = V_{RF} \cos \omega_{RF} t\left(a + 2b(V_{LO} \cos \omega_{LO} t + V_b)\right) \tag{7.24}$$

$$\begin{aligned} &= (a + 2bV_b)V_{RF} \cos \omega_{RF} t && \text{(signal)} \\ &\quad + bV_{LO}V_{RF} \cos(\omega_{LO} + \omega_{RF})t && \text{(sum)} \\ &\quad + bV_{LO}V_{RF} \cos(\omega_{LO} - \omega_{RF})t && \text{(difference frequency (IF))} \end{aligned}$$

7.4.3.1 Conversion loss

The conversion loss for ω_{RF} to $\omega_{LO}+\omega_{RF}$ and ω_{RF} to $\omega_{LO}-\omega_{RF}$ is now given by:

$$CL = \frac{1}{4R_S R_L(bV_{LO})^2} \tag{7.25}$$

This is identical to the result given in eqn. (7.13). The results for the input and output impedance and signal isolation will also be found to be identical.

The above is a very simple example to illustrate how the mixer performance can be determined when the device non-linearity is expressed as the first two terms of a power series. Such a device does not exist in practice, because although some two-terminal devices have a square law characteristic in the forward direction, they have very little conduction in reverse. The characteristic can then be represented by either a much higher order power series [5] or by splitting the forward and reverse parts of the characteristic, and then determining the Fourier series for the conductance waveform separately [6]. When the diode is being driven hard it can be considered as a sampler, switching between two states. This bilinear or switch model has been used extensively to analyse the performance of diode mixers and will be referred to in the forthcoming sections.

The references are divided into eight areas: those giving background information on mixer operation and performance [1–15], diode mixers [16–58], coupling structures [59–81], active FET mixers [82–109], resistive FET mixers [110–126], image-rejection and single-sideband mixers [127,128], sub-harmonic mixers [129–133], and distributed mixers and coupling structures [134–143]. Reference 16 onwards are given in chronological order for ease of identification.

Further details on the operation of various diode and FET mixer types, and the required coupling structures, will now be discussed separately.

7.5 Diode mixers

The Schottky barrier diode has been the primary non-linear device for microwave mixers. Figure 7.4 shows its equivalent circuit. Also indicated are the key equations defining the non-linearities, C_j and R_j, which are responsible for the mixing operation. The parasitic series resistance, R_s, has to be minimised, in order to achieve high diode cut-off frequency and low mixer conversion loss. This requires a low resistance current path within the diode structure, which is implemented by a highly doped active layer below the cathode.

In monolithic circuits, the fabrication process for Schottky diodes is generally not compatible with those for MESFETs, although fabrication processes incorporating both structures have been reported [35]. Consequently, implementing diodes using MESFET based processes usually involves some derivatives of the MESFET structure. For example, the gate finger can be utilised as the anode of a Schottky diode contact by connecting the drain and source fingers together to form the cathode. However, these diodes have high series resistance due to the absence of a highly doped active layer below the gate Schottky contact. This results in rather low diode cut-off frequencies, of around 200–300 GHz, compared to those realised with a planar diode process using a vertical structure, which can achieve cut-off frequencies of 1000–1500 GHz without recourse to sub-micron geometries. Thus, mixers realised using these sub-optimum diodes have higher conversion losses than those fabricated using a diode-based fabrication process.

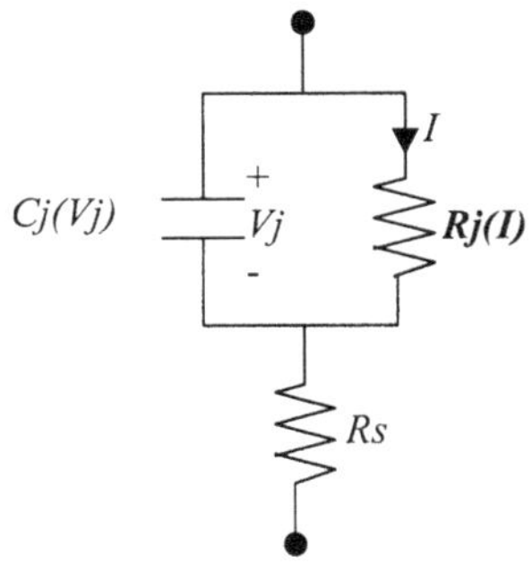

$$I = I_s(e^{\alpha V_j} - 1)$$

$$Cj = \frac{C_j(0)}{\left(1 - \frac{V_j}{\phi}\right)^{\gamma}}$$

Figure 7.4 Schottky diode equivalent circuit model

7.5.1 Diode mixer design considerations

We have seen in Section 7.1 that the use of non-linear elements for frequency conversion generates many mixing product frequencies. This is indeed the case in diodes, which are highly non-linear due to the exponential nature of their I–V characteristics. The objectives of diode mixer design is to recover the input signal at the desired frequency with minimum signal loss and noise performance degradation, and to minimise the signal level in other spurious mixing products.

7.5.1.1 Conversion loss

To achieve minimum conversion loss, the following two fundamental conditions must be satisfied:

1. minimum power from the input signal and mixed product frequencies is dissipated in the non-linear element(s)
2. all out-of-band mixed product frequencies are terminated reactively.

However, it should be noted that this is, in general, contrary to the requirements for low out-of-band mixing and intermodulation product levels, where it is 'good mixer practice' to terminate out-of-band products resistively [7]. An appropriate compromise must be made, depending on the system application. In a wideband ESM receiver, for example, low spurious signal levels are more important than low loss and noise figure, whereas these are not so critical in a narrowband satellite communications receiver, where they can be removed by filtering.

Condition (1) is obvious, and ideally, would be satisfied by a device which at no time during the LO cycle assumed an impedance with a resistive component. Although this condition is satisfied by a non-linear capacitor, the Manley–Rowe relationships [8,9] show that a parametric down-converter has a power loss which is equal to the ratio between the input and output frequencies in the unconditionally stable state. These points suggest that the optimum device would be one switching between two resistance values, which, for a zero loss mixer, tend to 0 and ∞, i.e. a short and open circuit. Saleh [2] presented a rigorous mathematical proof for this and showed that the optimum time-varying conductance or resistance waveform was a rectangular pulse or group of pulses.

Although it is possible to simulate this ideal switch at low frequencies using, for example, mechanical or micro-electromechanical switches, normally it is necessary to utilise the non-linear properties of a semiconductor or thermionic device. The Schottky diode has been used almost exclusively in low loss mixers at microwave frequencies. To obtain low loss, it is necessary to pump the diode strongly, so that the junction closely approximates a fully on or off state, and therefore dissipates minimum power. Under high drive conditions, a typical diode impedance is limited by the series resistance in the forward direction, and the junction capacitance in parallel with a high resistance, representing the leakage, in the reverse direction. The time varying impedance or admittance then tends to a rectangular pulse waveform, and a particular value of pump voltage or current and DC bias will determine how long the device conducts for each pump cycle. This two-state, or bilinear, model has been used extensively in the past to determine the optimum conditions for minimum loss in a specified embedding structure, as well as to investigate other mixer phenomena. Although such a model does provide considerable insight into optimum operating conditions, it has some limitations at low pump power levels. The main advantage of this model is the ease with which a solution for the conversion loss performance may be obtained for a specified embedding structure. Once the device parameters and embedding structure have been defined, the only variables are the source and load impedances and the conduction time per pump cycle or pulse duty ratio (PDR).

Condition (2) must be satisfied in order to avoid power loss at out-of-band frequencies, although there are cases where the difference between total and selective reactive termination is small. Since a reactive termination can vary from an open to a short circuit, it is informative to note the general effects of these. Figure 7.5, taken from a study by the second author [10], shows how the conversion loss of a mixer can be improved by progressively terminating more higher order mixing products in a short circuit. The main conclusions of this work on single loop circuits were as follows:

a) As more currents are terminated in a short circuit (current idler) the attainable minimum conversion loss decreases, the optimum pulse duty ratio decreases and the optimum source and load impedances increase. In the limit, where all higher order mixed products are terminated in a short circuit, we have a Y-mixer. In this circuit, with a high-Q device, the optimum operating conditions are very high source and load impedances and a very small pulse duty ratio, tending to 0; i.e. impulse conditions.

b) The converse of (a) is also true. In the limit, where all higher order mixed products are terminated in an open circuit (voltage idler), we have a Z-mixer. In this circuit, with a high-Q device, the optimum operating conditions are very low source and load impedances and a very large pulse duty ratio, tending to 1; i.e. inverse impulse conditions.

c) Following from (a) and (b), it is possible to select a pulse duty ratio and terminate the mixed products reactively but in different ways (i.e. s/c, o/c or somewhere in between) so that the minimum conversion loss is realised. Specific cases of this are the G- (and H-) mixers, where odd order products are terminated in an open (short) circuit and even order products are terminated in a short (open) circuit. For these two mixers, the optimum pulse duty ratio is 0.5.

d) In general, once reactive terminations have been obtained by correctly terminating the sum and image frequencies, there is little to be gained by attempting to recover energy from other mixed products, unless a high-Q medium for the embedding structure and a high-Q device are available. Consideration of several microstrip structures shows that the extra potential improvement in conversion loss achieved by terminating higher order mixed products reactively is usually more than cancelled out by the additional loss of the embedding circuit.

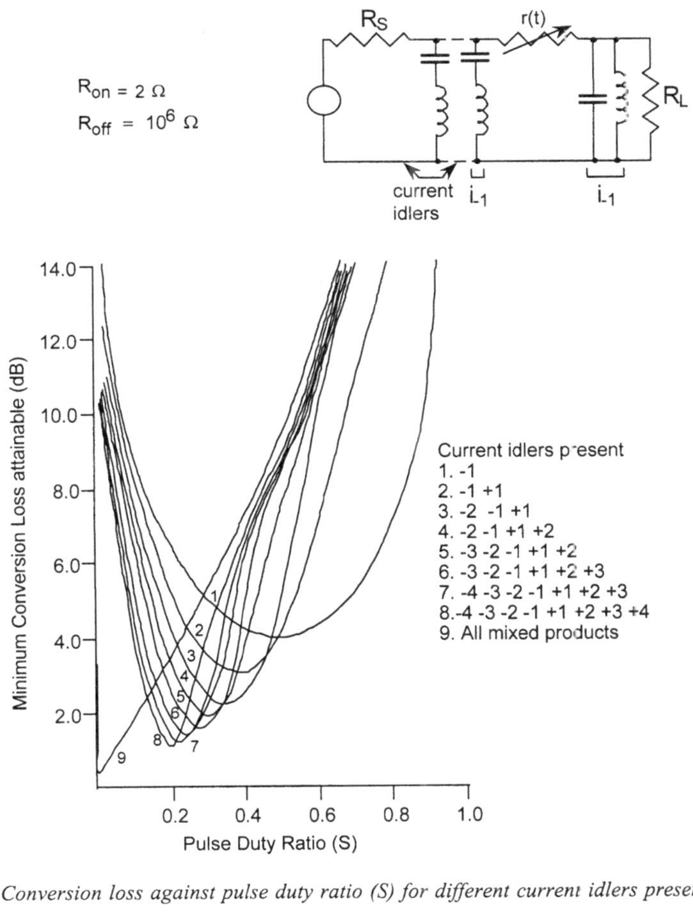

Figure 7.5 Conversion loss against pulse duty ratio (S) for different current idlers present in a mixer

e) In most practical cases, the non-linear device will have parasitic capacitance, inductance and resistance. The main parasitics in a Schottky barrier diode are the 'on' resistance, which is primarily due to the device series and junction resistance and the 'off' capacitance, which is due to the junction. The latter tends to progressively 'short circuit' higher order mixed products across the diode junction. Therefore, the Y-mixer is the single most commonly used circuit at microwave frequencies. Consideration of the noise performance shows that it is preferable to use short circuit terminations to avoid parametric amplification of noise from mixed products to the IF.

f) In mixers where high-Q embedding circuits cannot be realised (e.g. microstrip), the device's zero-bias junction capacitance should have a value such that the reactance at the signal frequency is typically twice the signal source impedance. Once this has been achieved, the series resistance should be made as small as possible.

7.5.1.2 Noise figure

The noise figure of a mixer is quite difficult to calculate for the general case. However, if it can be assumed that the dominant noise mechanisms are shot and thermal noise, i.e. pump noise and other device noise (such as $1/f$ noise) are assumed to be zero, the situation is simplified considerably.

It was shown by the second author [10,11], developing the work of Nemlikher and Strukov [12,13], that the noise figure, F_m, of a mixer using Schottky barrier diodes could be represented by the following equations:

$$t_{eq} = \frac{F_m - 1}{L_i - 1} \tag{7.26}$$

$$= \frac{p_r + \frac{n}{2} p_j}{p_r + p_j} \tag{7.27}$$

$$L_m = L_i L_{mm} \tag{7.28}$$

where t_{eq} is the mixer homogeneous equivalent noise temperature ratio
F_m is the mixer noise figure
L_i is the mixer intrinsic conversion loss
p_r, p_j are the normalised power coefficients which account for the proportions of power dissipated in the circuit resistances and device junction(s), respectively
n is the ideality factor of the diode (typically 1.1)
L_m is the mixer conversion loss
L_{mm} is the mixer mismatch loss at the input

This equation assumes the mixer is at 290 K. It should be noted that an alternative definition for the noise temperature ratio, t_m, of a mixer with total loss L_m is given by:

$$t_m = \frac{F_m}{L_m} \tag{7.29}$$

This definition is commonly used because it is compatible with Friis' formula for cascaded networks. Although it is one way to represent the noise performance of a mixer, or any lossy network in general, unfortunately it does not help the analysis or understanding of the problem because both types of loss are lumped together.

When the diode is weakly driven (i.e. the pump drive power is low), the signal and IF are perfectly matched and all out-of-band mixed products are terminated reactively, then:-

$$p_j \gg p_n$$

$$t_{eq} \rightarrow n/2$$

$$\therefore \quad F_m \rightarrow nL_i/2 + (2-n)/2 \tag{7.30}$$
$$\sim nL_i/2 \text{ when } L_i \text{ is large}$$

This situation is seldom of practical interest, unless the resistive losses are very low, because the conversion loss will be very high.

In most mixers operating under practical pumping conditions, p_r is much greater than p_j [10]. Indeed, under strong pumping conditions where the junction is made to switch hard between the 'on' and 'off' state, close to zero power is dissipated in the junction. Hence, p_j tends to zero and t_{eq} tends to 1:

$$p_j \rightarrow 0$$

$$t_{eq} \rightarrow 1$$

$$\therefore \quad F_m = L_i \tag{7.31}$$

The overall noise figure, F_t, of a mixer in cascade with an IF amplifier (having a noise figure F_{if}), can be found using the Friis formula for cascaded networks, viz:

$$F_t = F_m + (F_{if} - 1)\, L_m$$
$$= F_{if} L_m \tag{7.32}$$

7.5.2 Single-ended diode mixers

One of the most basic forms of mixer circuit is the single-ended diode mixer shown in Figure 7.6. It is often used as a basis for single-balanced mixers. It employs a single diode as the mixing element with passive networks to perform functions of signal isolation, impedance matching and presenting appropriate terminations for the LO harmonics and undesired mixing frequencies. Other circuit requirements include DC biasing and current return for the diode as well as any rectified current generated by the LO.

With simple mixers, the primary function of the passive networks is signal separation, while impedance matching requirements are secondary. Therefore, filters are generally used at the LO, RF and IF ports for signal isolation. This is only applicable when these frequencies are clearly separated. Moreover, high-Q filters are required to achieve good isolation. Therefore, single-ended diode mixers are rarely used, especially in MMICs, since filter structures cannot achieve high-Q performance. They are only found in millimetre-wave mixers, where high-Q waveguide filters are relatively small and scarce LO power only permits the use of a single device.

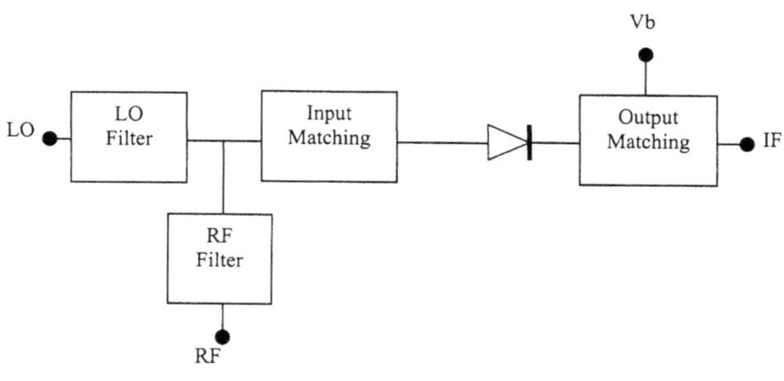

Figure 7.6 Single-ended diode mixer

The input and output matching networks are designed to conjugate match the LO, RF and IF port impedances to the diode. As the LO and RF are applied to the same diode terminal, it may not be possible to simultaneously match the LO to the diode large-signal impedance and the RF to the pumped diode input impedance. To minimise conversion loss and noise figure, the input matching network is usually designed to match the RF to the diode under typical pumping conditions.

In narrow band designs, the required terminations to be presented to the LO harmonics can be obtained using simple stubs. Figure 7.7 shows the microstrip implementation of the diode mixer input and output networks. The LO/RF input has a short circuit stub which is $\lambda g/4$ at the LO frequency. It presents a short circuit to the LO second harmonic frequency. The mixer IF output has two open circuit stubs: the $\lambda g/4$ open circuit stub presents a short circuit for the LO and its odd harmonics, whereas the $\lambda g/8$ open circuit stub presents a short circuit for the LO second harmonic and odd order multiples of it. In combination, they approximate a parallel tuned circuit providing short circuits to the LO harmonics.

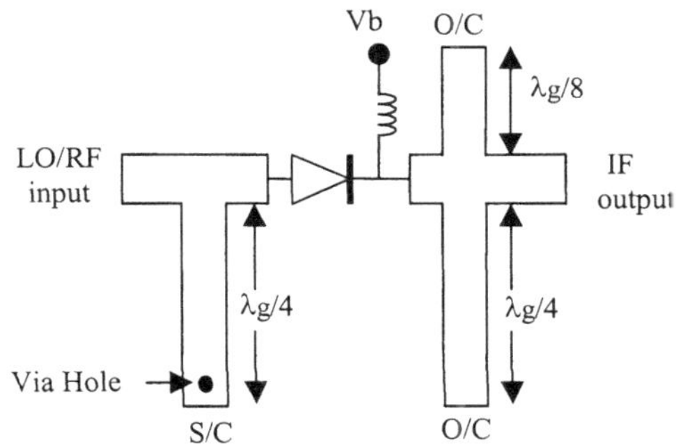

Figure 7.7 Microstrip implementation of a single-ended mixer

7.5.3 Single-balanced diode mixers

In single-balanced mixers, two signals are mutually isolated due to the circuit geometry rather than the use of filters. Usually, the LO and RF are mutually isolated and separated from the IF port by means of a low-pass filter. Figure 7.8 shows single-balanced mixers realised by combining two mixing diodes in parallel with a hybrid coupler. The LO and RF are applied to the isolated ports of the hybrids to achieve mutual isolation between these signals. The hybrid can be a 180° hybrid, such as the rat-race coupler, or a 90° hybrid, such as the branch line or Lange coupler. Also required are the low-pass filter for IF extraction, DC biasing and matching networks, which are not shown in Figure 7.8.

Besides inherent LO-RF isolation, single-balanced mixers also provide rejection of AM noise in the LO and suppression of some spurious mixing products. The mixing products that are suppressed depend on the relative phase relationships of the LO and RF in the two mixing diodes. These phase relationships are indicated in Figure 7.8 as θ_{LO1}, θ_{RF1}, and θ_{LO2}, θ_{RF2}. Taking into account these phase relationships, the voltages, V_1 and V_2, across the two diodes can be expressed as:

$$V_1 = V_{LO}\cos(\omega_{LO}t + \theta_{LO1}) + V_{RF}\cos(\omega_{RF}t + \theta_{RF1}) \tag{7.33}$$

$$V_2 = V_{LO}\cos(\omega_{LO}t + \theta_{LO2}) + V_{RF}\cos(\omega_{RF}t + \theta_{RF2}) \tag{7.34}$$

The currents, I_1 and I_2, in each diode can then be evaluated as a power series, using eqn. (7.8). As the two diode terminals are reversed, the total current flowing in the IF port is given by I_1-I_2. The resulting expansion for the IF current will indicate which of the mixing products, in the general form $\pm\, m\omega_{RF} \pm n\omega_{LO}$, are cancelled out at the IF port. For Figure 7.8*a*, where the LO is applied to the Δ port and the RF is applied to the Σ port of a 180° hybrid, all the mixing products corresponding to *m* even are suppressed. If the LO and RF ports are reversed, as in Figure 7.8*b*, then all the mixing products corresponding to *n* even are suppressed. For the 90° hybrid in Figure 7.8*c*, only the mixing products corresponding to m and n even are suppressed. In addition, the sum frequency, $\omega_{RF} + \omega_{LO}$, is also suppressed. For up-conversion applications, an IF balun is required in Figure 7.8*c*. In this case, the difference frequency, $\omega_{RF} - \omega_{LO}$, is suppressed. Besides differences in mixing product suppression, the use of 180° hybrids generally gives better LO and RF isolation, while 90° hybrids give better LO and RF port matching.

The anti-phase LO relationship in Figure 7.8*a* can also be achieved over a broad bandwidth using a LO balun. An RF filter is then required to apply the RF in phase. Figure 7.9 shows two possible locations for applying the RF. At low frequencies, a classical transformer can be used for the LO balun. When an LO transformer is used in Figure 7.9*a*, the virtual ground at the centre tap of the transformer can be used for applying the RF without an RF filter. As the LO and RF phase relationships are the same as Figure 7.8*a*, it also suppresses spurious mixing products corresponding to n even.

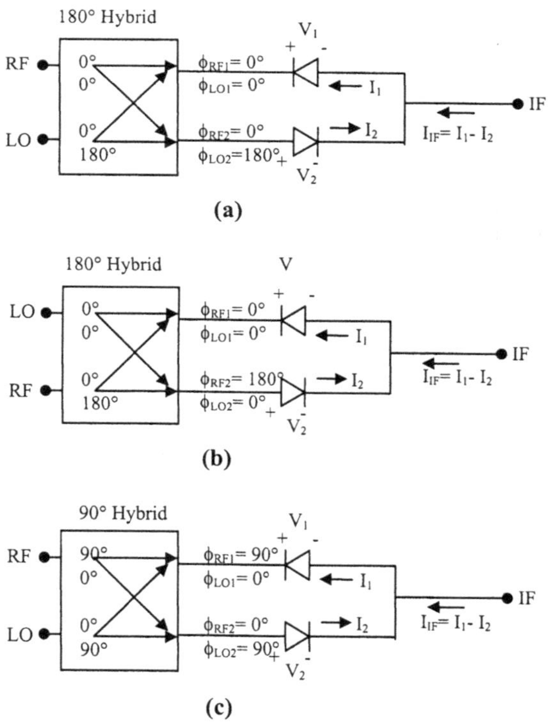

Figure 7.8 Single-balanced mixers using hybrids

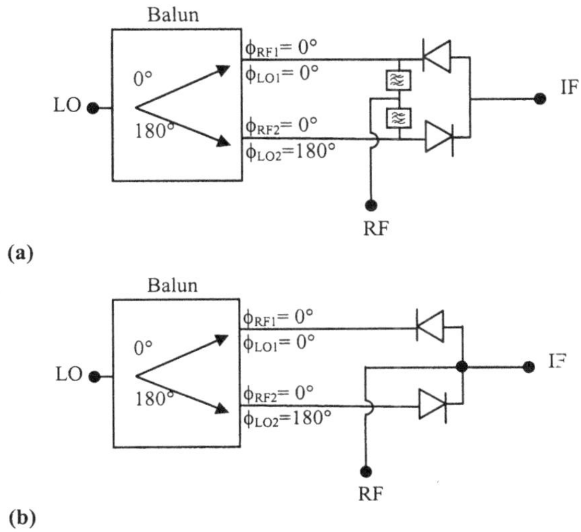

Figure 7.9 Single-balanced mixers using baluns

7.5.4 Double-balanced diode mixers

In double-balanced mixers, all the three mixer ports are mutually isolated, due to the balanced nature of the circuit geometry rather than the use of filters. As single-balanced mixers have inherent LO–RF isolation, the additional isolated IF port can be obtained by applying balanced LO and RF signals to two single-balanced mixers, as shown in Figure 7.10. Due to the anti-phase relationships of the LO and RF, the IF port presents a virtual ground to these signals, thus ensuring LO-IF and RF-IF isolation. If two single-ended mixers of Figure 7.9*a* are combined, the result is the ring diode mixer in Figure 7.11*a*. The three resulting LO baluns are replaced by a single LO balun. Similarly, combining two single-ended mixers of Figure 7.9*b* will result in the star diode mixer in Figure 7.11*b*. Also indicated in Figure 7.11 are the balanced outputs for the IF. For the star mixer, the IF can be extracted from the common node of the diodes. The DC grounds in the LO and RF baluns provide the IF current return. In the case of the ring mixer, high impedance stubs are required at the diode terminals for IF extraction and IF return. At low frequencies, where classical transformers can be used as LO and RF baluns, the centre taps in the transformers provide convenient points for IF extraction and IF return.

In addition to inherently isolated ports, double-balanced mixers also provide rejection of AM noise in the LO and suppression of certain mixing products. Following the same procedure as in single-balanced mixers, the relative LO and RF phase relationships in each diode are shown in Figure 7.12. Due to the diode orientations, the total IF current is given by I_1-I_2+I_3-I_4, where I_1 to I_4 are diode currents indicated in Figure 7.12. Expressing the resulting IF current as a summation of power series expansions will show that all the spurious mixing products corresponding to *m* or *n* even are suppressed.

Therefore, the use of balanced mixer configurations enhances the overall mixer performance in terms of port isolations, LO AM noise and spurious response suppressions. In addition, the input power handling capability also increases due to the increased number of mixing devices. The LO power required, however, also increases correspondingly. The enhancements offered by balanced mixers can be obtained in practical mixers if the diodes have identical properties and the hybrids or baluns are well balanced. The former requirement can be readily achieved in monolithic circuits. Some hybrid and balun structures, however, are not amenable for monolithic circuits and alternate forms of implementation have to be developed.

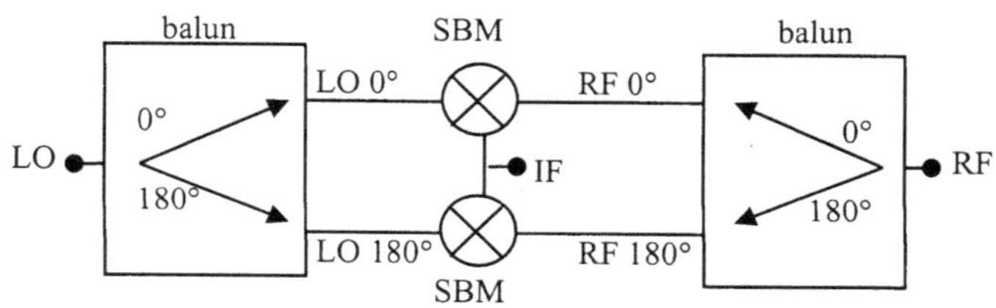

Figure 7.10 Double-balanced mixers as a combination of two single-balanced mixers (SBMs)

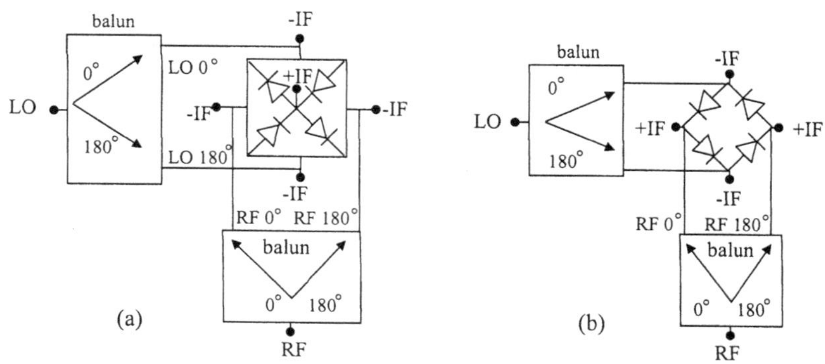

Figure 7.11 Double-balanced mixers derived from Figures 7.9a and 7.9b, respectively

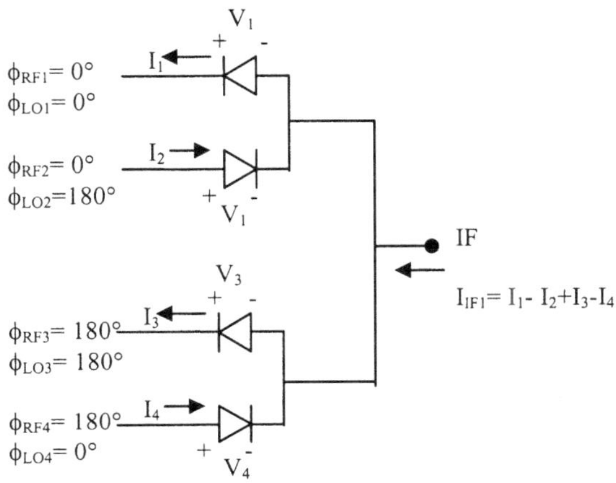

Figure 7.12 Phase relationships in double-balanced diode mixers

7.5.5 A 94 GHz diode mixer design example

The following paragraphs describe some of the steps involved in designing a 94 GHz mixer. Although the work described below makes rather ideal assumptions for some of the local oscillator and mixed product terminations, the overall procedure and method of design can be followed for other designs at different frequencies. The analyses presented here were made with the program DIODEMX, which is a pubic domain program written by Stephen Maas and is listed at the back of the first edition of [3].

A single diode mixer was first designed using the program DIODEMX. The general aim was to design a single-balanced mixer at 94 GHz, operating between source and load impedances of 50 Ω. Therefore, a single-ended mixer was designed to operate from a source impedance of 50 Ω driving an IF load impedance of 100 Ω (these two outputs are combined at the IF to give 50 Ω).

One of the fundamental problems in mixer design is that there are so many variables that it is difficult to see how to optimise the circuit. In this design approach, the concept of equivalent pulse duty ratio for the pump waveform is used. This is described in [10] and the key equations are given in Figure 7.13.

The basic assumptions made here are that the diode is voltage driven by a pump waveform of amplitude kV_p, and it operates at a point V_{dc} on the DC characteristic. This operating point must be derived from the values used in DIODEMX by subtracting the voltage drop across the series resistance; i.e. V_{dc} as defined here is V_{dc} (given by DIODEMX™) – I_{dc} × diode series resistance. The parameter k is a voltage tapping factor and accounts for the fact that only a portion of the pump voltage appears across the diode junction. When the voltage kV_p plus the diode bias voltage V_{dc} exceeds the turn on voltage, the diode turns on, and when this combination is less than V_t, the diode turns off. The on time per pump cycle is the pulse duty ratio, denoted by S. We now have to determine the turn-on voltage for a device with an I–V characteristic which is a continuous function. This can be determined by a series of calculations on the selected mixer circuit.

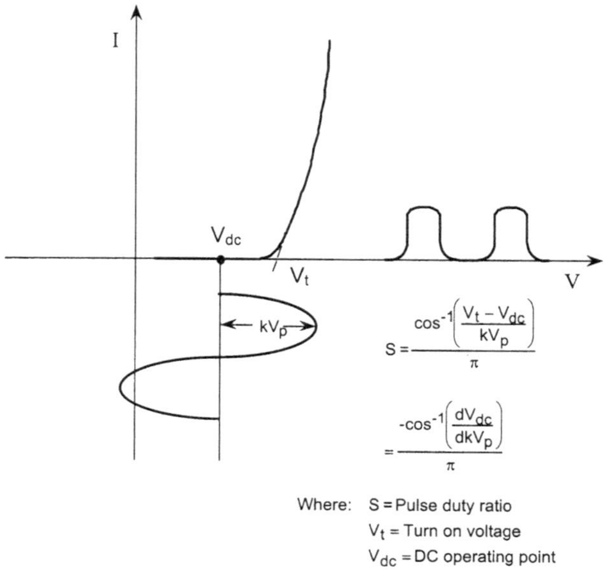

Figure 7.13 Approximation of pulse duty ratio

Examination of the equation for pulse duty ratio S given in Figure 7.13, shows that, given that V_t is fixed for a given device, there are a range of values for kV_p and V_{dc} which give the same value of S. Further investigation shows that if we plot kV_p against V_{dc} for a constant pulse duty ratio, we have a straight line with a negative slope. The slope of this line gives the pulse duty ratio and the point at which it intersects the DC bias voltage axis gives V_t. In general, k will not be known, and therefore we plot V_p against V_{dc}. Since a plot of kV_p against V_{dc} has the same slope and intersects the DC bias voltage axis at V_t, k can be found if the value of V_t is known. This is given approximately by the junction voltage, which causes a current to flow in the junction such that the small signal resistance is equal to the real part of the RF source impedance. However, since the pulse duty ratio is given by the slope of the plot, it is not necessary to determine the value of k, and a plot of V_p against V_{dc} is sufficient. We know that a given embedding structure requires a particular value of S to achieve minimum conversion loss. We can now compute a series of conversion loss curves for different DC operating points (note that this is the operating point on the I–V characteristic and not the external bias voltage). We know that the minimum value of conversion loss on each curve corresponds to the optimum pulse duty ratio, S_{opt}. Therefore, if we calculate the generator voltage magnitude from the pump power level at this minimum, we can plot this value of V_p together with the operating point voltage V_{dc}. Plotting a series of these results gives a straight line as mentioned above.

In the first stages of this work, an attempt was made to find these minima in conversion loss by plotting conversion loss against pump power for fixed external bias voltages. A typical result for the mixer under consideration is given in Figure 7.14. One problem in using this method is that the minimum is very poorly defined. The reason for this is that although the external bias voltage is fixed, the operating voltage on the I–V characteristic decreases as pump power is increased due to the opposing voltage developed across the diode's series resistance, caused by the increasing rectified current. What is happening in this case is that the operating point voltage is maintained at approximately the optimum value for a wide range of pump powers. This phenomenon can be deliberately exploited in some mixer designs where it is required to maintain a low conversion loss over a wide range of pump powers. In this case, the total series resistance in the DC bias circuit would be carefully selected to keep the operating point constant over the range of pump power levels required.

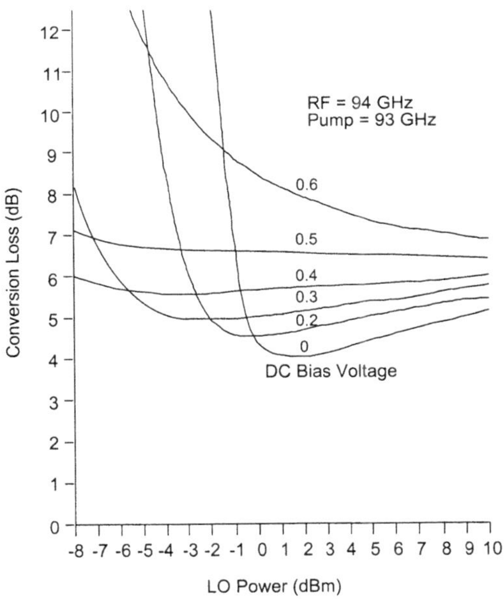

Figure 7.14 Conversion loss against pump power for different bias voltages

In view of this problem, conversion loss was plotted against the operating point voltage. Conversion loss curves were plotted for pump power levels corresponding to pump generator voltages of 0.4, 0.6, 0.8, 1.0 and 1.2 V, which correspond to power levels of -3.98, -0.46, 2.04, 3.98 and 5.56 dBm, respectively, as shown in Figure 7.15. The minima of these curves enabled five conversion loss minima to be obtained. These minima were obtained for the optimum value of pulse duty ratio. Therefore, five values of V_p against V_{dc} were plotted in Figure 7.16. These points showed a very good fit to a straight line and enabled the value for V_t/k and S_{opt} to be determined, as previously described. In this case they were found to be 0.65 and 0.1609, respectively.

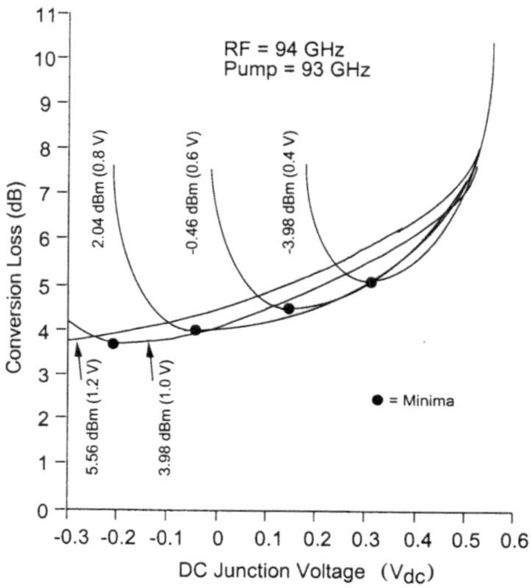

Figure 7.15 Conversion loss against DC junction voltage for different power levels

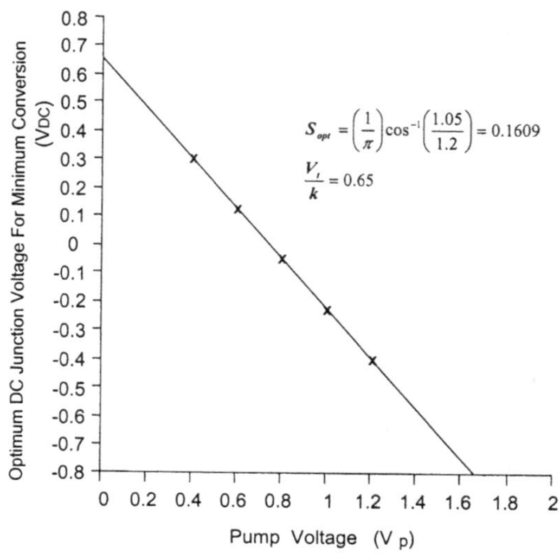

Figure 7.16 Optimum DC junction voltage against pump voltage

Following computation of the above results, the pulse duty ratios were calculated for each of the four highest power settings given above. This was done by running the program DIODEMX with the pump voltage fixed and sweeping the external DC voltage bias. For each value of external voltage bias and the calculated rectified current, the operating point voltage V_{dc} was determined. The values for V_{dc} and the fixed V_p enabled a value of pulse duty ratio to be found for each value of operating point voltage using the equation given in Figure 7.13. These values were used to plot four curves of conversion loss against pulse duty ratio for each of the four largest pump power levels, as shown in Figure 7.17. It can be seen that all four conversion loss minima occur at the same value of pulse duty ratio, as expected, but the actual values of the minima increase with decreasing pump power level. The reasons for this are that as the pump power is decreased, the time taken to turn the diode on and off increases, and that, in the on-state, the device has a higher impedance, thus causing increased signal power dissipation in the diode junction. The reason for the latter is that in the on-state the diode is passing a lower current, and since the differential resistance is inversely proportional to the current, the 'on' junction resistance is higher and hence dissipates more power and raises the conversion loss.

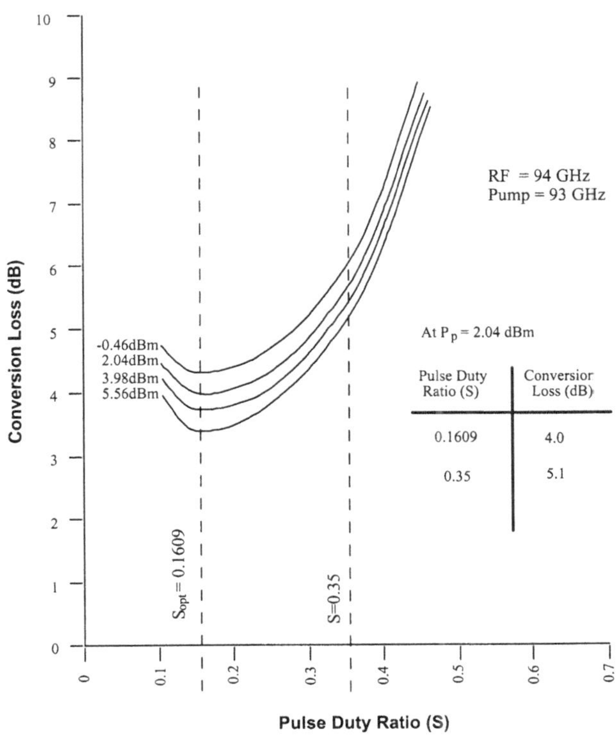

Figure 7.17 Conversion loss against pulse duty ratio

Having determined the equivalent pulse duty ratios to plot the conversion loss curves, the input and output impedance were plotted against pulse duty ratio, as shown in Figures 7.18 and 7.19. The impedance plots were drawn this way rather than on a Smith chart as it is easier to interpret and interpolate values. The curves given in Figures 7.17, 7.18 and 7.19 completely characterise the mixer circuit within the pump power range -0.46 to 5.56 dBm. The use of these figures to determine an optimum design will now be discussed.

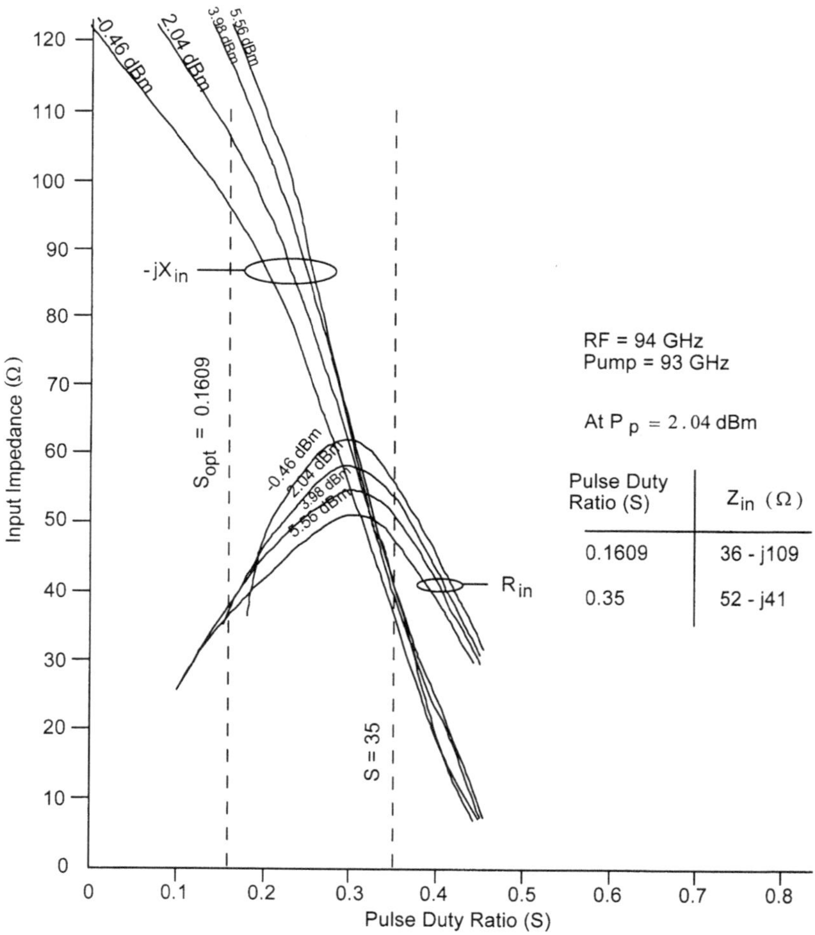

Pulse Duty Ratio (S)	Z_{in} (Ω)
0.1609	36 - j109
0.35	52 - j41

Figure 7.18 Input impedance against pulse duty ratio for different pump power levels

The first point to be resolved in the mixer design is how much pump power will be available. Clearly, the higher the pump power level, the lower the conversion loss and noise figure, but beyond a certain point the improvement in performance becomes negligible. Examination of Figure 7.17 suggests that a pump power level of 2 dBm would give a theoretical conversion loss of around 4 dB. However, from Figures 7.18 and 7.19, we see that for a pulse duty ratio corresponding to S_{opt} = 0.1609, the input impedance is 36 – j109.5 Ω and the output impedance is 482 + j105 Ω. The output impedance is clearly too high and, although this value will be halved when two single-ended mixers are combined in the balanced design, i.e. 241 – j52.5 Ω, it is still about twice the value required. However, these three figures enable a trade-off to be readily established. We first set the constraint that the output impedance must have a real part of about 100 Ω. Examination of Figure 7.19 shows that this can be achieved with a pulse duty ratio of 0.35, which gives an input impedance of 52 – j41 Ω and an output impedance of 117 + j12.5, and that, at this point, the conversion loss is only degraded to 5.1 dB.

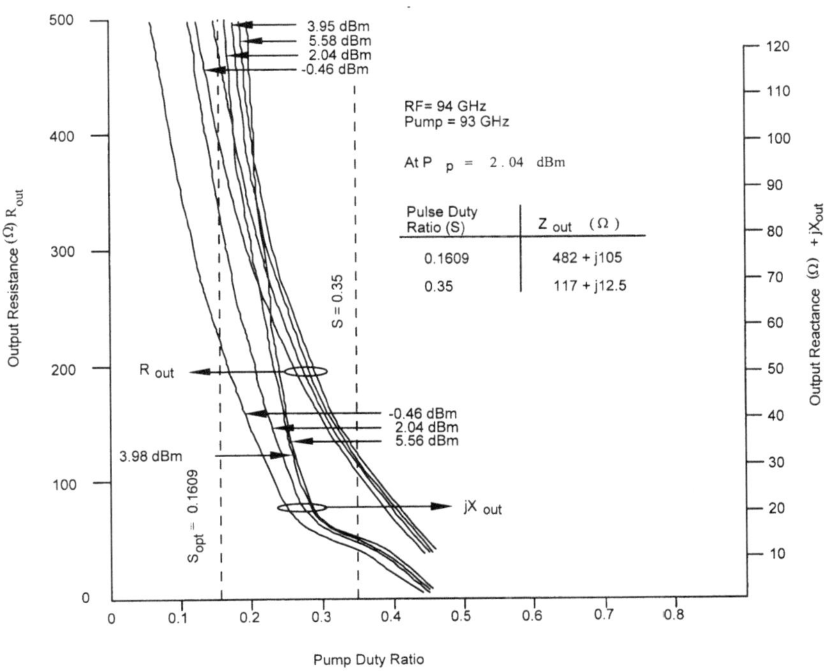

Figure 7.19 Output impedance against pulse duty ratio for different pump powers

Finally, we now wish to do a check on the mixer performance. The first step in this process is to determine the optimum value of external voltage bias. This is readily determined using the equation in Figure 7.7, rewritten to calculate V_{dc}, from which we find that for $V_p = 0.8$, $V_t/k = 0.65$ and $S = 0.35$, so $V_{dc} = 0.2868$. We can now sweep through this voltage using DIODEMX to produce a set of results on either side of the optimum bias voltage. The analysis shows input and output impedances of $51.8 - j41.2\ \Omega$ and $117 + j12.5\ \Omega$, thus confirming the values selected from Figures 7.12 and 7.13. It should be remembered that these are the input and output impedances, which require a conjugate match to achieve the conversion loss figure of 5.22 dB. Therefore, the embedding impedances should be $51.8 + j41.2\ \Omega$ and $117 - j12.5\ \Omega$ at the signal and IF, respectively. Having found the optimum operating conditions for this application, it is now necessary to design the matching circuit for the RF input. This can be realised by an additional open circuit stub. However, after insertion of this stub, it is necessary to iterate the design process to ensure that the operating conditions are still optimum. The reason for this is that when the matching stub is placed in the circuit, the terminations at other frequencies are also affected, thus making the arrival at an optimum condition an iterative process. The IF is matched by a lumped-element at the IF output.

In this example, we are assuming a pump power of 2 dBm will be available. It should be noted that when the two SEMs are combined with a 3 dB coupler to form a SBM, the power required will be 5 dBm, plus an extra amount to compensate for the loss in the coupler.

7.6 Coupling structures

Preceding sections have shown the importance of coupling structures such as 90° and 180° hybrid couplers, baluns, power splitters and combiners in balanced mixers. In MMICs, it is desirable to make the overall mixer circuit as small as possible. Generally, the coupling structures are the circuits that require the most area. This is particularly true for distributed circuits at low frequencies. Therefore, it is common practice to use either a folded version of the distributed circuit or a lumped approximation, in order to reduce the overall size. At millimetre-wave frequencies, distributed circuits are already quite small and commonly used. An alternative to the passive distributed and lumped-element coupling circuits is an active circuit. Active coupling circuits have the potential to be much smaller than the distributed or lumped circuits, but although they can have gain, the noise performance of most circuits investigated to date is poorer. Often, the asymmetry of an active balun or splitter means that the noise figure to each output port is not the same, even though the gains may be equally balanced. This can be quite acceptable in some receiver architectures where the mixer is preceded by a low noise amplifier. However, although insertion of a low noise amplifier in front of a mixer can decrease the overall noise figure of the combination, it also decreases the dynamic range. Recently, balanced oscillators [79], which have two equal amplitude outputs in anti-phase, have been demonstrated at microwave frequencies. They have the potential of directly generating the anti-phase LO signals required in balanced mixers, without the use of baluns.

7.6.1 90° and 180° hybrids

Examples of the 90° hybrid are the branch-line and the Lange coupler. The rat-race coupler is an example of a 180° hybrid. A detailed description of these hybrid couplers can be found in Chapter 3. When employing these hybrids for the balanced mixer applications, shown in Figure 7.8, it should be noted that the hybrid ports are generally not matched at the diode terminals. Unless impedance matching networks are employed, there will be LO and RF signals reflected at the diode terminals. This has various effects on the mixer LO-RF isolation and the LO and RF input return losses, depending on the type of hybrid employed.

For the 90° hybrid in Figure 7.8*c*, if the diode ports are not matched but have equal reflection coefficients, the LO (RF) signals reflected at the diode terminals will add constructively at the RF (LO) input and cancel out at the LO (RF) input. Thus, the LO power is reflected into the RF input and vice versa. This results in poor isolation between the LO and RF but good matching at the LO and RF ports.

The opposite occurs in the 180° hybrids in Figures 7.8*a* and 7.8*b*. In this case, any LO (RF) signal reflections at the diode terminals will add constructively at the LO (RF) port and cancel out at the RF (LO) input. Therefore, any diode mismatch results in LO power being reflected back to the LO input port and RF power being reflected back to the RF input port. Thus, the mixer will have poor matching at the LO and RF ports but good isolation between the LO and RF.

7.6.2 Passive baluns

A balun provides a transition between an unbalanced and a balanced transmission line. Intuitively, an unbalanced transmission line carries signals between a conductor and the ground plane, whereas a balanced transmission line carries the signal between two conductors. The balanced output can be treated as two separate ports with anti-phase signals. Thus, the balun can be used as an anti-phase power splitter. However, as a lossless, passive three-port network, the balun cannot be matched at all ports and the two outputs are not isolated. Because the anti-phase relationship is due to a mode transition, it exhibits broadband performance.

Unfortunately, most commonly used balun structures are not amenable to monolithic integration. They usually require metallisation on both sides of a suspended substrate or have some other non-planar structure. Various balun configurations have been reported for applications in MMICs [65,68,71]. Among them, the planar version of the Marchand balun [59] is perhaps one of the most attractive, due to its planar structure and wideband performance.

Figure 7.20 shows the schematic diagram of the planar Marchand balun. It provides anti-phase outputs to load terminations Z_1, from an unbalanced input with source impedance Z_0. It consists of two coupled sections, each of which is one quarter-wavelength long at the centre frequency of operation. The coupled sections may be realised using microstrip coupled lines [53], Lange couplers [71], multi-layer coupled structures [76] or spiral coils [75,77].

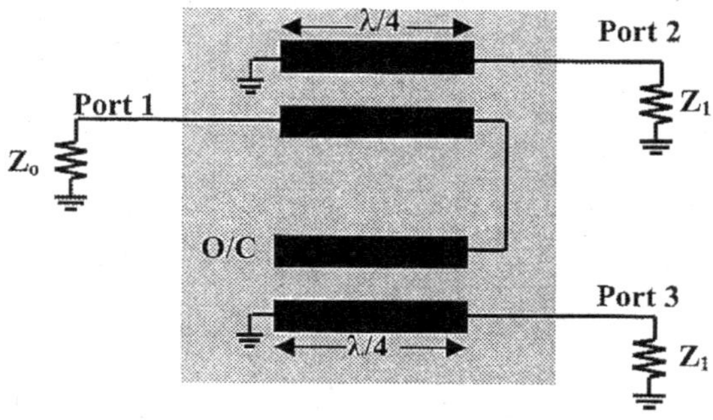

Figure 7.20 Schematic diagram of planar Marchand balun

These baluns are usually designed through full-wave electromagnetic circuit simulations [68] or lumped-element models [77]. Various synthesis techniques have also been reported. Tsai and Gupta [70] use the design curves of the original Machand balun [60] to synthesise a planar Marchand balun based on coupled line equivalent circuit models. Schwindt and Nguyen [74] apply a circuit synthesis technique based on the analytically derived scattering parameters of the multilayer planar Marchand balun. The planar Marchand balun can also be analysed as a combination of two identical coupled sections. This simplifies the balun design to designing couplers with the appropriate coupling factor. This approach also provides valuable insight into the balun operation, relating the coupling factor of the coupler to the balun input and output impedances. This results in an effective technique for designing impedance-transforming baluns [81].

Based on this approach, the required coupling factor, C, for equal power transfer to each output port is given by:

$$C = \frac{1}{\sqrt{\frac{2Z_1}{Z_o} + 1}} \tag{7.35}$$

where Z_0 and Z_1 are the source and load impedances, respectively.

When eqn. (7.35) is satisfied, the balun is matched at the input and has transmission coefficients of -3dB with opposite phase. The outputs, however, are not matched or isolated and both have a value of -6dB. To achieve perfect output port matching and isolation, some form of resistive network needs to be added between the output ports, just like the Wilkinson power divider. Figure 7.21 shows a perfectly matched balun with a resistive network consisting of a half-wavelength transmission line terminated with Z_1 at both ends.

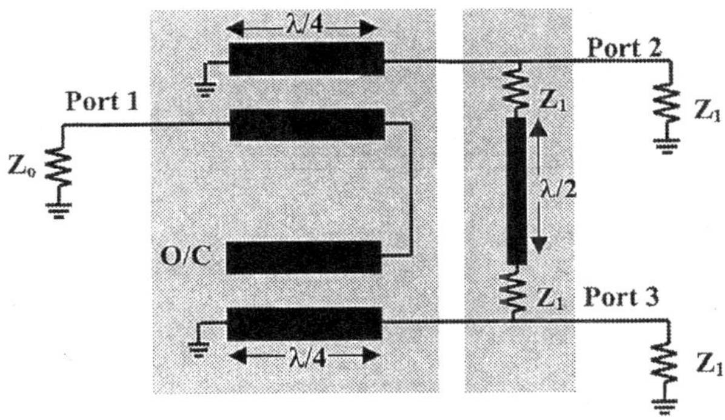

Figure 7.21 Schematic diagram of perfectly-matched planar Marchand balun using a resistive network

7.6.2.1 Coupled-line Marchand balun

Figure 7.22 shows an example of a planar Marchand balun, realised using CPW parallel coupled lines [80]. The circuits are fabricated on a 400 μm GaAs substrate using the millimetre-wave IC fabrication facilities at the University of Glasgow. To enable accurate characterisation using the Agilent Technologies 8510XF RFOW system at the University of Surrey, three baluns were fabricated and measured as two-ports (with each spare port terminated on-chip in turn). The balun insertion loss and phase error performance are shown in Figure. 7.23, and the circuit exhibits excellent bandwidth (40 to 90 GHz).

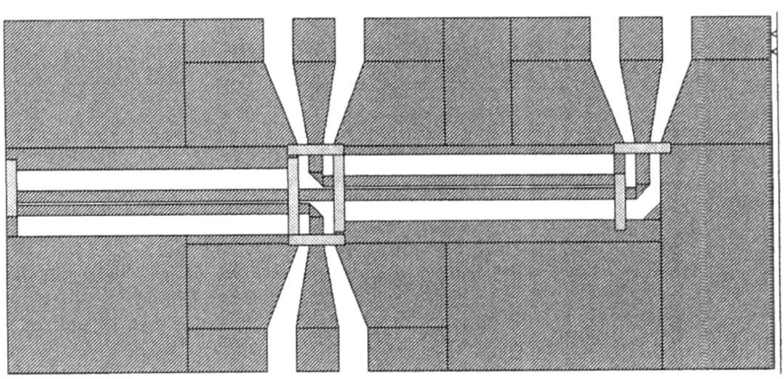

Figure 7.22 Layout of a CPW planar Marchand balun

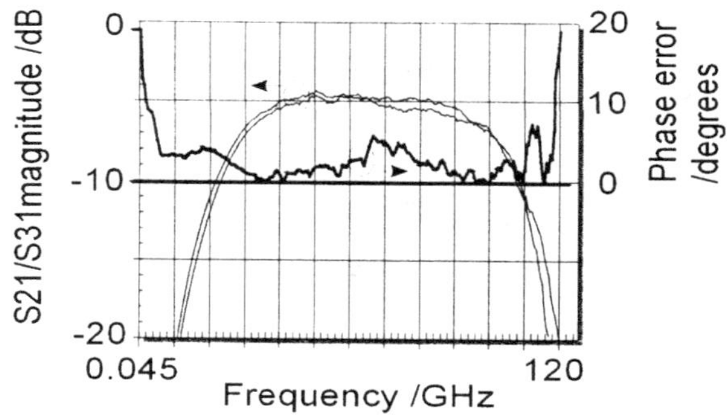

Figure 7.23 Measured performance of CPW planar Marchand balun

7.6.2.2 Lange coupler Marchand balun

According to eqn. (7.35), tight couplers are required for baluns transforming to low impedances. This can be achieved by employing Lange couplers in place of parallel-coupled lines. Figure 7.24 shows an example of a Lange coupler planar Marchand balun. It has an average insertion loss of 1.5 dB, with amplitude and phase balances of less than 0.2 dB and 5°, respectively, over the 17 to 34 GHz frequency band.

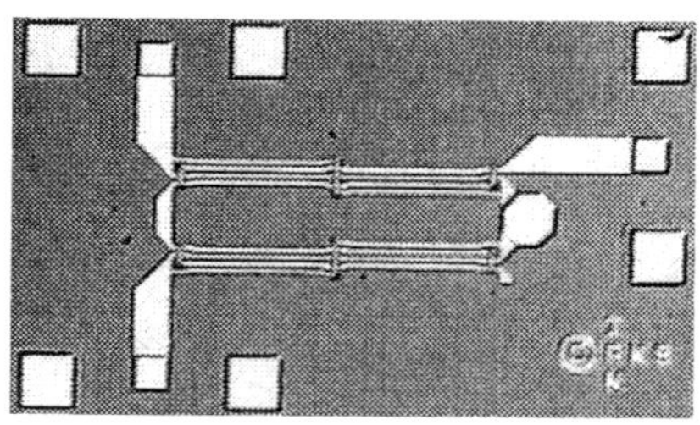

Figure 7.24 A Lange coupler planar Marchand balun (0.8 × 1.4 mm²)

7.6.2.3 Spiral transformer Marchand balun

The low frequency lumped-element baluns usually require a large number of components. To maintain high yield in the monolithic process, it is highly desirable to maintain a small circuit area and a low component count. This can be achieved by realising a planar Marchand balun using spiral transformers [78].

The use of planar spiral transformers was pioneered by Honeywell and Pacific Monolithics as a monolithic implementation of the magnetically coupled coil transformer [61]. Capacitive coupling, which also exist between the spirals, presents a serious parasitic in the transformer. As a result, the transformer performs poorly as a classical centre-tapped transformer balun. However, studies of the planar spiral balun transformers have illustrated that they can be effectively modelled as coupled–line elements [62,64]. Thus, the transformer can be employed as quadrature couplers in a Marchand balun. This results in a simple, compact design with readily predictable performance.

A photograph of a spiral transformer balun is shown in Figure 7.25. It consists of two oppositely wound twin-coil transformers connected in series. The input coil is terminated with an open circuit, while the output coils are terminated in a short circuit using a via hole. The spiral transformer has 4.5 turns with a track width and gap of 12 μm and 8 μm, respectively. The balun was designed for 4 GHz operation and measures 0.7×1.5 mm^2.

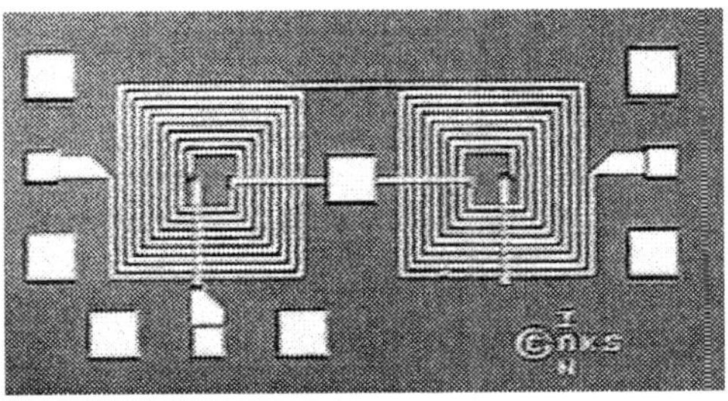

Figure 7.25 A spiral transformer balun (0.7 X 1.5 mm²)

Figure 7.26 shows the insertion loss and the phase error of the balun. The insertion loss of the two outputs track each other closely, within 0.2 dB, up to 6GHz. The insertion loss at the centre frequency is less than 0.5 dB at mid band and the phase error is within 10°, from 2 to 6 GHz.

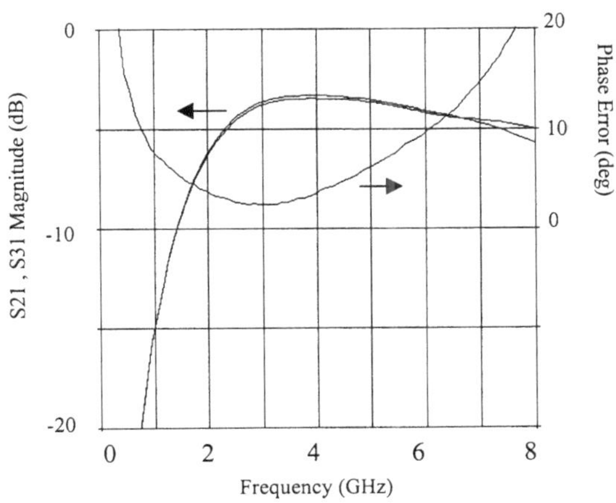

Figure 7.26 Measured insertion loss and phase error of a spiral transformer balun

The lower and upper limits of the balun's operating bandwidth are determined, respectively, by the inductance and the resonant frequency of the transformers. Thus, the operating bandwidth can be increased by using thinner tracks and a smaller number of turns for the spiral transformers, while maintaining the same electrical length. However, this will be accompanied by higher insertion loss and larger chip areas. Therefore, compromises need to be made to achieve the optimum balun performance.

7.6.2.4 Wilkinson divider with high-pass and low-pass sections

An alternative way to construct an in-phase or anti-phase equal amplitude split is to use the Wilkinson power divider. The in-phase splitter is a standard Wilkinson divider, which can be realised in distributed or lumped form. A lumped-element anti-phase splitter can be realised by following one output with a high-pass filter structure and the other by a low-pass filter structure. The distributed form can, for homogeneous dielectrics, have exact anti-phase tracking over infinite bandwidth. The inhomogeneous microstrip version can have very broadband performance. This circuit can be meandered to reduce the size. The lumped-element splitter would be realised using spiral inductors and MIM capacitors, but would tend to have narrower bandwidth than the distributed form. These passive structures are described in Chapter 3.

7.6.3 Active power splitters and combiners

The most basic form of an active power divider consists of a pair of transistors, with their inputs directly connected but with separate outputs. This is an excellent method, providing gain, near perfect amplitude and phase balance, as well as good reverse isolation and isolation between the two (or more) outputs. The basic power combiner consists of a pair of transistors with separate inputs and a common output. However, strictly speaking, this is not a true power combiner because two input signals at different frequencies would create intermodulation products, as the transistor outputs are not isolated.

It has also been demonstrated that MESFETs can be integrated into CPW/slotline structures to realise active power dividers and combiners (referred to as line unified or LU-FETs [66,67]), as illustrated in Figure 7.27.

7.6.4 Active baluns

The passive balun has the advantage that it can be used for both power dividing and power combining applications, and it is ideal for use at millimetre-wave frequencies. Unfortunately, they can be prohibitively large for applications below X-band and suffer from excessive levels of insertion loss. As a result, baluns have been developed to overcome both of these shortcomings. Active baluns are unilateral and must be designed as either power splitters or power combiners. A number of active balun structures for mixers have been investigated and published in the literature.

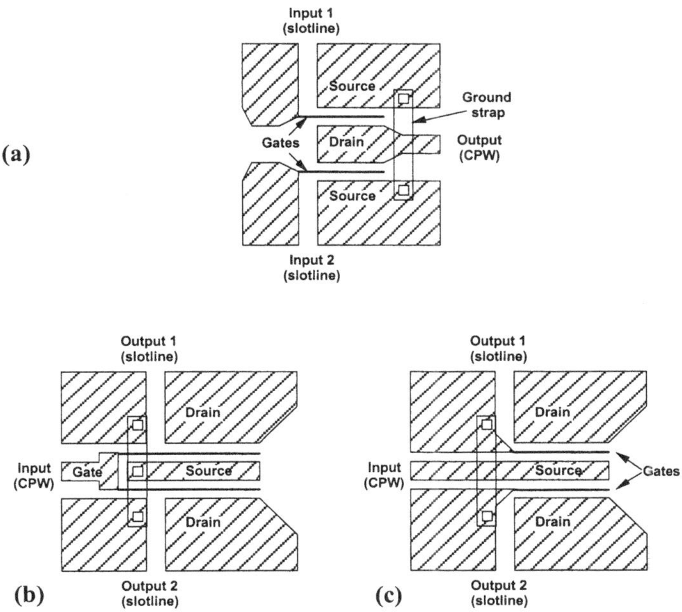

Figure 7.27 LU-FET examples: (a) In-phase power combiner and (b) and (c) power dividers

An example of a simple active balun power combiner was demonstrated in monolithic form by Jouanneau-Douard *et al.* [69]. The circuit design is based on a totem pole arrangement of two MESFETs: one in common-source configuration and the other in common-drain configuration. The two balanced input signals are applied to the two gates, and the unbalanced output signal is taken from the common hot ends of the transistors. The measured results of this compact implementation demonstrated an amplitude balance of ±0.5 dB and a phase balance of ±10°, across the 1 to 18 GHz frequency range.

The simplest technique for implementing an active balun power divider is to take the balanced output from the source (non-inverting) and drain (inverting) terminations of a single MESFET, as illustrated in Figure 7.28*a*. The main drawback of this technique is that the output port impedances and parasitics are unequal, causing significant amplitude and phase imbalances. A technique for overcoming these problems by cross-connecting identical active baluns, as shown in Figure 7.28*b*, has been investigated and successfully implemented by Goldfarb *et al.* [72].

Another technique for implementing an active balun power divider is to use a differential push-pull amplifier, as illustrated in Figure 7.28*c*. With this technique, imbalance can be introduced because the constant-current source transistor is not connected to a virtual ground [73]. However, it has been successfully employed in an L-band double-balanced mixer [63].

The previous two active balun power dividers are generally used in low frequency applications. A high frequency implementation has been demonstrated using a common-gate/common-source pair, illustrated in Figure 7.28*d*. This technique has been successfully employed in a 3 to 18 GHz monolithic double-balanced mixer [117].

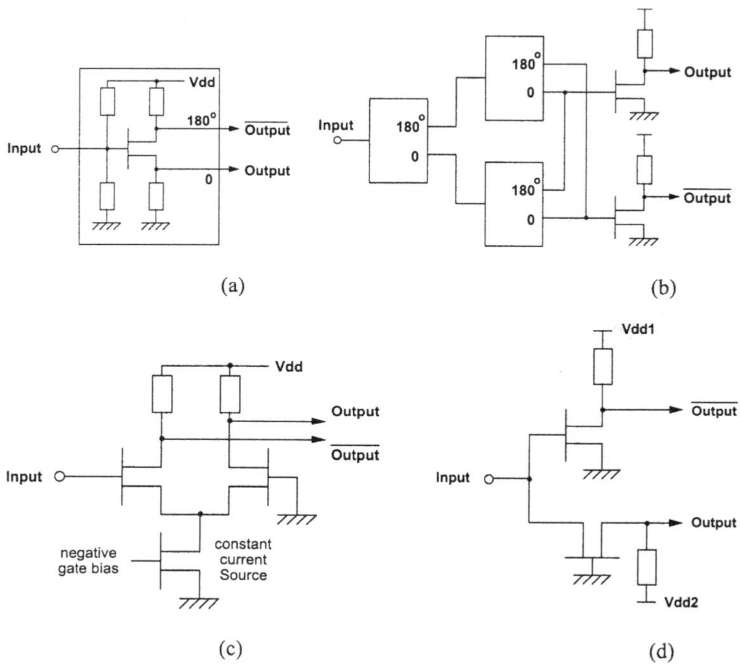

Figure 7.28 Active baluns: (a) source/drain, (b) cross-coupled, (c) push-pull and (d) common-source/common-gate FET pair

7.6.5 Balanced oscillators

Instead of passing a signal through a balun to obtain anti-phase signals, it is also possible to generate these anti-phase signals directly from a balanced oscillator. The anti-phase relationship of the oscillator outputs is a result of the inherent symmetry in the oscillator circuit, rather than the phase shifting characteristics of transmission line and filter structures. Thus, exact amplitude and phase balance can be readily obtained, especially in monolithic circuits. Monolithic balanced oscillators have been demonstrated at K and Ka-band [79].

7.7 Active FET mixers

FETs are widely used in MMIC mixers, as many monolithic foundry processes are based on the MESFET structure. Consequently, FET mixers can be fabricated on the same chip with other FET-based circuits, such as amplifiers, oscillators and switches, to form integrated subsystems. In addition, active FET mixers offer the potential to achieve conversion gain. Generally, they also require less LO power, have higher dynamic range and give comparable noise performance to diode mixers.

A study by Fairburn *et al.* [14] identified several mixing actions using the active mode in a single-gate FET. They can be attributed to the various non-linear elements in the active FET. Figure 7.29 shows the large-signal equivalent circuit of an active FET. The dominant non-linear elements are the gate-source capacitance, transconductance and drain conductance. Depending on the biasing conditions, each of these elements can be employed for frequency mixing. According to which FET terminal the LO is applied to, these active FET mixers can be classified as gate, drain and source mixers. Figure 7.30 shows the biasing points and the I–V trajectories of the pumped device, for each mixer type.

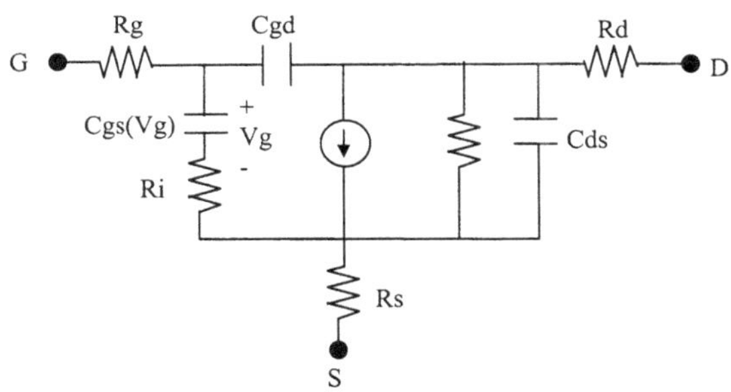

Figure 7.29 The non-linear intrinsic elements in the active FET

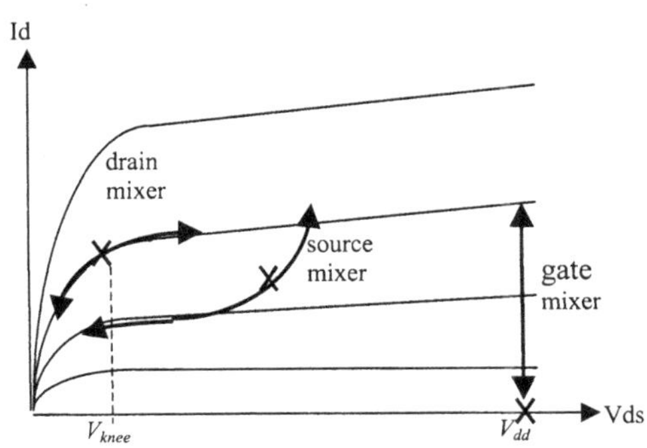

Figure 7.30 I–V trajectories and bias points of the gate, drain and source pumped mixers

7.7.1 Gate mixer

The LO voltage is applied to the gate and the time-varying transconductance is responsible for frequency conversion. The gate is biased near pinch-off and the drain biased at V_{dd} to provide the best conversion response. Under these conditions the instantaneous values of the other non-linear elements in the model are approximately constant.

7.7.2 Drain mixer

The non-linearity at the knee of the I–V characteristic is used for this mixing action. Therefore, the optimum drain-source bias is when $V_{ds}=V_{knee}$. The LO is applied to the drain and the dominant contribution to frequency conversion is the variation of the drain-source resistance.

7.7.3 Source mixer

In this mode, the FET is being used in a common-drain configuration. The FET is biased in the middle of the I–V characteristic and the LO voltage is applied at the source, to modulate both V_{gs} and V_{ds}. This is the most complicated mode of mixing from the point of view of analysis, since all the non-linear elements within the FET model are time-varying.

7.7.4 Active FET mixer design considerations

Many of the techniques used in amplifier designs can be applied to mixer design. The aim of FET mixer design is to achieve optimum RF performance. In most receiver applications, this is achieved when minimum noise figure is obtained with adequate conversion gain. This is different from the diode mixer, where maximum conversion efficiency is usually sought. It is also important that the mixer is 'well-behaved'. 'Well-behaved' is a concept which is not as precisely defined as one might like. It means that the mixer does not have any unusual properties that compromise its practical use in a system. Specifically, it must be stable under small-signal and large-signal conditions, its performance parameters must vary gracefully with pump level and environmental temperature, it should be approximately unilateral (i.e. have high enough reverse loss), and its performance should not be unduly sensitive to source or load impedances.

These properties are best realised by a gate mixer: one in which the time-varying transconductance is the dominant contributor to frequency conversion and the variation of other parameters is minimal. The effect of harmonically varying gate-source capacitance and drain conductance must be minimised. Additionally, the transconductance swing must be maximised for a given device in order to minimise noise figure, optimise gain (for a given set of source and load impedances) and to minimise gain variation with pump level. The most important requirement in achieving these conditions is that the FET is kept in its current saturation region throughout the pump cycle, as shown in Figure 7.30. Full saturation can be achieved by ensuring that the variation of $V_{ds}(t)$ under pumping is minimal; it is best if it remains constant at its DC value. This condition is achieved by short circuiting the drain at all pump harmonics; then the drain current, which may have a fairly high peak value, cannot cause any drain-source voltage variation. The RF voltage across the gate-drain capacitance is then minimal, so feedback is minimal and large-signal stability is good. In addition, the gate should also be short-circuited at the IF frequency to avoid instability.

The various mixer configurations based on the gate mixer will be described in this section. The general operation and characteristics will be presented. More detailed information on the design of FET mixers can be found in [3].

7.7.5 Single-ended FET mixers

Figure 7.31 shows the main building blocks of the single-ended FET mixer. Operating as a gate mixer, the FET is biased at pinch-off and the LO is applied to the gate to modulate the transconductance between zero and the maximum value. To ensure that the transconductance varies between these two limits, the drain needs to present a short circuit to the LO and all its harmonics. The RF is then applied at the gate to the time-varying transconductance. As in the single-ended diode mixer, a filter diplexer is required at the LO/RF input to isolate the LO and RF ports. The IF generated is extracted from the drain. An IF short circuit is required at the gate to prevent the FET from amplifying any IF noise, which may lead to instability. In general, no attempt is made to enhance the performance by feeding back higher order mixing products as with the diode mixer, as this can lead to instabilities and oscillation unless it is carefully controlled. Impedance matching networks are employed between the LO/RF and IF ports and the active device. The FET input and output impedances are a function of the LO power, as well as biasing conditions. They can be estimated using a non-linear simulator. A typical circuit configuration for implementing the required matching networks and terminations of a single-ended FET mixer is shown in Figure 7.32.

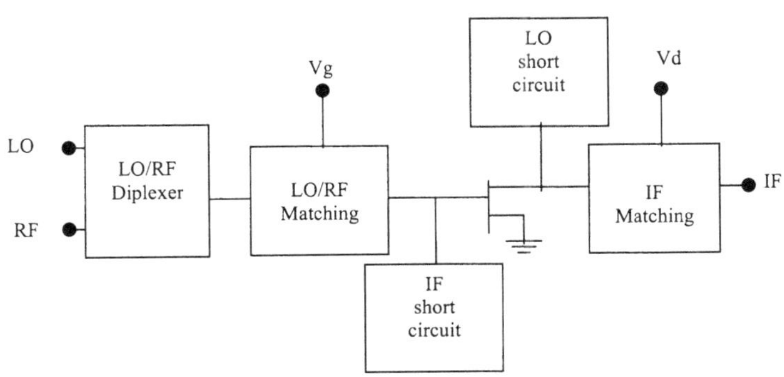

Figure 7.31 Single-gate single-ended FET mixer

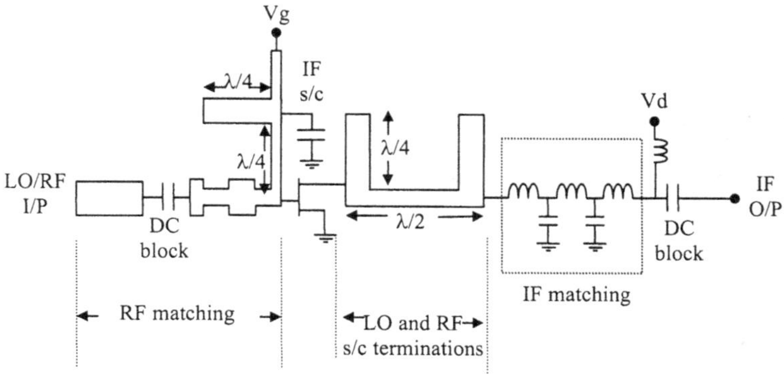

Figure 7.32 Schematic diagram of a typical FET single-ended mixer

The dual-gate FET provides two separate gates for the LO and RF. This simplifies the mixer design by eliminating the need for a coupling structure for the and RF. It also allows separate LO and RF matching networks. The introduction of the second gate, however, complicates the analysis of the FET mixer. Generally, it can be treated as a series combination of two FETs, as shown in Figure 7.33. Individual gate bias, V_{g1} and V_{g2}, can be applied to each gate, but the sum of the two drain source voltages, V_{d1} and V_{d2}, is equal to the applied V_{ds}. The two FETs also have a common drain current. For a stable operating point, one of the FETs is biased in the saturation mode, while the other is in the linear region. For efficient mixer operation, the lower FET is biased in the linear region to act as a drain mixer, while the upper FET is biased in the saturation region to operate as a source follower. The LO is applied to the upper FET, to generate time-varying waveforms for G_m and G_{ds} in the lower FET. Frequency mixing occurs in the lower FET, in which the RF is applied to the gate, and the IF is generated at the drain. By incorporating an IF short circuit in the LO matching network, the upper FET can also act as a common-gate amplifier for the IF. Like the single-ended mixer, the IF matching network is generally designed to present short circuit terminations to the LO and RF.

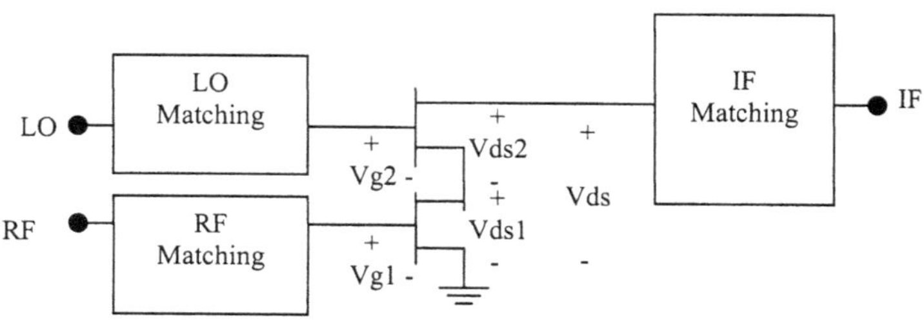

Figure 7.33 Dual-gate single-ended FET mixer

7.7.6 Single-balanced FET mixers

Single-balanced FET mixers can be realised by combining two single-ended FET mixers in parallel using hybrids, as shown in Figure 7.34. Like single-balanced diode mixers, they exhibit inherent isolation between LO and RF, suppression of LO AM noise and rejection of certain spurious responses. However, the FET terminals cannot be 'reversed' as with the diode. Therefore, an IF hybrid is required to combine the out-of-phase IF generated at the drains of the two FETs.

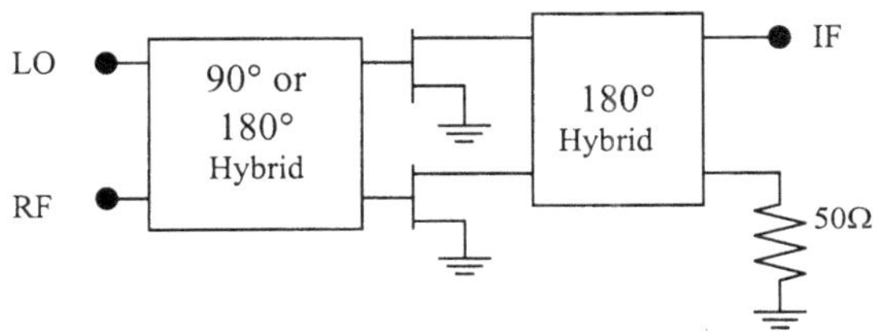

Figure 7.34 Single-balanced FET mixer using hybrid couplers

The isolated gates of the dual-gate FET allow the realisation of single-balanced mixers using baluns in place of hybrids. Figure 7.35 shows a dual-gate FET single-balanced FET mixer, in which the RF is applied in-phase, while the LO is applied out-of-phase using a LO balun. Similarly, the IF hybrid can also be replace by an IF balun, which generally exhibits broadband performance.

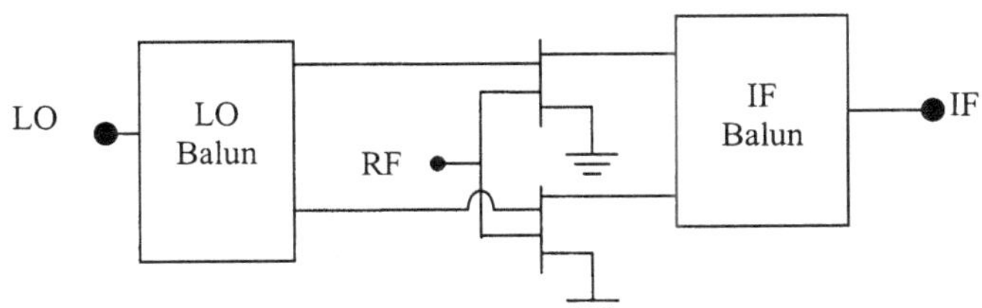

Figure 7.35 Dual-gate single-balanced FET mixer

Figure 7.36 shows a single-balanced mixer employing 'stacked' single-gate FETs. This circuit is an adaptation of the analogue multiplier using bipolar transistors. In the analogue multiplier, a differential input is applied to the upper transistors, while a single-ended input is applied to the lower transistor. The differential current through the upper transistors is then proportional to the product of the two input voltages. Based on this approach, millimetre-wave mixers employing PHEMTs have been demonstrated [109]. Operating as an active FET mixer, the LO is applied in anti-phase to the upper FETs, while the RF is applied to the lower FET. The differential IF generated by voltage multiplication process is combined by an IF balun. For optimum conversion efficiency, the two upper FETs are biased in the linear region (to operate as FET switches, to be switched 'on' and 'off' alternately by the differential LO voltage), while the lower FET is biased in the saturated region (to function as an RF amplifier). The differential LO voltage results in a virtual ground at the connection point of the upper FETs, providing inherent isolation between the LO and RF. Other attributes, like LO AM noise rejection and spurious suppression, also apply to this single-balanced mixer.

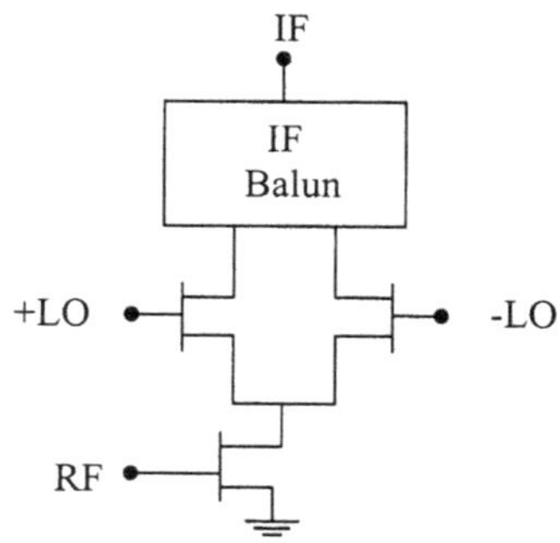

Figure 7.36 Single-balanced FET mixer using 'stacked' FETs

7.7.7 Double-balanced FET mixers

As with diode mixers, two balun-based single-balanced mixers can be combined to realise a double-balanced active FET mixer, as in Figure 7.10. Figure 7.37 shows the result of combining the two dual-gate mixers of Figure 7.35. The phase relationships of the LO, RF and IF in each mixing element is the same as the double-balanced diode mixer in Figure 7.12. Therefore, they have the same port isolation, LO AM noise and spurious mixing production suppression characteristics. In addition, the isolated gates of the dual-gate FET also provide additional isolation between the LO and RF.

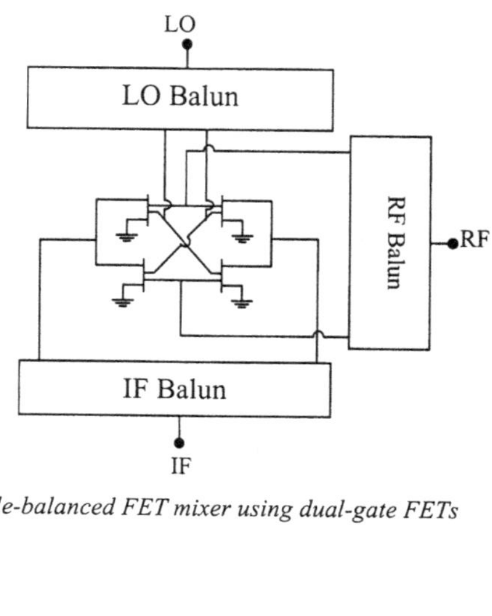

Figure 7.37 Double-balanced FET mixer using dual-gate FETs

Combining the single-gate FET mixers in Figure 7.36 results in the double-balanced mixer shown in Figure 7.38. This mixer is analogous to the BJT Gilbert cell, which is commonly employed in low microwave frequency active mixers, due to its excellent performance.

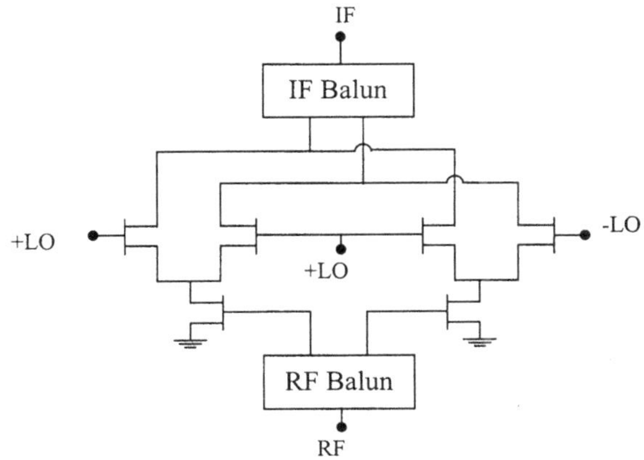

Figure 7.38 Double-balanced FET mixer based on the Gilbert cell

7.8 Resistive FET mixers

As discussed in Section 7.1, the use of non-linear elements for frequency mixing results in spurious mixing products and intermodulation. Instead, if a voltage- or current-controlled circuit element, whose transfer function varies linearly as in eqn. (7.4), is employed, a mixer without any spurious responses can be realised. A circuit element that approximates this behaviour does exist: it is the channel conductance of an unbiased FET. A mixer based on the unbiased FET is referred to as the resistive FET mixer and was first described by Maas [110].

Figure 7.39 shows the equivalent circuit of the FET at low drain bias voltages. Also shown is the variation of channel conductance, G_d, with the gate bias, V_{gs}, for a typical PHEMT device. The channel conductance varies relatively linearly over a specific range of gate voltage. This is in contrast to the strongly non-linear active FET transconductance and the Schottky diode junction resistance, which varies exponentially with applied voltage. Thus, when the LO is applied to the gate, a time-varying conductance with very low LO harmonics is generated. This results in a mixer with low spurious mixing products. Other advantages include low DC power consumption and natural separation of LO and RF/IF ports. In FET-based monolithic circuits, the resistive FET is often a better option than implementing Schottky diodes using derivatives of the FET structure. Resistive mixers have been demonstrated up to F-band using InAlAs/GaInAs/InP HEMTs [123].

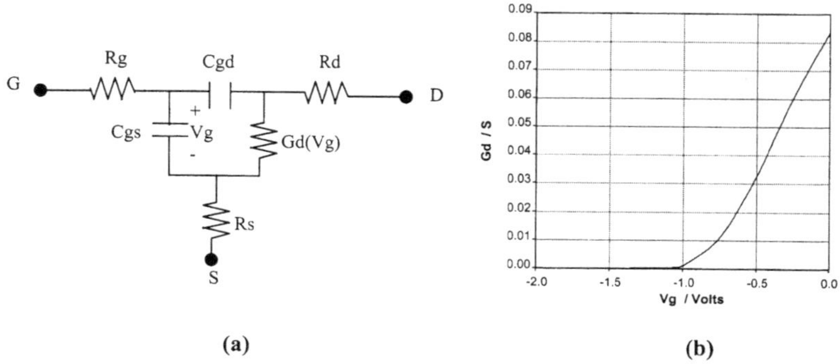

Figure 7.39 The resistive FET: (a) equivalent circuit and (b) variation of G_d with V_{gs}

7.8.1 Single-ended resistive FET mixer

Figure 7.40 shows the block diagram of a single-ended resistive FET mixer [110]. The channel resistance of the FET is modulated by the LO at the gate, while the RF is applied to the drain. The IF is usually extracted from the drain through a low pass filter. A biasing voltage is applied to the gate to achieve optimum performance. A major design consideration is the coupling between the LO and RF through C_{gd}, which is relatively high at low drain bias voltages. Therefore, bandpass filters are required for the LO and RF ports for isolation.

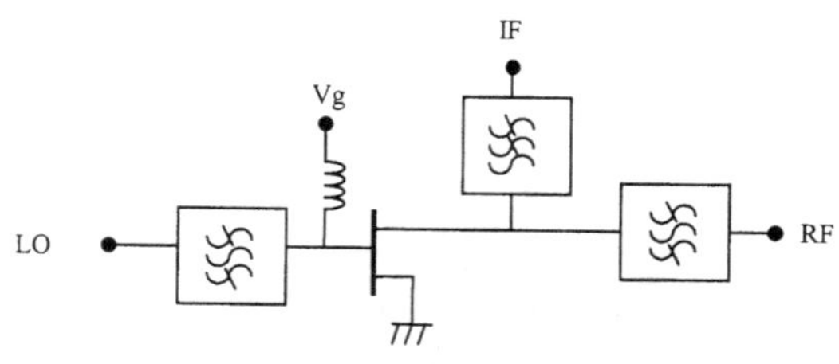

Figure 7.40 Single-ended resistive FET mixer

7.8.2 Single-balanced resistive FET mixer

Single-balanced resistive FET mixers [111,118] can be realised by combining two single-ended mixers with a LO or RF balun, as shown in Figures 7.41*a* and 7.41*b*. The use of an LO or RF balun corresponds to applying the LO or RF to the Σ port in active FET mixers with 180° hybrids. They suppress the same spurious mixing products. The inherent isolation between the LO and RF ports, and LO AM noise rejection characteristics are also similar.

An alternative approach to achieve inherent LO and RF isolation is the use of dual-quadrature hybrids, as shown in Figure 7.41*c* [113]. There are two signal paths between the LO and RF ports: one though the coupled ports and the other through the direct ports of the two 90° hybrids. Therefore, these signal paths have a phase difference of 180°, resulting in signal cancellation at the LO and RF ports. As the reflected LO and RF signals are terminated in the 50 Ω loads of the 90° hybrids, the VSWR of the LO and RF ports also improves.

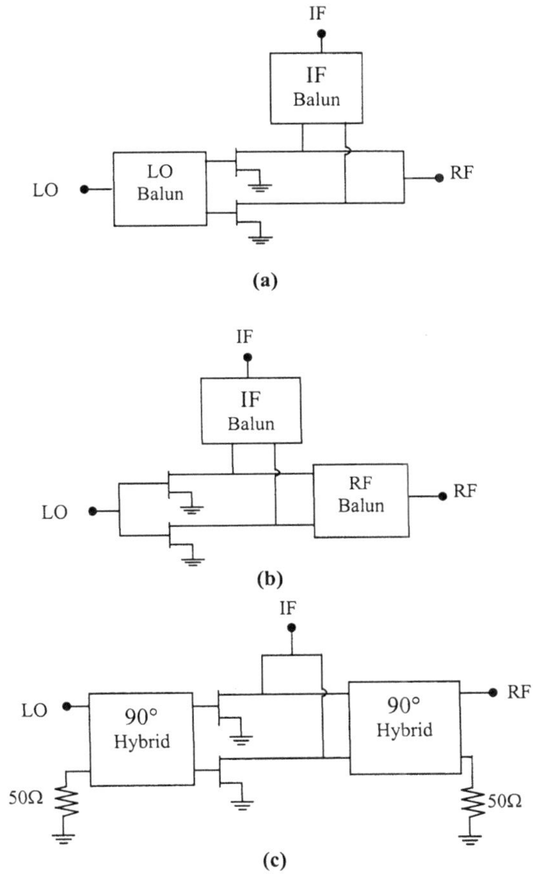

Figure 7.41 Single-balanced resistive FET mixers

7.8.3 A 60 GHz resistive FET mixer design example

In this example, resistive FET mixers are developed for direct conversion receivers, which downconvert the RF signal directly to baseband. This receiver architecture eliminates the IF circuitry, including IF filters. This results in reduced design complexity and minimal hardware. Thus, they are very attractive for small size terminals in mobile communications. However, these receivers can suffer from small instantaneous dynamic range due to undesired baseband intermodulation products. The resistive FET mixer, which exhibits low intermodulation, is especially suited for direct conversion applications [125].

Another advantage of using a resistive FET mixer for direct conversion is that it exhibits low $1/f$ noise. This is because minimal current passes through the device when it is pumped by the LO voltage at the gate. This is unlike the diode mixer, where the small-signal junction conductance is proportional to the diode current. To generate a time-varying conductance for efficient frequency conversion, currents of several mA have to pass through the diode during each LO pump cycle, leading to high $1/f$ noise.

For downconversion to baseband, the RF and LO frequencies are very close together and it is not feasible to separate them through filtering. The use of hybrids for RF and LO injection will result in large chip areas. In this design example, the alternative topology shown in Figure 7.42 is employed [114]. In this mixer circuit, the three different terminals of the PHEMT are used for the RF, LO and IF signals to achieve port isolations. Inductive feedback is applied between the gate and drain to form a resonance with the gate-drain capacitance, minimising any coupling between the LO and RF. The IF generated by the time-varying channel resistance is extracted through a low pass filter from the source instead of the drain. A DC ground is included at the drain to ensure zero drain bias and to present a short-circuit for the IF. The IF filter is designed to present a virtual ground to both the LO and RF. Simple matching networks are included for the LO and RF ports. This results in a compact monolithic design with good isolations between the three mixer ports.

Figure 7.43 shows a microphotograph of the fabricated single-ended mixer, measuring 1 × 1 mm^2. The mixer is designed around a 2 × 60 µm PHEMT device. The LO matching consists of an open-circuit stub, while the RF matching employs a short-circuit stub, which also provides the required DC ground. The IF is extracted from the source through a high impedance line. As the RF and LO frequencies are close together, a simple quarter-wavelength open-circuit stub is placed at the source to present a virtual ground to the LO and RF signals. Coupling between the LO and RF through the gate to drain capacitance is minimised by forming a parallel resonance with an inductive loop. A series blocking capacitor in the loop prevents DC shorting between the gate and drain. Reverse bias is applied to the gate through a mesa resistor. The value of the gate voltage is somewhat less than the pinch-off voltage for maximum channel resistance variation when the LO is applied.

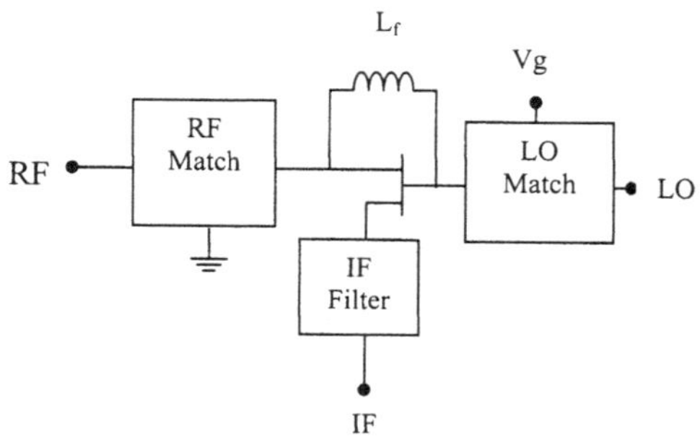

Figure 7.42 Single-ended resistive FET mixer for direct conversion

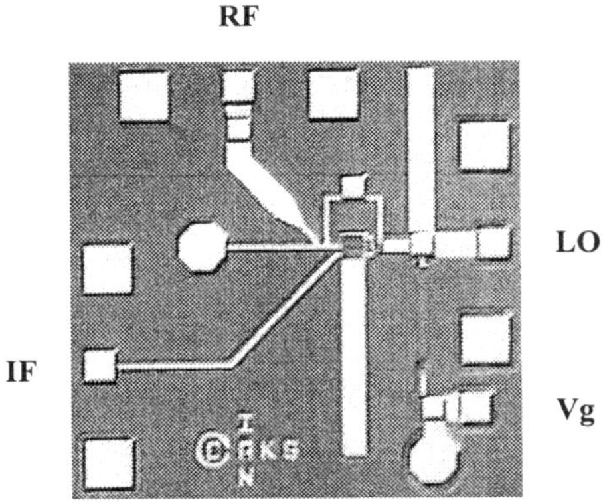

Figure 7.43 Microphotograph of the single-ended resistive FET mixer (1 × 1 mm²)

Based on the use of dual quadrature hybrids, a single-balanced resistive FET mixer is realised by employing two of the previously described 60 GHz mixers. Figure 7.44 shows a photograph of the fabricated mixer, measuring 1.3 × 1.6 mm². The LO and RF are applied to the two devices in quadrature using Lange couplers. The IF generated by the devices, which are in-phase, are combined directly using high impedance transmission lines.

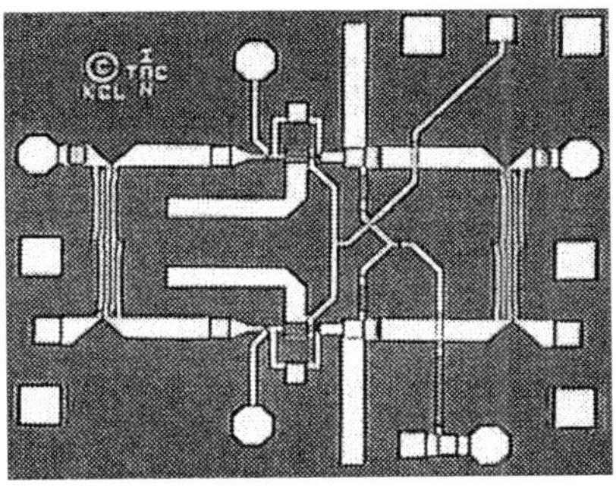

Figure 7.44 Microphotograph of a single-balanced resistive FET mixer (1.3 × 1.6 mm²)

The mixers were measured on-wafer using V-band CPW probes. The performance of the two mixers is summarised in Table 7.1. The conversion loss was measured as the RF is swept above and below the fixed LO frequency of 60 GHz. The single-ended mixer exhibits a conversion loss of 9 ± 0.5dB for a DC to 0.3 GHz IF, and 10 ± 1dB for higher IFs up to 1GHz. The single-balanced mixer achieves a conversion loss of 8.5 ± 0.5 dB for a DC to 1 GHz IF. These measurements were performed at a gate bias of -0.6 V and a LO power level of 10 dBm.

Table 7.1 Measured performance of the direct-conversion resistive mixers

	Single-ended mixer	Single-balanced mixer
Conversion loss	10 ± 1 dB	8.5 ± 0.5 dB
LO-RF isolation	30 dB	32 dB
LO return loss	-6 dB	-10 dB
RF return loss	-5 dB	-15 dB
IP2	25 dBm	28 dBm
IP3	28 dBm	30 dBm

7.8.4 Double-balanced resistive FET mixer

As illustrated in the 60 GHz resistive mixer design example, by providing an IF short-circuit at the drain the IF can be extracted from the source instead of the drain. This is also the case in the double-balanced resistive mixer, shown in Figure 7.45 [112,119]. The RF balun provides a virtual ground at the drain for IF and LO. Similarly, the LO balun provides a virtual ground for the RF and IF, while the IF balun provides a virtual ground for the LO and RF. Thus, all the three mixer ports are mutually isolated. It also has the same advantages as previously described for double-balanced mixers.

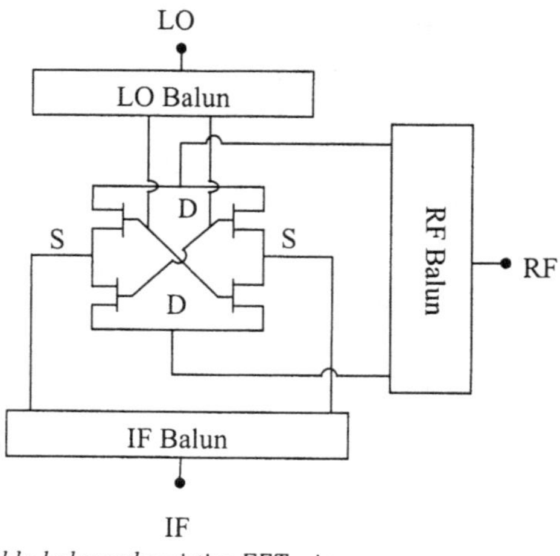

Figure 7.45 Double-balanced resistive FET mixer

Figure 7.46 shows an example of a double-balanced resistive mixer MMIC [126]. It is employed in a millimetre-wave upconverter for point-to-multipoint distribution systems. The mixer employs four 0.25 μm AlGaAs/InGaAs PHEMTs. The LO, RF and IF baluns used are similar to those described in Section 7.1.2. The mixer conversion loss was limited by the amount of LO power available. It exhibits 16 dB conversion loss with 10 dBm of LO power at 38 GHz. However, by increasing the LO power to 17 dBm, the conversion loss reduces to 9 dB in the upconverter.

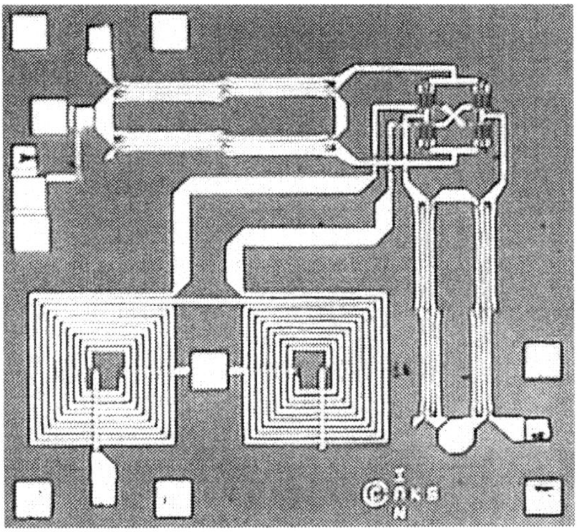

Figure 7.46 A millimetre-wave double-balanced resistive mixer (3 × 3.2 mm²)

7.9 Other mixer structures

7.9.1 Image-rejection and single-sideband mixers

For a given LO frequency, there are two different RFs, $(\omega_{LO} + \omega_{IF})$ and $(\omega_{LO} - \omega_{IF})$, that can result in the same IF. Depending on the application, one of these frequencies is the desired RF frequency, while the other is the image frequency. In receiver applications, this implies that there are two frequencies that will down-convert to the same IF. In the case of up-conversion, two RF frequency bands, referred to as the upper and lower sidebands, are generated. In many applications, these frequencies are close together and it is not feasible to isolate them using filters. Instead, hybrids are used to isolate them by introducing different phase shifts to the two frequencies in the mixer circuit.

Figure 7.47*a* shows the block diagram of the phasing-type image-rejection mixer. It consists of two identical mixers, usually double-balanced, which are fed with phase-quadrature RF input signals and an in-phase LO. The two downconverted sideband signals can then isolated at the outputs of the 90° hybrid, with the unwanted sideband terminated in a resistive load.

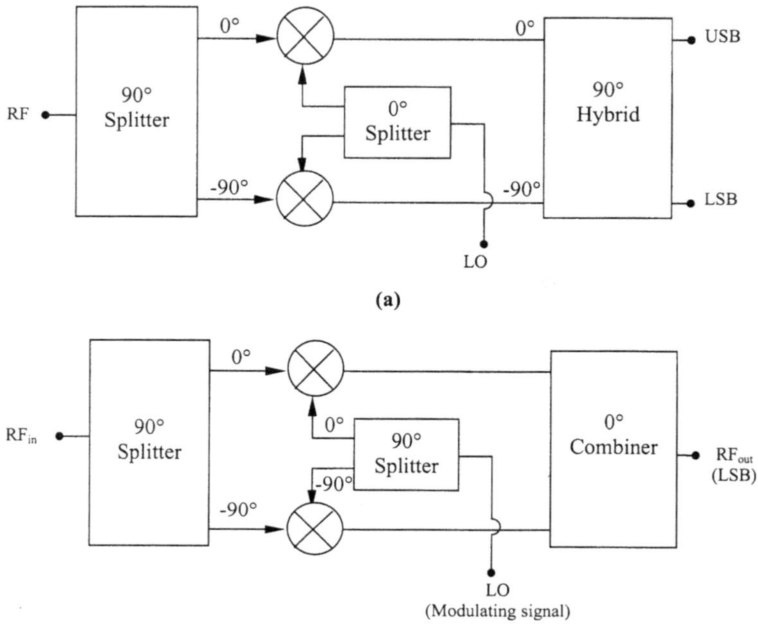

(a)

(b)

Figure 7.47 (a) Image-rejection mixer block diagram and (b) single-sideband modulator block diagram

The block diagram of the phasing-type single-sideband modulator is shown in Figure 7.47*b*. The RF input signals and the modulating signals are fed in phase-quadrature, while the mixer outputs are combined with an in-phase power combiner. As shown, the single-sideband modulator produces only the lower sideband. If the modulating frequency's 0 and 90° signals are switched then the upper sideband is produced.

In recent years, monolithic small-shift frequency translators have been demonstrated using single-sideband modulators of this type. One notable example was demonstrated by Parisi [127]. Here, single-balanced mixers were employed in an all lumped-element design. With an input carrier frequency of f_0=6.4 GHz and a maximum modulating frequency of 0.5 GHz, a conversion loss of 7 dB, carrier suppression of 20 dB and image suppression of 18 dB were reported. A more recent single-sideband modulator example was demonstrated by Baree *et al.* [128]. Here, compact common-gate/common-source balanced FET mixers were employed with a combination of distributed and lumped-element couplers. With an input frequency covering 4.25 to 13.75 GHz and a fixed LO of 2 GHz, a conversion gain of approximately 5 dB and an upper-sideband suppression of 20 dB were maintained.

7.9.2 Sub-harmonically pumped mixers

Most mixers use the fundamental of the LO to produce an IF. Harmonic mixers deliberately use a harmonic ($n > 1$) of the LO. This mixer type is especially attractive at millimetre-wave frequencies, where the cost of fundamental LO power and the complexity of phase-locking increases rapidly with frequency.

Figure 7.48 shows the schematic diagram of a resistive harmonic mixer. It consists of two resistive FETs, whose drains are connected together, while the gates are pumped by in anti-phase through a LO balun. The RF is applied to the drains through a high pass filter. The channel resistances of the FETs are modulated by the LO frequency. The parallel combination of the two anti-phase conductance waveforms has a frequency which is twice the LO frequency. The IF generated by the mixing process is extracted at the drain, through a low pass filter. The anti-phase LO relationship also provides rejection of undesired LO leakage through the gate-drain feedback capacitances.

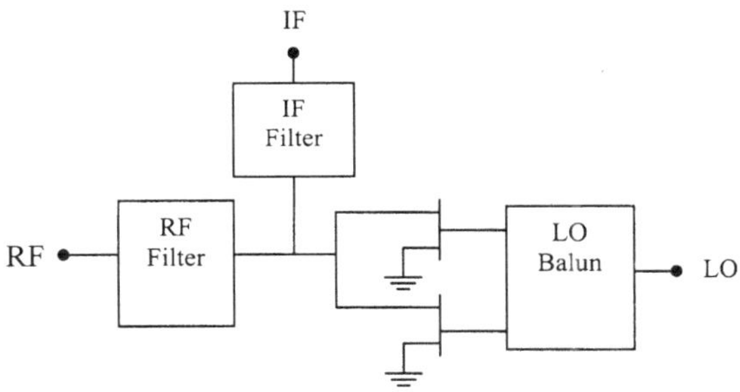

Figure 7.48 Subharmonically-pumped resistive FET mixer

Figure 7.49 shows an example of a resistive harmonic mixer employing 0.25 μm gate length AlGaAs/InGaAs PHEMTs [133]. The circuit uses two 2 × 60 μm PHEMTs. The RF and IF filters are implemented using high impedance lines and MIM capacitors. The LO balun was implemented as a Marchand balun using two Lange couplers. The gate bias for the PHEMTs are applied to the LO balun through a mesa resistor. The mixer downconverts an RF from 35-40 GHz to an IF of 1-5 GHz, with a fixed LO of 17.5 GHz.

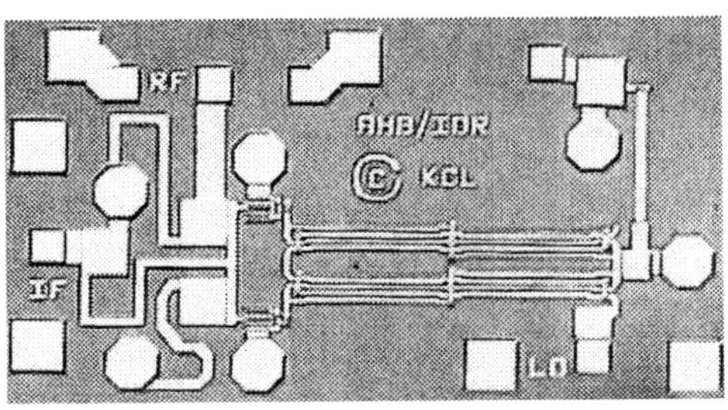

Figure 7.49 Microphotograph of a millimetre-wave subharmonically-pumped resistive FET mixer ($1.1 \times 2.1 mm^2$)

7.9.3 Distributed FET mixers and coupling structures

The distributed (or travelling-wave) amplifier, described in Chapter 5, achieves ultra-wideband operation by incorporting the transistor input and output capacitances into so-called 'artificial transmission lines', which are in fact low-pass ladder networks. The beauty of this technique is that matching can easily be achieved over wide bandwidths, such as 1 to 20 GHz. As a result, the travelling-wave matching technique has found applications in mixers, active baluns, combiners and splitters.

The distributed FET mixer was first demonstrated by Tang and Aitchison [134,135] in hybrid form, using single-gate FETs with a passive coupler to combine the RF and LO signals. Dual-gate FETs can also be used so that the coupler is not needed [136]. The mixer then has a gate-line for the RF, a gate-line for the LO and a drain-line for the IF, as shown in Figure 7.50. Because this has some in-built LO/RF isolation, it is very suitable for MMIC implementation.

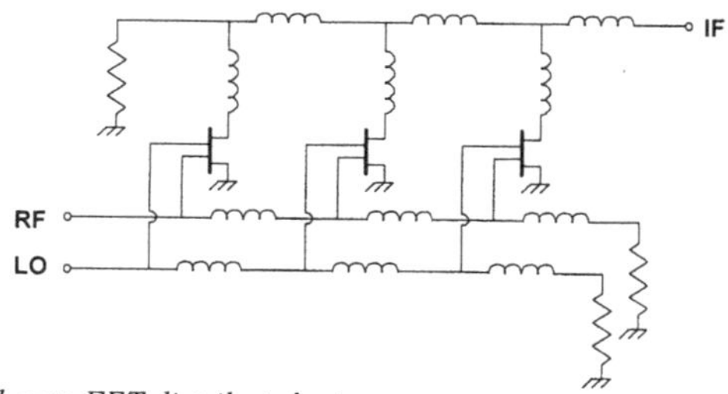

Figure 7.50 Dual-gate FET distributed mixer

Although dual-gate FETs provide a convenient means of implementing distributed mixers, they may not be readily available. Alternatively, a cascode of two single-gate FETs can be used to mimic the dual-gate FET. Figure 7.51 shows an example of a cascode distributed mixer using four sections. It consists of two tiers of FETs in cascode, with separate artificial transmission lines for RF, LO and IF. The RF and LO are applied to the lower and upper gate lines, respectively, while the IF is extracted from the drain line. For a fixed IF of 1 GHz, it achieves 3 dB conversion gain over an RF range of 2 to 20 GHz [142].

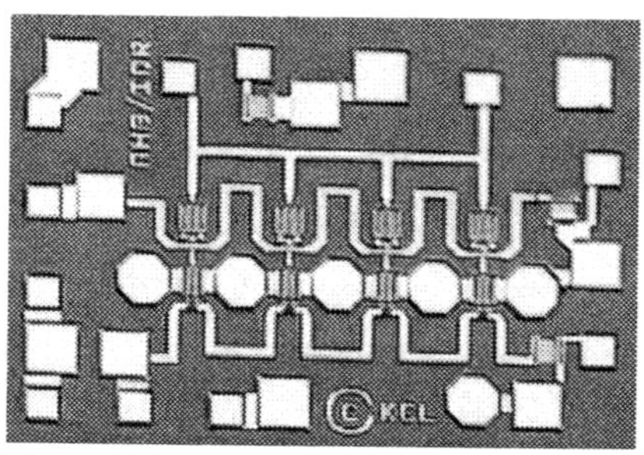

Figure 7.51 A cascode-FET distributed MMIC mixer (1.3 × 1.9mm²)

The distributed mixer can also be implemented using resistive FETs to incorporate low intermodulation properties with wideband performance. Operating as a resistive mixer, the channel resistance of the FETs is modulated by the LO signal along the gate line, while the RF signal is applied along the drain line. The low frequency IF generated is extracted at the terminated end of the drain line, through a low pass filter. A 2-18 GHz MMIC prototype, shown in Figure 7.52, achieves 10 dB conversion loss and IP3 of 28 dBm, with 10 dBm of LO power [143].

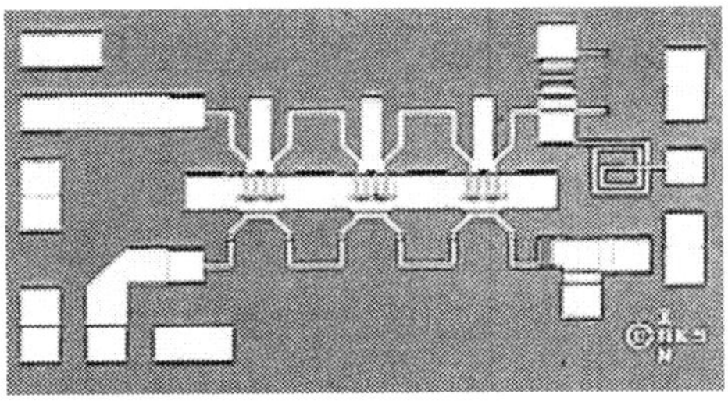

Figure 7.52 A resistive FET distributed MMIC mixer (1.1 × 2.1mm²)

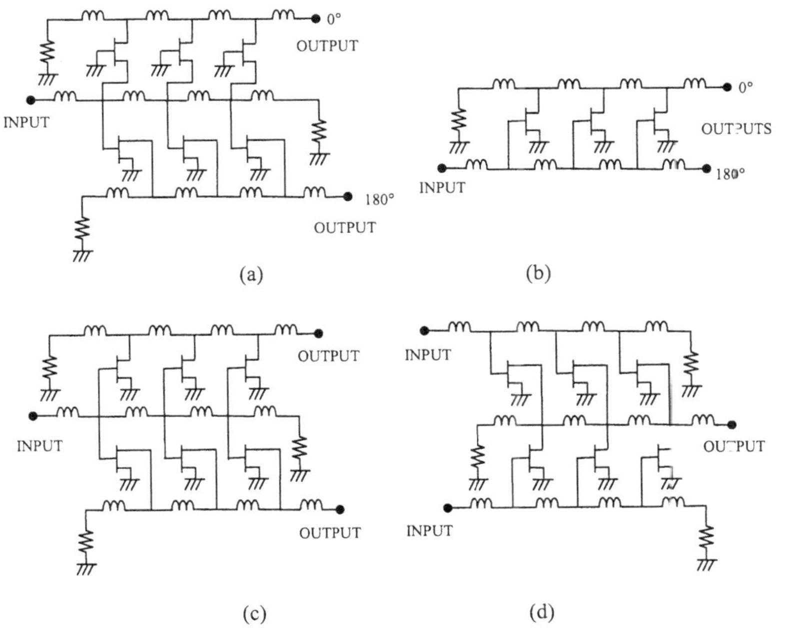

Figure 7.53 Distributed coupling structures: (a) active balun with common-source/common-gate FETs, (b) gate-line termination balun, (c) power splitter and (d) combiner

A number of distributed active coupling structures for mixers have also been investigated and published in the literature. Some of the most notable work performed was by Robertson [140], who investigated practically the noise properties as well as the small- and large-signal characteristics. The most common technique for distributed active baluns is the use of common-gate and common-source FET pairs, on a common input line [139], as shown in Figure 7.53*a*. Distributed active baluns using a gate-line termination technique, as shown in Figure 7.53*b*, have also been demonstrated [141]. A distributed active splitter [137] has a single gate-line and two drain-lines with the FET connected to a common drain-line, as shown in Figure 7.53*c*. The distributed active combiner [138] has two gate-lines with the FETs connected to a common drain line, as shown in Figure 7.53*d*. In some respects this is not a true power combiner, because input signals at different frequencies would suffer intermodulation via the drain-voltages of the FETs. Nevertheless, this effect is unimportant for many applications, and this combiner has been used successfully in distributed mixers.

7.10 Some final comments and observations

A vast amount of theoretical and practical work has been published in the field of mixer design using diodes, FETs and other more complex structures. In recent years, much of the experience gained, together with new ideas, has been applied to realising the circuits in monolithic form. Although many of the active forms using FETs look promising and show good small signal-performance, it should be noted that the noise performance and dynamic range are often rather poor. If a publication does not give the noise figure, then it is probably bad! The best mixer conversion performances and noise figures appear to be obtained when switching devices are used together with distributed (sometimes folded to reduce the size) or lumped-element coupling structures. In this type of circuit, one must then consider carefully whether it would be more cost effective to put the coupling structure on a separate MIC substrate, rather than use costly GaAs real estate for a purely passive circuit. This is especially true if multilayer multi-chip module technology is used for the complete transceiver module.

The key parameters of the most notable diode, active and resistive FET mixers are summarised in Tables 7.2, 7.3 and 7.4, respectively. Note that the figures in square brackets are not given in the references, and are based on the assumption that the single-sideband noise figure is equal to the conversion loss, and that the single-sideband noise figure is 3 dB higher than the double-sideband noise figure. They are given for comparison purposes only. This assumption can only be made when the baluns or coupling circuits are passive. The mixers in Table 7.2 are all diode mixers. A mixer is considered to be a diode mixer when the part which is switching and performs the frequency conversion process is a diode. Therefore, a mixer which uses FETs to form active baluns has been included in this table. The references in these tables cover only mixers where performance was given in terms of conversion loss or noise figure for the mixer alone. Therefore, references have been excluded which do not allow the performance of the mixer alone to be calculated, or only give a loss/gain figure for a mixer and amplifier combination.

Ref.	Company							
16	Siemens	15	70 (MHz)	8.5	[5.5]	10 inc. 1.5 dB IF [8.5]	130	Microstrip SBM with branch-line coupler.
17	Honeywell	30-40	5-500 (MHZ)	[7.5]	6.0 (31-39 GHz) inc. 1.5dB IF [4.5]	[7.5]	660	Suspended stripline SBM in WG. pump power 6mW, no dc bias.
18	GEC	8-12	1-500 (MHz)	4.95	[1.95]	[4.95]	-	Theoretical figure. Microstrip SBM with Lange couplers.
19	Honeywell	93-97	5-500 (MHz)	[7.8]	6.3 inc. 1.5dB IF [4.8]	[7.8]	600	Cross bar SBM in WG.
20	Hughes	75-110	1-9	9.0 av, 7.5 @ 76 GHz	[6.0] [4.5]	[9.0] [7.5]	3000	Cross bar SBM in WG.
21	Honeywell	80-105	-	[7.1]	5.6 @ 94.5 GHz inc. 1.5 dB IF [4.1]	[7.1]	700	Microstrip SBM with rat race coupler.
23	Hughes	73.6-83.6	8-18 (GHz)	8.0, 4.6 @ 91.1 GHz	[5.0] [1.6]	[8.0] [4.6]	-	CPW & slotline SBM.
24	GE	26-40	1	6.0	[3.0]	[6.0]	640	CPW & CPS SBM with bow tie transition.
25	MIT Lincoln Labs.	34-36	1-2	6.5 +/- 0.5 (34-36 GHz)	[3.5 +/- 0.5]	[6.5 +/- 0.5]	-	Microstrip SBM with branch-line coupler.
27	GE	30	1	6.0	[3.0]	[6.0]	637	CPW and CPS SBM using Mott diodes.
28	Hughes	27-30	3.5-6.5	10.5	[7.5]	[10.5]	550	Microstrip SBM with Lange coupler. Part of complete RX.
29	Honeywell	27-30	5.5-8.0	8.0 @ 30 GHz	[5.0]	[8.0]	>1000	Microstrip SBM with modified rat race coupler.
30	Pacific Monolithics	3.7-4.2	0.1-1.8	-	-	-	-	Complete RX with RF & IF LNAs. DBM. NF=4.0dB.
33	Hughes	75-88	4-17	6.8-10.0	[3.8-7.0]	[6.8-10.0]	450-560	Microstrip SBM with Lange coupler.
34	T I	2-18	500 (MHz)	12.0 @ 10 GHz	[9.0]	12.0-15.0	-	DBM with FET baluns.
35	AEG	50-70	-	6 @ 60GHz	3.3 @ 60 GHz	[6.3]	2300	Microstrip SBM with modified branch-line coupler.
36	Thomson	7.47	30 (MHz)	6.7	[3.7]	[6.7]	-	Lumped element rat race coupler.
37	Telefunken Sys'technik	94	1	8 @ 94 GHz	6 @ 94 GHz	[9.0]	2000	Microstrip SBM with branch-line coupler.
38	TRW	7-10	5	10.0	[7.0]	[10.0]	-	MESFET technology DBM using lumped element high & low pass networks for balun. High IP3
39	TRW & ESG	30-40	20-100 (MHz)	5.0 @ 35 GHz	[2.0]	[5.0]	-	Complete downconverter in HEMT technology. Microstrip SBM with Lange couplcr.

Table 7.2 Continued

41	Martin Marietta & GAMMA	85-100	-	11.8 @ 94 GHz	11.3 @ 4 GHz	[14.3]	-	HEMT technology. Microstrip SBM with branch-line coupler.
42	TRW & ESG	92-96	1-8	<9.0 (90-8 GHz) 7.5 @ 94 GHz	[<6.0] [4.5]	[<9.0] [7.5]	370	PHEMT technology downconverter with 2 stage RF LNA. Microstrip SBM with rat race coupler. 5.3dB gain & 6.8dB NF @ 94GHz.
43	Varian	6-20	2-7	6.2-9.8	[3.2-6.8]	[6.2-9.8]	-	CPW, CPS & slotline SBM.
45	Telefunken Sys'technik	94	0.1-1.8	6.0 @ 94 GHz	4.0	[7.0]	-	MESFET technology downconverter. Microstrip SBM with Lange coupler.
46	Raytheon *et al.*	35	-	19 @ 35 GHz	8.7 @ 35 GHz	[11.7]	-	Quasi-optical millimetre-wave monopulse RX with slot ring SBMs.
47	THORN EMI & Alpha	35	100 (MHz)	4.6	[1.6]	[4.6]	1058	Microstrip SBM with sum enhancement. Rat race mixer with vertical diode.
47	THORN EMI & Alpha	35	100 (MHz)	5.7	[2.7]	[5.7]	1058	Microstrip SBM with Lange coupler with vertical diode.
48	Nonlinear Consulting & TRW	26-40	DC-10	5.0-10.0	[2.0-7.0]	[5.0-10.0]	235	Planar Marchand baluns. DBM with diodes in star configuration.
49	Northrop	16-21	200 (MHz)	8.0-11.0	-	-	-	PHEMT technology. Microstrip DBM.
50	TRW	10-18	-	-	-	-	-	HEMT technology image rejection RX. Microstrip SBM with Lange couplers. 7-10dB gain.
51	COMSAT	91-96	50 (MHz)	8.5	-	-	-	LNA SBM combination. CG=7dB & NF=7.3dB.
52	TRW, Hughes & Hexwave	20.2-21.2	1.55	-	-	6.0	-	LNA SBM combination. CG=15dB & NF=6.0dB.
53	TI	6-18	500 (MHz)	<8.0	[<5.0]	[<5.0]	-	DBM with novel form of Marchand balun.
54	TRW	18-22	6-10	5.0-6.0	[2.0-3.0]	[5.0-6.0]	-	HBT diode DBM.
55	TRW	43-46	16-21	<6.0	[<3.0]	[<6.0]	1.02THz	HEMT diode DBM with Lange couplers.
56	TRW	60-63	8.1	-	-	-	-	HEMT diode SBMs using modified rat race couplers. RF & IF LNAs. CG >10dB.
57	Nonlinear Technologies & TRW	18-40	DC-10	5.5-9.0	-	-	-	Planar Marchand baluns. DBM with diodes in ring configuration.
58	Hittite	5-20 6-11 4-8 1.8-5 1-2	dc-3.0 dc-2.0 dc-2.0 dc-2.0 dc-1.0	10.0 10.0 10.0 11.0 11.0	[7.0] [7.0] [7.0] [8.0] [illegible]	[10.0] [10.0] [10.0] [11.0] [illegible]	- - - -	DBMs with folded passive baluns or planar transformers. 25-35dB pump-RF isolation. P_p=13-20dbm. P_{1dB}=10dbm. IP3=20dbm.

Table 7.3 A comparison of different monolithic active mixers using FETs, HEMTs and pHEMTs

Ref.	Company	RF (GHz)	IF (GHz)	CG (dB)	DSB NF (dB)	SSB NF (dB)	Gate Length (μm)	Comments
82	CISE & Alcatel	11.7-12.4	-	-7.5	-	-	-	Dual gate FET SEM.
83	Matsushita	100-800 (MHz)	100 (MHz)	6.0-8.0	-	-	1.0	DBM using dual gate FETs. P_p=10dBm.
84	GE	6-18	0.1-2.0	-8.0	-	-	0.5	DBM using single gate FETs. Theoretical figure based on switching model.
86	TI	2-18	0.2-1.4	0--6.5 (0.1-1.4 GHz IF)	15.0-18.0	-	0.5	4 stage distributed SEM realised with dual gate FETs.
87	TRW	6-10	3	7+/-1.5 (6-10 GHz)	13.0	-	0.5	Dual gate SBM. Complete RX with RF & IF LNAs. P_p=10dBm.
88	Plessey	6-4	2-7	5.0	13.0	-	1.0	Dual gate common source SBM. Complete downconverter with RF & IF LNAs.
89	Avantek	0.1-6	70 (MHz)	8-15 (RF=2 GHz, IF=250 MHz)	-	15.5-16 (RF=2 GHz & IF=250 MHz)	-	Silicon bipolar active mixer using a Gilbert cell.
95	Varian & Litton	23-40	2-13	0-5	-	-	0.25	2 stage distributed SEM realised with single gate FETs. Pump and RF to gate via external combiner
96	Sharp	11.7-12.2	1.0-1.5	>0 (11.7-12.2 GHz)	<6.5 (11.7-12.2 GHz)	-	-	Drain pump injection SBM. Downconverter for broadcast satellite applications. P_p=8dBm.
98	TRW & UCLA	DC-5	8-10	-4.0 @ 2.5 GHz	-	-	0.5	DBM using dual gate FETs. Upconverter chip with RF & IF LNAs.
101	Pacific Monolithics	34.1-36	0.1-2	-4.0 @ 3 GHz	-	-	-	PHEMT mixer fed via drain. Complete RX chip with RF & IF LNA. P_p=-7dBm.
102	Comms. Res. Centre.	19.5-22	4	<-3.0	-	-	0.25	PHEMT process. SBM. Pump and RF to gate via external diplexer.
103	Comms. Res. Centre.	28.5-31	4	1-3.7	-	-	0.25	PHEMT process. Same as above.
104	Fujitsu	58-62	140 (MHz)	-7.5 @ 59.86 GHz	-	-	0.15	SBM active drain mixer with Lange couplers in an image rejection configuration.
105	M/A COM & Anadigics	2.5	100 (MHz)	0.9	-	15.0	-	Silicon MMIC Gilbert cell mixer with 20dB improvement in dynamic range.
109	Plextek	41-44	5.2	6 @ 44 GHz	-	-	0.25	PHEMT MMIC Gilbert cell type mixer

Table 7.4 A comparison of different monolithic resistive mixers using FETs, HEMTs and pHEMTs

Ref	Company	RF (GHz)	IF (GHz)	CL (dB)	DSB NF (dB)	Input IP3 (dBm)	Gate Length (μm)	Comments
113	TriQuint	10-14	DC-1.5	10 @ 12 GHz	-	19	-	SBM with Lange couplers for LO and RF ports. Pp=10 dBm
114	Honeywell	27-37	10-100 (MHz)	9	7.5		0.25	SEM with resonant loop from drain to gate for isolation. Pp=9 dBm.
115	TRW	26-29	DC-2	7-9		>24	0.2	PHEMT SBM with Lange couplers for LO and RF ports. Pp=13 dBm
116	Alcatel	12.7-15.5	50 (MHz)	-6.2--6.9	-	-	0.25	HEMT process - cold shunt HEMT mixer cell SBM.
117	Raytheon	4-18	1-2	<12.0		>18 dBm	0.25	PHEMT SBM with active pump balun & passive RF balun. Po=7 dBm
118	TRW	93.2	0.2	12.8			0.1	HEMT SBM with Lange couplers for LO and RF ports, external hybrids for IF ports Pp=8 dBm
120	TRW	DBM: 3-18 SBM: 2-16	0.75-1.25	DBM: 7.5-9 (4-13 GHz) 7.5-11 (3-18 GHz) SBM: 8-10 (4-15 GHz) 8-11 (2-16 GHz)	-	DBM: 19.5-27.5 dBm SBM: 20-28.5 dBm	0.2-0.25	HEMT DBM.: Active pump balun, passive RF & IF baluns. Pp=16 dBm HEMT SBM: Active pump balun, passive RF balun & passive IF divider. Pp=16 dBm
121	TRW	93.5-97.5	1.5	9			0.1	MODFET IRM. Pp=8 dBm
123	Chalmers University	112.5	2.5	9	-	-	0.15	SEM using InAlAs/InGaAs/InP HFET. Pp = 4 dBm
124	Mitsubishi	55-60	0.4-5.4	8.4-12.5	-	-	0.15	CPW SEM. Pp = -2 dBm.

7.11 References

7.11.1 General

1 TUCKER, D. G.: 'Circuits with periodically-varying parameters' (Macdonald and Co. Ltd., London, 1964)

2 SALEH, A. A. M.: 'Theory of resistive mixers' (Research Monograph no. 64) (MIT Press, Cambridge, MA and London, 1971)

3 MAAS, S. A.: 'Microwave mixers' (Artech House, Dedham, MA, 1986)

4 MAAS, S. A.: 'Non-linear microwave circuits' (Artech House, Norwood, MA, 1988)

5 FAYOS, J. R., and NIGHTINGALE, S. J.: 'Determination of the transfer function of a non-linear circuit for prediction of intermodulation characteristics', *Proceedings of IEEE MTT-S International Microwave Symposium*, Baltimore, 1986, pp. 499–502

6 HOWSON, D. P., and GARDINER, J. G.: 'High-frequency mixers using square-law diodes', *The Radio and Electronic Engineer*, 1968, pp. 311–316

7 WILL, P.: 'Reactive loads – the big mixer menace', *Microwaves*, 1971, pp. 38–42

8 MANLEY, J. M., and ROWE, H. E.: 'Some general properties of non-linear elements: Part (1) General energy relations', *Proceedings of the IRE*, 1956, **44** (7), pp. 904–913

9 ROWE, H. E.: 'Some general properties of non-linear elements: Part (2) Small signal theory', *Proceedings of the IRE*, 1958, **46** (5), pp. 850–860

10 NIGHTINGALE, S. J.: 'Loss and noise characteristics of microwave mixers'. PhD thesis, University of Canterbury, Kent, 1980

11 NIGHTINGALE, S. J., and SCHIEK, B.: 'Direct and accurate measurements of mixer equivalent noise temperatures', *Proceedings of 8th EuMC*, Paris, 1978, pp. 712–716

12 NEMLIKHER, Y. A., and STRUKOV, I. A.: 'The impedance and transfer characteristics of resistive frequency converters using Schottky barrier diodes', *Radio Engineering and Electronic Physics*, 1974, **19** (1), pp. 111–118

13 NEMLIKHER, Y. A., and STRUCKOV, I. A.: 'Noise characteristics of frequency converters using Schottky barrier diodes', *Radio Engineering and Electronic Physics*, 1974, **19** (1), pp. 119–126

14 FAIRBURN, M., MINNIS, B. J., and NEALE, J.: 'A novel monolithic distributed mixer design', *Proceedings of the IEE Colloquium on Microwave and Millimetre Wave Monolithic Integrated Circuits*, Digest no. 1988/117, 1988

15 LIECHTI, C. A.: 'Down converters using Schottky barrier diodes', *IEEE Transactions on Electron Devices*, 1970, **ED-17** (11), pp. 975–83

7.11.2 Diode mixers

16 RISTOW, D., ENDERS, N., and KNIEPKAMP.: 'A monolithic GaAs Schottky barrier diode mixer for 15 GHz', *Proceedings of EuMC*, Paris, 1978, pp. 707–711

17 CONTOLATIS, A., CHAO, C., JAMISON, S., and BUTTER, C.: 'Ka-band monolithic GaAs balanced mixers', *Proceedings of IEEE Microwave and Millimeter Wave Monolithic Circuits Symposium*, Dallas, 1982, pp. 28–30
18 JACOMB-HOOD, W. W., SMITH, J. R., and HIPWOOD, L. G.: 'High quality X-band monolithic diode mixer', *Proceedings of IEEE Microwave and Millimeter Wave Monolithic Circuits Symposium*, Boston, 1983, pp.1–4
19 CHAO, C., CONTOLATIS, A., JAMISON, S., and BUTTER, C.: '94 GHz monolithic GaAs balanced mixers', *ibid.*, pp. 50–53
20 YUAN, L. T.: 'A W-band monolithic GaAs crossbar mixer', *Proceedings of IEEE Microwave and Millimeter Wave Monolithic Circuits Symposium*, San Francisco, 1984, pp. 67–69
21 BAUHAN, P., CONTOLATIS, T., ABROKWAK, J., and CHAO, C.: '94 GHz planar GaAs monolithic balanced mixer', *ibid.*, pp. 70–73
22 KERMARREC, C., FAGUET, J., VANION, B. *et al.*: 'The first GaAs fully integrated microwave receiver for DBS applications at 12 GHz', *Proceedings of EuMC*, Liège, 1984, pp. 749–754
23 YUAN, L. T., and ASHER, P. G.: 'A W-band monolithic balanced mixer', *Proceedings of IEEE Microwave and Millimeter Wave Monolithic Circuits Symposium*, St Louis, 1985, pp. 71–73
24 NIGHTINGALE, S. J., UPTON, M. A. G., MISHRA, U. K., PALMATEER, S. C., and SMITH, P. M.: 'A 30 GHz monolithic single balanced mixer with integrated dipole receiving element', *ibid.*, pp. 74–77
25 CHU, A., COURTNEY, W. E., MAHONEY, L. J., MANFRA, M. J., and CALAWA, A. R.: 'Dual function mixer circuit for millimeter wave transceiver applications', *ibid.*, pp. 78–81
26 KERMARREC, C.: 'GaAs MMICs for mass production 12 GHz down converters', *Proceedings of EuMC*, Paris, 1985, pp. 87–95
27 NIGHTINGALE, S. J., UPTON, M. A. G., MISHRA, U. K., PALMATEER, S. C., and SMITH, P. M.: 'An EHF coplanar monolithic single balanced mixer using Mott diodes', *ibid.*, pp. 635–640
28 LIU, L. C. T, LIU, C. S., KESSLET, J. R., and WANG, S. K.: 'A 30 GHz monolithic receiver', *Proceedings of IEEE Microwave and Millimeter Wave Monolithic Circuits Symposium*, Baltimore, 1986, pp. 41–44
29 BAUHAN, P., CONTOLATIS, A., SOKOLOV, V., and CHAO, C.: '30 GHz balanced mixers using an ion-implanted FET-compatible 3-inch GaAs wafer process technology', *ibid.*, pp. 45–49
30 PODELL, A. F., and NELSON, W. W.: 'High volume, low cost, MMIC receiver front end', *ibid.*, 1986, pp. 57–59
31 GELLER, B. D., and ABITA, J. L.: 'An integrated DBS receiver front-end using a monolithic frequency converter', *Proceedings of EuMC*, Dublin, 1986, pp. 613–618
32 RAMACHANDRAN, R., MOGHE, S., HO, P., and PODELL, A.: 'An 8-15 GHz GaAs monolithic frequency converter', *Proceedings of IEEE Microwave and Millimeter Wave Monolithic Circuits Symposium*, Las Vegas, 1987, pp. 31–34
33 TRINH, T. N., WONG, W. S., LI, D., and KESSLER, J. R.: 'Ion implanted W-band monolithic balanced mixers for broadband applications', *ibid.*, pp. 89–92

34 PAVIO, A. M., HALLADAY, R. H., BINGHAM, S. D., and SAPASHE, C. A.: 'Broadband monolithic single and double ring active/passive mixers', *Proceedings of IEEE Microwave and Millimeter Wave Monolithic Circuits Symposium*, New York, 1988, pp. 71–74

35 ADELSEK, B., COLQUHOUN, A., DIEUDONNÉ, J.-M. *et al.*: 'A monolithic 60 GHz diode mixer in FET compatible technology', *Proceedings of IEEE Microwave and Millimeter Wave Monolithic Circuits Symposium*, Long Beach, 1989, pp. 91–94

36 TELLIEZ, I., CHAUMAS, P., RUMELHARD, C., and PATAUT, G.: 'GaAs monolithic balanced mixer for C-band direct demodulation receiver', *Proceedings of EuMC*, Wembley, 1989, pp. 731–736

37 ADELSECK, B., DIEUDONNE, J.-M., SCHMEGNER, K. E., COLQUHOUN, A., EBERT, G., and SELDERS, J.: 'A monolithic 94 GHz balanced mixer', *Proceedings of IEEE Microwave and Millimeter Wave Monolithic Circuits Symposium*, Dallas, 1990, pp. 111–114

38 TON, T. N., DOW, G. S., CHEN, T. H., LACON, M., LIN, T. S., BUI, S., and YANG, D.: 'An X-band monolithic double double-balanced mixer for high dynamic range receiver application', *ibid.*, pp. 115–118

39 MAOZ, B., REYNOLDS, L. R., and OKI, A.: 'FM-CW radar on a single GaAs/AlGaAs HBT MMIC chip', *Proceedings of IEEE Microwave and Millimeter Wave Monolithic Circuits Symposium*, Boston, 1991, pp. 3–6

40 YONAKI, J., CARANDANG, R., ALLEN, B. A. *et al.*: '35 GHz InGaAs HEMT MMIC downconverter', *ibid.*, pp. 47–50

41 TRIPPE, M. W., WEINREB, S., DUNCAN, S. W. *et al.*: 'mm-wave MIMIC receiver components', *ibid.*, pp. 51–54

42 CHANG, K. W., WANG, H., CHEN, T. H. *et al.*: 'A W-band monolithic pseudomorphic InGaAs HEMT downconverter', *ibid.*, pp. 55–58

43 EISENBERG, J., PANELLI, J., and OU, W.: 'A new planar double-double balanced MMIC mixer structure', *ibid.*, pp. 69–72

44 DÄMBKES, H., and SCHMIDT, L. P.: 'MMIC technology in Europe for millimetre-wave applications', *Proceedings of GaAs '92 Symposium*, Noordwijk, 1992 session 1B, paper 1

45 DIEUDONNE, J.-M., RITTMEYER, R., ADELSEK, B., and COLQUHOUN, A.: 'A 94 GHz monolithic downconverter in a MESFET technology', *Proceedings of IEEE Microwave and Millimeter Wave Monolithic Circuits Symposium*, Alburquerque, 1992, pp. 69–72

46 GINGRAS, R. L., DRUBIN, C., COLE, B. *et al.*: 'Millimeter-wave slot ring mixer array', *ibid.*, pp. 105–107

47 LANG, R. J., BHUMBRA, S. S., NIGHTINGALE, S. J., and McDERMOTT, M. G.: 'Low conversion loss Ka-band GaAs monolithic diode mixers', *Proceedings of EuMC*, Helsinki, 1992, pp. 735–740

48 MAAS, S. A., and CHANG, K. W.: 'A broadband planar doubly balanced monolithic Ka-band mixer', *Proceedings of IEEE Microwave and Millimeter Wave Monolithic Circuits Symposium*, Atlanta, 1993, pp. 53–55

49 FUDEM, H., MOGHE, S., DIETZ, G., and CONSOLAZIO, S.: 'A highly integrated MMIC K-band transmit/receive chip', *ibid.*, pp. 119–122

50 KATZ, R., AUST, M. V., KASODY, R. *et al.*: 'A highly compact wideband GaAs HEMT X-Ku band image-reject receiver MMIC', *ibid.*, pp. 131–134

51 HUANG, H. C., LAUX, P., BASS, J. F. *et al.*: 'A W-band multifunction MMIC', *Proceedings of IEEE Microwave and Millimeter Wave Monolithic Circuits Symposium*, San Diego, 1994, pp. 37–40
52 LESTER, J. A., HUANG, P., AHMADI, M. *et al.*: 'HEMT MMIC chip set for low cost miniaturised EHF satcom transceiver', *ibid.*, pp. 85–88
53 BRINLEE, W. R., PAVIO, A. M., and VARIAN, K. R.: 'Novel planar double-balanced 6-18 GHz MMIC mixer', *ibid.*, pp. 139–142
54 KABAYASHI, K. W., KASODY, R., OKI, A. K., DOW, S., ALLEN, B., and STREIT, D. C.: 'K-band double balanced mixer using GaAs HBT THz Schottky diodes', *ibid.*, pp. 209–212
55 AUST, M. V., ALLEN, B., DOW, G. S., KASODY, R., BIEDENBENDER, M., and WANG, N.: 'A low noise high gain Q-band monolithic HEMT receiver', *ibid.*, pp. 217–220
56 MINOT, K., AUST, M., KASODY, R. *et al.*: 'A MMIC chip set for V-band crosslink communication systems', *ibid.*, pp. 221–224
57 MAAS, S. A., YAMADA, F. M., OKI, A. K., MATOVELLE, N. and HOCHULI, C.: 'An 18-40 GHz monolithic ring mixer', *Proceedings of IEEE Radio Frequency Integrated Circuits Symposium*, 1998, pp. 29–32
58 Hittite Microwave Corporation, 21 Cabot Road, Woburn, MA 01801, USA

7.11.3 Coupling structures

59 MARCHAND, N.: 'Transmission line conversion transformers', *Electronics*, 1944, **17** (12) pp.142–145
60 CLOETE, J. H., 'Graphs of circuit elements for the Marchand balun', *Microwave Journal*, 1981, **24**, pp. 125–128
61 FERGUSON, D. BAUHAHN, P. KEUPER, J. *et al*,: 'Transformer coupled high-density circuit technique for MMIC', *IEEE MTT-S International Microwave Symposium Digest*, 1984, pp. 34–36
62 WIEMER, L. JANSEN, R. H. ROBERTSON, I. D., and SWIFT, J. B.: 'Computer simulation and experimental investigation of square spiral transformers for MMIC applications', *IEE Colloquium on Computer Aided Design of Microwave Circuits Digest*, 1985, pp. V1–5
63 JEAN, P., PAUKER, V., and DAUTRICHE, P.: 'Wide-band monolithic GaAs phase detector for homodyne reception', *IEEE MTT-S Internatiomnal Microwave Symposium Digest*, 1987, pp. 169–171
64 BOULOUARD, A., and LE ROUZIC, M., 'Analysis of rectangular spiral transformers for MMIC applications', *IEEE Transactions on Microwave Theory and Techniques*, 1989, **MTT-37**, pp. 1257–1260
65 PAVIO, A. M. AND KIKEL, A. 'A monolithic or hybrid broadband compensated balun', *IEEE MTT-S International Microwave Symposium Digest*, 1990, pp. 483–486
66 TOKUMITSU, T., HARA, S., TAKENAKA, T., and AIKAWA, M.: 'Divider and combiner line-unified FET's as basic circuit function modules - Part 1', *IEEE Transactions on Microwave Theory and Techniques*, 1990, **MTT-38** (9), pp. 1210–1217

67 TOKUMITSU, T., HARA, S., TAKENAKA, T., and AIKAWA, M.: 'Divider and combiner line-unified FET's as basic circuit function modules - Part 2', *IEEE Transactions on Microwave Theory and Techniques*, **MTT-38** (9), pp. 1218–1226

68 CHEN, T. H., CHANG, K. W., WONG, H., DOW, G. S., LIU, L. C. T., BUI, S. B. T., and LIN, T.S.: 'Broadband monolithic passive baluns and monolithic double-balanced mixer', *IEEE MTT-S International Microwave Symposium Digest*, Boston, 1991, pp. 861–864

69 JOUANNEAU-DOUARD, M., BROUZES, H., BIONAZ, S., and LEVY, D.: 'A 1 to 18 GHz out of phase combiner', *IEEE Microwave and Millimeter-Wave Monolithic Circuit Symposium Digest*, 1992, pp. 83–86

70 TSAI, C. M. and GUPTA, K. C.: 'A generalized model for coupled lines and its applications to two-layer planar circuits', *IEEE Transactions on Microwave Theory and Techniques*, 1992, **MTT-40**, pp. 2190–2099

71 TSAI, M. C.: 'A new compact wideband balun', *Proceedings of IEEE Microwave and Millimetre Wave Monolithic Circuit Symposium*, 1993, pp. 123–125

72 GOLDFARB, M. E., COLE, P. J. B., and PLATZKER, A.: 'A novel MMIC biphase modulator with variable gain using enhancement-mode FETs suitable for 3 V wireless applications', *IEEE Microwave and Millimeter-Wave Circuits Symposium Digest*, 1994, pp. 99–102

73 MAAS, S.: 'MMIC mixers for commercial applications', *Proceedings of IEEE MTT-S Workshop on Designing RF Signal Processing MMICs for Commercial Communication Systems*, 1994, pp. 99–106

74 SCHWINDT, R., and NGUYEN, C.: 'Computer-aided analysis and design of a planar multilayer Marchand balun', *IEEE Transactions on Microwave Theory and Techniques*, 1994, **MTT-42**, pp. 1429–1434

75 GOKDEMIR, T., ECONOMIDES, S. B. KHALID, A. A. REZAZADEH, A., and ROBERTSON, I. D. 'Design and performance of GaAs MMIC CPW baluns using overlaid and spiral couplers', *IEEE MTT-S International Microwave Symposium Digest*, 1997, pp. 401–404

76 NISHIKAWA, K., TOYODA, I., and TOKUMITSU, T., 'Compact and broad-band three-dimensional MMIC balun', *IEEE Transactions on Microwave Theory and Techniques*, 1999, **MTT-47**, pp. 96–98

77 YOON, Y. J. *et al.*,: 'Design and characterization of multilayer spiral transmission-line baluns', *IEEE Transactions on Microwave Theory and Techniques*, 1999, **MTT-47**, pp. 1841–1847

78 ANG, K. S., ECONOMIDES, S. B., NAM, S., and ROBERTSON, I. D.: 'A compact MMIC balun using spiral transformers', *Proceedings of Asia Pacific Microwave Conference*, Singapore, 1999, pp. 655–658

79 ANG, K. S., UNDERHILL, M. J., and ROBERTSON, I. D.: 'Balanced monolithic oscillators at K and Ka-band', *IEEE Transactions on Microwave Theory and Techniques*, 2000, **48** (2), pp. 187–193

80 ANG, K. S., ROBERTSON, I. D., ELGAID, K., and.: THAYNE, I. G.: '40 to 90GHz impedance transforming CPW Marchand balun', *IEEE MTT-S International Microwave Symposium Digest*, Boston, 2000, pp1141–1144

81 ANG, K. S., and ROBERTSON, I. D.: 'Analysis and design of impedance-transforming Marchand baluns', *IEEE Transactions on Microwave Theory and Techniques*, 2001, **MTT-49** (2), pp. 402-406

7.11.4 Active FET mixers

82 BASTIDA, E. M., CAVALIERI D'ORO, E., DONZELLI, G. P., FANELLI, N., FAZZINI, G., and SIMONETTI, G.: 'A monolithic 800 MHz bandwidth DBS front end receiver for mass production', *Proceedings of EuMC*, Liège, 1984, pp. 755–760

83 KANAZAWA, K., KAZUMURA, M., NAMBU, S., KANO, G., and TERAMOTO, I.: 'A GaAs double-balanced dual-gate FET mixer IC for UHF receiver front-end applications', *Proceedings of IEEE Microwave and Millimeter Wave Monolithic Circuits Symposium*, St. Louis, 1985, pp. 63–65

84 NIGHTINGALE, S. J., UPTON, M. A. G., and DANDEKAR, N. V.: 'The design and performance of a double balanced mixer using FETs', *IEEE MTT-S International Microwave Symposium Digest*, St. Louis, 1985, pp. 717–720

85 NIGHTINGALE, S. J., UPTON, M. A. G., DANDEKAR, N. V., and KONG, W. M.: 'The design and performance of a wideband monolithic double balanced mixer using FETs', *Proceedings of EuMC*, Paris, 1985, pp. 940–945

86 HOWARD, T. S., and PAVIO, A. M.: 'A distributed monolithic 2-18 GHz dual-gate FET mixer', *Proceedings of IEEE Microwave and Millimeter Wave Monolithic Circuits Symposium*, Las Vegas, 1987, pp. 27–30

87 YANG, D. C., ESFANDIARI, R., LIN, T. S., and O'NEILL, T.: 'Wideband GaAs MMIC receiver', *ibid.*, pp. 101–103

88 HARVEY, A. R., COTTON, F. J., PENNINGTON, D. C., EATON, R. M., and COOPER, P. D.: 'A low noise GaAs MMIC satellite downconverter for the 6-4 GHz band', *ibid.*, pp.139–142

89 WHOLEY, J., KIPNIS, I., and SNAPP, C.: 'Silicon bipolar double balanced active mixer MMICs for RF and microwave applications up to 6 GHz', *Proceedings of IEEE Microwave and Millimeter Wave Monolithic Circuits Symposium*, Long Beach, 1989, pp. 133–137

90 ROBERTSON, I. D., and AGHVAMI, A. H.: 'A practical distributed FET mixer for MMIC applications', *IEEE MTT-S International Microwave Symposium Digest*, Long Beach, 1989, pp. 1031–1032

91 ROBERTSON, I. D.: 'Mixer designs for GaAs technology', in HAIGH, D., and EVERARD, J. (Eds): *GaAs technology and its impact on circuits and systems*, (Peter Peregrinus Ltd., 1989), Chap. 9

92 PHILIPPE, P., and PERTUS, M.: 'A 2 GHz enhancement mode GaAs down converter IC', *Proceedings of IEEE Microwave and Millimeter Wave Monolithic Circuits Symposium*, Boston, 1991, pp. 61–64

93 ROBERTSON, I. D., and AGHVAMI, A. H.: 'A novel 1-15 GHz matrix distributed mixer', *Proceedings of 21st European Microwave Conference*, Stuttgart, 1991, pp. 489–494

94 ROBERTSON, I. D., and AGHVAMI, A. H.: 'A compact X-band GaAs monolithic balanced FET mixer', *Proceedings of European GaAs Applications Symposium*, session 3A, paper 2

95 MAJUDI-AHY, R., NISHIMOTO, C., RUSSELL, OU, W., BANDY. S., and ZDASIUK, G.: '23-40 GHz InP HEMT MMIC distributed mixer', *IEEE MTT-S International Microwave Symposium Digest*, 1992, pp. 1063–1066

96 SAKUNO, K., YOSHIMASU, T., MATSUMOTO, N. *et al.*: 'A miniature low current GaAs MMIC downconverter for Ku-band broadcast satellite applications', *Proceedings of IEEE Microwave and Millimeter Wave Monolithic Circuits Symposium*, Albuquerque, 1992, pp. 101–104
97 BONATO, G. L., and BÓVEDA, A.: 'GaAs monolithic image rejection down-converter for point-to-multipoint communication systems', *ibid.*, pp. 131–134
98 NEILSON, D., ALLEN, B., KINTIS, M., HOPPE, M., and MAAS, S.: 'A broadband upconverter IC', *ibid.*, pp. 163–165
99 DEVLIN, L. M., BUCK, B. J., CLIFTON, J. C., DEARN, A. W., and LONG, A. P.: 'A 2.4 GHz single chip transceiver', *Proceedings of IEEE Microwave and Millimeter Wave Monolithic Circuits Symposium*, Atlanta, 1993, pp. 2326
100 KAMITSUNA, H., and OGAWA, H.: 'Monolithic image rejection optoelectronic up-converters that employ the MMIC process', *ibid.*, pp. 75–78
101 RAMACHANDRAN R., WOO C., NIJJAR, M., and FISHER D.: 'A fully integrated 35 GHz MMIC receiver with on-chip LO', *ibid.*, pp. 93–95
102 FEUDALE, M. A., SURIANI, A. M., McPARTLIN, M. J., and OLSEN, E. A.: 'MMICs for satellite Ku band TLC repeaters', *ibid.*, pp. 229–232
103 VERVER, C. J., and STUBBS, M. G.: 'Development of 30 and 20 GHz mixers for miniaturised personal communications systems', *ibid.*, pp. 269–272
104 SAITO, T., HIDAKA, N., OHASHI, Y., SHIMURA, T., and AOKI, Y.: '60 GHz MMIC downconverter using an image-rejection active HEMT mixer', *Proceedings of IEEE Microwave and Millimeter Wave Monolithic Circuits Symposium*, San Diego, 1994, pp. 77–80
105 MONIZ, J. M., and MAOZ, B.: 'Improving the dynamic range of Si MMIC Gilbert cell mixers for homodyne receivers', *ibid.*, pp. 103–106
106 WOO, C., RAMACHANDRAN, R., BURNS, L., MARASCO, K., and FISHER, D.: 'A frequency agile X-band homodyne GaAs MMIC transceiver with a synthesized phase locked source for automotive collision avoidance radar', *ibid.*, pp. 129–132
107 KOBAYASHI, K. W., TRAN, L. T., BUI, S., OKI, A. K., STREIT, D. C., and ROSEN, M.: 'InAlAs/InGaAs HBT X-band double-balanced upconverter', *IEEE Journal of Solid-State Circuits*, 1994, **29** (10), pp. 1238–1243
108 KIM, M., HACKER, J. B., SOVERO, E. A., DEAKIN, D. S., and HONG, J. H.: 'A millimeter-wave multifunction HEMT mixer', *IEEE Microwave and Guided Wave Letters*, 1999, **9** (4), pp. 1154–1156
109 DEARN, A. W., and DEVLIN, L. M.: 'A mm-wave monolithic Gilbert-cell mixer', *Proceedings of IEEE Radio Frequency Integrated Circuits Symposium*, Boston, 2000

7.11.5 Resistive FET mixers

110 MAAS, S.: 'A GaAs MESFET mixer with very low intermodulation', *IEEE Transactions on Microwave Theory and Techniques*, 1987, **MTT-35**, pp. 425–429

111 MAAS, S.: 'A GaAs MESFET balanced mixer with very low intermodulation', *IEEE MTT-S International Microwave Symposium Digest*, 1987, pp. 895–897

112 S. WEINER, D. NEUF and S. SPOHRER, '2 to 8 GHz double balanced MESFET mixer with +30 dBm 3rd order intercept', *IEEE MTT-S International Microwave Symposium Digest*, 1988, pp 1097–1100

113 KRUGER, D. A.: 'Monolithic dual-quadrature mixer using GaAs FETs', *Microwave Journal*, 1990, pp. 201–206

114 GEDDES, J., BAUHAHN, SWIRHUN, S.: 'A millimeter wave passive FET mixer with low 1/f noise', *IEEE MTT-S International Microwave Symposium Digest*, 1991, pp. 1045–1047

115 CHANG, K. W., CHEN, T. H., BUI, S. B. T., LIU, L. C. T., and NGUYEN, L.: 'High performance resistive EHF mixers using InGaAs HEMTs', *IEEE MTT-S International Microwave Symposium Digest*, 1992, pp. 1409–1412

116 HÜBNER, M., and RATTAY.: 'K- and Ku-band MMICs for radio link communications', *Proceedings of GaAs '94*, Torino, 1994, pp. 59–62

117 TSAI, M. C., SCHINDLER, M. J., STRUBLE, W., VENTRESIA, M., BINDER, R., WATERMAN R., and DANZILIO D.: 'A compact wideband balanced mixer', *IEEE Microwave and Millimeter-wave Monolithic Circuit Symposium Digest*, 1994, pp.135–138

118 K. W. CHANG, E. W. LIN, H. WANG, K. L. TAN and W. H. KU, 'A W-band monolithic, singly balanced resistive mixer with low conversion loss', *IEEE Microwave and Guided Wave Letters*, 1994, **4** (9), pp. 301–302

119 FLAVIIS, F. D., and MAAS, S. A.: 'X-band doubly balanced resistive FET mixer with very low intermodulation', *IEEE Transactions on Microwave Theory and Techniques*, 1995, **MTT-43**, pp. 457–460

120 T. H. CHEN *et al.*, 'Broadband Single- and Double-Balanced Resistive HEMT Monolithic Mixers', *IEEE Transactions on Microwave Theory and Techniques*, 1995, **MTT-43**, pp. 477–484

121 LIN, E. W., KU, W. H.: 'Device considerations and modeling for the design of an InP-based MODFET millimeter-wave resistive mixer with superior conversion efficiency', *IEEE Transactions on Microwave Theory and Techniques*, 1995, **43** (8), pp. 1951–1959

122 YHLAND, K., RORSMAN, N., and ZIRATH, H. G.: 'Novel single device balanced resistive HEMT mixers', *IEEE Transactions on Microwave Theory and Techniques.*, 1995, **MTT-43**, pp. 2863–2867

123 KARLSSON, C., RORSMAN, N., and ZIRATH, H.: 'A monolithically integrated F-band resistive InAlAs/InGaAs/InP HFET mixer', *IEEE Microwave and Guided Wave Letters*, 1995, **5** (11), pp. 394–395

124 KASHIWA, T., KATOH, ISHIDA, T., KOJIMA, Y., YAMAMOTO, Y., KOMARU, M., and MITSUI, Y.: 'A V-band monolithic InP HEMT resistive mixer with low LO power requirements', *IEEE MTT-S International Microwave Symposium Digest*, 1997, pp 369–372

125 ANG, K. S., CHONGCHEAWCHAMNAN, M., and ROBERTSON, I. D.: 'Monolithic resistive mixers for 60GHz direct conversion receivers', *Proceedings of IEEE Radio Frequency Integrated Circuits Symposium*, Boston, 2000, pp. 35–38

126 ANG, K. S., and ROBERTSON, I. D.: 'A monolithic double-balanced upconverter for m-wave point-to-multipoint distribution systems', *Proceedings of EuMC*, Paris, 2000

7.11.6 Image-rejection and single-sideband mixers

127 PARISI, S. J.: 'Monolithic, lumped element, single sideband modulator', *IEEE MTT-S International Microwave Symposium Digest*, Albuquerque, 1992, pp 1047–1050.

128 BAREE, A. H., ROBERTSON, I. D., and BHARJ, J.S: 'A compact wideband MMIC image-rejection mixer chip, *Proceedings of IEEE Asia-Pacific Microwave Conference*, 1994, pp 241–244.

7.11.7 Sub-harmonic mixers

129 BEMKOPF, P., and TAJIMA, Y.: 'A monolithic Ka-band sub-harmonically pumped frequency converter', *Proceedings of IEEE Microwave and Millimeter Wave Monolithic Circuits Symposium*, Boston, 1991, pp. 43–46

130 PAYNE, D., HICKS, R., LAMPORT, R., STEWART, E., VAN LEEUWEN, B., and RAFFAELLI, L.: 'A GaAs monolithic V-band receiver for space communications', *Proceedings of European GaAs Applications Symposium*, 1992, session 1B, paper 2

131 ZIRATH, H., ANGELOV, I., and RORSMAN, N.: 'A millimeterwave subharmonically pumped resistive mixer based on hererostructure fied effect transistor technology', *IEEE MTT-S International Microwave Symposium Digest*, Albuquerque, 1992, pp 599–602.

132 ZIRATH, H., ANGELOV, I., RORSMAN, N., and KARLSSON, C.: 'A W-band subharmonically pumped mixer based on pseudomorphic heterostructure field effect transistor technology', *IEEE MTT-S International Microwave Symposium Digest*, 1993, pp 341–344.

133 ANG, K. S., BAREE, A. H., NAM, S, and ROBERTSON, I. D.: 'A millimeter-wave monolithic sub-harmonically pumped resistive mixer', *Proceedings of Asia-Pacific Microwave Conference*, Singapore, 1999, pp. 222–225

7.11.8 Distributed mixers and coupling structures

134 TANG, O. S. A., and AITCHISON, C. S.: 'A practical microwave travelling-wave MESFET gate mixer', *IEEE MTT-S International Microwave Symposium Digest*, St. Louis, 1985, pp. 605–608

135 TANG, O. S. A., and AITCHISON, C. S.: 'A very wide-band microwave MESFET mixer using the distributed mixing principle', *IEEE Transactions on Microwave Theory and Techniques*, 1985, **MTT-33** (12), pp. 1470–1478

136 HOWARD, T. S., and PAVIO, A. M.: 'A distributed 1-12GHz dual-gate FET mixer', *IEEE MTT-S International Microwave Symposium Digest*, 1986, pp. 329–332.

137 BARTA, G. *et. al.*: 'Surface-mounted GaAs active splitter and attenuator MMICs used in a 1-10 GHz leveling loop', *IEEE Transactions on Microwave Theory and Techniques*, 1986, **MTT-34** (12), pp. 1569–1575

138 GERARD, R. E. J., 'Multisignal amplification', US patent no. 4423386, The Marconi Company Limited, Chelmsford, England.

139 PAVIO, A. M. *et. al.*: 'Double balanced mixers using active and passive techniques', *IEEE Transactions on Microwave Theory and Techniques*, 1988, **MTT-36**, pp. 1948–1957

140 ROBERTSON, I. D.: 'Novel techniques for GaAs MMIC receivers', PhD thesis, King's College, London, 1991

141 ROBERTSON, I. D., and AGHVAMI, A. H.: 'A novel wideband MMIC active balun', *Proceedings of 20th European Microwave Conference*, Stuttgart, 1991, pp. 489–494

142 ANG, K. S., and ROBERTSON, I. D.: 'Multiple-FET mixer analysis with applications to the cascode distributed mixer', *Proceedings of High Frequency Postgraduate Student Colloquium*, University of Leeds, UK, 1999, pp. 22–27

143 ANG, K. S., NAM, S., and ROBERTSON, I. D.: 'A 2 to 18 GHz monolithic resistive distributed mixer', *Proceedings of European Microwave Conference*, Munich, Germany, 1999, pp. 222–225

Chapter 8

Switches and attenuators

J. S. Joshi

8.1 Introduction

Microwave switches are essential elements for a number of widely varying applications, ranging from highly sophisticated space communications systems to more common applications simply requiring the switching of an RF signal from one path to another (e.g. for the purpose of providing redundancy or antenna diversity). Although the specific requirements for each application have to be tailored to individual needs, the essential characteristics of the switches remain the same. There is a continuing need for high performance, small and versatile microwave switches for use in a range of RF commercial systems such as cellular GSM, UMTS, cable modems, direct broadcast and other point-to-point and point-to-multi-point broadcast systems and applications. With the increasingly complex nature of information transmission the demand for bandwidth rises, and so also the need to accommodate these functions in smaller outlines and ever decreasing power consumption requirements. They allow routing of microwave signals through desired paths in the system and sub-system, and enable the realisation of different schemes for signal combinations necessary for system implementation.

Today there are phased-array antennas, smart antennas (capable of providing shaping and electronic control of the main beam) and modern communications systems that require amplitude control and gain equalisation in different paths. This means that there are increasing requirements for gain setting and control circuits. These circuits are generally of digital architecture and provide binary weighted attenuation control to achieve the required performance objectives, or they can be analogue in nature.

In this chapter, we will first review the GaAs FET switch mechanism and the development of an equivalent circuit for switching operation. The advantages of monolithic switching operation will be described along with some of its limitations. The second part of this chapter deals with different schemes for the realisation of GaAs MMIC variable attenuators. A number of techniques will be examined and the performance characteristics offered by selected configurations compared.

8.2 GaAs FET MMIC Switches

The implementation of solid-state PIN diode switches several decades ago resulted in a significant improvement in the mass, size and switching speed of microwave switches compared with their then available electromechanical counterparts. Compact PIN diode circuits can be easily realised in hybrid integrated circuits. This enabled their widespread use in a range of different applications. PIN diodes can provide faster switching speeds and can handle medium to large RF power levels. The biasing requirements of PIN diodes are such that in the on-state they require high forward current, and in the off-state they require large reverse bias voltage across the semiconductor junction. Their switching characteristics are highly dependent on the individual characteristics of the PIN diode. In recent years these PIN diode switches have been increasingly replaced by GaAs FET based monolithic switches, especially for low to medium power applications. Medium to high power GaAs FET MMIC switches have also been demonstrated. In recent years, the realisation of single pole multithrow switches, with on-chip control circuits and in low cost packages, have made possible the realisation of new architectures for wireless communications type applications. GaAs FET MMIC switches have also been demonstrated for high linearity and power handling requirements.

The inherent advantages of GaAs FET-based switches over conventional PIN diodes are simplified bias networks, negligibly small DC power requirements of the FET switch itself, simplified design of driver circuits, faster switching speeds, etc. Furthermore, monolithic implementation of such switches means that increased functionality can be designed within a single chip measuring only a few square millimetres. Monolithic implementation has resulted in some very high performance and extremely complex circuits being realised on very small chips, with application-friendly control architecture.

8.2.1 GaAs FET switching mechanism

The FET switch is a three-terminal device, in which the gate bias voltage V_g controls the states of the switch. The FET acts as voltage-controlled resistor, in which the gate bias controls the drain-to-source resistance in the channel. The intrinsic gate-to-source and drain-to-gate capacitances and device parasitics limit the performance of the FET switch at higher frequencies

In a typical switching mode, a high impedance state is produced when the negative gate-source bias is larger in magnitude than the pinch-off voltage, V_p, (i.e. $|V_{gs}| > |V_p|$). When a zero bias gate voltage is applied across the gate terminal, the FET switch is in a low impedance state. These two linear operational regions of the FET are shown graphically in Figure 8.1. It is important to note that in either state virtually no DC power is required by the FET switch. Thus, for all practical purposes, from a power consumption point of view these FET switches can be considered as passive devices. This negligible DC power requirement of the FET switch results in a considerable simplification of its driver circuitry. Being a planar structure, with the gate stripe located between the source and drain terminals, the FET switch acts as a bi-directional switch.

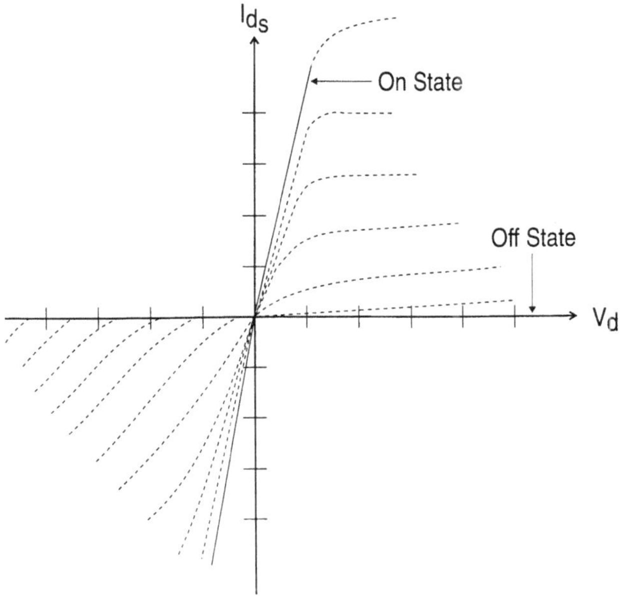

Figure 8.1 Linear operating regions of a FET switch

Figure 8.2 shows the simplified cross-section of a conventional FET structure. Under the zero gate bias condition, the channel region is virtually open, except for the zero field depletion layer thickness. Thus, for current levels less than the saturated channel current, I_{dss}, the FET can be modelled as a linear resistor. When a negative gate-source voltage is applied, such that $|V_{gs}| > |V_p|$, the channel region is completely depleted of free charge carriers. Under these conditions, the FET can be modelled by series and parallel combinations of resistors and capacitors. The parts of the FET associated with resistive and capacitive regions are shown in Figure 8.2. Assuming that the gate termination represents a high RF impedance at the frequency of operation, the off-state equivalent circuit can be expressed as a parallel combination of a resistor and a capacitor, as shown in Figure 8.2. For conditions when $1/\omega C_g >> r_g$, the effective drain-to-source capacitance is simply $(C_{sd} + C_g/2)$ and the effective drain resistance is the parallel combination of r_d and $2/(\omega^2 C_g^2 r_g)$. The figure-of-merit for a switched FET can be expressed as a ratio of its effective off-state resistance to its on-state resistance [1,2].

The physical parameters of the device strongly influence the equivalent circuit parameters of the switched FET. The important device physical parameters are the channel geometry, gate length, channel doping density and pinch-off voltage. The metallisation scheme, gate recess structure, etc. influence the on-state resistance of the FET switch; the off-state impedance depends on the source-drain capacitance C_{sd}, its parallel resistance r_d, and the drain-to-gate and source-to-gate capacitance C_g and its associated series resistance r_g.

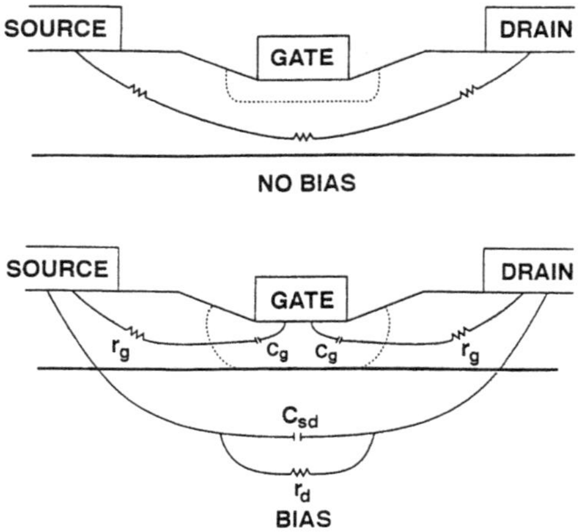

Figure 8.2 Cross-section of a FET switch with resistive and capacitive regions

The capacitance C_{sd} represents the fringing capacitance between the source and drain electrodes. Its value mainly depends on the spacing between them, and is virtually independent of device parameters such as channel doping and pinch-off voltage. Due to the symmetry of the device, the drain-to-gate and gate-to-source capacitances are equal and are represented by C_g. The gate pad capacitance to ground, if significant, needs to be added to the equivalent circuit. A simple but reasonable approximation of C_g is half the gate capacitance with the channel region fully depleted. C_g is thus dependent on both the channel doping and the pinch-off voltage of the device. The resistances that contribute to the off-state impedance are r_d and r_g, as shown in Figure 8.2. Resistance r_d is in parallel with C_{sd} and represents the RF losses associated with the source-drain capacitance. Its value can be arrived at from a reasonable Q value for the capacitance. Resistance r_g represents the resistance of the channel outside the pinch-off region in the of-state.

The switching speed of the FET switch can also be calculated from the equivalent circuit parameters. Assuming that an internal gate resistance of 1.5 kΩ is included in series with the gate terminal and if 10 pF is used in the low-pass bias circuit filter, the charging time constant is determined mainly by these two values, and works out to be 15 ns. Reduction in switching times are quite possible by optimising the gate bias circuit design.

8.2.2 Switched FET equivalent circuit

It should be noted that the physical basis of these device parameters is applicable to discrete as well as MMIC FET devices. However, for monolithic circuit implementation special processing steps can be implemented by the GaAs foundries (e.g. ion-implantation, etc.), in order to achieve FET characteristics specially dedicated to switching applications. Some foundry processes enable the fabrication of not only the RF switching function but also the associated driver circuitry on the single chip.

To develope a model for switching FETs, a range of device patterns of varying unit and total gate widths and different numbers of gate fingers are processed using the standard switch-specific GaAs MMIC process. These processed switch circuits are then RF-on-wafer probed at on-state (V_{gs} = 0 V) and off-state (V_{gs} = -5 V) over a wide frequency range. From these measurements a multi-element physical characteristics compatible equivalent circuit model is generated. This model is made scalable by introducing a physically meaningful dependency of all the parameters on the gate widths and the number of gate fingers. This equivalent circuit, along with scaling rules for an MMIC process from a European foundry, is shown in Figure 8.3.

The insertion loss of a four-finger gate geometry series FET switch in the on-state is shown in Figure 8.4 over a 1 to 20 GHz frequency range as a function of total gate periphery. As expected, the insertion loss decreases as total gate periphery increases. Figure 8.5 shows the performance characteristics of a series FET switch in the off-state. Due to increased capacitive coupling, the isolation in the off-state for larger gate periphery devices is lower than for smaller gate periphery devices. The switching characteristics are principally controlled by the total gate width of the transistor. The actual number of gate fingers has a marginal influence.

In the on-state, the low channel resistance of the FET is the dominating parameter and this results in a virtually independent frequency response. In the off-state, however, the equivalent capacitance of the circuit dominates and this shows strong frequency dependence. Therefore, insertion loss for the series FET and isolation for the shunt FET configurations, both in the on-state, show frequency invariant behaviour. On the other hand, isolation for the series FET and insertion loss for the shunt FET configurations, both in the off-state, show frequency dependence. Increasing the size of the FET reduces its on-state resistance and thus reduces insertion loss. However, it also results in increased gate-source and gate-drain capacitance and increased source-drain fringing capacitance, which limit the switch performance in the off-state.

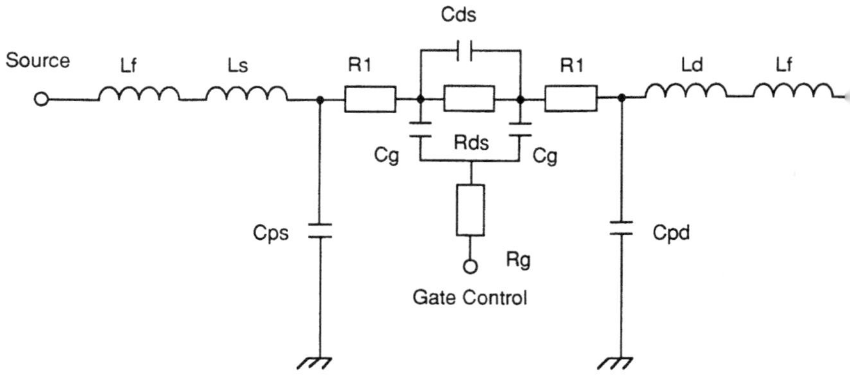

(a)

	ON	OFF
R1	0.52 + 750/**W**	1.21 + 1910/**W**
Rds	0.20 + 583/**W**	Infinite
Cds	Zero	O.11**W** + N + 5
Cg	**W** + 1.25N	0.17**W** + 1.25N

Cpd=1.2N + (0.057 + 0.015N)*w*

Cps=1.23N + (0.057 + 0.009N)*w*

Lf =80 pH

Ld= [1.83N^2 -18N + *w*(0.873 - 0.095N)] / [K + 1]

Ls = K x Ld

where K =1 for N=1; K =2 for N=2; K =1.5 for N=4; K =1.25 for N=8

In general: K= 1 + 2/N for N >1 and K =1 for N = 1

W is the total gate width in microns
w is the unit gate width in microns
N is the number of gate fingers

Capacitances ere in femtoFarads
Inductances are in picoHenries
Resistances are in Ohms

(b)

Figure 8.3 Philips D07M switched FET foundry model: (a) equivalent circuit and (b) scaling rules [3]

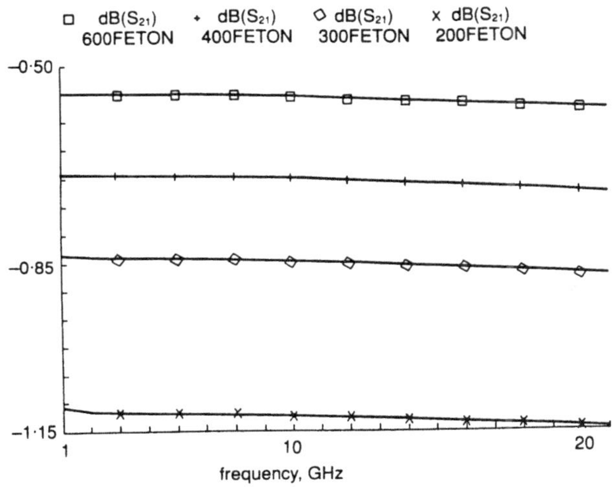

Figure 8.4 Insertion loss against frequency for a series FET switch in the on-state

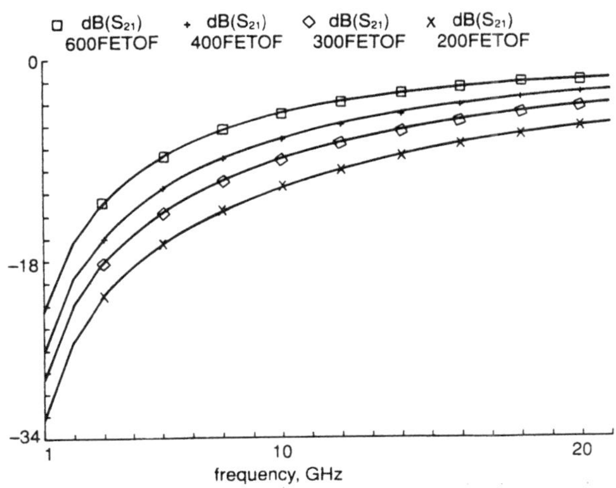

Figure 8.5 Insertion loss against frequency for a series FET switch in the off-state

8.2.3 Implementation of MMIC switches

FET switching circuits can be designed in the same way as conventional PIN diodes by using on- and off-state equivalent circuits as required in the overall circuit configuration. The FET can be in series or in shunt configuration with respect to the transmission lines to which these switches are connected. Before going into the details of these switch configurations, it is worthwhile to note some important differences between the conventional PIN diode-based switches and FET-based MMIC switches. Since the FET is a three-terminal device the switching occurs only through the gate bias control voltage — no other voltage is necessary for switch operation. The RF transmission lines do not carry any DC voltage, and therefore, there is no necessity for using DC blocking capacitors between the various FET switching elements. This represents a very significant advantage in lowering the component count for MIC circuits and results in a simpler configuration for MMIC circuits.

In the two switching states the gate junction is either reverse biased or is biased at zero gate-source bias voltage. In both these states the gate draws negligible current, which greatly simplifies the control circuit design requirements. In the off-state of the FET switch, the gate-drain and gate-source capacitances are equal, as both the source and drain terminals are at ground potential. As a result, the gate and drain terminals are not isolated from each other and the RF impedance of the gate bias circuit affects the equivalent drain-source impedance. In the limit, when the gate terminating impedance is kept very high by including a built-in 1.5 kΩ resistor, for example, the equivalent drain-source capacitance can be approximated as $C_d + C_g/2$. The effect of this capacitive shunting should be included in the design of off-state circuits. Above 10 GHz, this capacitive shunting effect in the off-state may result in the off-state impedance of the FET being much smaller than the value of r_d. It may be necessary to resonate out the capacitance with a shunt inductor across the source-drain terminals at the operating frequency of interest. This can be implemented in a number of ways, ranging from a simple shunt inductor [4] to a distributed type structure as used by Ayasli *et al.* [5].

The performance of the simple series and shunt monolithic FET switches, shown in Figures 8.4 and 8.5, respectively, may be adequate for certain limited applications, but generally a combination of series-shunt configurations are needed to achieve the desired low insertion loss, high isolation and good return loss characteristics.

Reflective and absorptive single pole single throw (SPST) and single pole double throw (SPDT) switches are the basic building block components that can be used for more complex configurations. For moderate frequencies of up to 8 GHz and for the wireless applications, which are generally below 4 GHz, a number of switches have become available in the last few years in low cost plastic packages. Single pole multithrow switches with up to eight outputs and with integral driver circuitry have also been demonstrated, and are now commercially available from at least one supplier. Smuk and Shifrin [6] reported monolithic multi-throw switches operating up to 4 GHz with built in decoder/driver logic circuitry in 1997. Switches with both reflective and terminated topology were demonstrated. The reflective topology provides slightly lower loss compared to the absorptive one. The MMIC process used had both the depletion as well as the enhancement FET capability. The depletion mode process was used for the RF switches and the enhancement mode process was used for the low power decoder/driver logic circuits, which helped to minimise power consumption. The decoder/driver circuit had three control lines to select one of the six outputs or an all-off state. Conventional control of an SPNT switch requires N complementary pair of driver control lines to bias the series/shunt FET switches. By using a 3:6 decoder circuit in this chip the number of input/output control lines were reduced by 9. This resulted in significant reduction in chip size, which in turn improved RF performance by minimising grounding inductance, stray coupling to the control lines, etc. A lesser number of control lines also results in easier integration with the PCB circuit. The chip measures 1.0×1.5 mm^2 and was fabricated using a standard 1 μm gate-length enhancement depletion process. For frequencies less than 2 GHz the insertion loss of the reflective switch is less than 0.5 dB. The driver decoder power consumption was 20 mW (4 mA at –5 V). The 1 dB input power compression point was within the range of 20 to 24 dBm. The die was also packaged in SOIC and QSOP plastic packages. The packaged chip had increased insertion loss and reduced isolation and return loss as expected. Similar circuits and even higher order switches are now being offered by the same company for large volume commercial applications [7].

8.2.4 High isolation switches

It is difficult to achieve high isolation characteristics with low insertion loss. The reason for this is that when the operating frequency increases, the isolation characteristics deteriorates due to the capacitive component of the switched FET in the off-state. The conventional approach has been to resonate the capacitive component at the desired frequency band. These techniques only provide isolation of the order of 40 dB, which may not be sufficient for some system applications. Imai, Minakawa and Okazaki [8] reported a high isolation and low insertion loss FET switch. A conventional series-shunt-series FET switch configuration was augmented by using an *R-C-R* T-network in parallel. In the off-state the switched FET parasitics and the *R-C-R* network act as a twin-T circuit, as illustrated in Figure 8.6. If the values of resistance and capacitance of the *R-C-R* T-network are selected to satisfy:

$$RC = 2R_{on2}\,(C_{off1} + C_{off3}) \tag{8.1}$$

then the composite circuit becomes a band rejection filter with an angular frequency

$$\omega_0 = 1/(\,2.C_{off1}\,C_{off3}\,R_{on2}\,R)^{1/2} \tag{8.2}$$

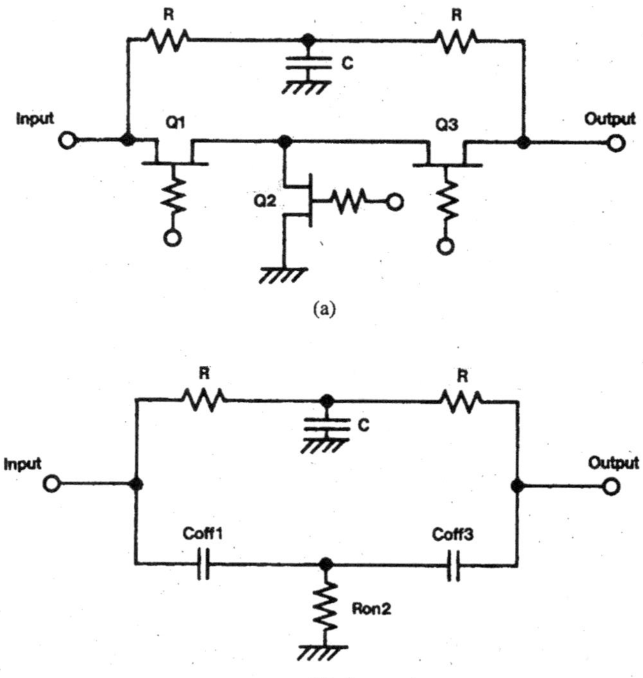

Figure 8.6 Circuit configuration for high isolation switch (a) R-C-R T-network and (b) its equivalent circuit in the off-state [8] (©1996 IEEE)

Thus, from the previous two equations, the values of R and C can be determined for the desired band rejection frequency. The authors presented a comprehensive mathematical analysis and computer simulation of the conventional switch configuration and their twin-T version. Measured results of two MMIC switches centred at 2 and 6 GHz are shown in Figure 8.7, which clearly demonstrate the usefulness of their technique. The two circuits had isolation greater than 60 dB. The FETs used in the chip had a total gate width of 800 μm and the chip size was 1.28×0.78 mm^2.

Another technique that has been demonstrated by Mizutani and Takayama [9] for a DC to 60 GHz switch uses what the authors call a shunt connected distributed FET (SCDF) technique. A shunt distributed FET was expressed as the infinite combination of a unit shunt FET of 0.15 μm gate HJFET and a unit series transmission line of the drain electrode. If the gate bias circuit is isolated with a large resistor, the unit FET in the open channel state is expressed as a simple resistor and in pinched-off state as a capacitor. The design parameters for the unit cell structure of 100 μm gate width were extracted as a unit resistance of 8 Ω for the open channel state and the unit capacitance of 30 fF for the pinched off-state. The capacitance and inductance for the unit 100 μm long transmission line composed of the drain electrode were 20 fF and 50 pH, respectively.

The *S*-parameters for the on-and off-state transmission lines containing the switched FETs were derived using standard formulations. Computer simulations showed an on to off ratio better than 20 dB and higher isolation at increasing frequencies up to 200 GHz! A shunt SCDF switch was fabricated using a 0.15 μm HJFET MMIC process that consisted of a single gate finger of 400 μm width. Input and output ports were attached at both ends of the drain electrode. The chip size was 0.4 × 0.07 mm². Maximum insertion loss of 1.37 dB and a minimum isolation of 22 dB were measured, which increased to 40 dB at 60 GHz. The distributed nature of the circuit was able to provide increasing isolation with frequency, unlike the conventional lumped element circuits.

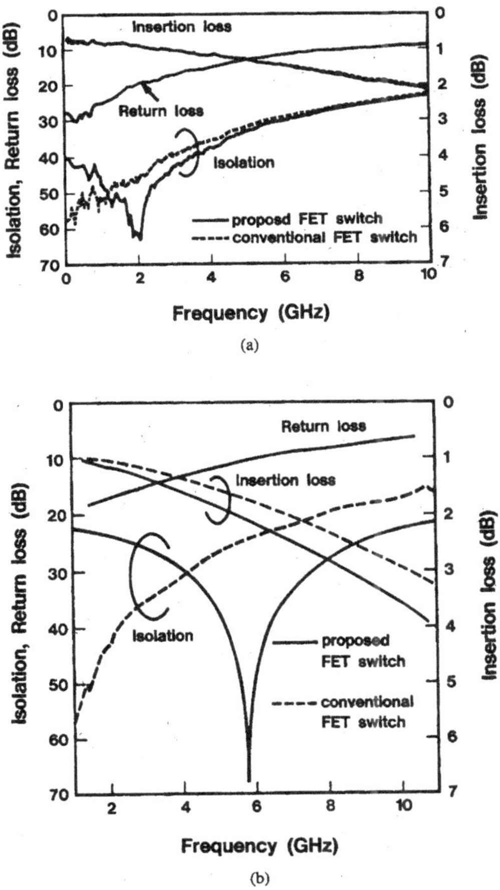

Figure 8.7 Experimental results of (a) conventional and high isolation MMIC switch tuned to 2 GHz and (b) tuned to 6 GHz [8] (©1996 IEEE)

MMIC switches capable of handling an RF power of more than 100 W at high switching speeds have also been reported by Katzin *et al.* [10]. A low-loss inductive gate bias network structure was developed that allowed high power RF switching. These were used to demonstrate S- and C-band SPDT switches capable of handling up to 300 W of CW RF power.

8.3 Digital attenuators

Microwave communications systems are continually demanding high performance and sophisticated components and equipment for increasingly complex user requirements. Components for gain setting and control functions have applications in a number of areas. Such a component is useful for multichannel communication systems where the gains through different channels need equalisation. Digitally controlled attenuator circuits are also applicable to phased array antennas, 'smart' antennas (where a selective coverage pattern is required) and other similar applications. This component is also finding increasing use in different wireless applications such as GSM, PCS, 3G, etc.

Prior to the advent of GaAs monolithic microwave integrated circuits, such complex circuit functions, even at microwave frequencies around 10 GHz, could only be realised using PIN diodes as switching elements in MIC configurations. These were bulky and required complex driver circuits for the PIN diodes. GaAs MMIC-based attenuator circuits can now be realised in an extremely small outline, have low mass, require negligible DC power consumption and have excellent performance characteristics. In fact, it can be safely stated that the realisation of such components in GaAs technology has made the implementation of complex new system architectures possible.

The attenuating function can be realised in analogue or digital form. Analogue implementations are briefly described in Section 8.5, but the majority of attenuators have been implemented in digital form. A digital attenuator consists of a number of cascaded units, each containing an individual bit that can be switched in or out to achieve the desired value of attenuation. These cascaded bits are usually binary weighted with the least significant bit (LSB) level as the minimum resolvable level. The most significant bit (MSB) determines the upper level of attenuation. Minimum attenuation is obtained when all the individual bit states are turned off, while at the maximum attenuation level all the individual bit states are turned on. The minimum attenuation value is also called the zero- or reference-state insertion loss of the attenuator.

In the wireless industry the control of signal strength, whilst maintaining functional integrity of the system in both the transmit and receive paths, is extremely important. Digitally controlled GaAs MMIC attenuators have been specifically developed to meet these needs. High volume and low cost targets have resulted in these now being available in low cost plastic packages. Some vendors offer a universal design covering DC to 4 GHz frequency range, whilst others have optimised designs for different sub-sectors within this range.

8.3.1 Design approaches

In the design of such digital attenuators the primary goal is to accurately meet the attenuation value requirements of each individual bit across the desired frequency range. In some cases, like the phased array antenna, the system requirements for these attenuators are such that the insertion phase through the attenuator should ideally be invariant for all possible attenuation states. This phase invariant performance makes their application easier, as no other parameters need to be adjusted to account for any finite insertion phase change for different states. This does away with any need for re-adjusting the insertion phase of the system every time the attenuator setting is changed. The input and output impedance matching of such digital attenuators have to be very good, so as not to cause any insertion loss and insertion phase ripples across the band of interest when cascaded with other components like phase shifters, amplifiers, etc. In fact, achieving excellent input and output matches for an individual attenuation bit itself is vitally important for the overall performance of the attenuator, as any mismatch interactions within the chip itself will worsen the amplitude and phase ripple performance. The zero state attenuation value of the digital attenuator should be designed to be as low as possible. For most applications additional attenuation directly impacts on the link budget. It is important to select a device with minimum possible insertion loss. Another important requirement is power handling, which is normally specified as input 3rd order intercept point (IP3) For wireless applications smaller size and easy to implement control circuitry are also important considerations. Needless to say, all these requirements cannot be simultaneously fulfilled and considerable trade-off is involved in the design of such state-of-the-art GaAs MMIC digital attenuators. In the following sections we consider individual design approaches in detail.

8.3.2 Segmented dual-gate FET technique

An efficient way to achieve equi-phase but differential gain paths at microwave frequency is to use GaAs dual-gate FET devices. In this arrangement the gate 1 and drain ports of the dual-gate FET are used as input and output ports, and gate 2 is used as a control terminal. Gate 1 is biased for the required gain level with gate 2 biased at saturation (the on-state). When gate 2 bias is increased towards pinch-off, the gain of the dual-gate FET decreases while the transmission phase stays constant. In practice, however, gain variation is limited to a value of about 10 dB, after which the change in transmission phase becomes excessive [11]. The transmission phase characteristics of the dual-gate FET are dependent on the device geometry, its material and processing parameters, etc.

For GaAs MMIC implementation, the dual-gate FET approach was modified by Hwang *et al.* [11] to produce segmented dual-gate FET (SDG FET) circuits. Figure 8.8 shows the schematic diagram of a segmented dual-gate FET circuit. The technique was investigated further by Naster and Hwang [12], Snow, Komiak & Bates [13] and Siweris [14]. The implementation has been at L- and C/X-band frequencies. In SDG FETs, gate 1 remains as before, but gate 2 is segmented into independently switched gates of different widths, allowing individual units of dual-gate FETs to be switched on or off for different gain settings. Since each dual-gate FET segment is only operated in either the on- or off-state, it is possible for the gain to be linearly scaled with the gate width. A variable gain amplifier with 5-bits of precision control results from the simple segmenting of the gate 2 sections into a binary form (i.e. 1:2:4:8:16). The DC power consumption, processing yield, chip size and dynamic range requirements determine the size of the most significant bit. The size of the least significant bit is thus automatically fixed from the overall bit requirements. In some cases the required gate 2 width for the LSB may be too small for practical realisation. Naster and Hwang [12] realised effective gate 2 widths of 5 and 10 μm by using a 4:1 capacitive voltage divider network with 20 and 40 μm gate 2 width segments. This segmented dual-gate technique has also been used by Devlin *et al.* [15] at Ku-band frequencies to realise a combined digital gain and phase control MMIC circuit. In this implementation, non-binary weighting was used for the 6-bit digital gain control function. A single chip unit containing both digital gain and phase control was designed and tested and had excellent performance characteristics.

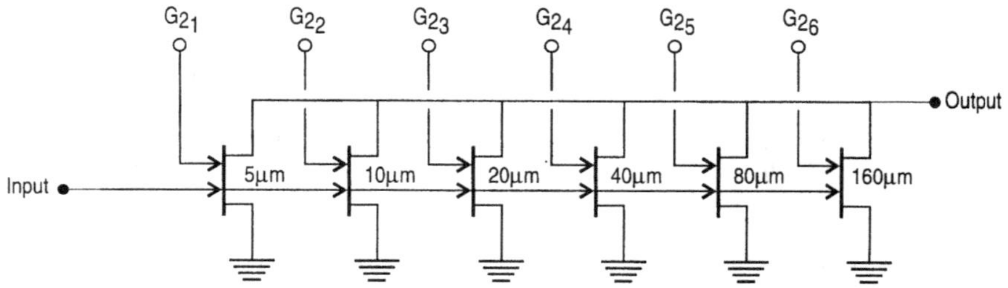

Figure 8.8 Schematic diagram of a segmented dual-gate FET attenuator

The advantages of the segmented dual-gate FET approach are small chip size, the inherent gain, broadband performance, etc. For the realisation of LSB units, where smaller gate widths are required, the gate scaling model may need to be radically modified. Such techniques are, however, only available to designers with access to in-house foundries, where individual optimisation is possible, and cannot be implemented by the majority of MMIC designers who only have access to commercial foundries with standard processes.

8.3.3 Switched attenuators

In the switched attenuator type, two SPDT switches are used to change the signal path between a through or reference line and an attenuator as shown in Figure 8.9. For attenuation values in excess of 4 dB, the attenuator can be formed by using a T- or π-type resistor network using either GaAs mesa resistors, or by using the dedicated nichrome resistors available in the foundry process. The schematic diagram of a broadband SPDT switch, as used by Gupta [16], is shown in Figure 8.10. As the switch does not contain any capacitive elements there is no lower frequency limitation on the performance. The upper frequency limit on the useability of the configuration is determined by the isolation of the SPDT switch itself. The isolation of the switch in the off-state determines the leakage of the RF signal to the unwanted path, and needs to be kept as high as possible to minimise any in-band ripple. The input and output impedance matching of individual bits are generally determined by the SPDT switch itself. Phase tracking between the two states can be easily achieved for larger bit values by equalising the reference path length to the resistive path length. The switched attenuator technique is attractive for the implementation of larger values of attenuation (> 4 dB). Being a symmetrical structure, its overall performance is less susceptible to the foundry process parameter variations and performance variations due to temperature are minimised.

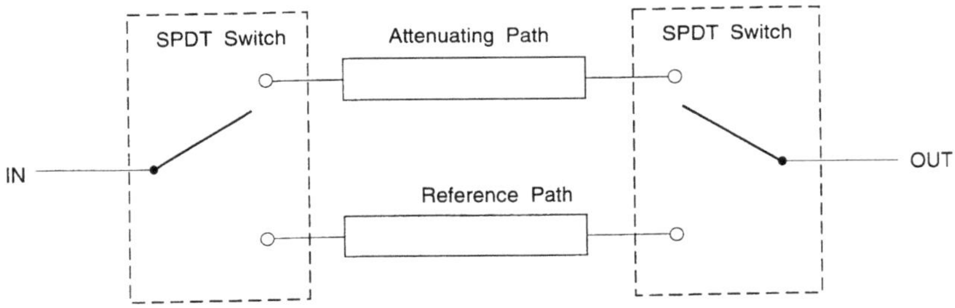

Figure 8.9 Schematic diagram of a switched attenuator

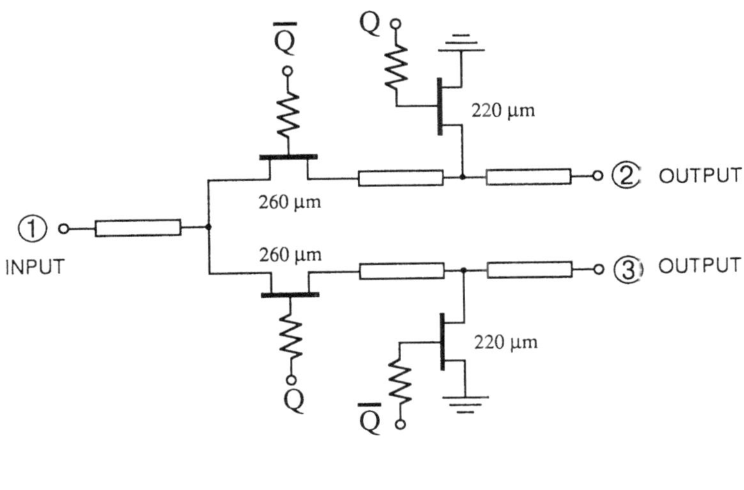

Figure 8.10 Schematic diagram of a SPDT switch [16]

A similar SPDT switch configuration has been used by Anderson [17], for the design of a constant phase MMIC digital attenuator, using the Philips D07M foundry. The insertion loss of the switch is less than 1.0 dB up to 5.5 GHz and increases with frequency, and the isolation is better than 28 dB for frequencies up to 10 GHz, as shown in Figure 8.11.

Apart from many implementations at lower frequencies [17], this technique has even been implemented at 44 GHz by Sjogren and others [18] using HEMT MMIC technology. They used a series-shunt SPDT switch configuration on a 0.15 μm HEMT process. The series switch had inductive parallel feedback to provide a compact circuit layout, while the shunt switch was used to provide additional isolation. One of the major drawbacks of using this technique for realisation of multibit MMIC attenuators is that the insertion loss of each pair of SPDT switches adds cumulatively to the zero or reference state insertion loss of the attenuator chip, thus resulting in higher overall insertion loss.

8.3.4 Switched scaled FETs

The switched scaled FET technique uses different sizes of FETs in two paths connected between the input and output ports. Instead of using external resistive networks to provide the desired attenuation, the difference in switched FET geometry on-state resistance values are used to achieve the selected attenuation. Hence the technique can be used for smaller attenuation bits of up to 2 dB. T-networks consisting of two passive FETs are used, as switchable series resistive elements and a shunt resistor to ground. Wider FETs are used in the reference path to minimise the reference state insertion loss. The attenuation ratios for each bit can be changed by adjusting the device widths and shunt resistor values. For the switched scaled FET configuration, additional short transmission-line segments need to be included to equalise the attenuating path length and the reference path. This phase equalisation technique is usually confined to small values of attenuation (< 4 dB, say) because for larger attenuation values the difference in FET sizes becomes excessive and makes it difficult to achieve phase matching between the two paths. Again, the symmetrical configuration makes the design more tolerant to process parameter variations, and the electrical performance is less sensitive to temperature variation. Figure 8.12 shows a schematic diagram of the switched scaled FET configuration. Although a T-type network topology is shown here, it is possible to realise a configuration using a π-network also. In the limiting case, when only a series FET is used in one of the arms, the reference state insertion loss can be kept to a minimum, as will be seen in Section 8.3.6.

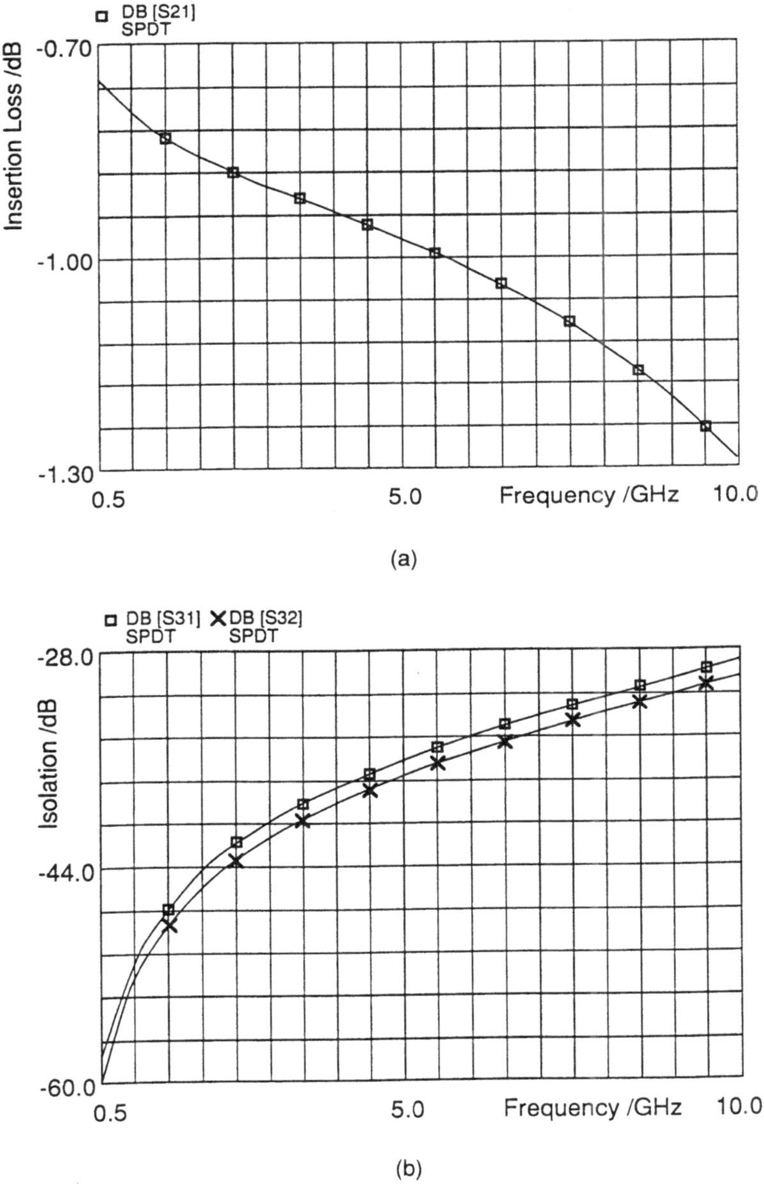

Figure 8.11 SPDT switch performance [17]: (a) insertion loss and (b) isolation

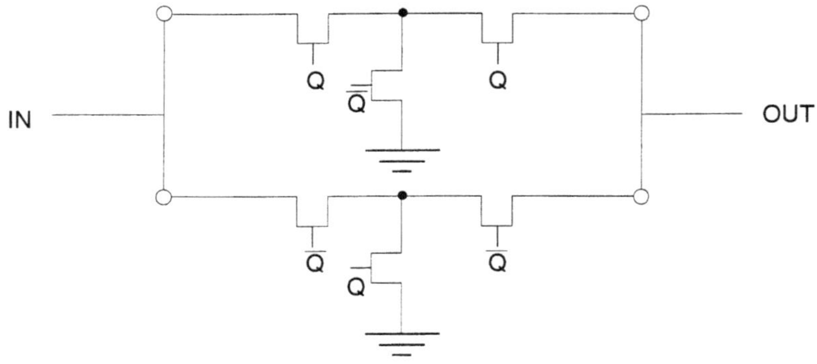

Figure 8.12 Schematic diagram of the switched scaled FET configuration

8.3.5 Switched bridged-T attenuator

The switched bridged-T attenuator configuration consists of a classical bridged-T attenuator with a shunt and series FET, as shown in Figure 8.13. Switched FETs are placed across the bridge resistor R_1 and in series with the shunt resistor R_2. These two FETs are switched on or off to switch between the zero state and the attenuation state, whose value is determined by the bridged-T attenuator. The bridged-T attenuator inherently provides good input and output matches. The attenuation value in dB is given by:

$$\text{Attenuation} = 20\ \text{Log}(R_2/(R_2+Z_0)) = 20\ \text{Log}(Z_0/(R_1+Z_0)), \text{ where } Z_0^2 = R_1R_2 \quad (8.3)$$

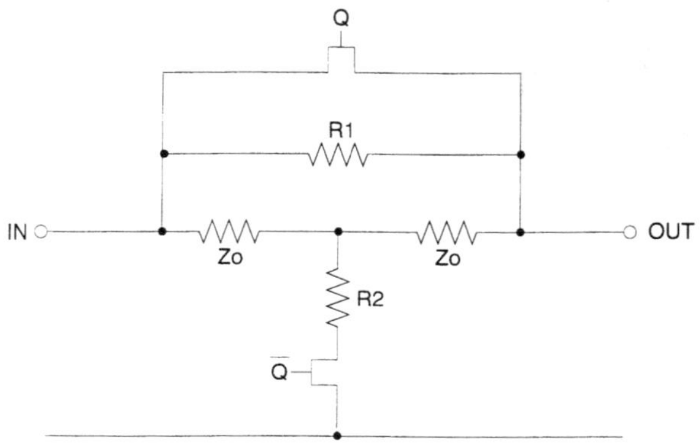

Figure 8.13 Switched bridged-T attenuator configuration

The performance of the individual bit is determined by the FET characteristics in the on- and off-states, and by the realisability limit on the required resistance values and the resistors' parasitics. Bayruns *et al.* [19] used this technique to realise a 5-bit digital attenuator with integral TTL-compatible switch drivers for operation up to 1.6 GHz. Instead of using a classical arrangement for the shunt resistor with series FET in the bridged-T, a parallel combination of a fixed resistor to ground and a switched FET in series with a resistor was used. This helped to keep the shunt FET from saturating at large input power levels. The contribution of the FET on- and off-state resistances was taken into account in the design of individual attenuation bits. Compared with the switched attenuator configuration, the zero state insertion loss for the bridged-T configuration is lower. Also, the chip size for a multibit attenuator is smaller when the bridged-T type structure is used. For the 5-bit attenuator realised by Bayruns *et al.* the insertion loss was only 3 dB and the chip size was only 1.8 × 1.8 mm^2. The bridged-T resistor network provides excellent input and output matches for an individual bit, and therefore, the effect of interaction between bits on in-band attenuation is minimal.

For lower attenuation bits, the resistor in the bridged-T can be replaced by FETs with the correct on-state resistance. This configuration is called a switched-FET bridged-T attenuator.

8.3.6 Switched T- and π-attenuators

The switched T- and π-attenuators are similar in principle to the above arrangement, except the topology of the attenuator is in classical T- or π-form and the individual resistive elements can be switched in or out of the circuit. This technique can also be termed a switched resistor method. Figure 8.14 shows a schematic representation of such an attenuator configuration. Bedard and Maoz [20] demonstrated a design for operation up to 10 GHz using a π-type configuration. Although similar to the bridged-T attenuator configuration, this technique offers lower insertion loss than the switched attenuator design as it uses much smaller FETs. As smaller FETs have lower parasitic capacitances this technique has good high frequency capability. However, unlike the bridged-T attenuator design that provides inherently good input and output matches for the individual bit, the designers who adopt this technique have to ensure that good port matches are achieved at the individual bits, and that any mismatches do not affect the overall performance of the attenuator chip. Bedard and Maoz [20] realised a 5-bit design with a 1 dB LSB bit. The 4 and 8 dB bits were realised in the classical topology. However, for the 16 dB bit, computer simulations showed that the inductances associated with the shunt resistors and the capacitances of the FETs were limiting the attenuation flatness at the high frequency end. To resolve this problem the 16 dB bit was designed as a cascade of two 8 dB bits. For smaller attenuation values of 0.5 to 2 dB, the classical T- or π-form can be replaced by either a series resistor shunted by a switched FET or a shunt resistor in series with a switched FET. The latter scheme was used in [20] to realise 1 and 2 dB bits. The measured results of the chip show that the reference or zero state insertion loss was less than 5 dB up to 10 GHz. Good accuracy for individual attenuation states was obtained but the relative phase shift for each attenuation state was as high as

60°. For smaller attenuation bits, resistors can be replaced by FETs with the correct on-state resistance value. This arrangement can be termed a switched-FET T- (or π-) attenuator.

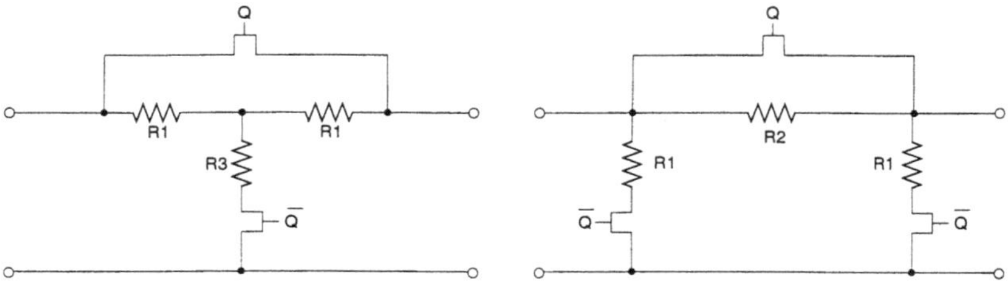

Figure 8.14 Switched T- and π-attenuator configurations

Neither the bridged- nor the switched-T type attenuators have the degree of symmetry of the switched attenuator and switched scaled-FET attenuator. This could be a disadvantage as the performance of a symmetrical configuration will be less susceptible to variations in MMIC process parameter variation, and the actual performance of the chip would be less sensitive to ambient temperature variations. This configuration, however, offers the possibility of minimising zero-state insertion loss. Krafsicik *et al.* [21] demonstrated a modified π-type embedded switched attenuator circuit topology. They used two parallel FETs in the series arm to reduce the line length parasitic for the high attenuation bits (> 4 dB). For the lower attenuation states an additional capacitor was added across the series π-resistor to adjust the attenuation flatness with frequency. Reference state insertion loss of the attenuator was less than 5 dB from 2 to 16 GHz. Input and output VSWRs were measured to be less than 1.5:1 for all states.

8.3.7 Linearity considerations

In the recent years there has been a growing demand for microwave digital and analogue attenuators for signal conditioning and control for the burgeoning requirements of the wireless communications industry. Both digital and analogue attenuators are being used for this purpose. This requirement has resulted in a number of GaAs vendors offering high performance components for different communications application (GSM, PCS, UMTS, etc.) and frequency bands. The bulk of these large volume applications are below 4 GHz and a number of digital attenuator chip sets for up to 5 bit configuration with 0.5 or 1 dB LSB are currently being offered. These are usually mounted in plastic packages rather than more expensive ceramic packages. For such signal level control applications, minimisation of reference state insertion loss is a key requirement. Another requirement from system considerations is linearity. As the attenuator is part of the RF chain, the designer needs to ensure that there is sufficient linearity margin in the component. The linearity of the attenuator can be defined in a similar way to an amplifier, that is 1 dB compression point when the insertion loss of the attenuator increases by 1 dB. Usually this parameter is defined at the input port rather than the output port like an amplifier. Some vendors provide 3rd order intercept point data based on two carrier measurements.

The conventional way of maintaining linearity is to size the gate width of the FETs accordingly. A technique for improving the linearity of the FET structure has been demonstrated by Miyatsuji and others [22]. A conventional GaAs FET is designed to have a low on-state resistance to obtain low insertion loss. Such a FET shows a steep I_d–V_{gs} characteristic, which causes distortion, especially when the FET is biased near its threshold voltage. The FETs with higher threshold voltages and smaller gate widths show a less steep pinch-off characteristic at the expense of higher on-state resistance. They use a squeezed gate FET geometry structure, in which two different gate geometry lengths are used. This technique maintains the higher pinchoff characteristics without increasing the on-state resistance significantly. The schematic of the FET is simply a parallel connection of the two gate width FETs. Third order intermodulation measurements show an in improvement of 10 dB compared to a constant length FET geometry, which occurs when the wider gate width is nearly 30% of the total gate width.

8.4 Digital attenuator design example

The selection of different attenuator topologies for the design and realisation of a multi-bit attenuator can be illustrated best by example. The Philips D07M standard foundry process was used for this design [23]. The model for the switching FET, developed by the foundry, was described in Section 8.2.2 (Figure 8.3). Initially, individual attenuation bits were designed and their performance evaluated for a cascaded 5-bit attenuator. A binary weighting arrangement with a 1 dB LSB was required. The chip was intended to be a generic functional block for applications up to 6.5 GHz.

8.4.1 Higher attenuation bits

For high attenuation states (i.e. the 4, 8 and 16 dB bits), the switched attenuator, bridged attenuator and switched resistor T-networks were investigated. For low attenuation bits, such as 1 and 2 dB, the switched scaled-FET, bridged FET T- and switched FET T-configurations were investigated.

The performance of the SPDT switches used in the switched attenuator type design is shown in Figure 8.11. The optimised performance characteristics for an 8 dB bit are shown in Figure 8.15. The reference state insertion loss is -1.9 ±0.2 dB/bit, the attenuation bit error is less than 0.02/-0.0 dB/bit, and the input and output matches are better than 22 dB. The path lengths in the two paths have been optimised to give an insertion phase error of less than 0.02° between the reference state and the attenuation state. The results for an optimised 8 dB attenuation bit using a switched resistor T-network design showed that the error was 0.2 dB. The differential phase shift was ±0.05°, and with 15 dB return losses.

8.4.2 Lower attenuation bits

For smaller bit attenuation values of up to 4 dB the techniques investigated were the switched scaled FET, bridged FET T- and switched FET T- networks. The optimised switched scaled FET circuit performance for a 2 dB attenuation bit is shown in Figure 8.16.The attenuation setting error is less than +0.1/-0.05 dB and port matches better than 25 dB over the frequency band. The phase error between the reference and attenuation states is +0.15/-0.4°. The techniques is not suitable for the 4 dB bit as it is difficult to minimise the phase error between the two paths.

The optimised design for a 2 dB bridged-T attenuator bit showed that in the desired frequency range of 1.5 to 6.5 GHz the reference state insertion loss was 1.4 dB. The attenuation bit error was less than -0.02 dB and the input and output matches were better than 25 dB. The insertion phase error was optimised to give a phase error of less than ±1° between the reference state and the attenuation state.

The optimised results for a 2 dB bit using thge switched FET T-network technique showed that the attenuation error is ±0.1 dB and port matches are only 15 dB. The phase difference between the reference and attenuation path is +0.4/-2.0°. The performance of the circuit was over a narrower frequency band and could not be implemented for a 1 dB bit.

Table 8.1 summarises the results of simulation and optimisation for the individual MMIC attenuator bits. It can be seen that for larger attenuation bits (> 4 dB) the switched attenuator or the switched resistor-T technique can be used. The bridged-T technique has poor port matches and limited operating frequency. On further examination, it can be seen that the attenuation errors and port matches of the switched resistor-T configuration are poorer than the switched attenuator design. For smaller bits, the switched scaled FET or the bridged FET T-techniques can be adopted. However, the bridged FET T-design has larger phase error than the former. Moreover, the configuration is not symmetrical and the individual bit performance might be subject to process parameter variations. It can be concluded that the switched attenuator and switched scaled FET techniques provide the best overall performance for the digital attenuator. The attenuation setting accuracy per bit is only ±0.07 dB and the phase error per bit is less than 0.4°. Both the input and output matches are better than 22 dB, thus making it easier to cascade individual bits without giving rise to any amplitude or phase ripples due to mismatch.

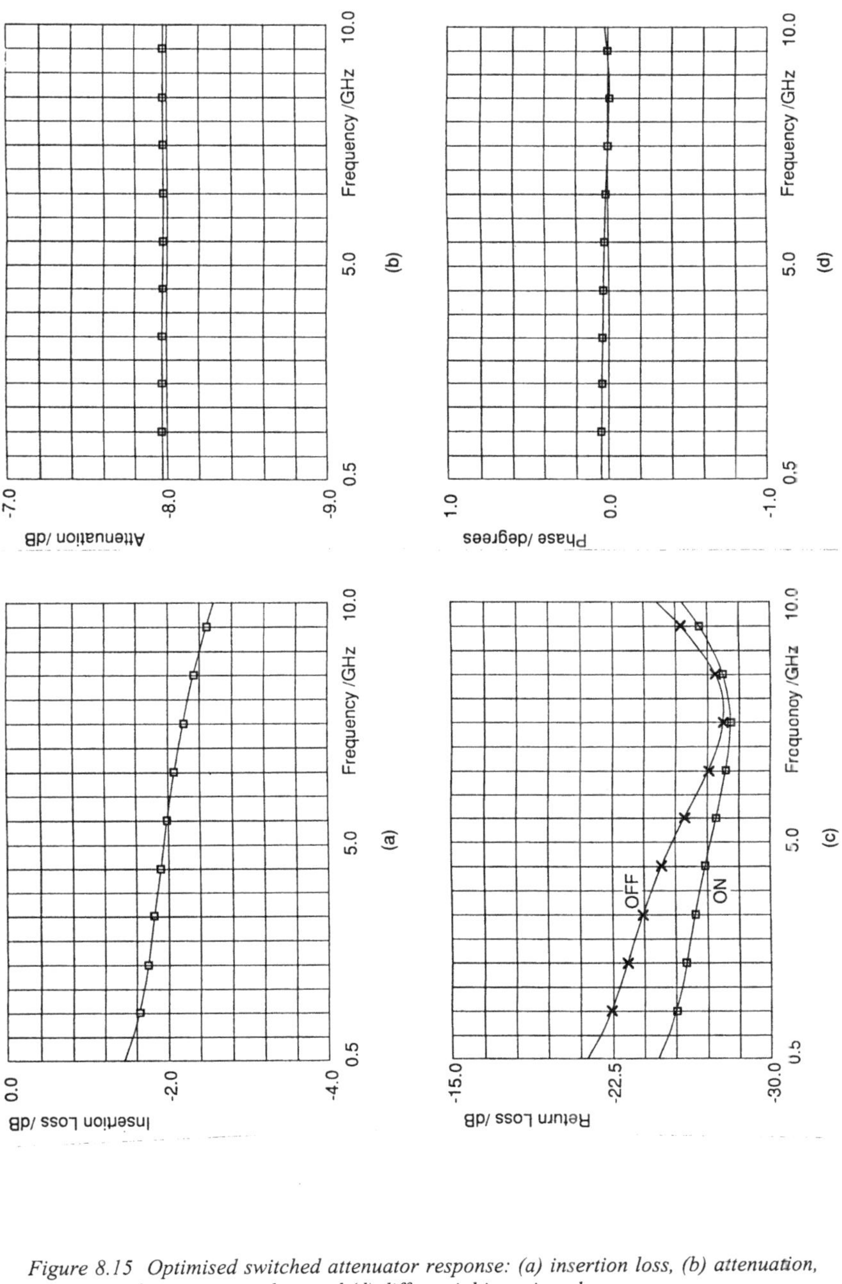

Figure 8.15 Optimised switched attenuator response: (a) insertion loss, (b) attenuation, (c) input and output return loss and (d) differential insertion phase

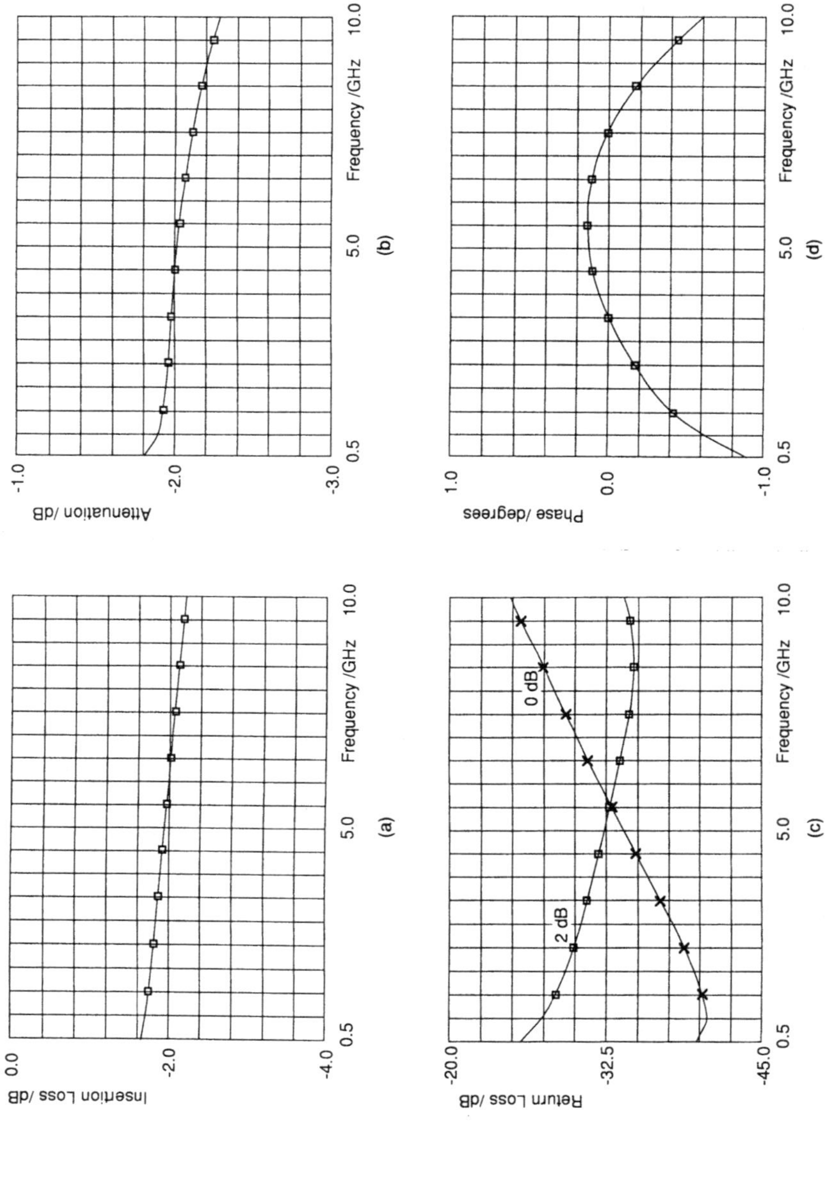

Figure 8.16 Optimised switched scaled FET circuit performance: (a) insertion loss, (b) attenuation, (c) input and output return loss and (d) differential insertion phase

Table 8.1 Computer simulation results

Atten uation value (dB)	Design approach	Attenuation error (dB)	Phase error (deg)	Frequency (GHz)	Match (dB)	Insertion loss per bit (dB)	Comments
4 8 16	Switched attenuator	+0.02/-0.0/bit	+0.04/ -0.0/bit	1.5 to 6.5	>22	1.9±0.2	Resistors limit attenuator values. Large area
4 8	Bridged-T attenuator	±0.1/bit	±1	< 2.0	>16	2±0.2	Resistors limit values
8 16	Switched resistor	0.2/-0.7/bit	0.05/ -0.3/bit	1.5 to 6.5	>15	1.9±0.15	Not symmetrical
1 2	Switched scaled FETs	±0.07/bit	0.2/ -0.4/bit	1.5 to 6.5	>28	1.95±0.15	FET geometries limit values
1 2	Bridged FET-T	±0.05/bit	0.65/ -0.5/bit	1.5 to 6.5	>24	1.15±0.15	Not symmetrical
2 4	Switched FET-T	0.3/bit	0.3/ -0.5/bit	1.5 to 5	>14	2.25±0.15	Not symmetrical

8.4.3 Overall chip design and layout

Having selected the individual attenuation bits, a cascading arrangement was optimised. Due to the better matches of the switched scaled FET bits, it was decided to insert them between the switched attenuator type bits. The 4 dB switched attenuator was the first element in the uni,t followed by the 1, 8, 2 and 16 dB bits. This arrangement provides better power handling for the chip. The simulated performance of the cascaded chip, including all modifications for layout, is shown in Figure 8.17 from 0.5 to 10 GHz. The desired frequency range was 1.5 to 6.5 GHz. Simulations show that the attenuation setting error for all possible states is less than 0.2 dB (less than one-quarter LSB). The insertion phase variation for all possible attenuator states is less than 1.5° over the frequency band of interest.

The chip was laid out using standard commercial layout software. Figure 8.18 shows a microphotograph of the 5.0 × 2.5 mm² chip. The RF input and output are at opposite ends with complimentary pairs of control lines along one edge. Great care was taken to avoid coupling across the attenuators, to minimise insertion loss and to reduce the number of through GaAs via-holes by the use of via sharing.

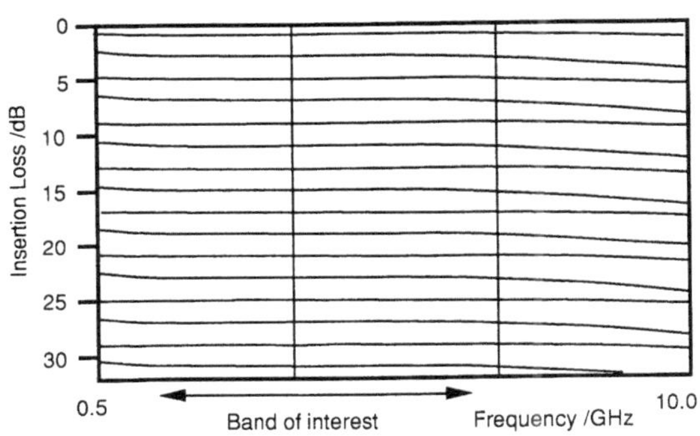

Figure 8.17 Simulated performance of the cascaded attenuator chip

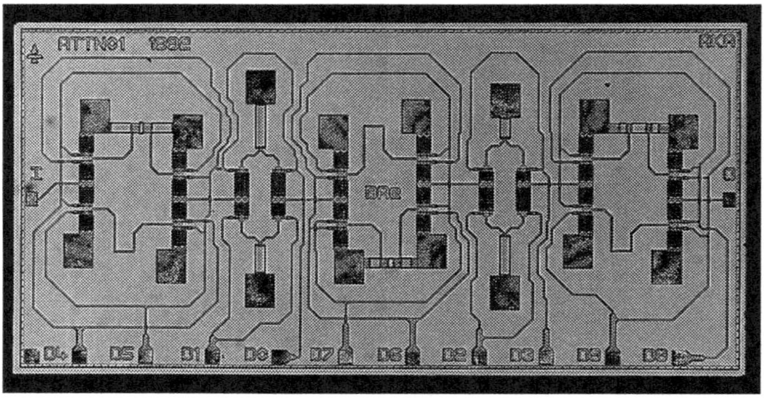

Figure 8.18 Microphotograph of the cascaded attenuator chip

8.4.4 Measured results

The measured attenuation states are shown in Figures 8.19 to 8.22. The zero state insertion loss was 8 dB up to a maximum operating input power of +21 dBm, as seen in Figure 8.19. The attenuation settings relative to the zero state are shown in Figure 8.20; peak setting errors are +0/-0.88 dB. These results show good correlation with predictions. The setting errors for all individual attenuation bits are negligible except for the 2 and 16 dB bits. For these the errors are 0.16 dB and 0.1 dB, respectively, in the band of interest. However, it is considered that the finite leakage in the test fixture itself may be a contributory factor for the 16 dB state. Normally, the cumulative errors for the 31 dB state (when all the individual bits are on) should be equal to the sum of individual bit errors. This has been observed at all the frequencies except at 3.5 GHz where the worst case attenuator setting errors occur in the higher attenuation states (> 16 dB).

The input and output port matches are better than 13.5 dB for all states, as shown in Figure 8.21. The differential insertion phase between states is shown in Figure 8.22, which shows that the peak errors are +3°/-1.6°. There is a phase error associated with the 16 dB bit caused by coupling across the attenuator. This accounts for the two clusters of traces: the cluster about 0° is for states below 16 dB and the cluster about +2° is for states where the 16 dB bit is switched on.

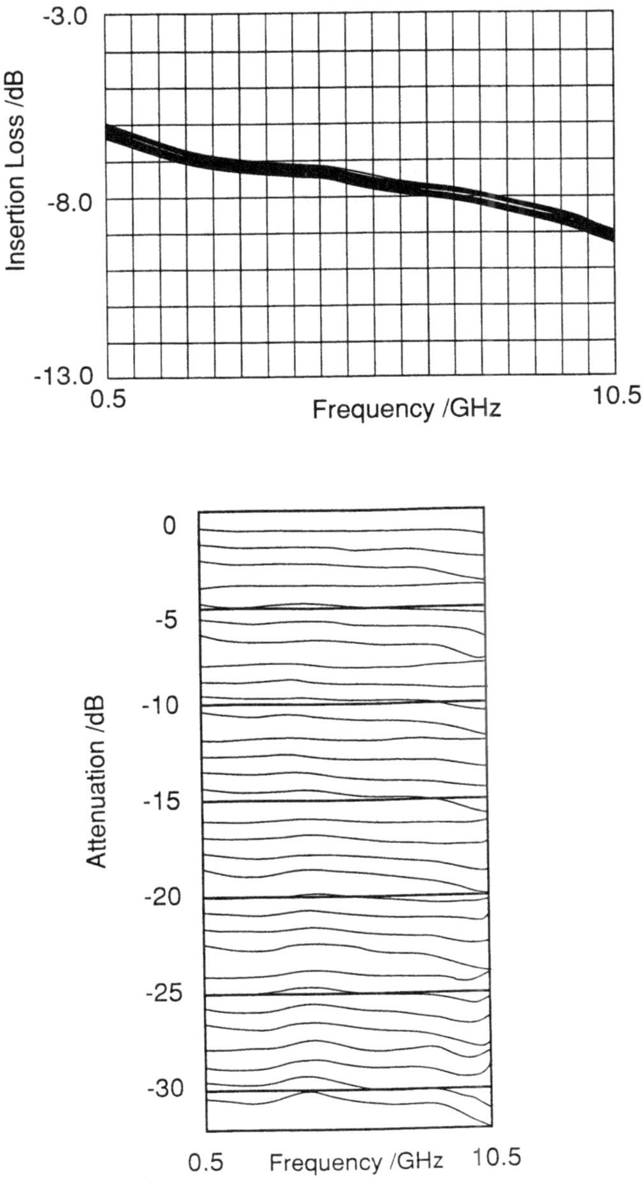

Figure 8.20 Attenuation settings relative to the zero state

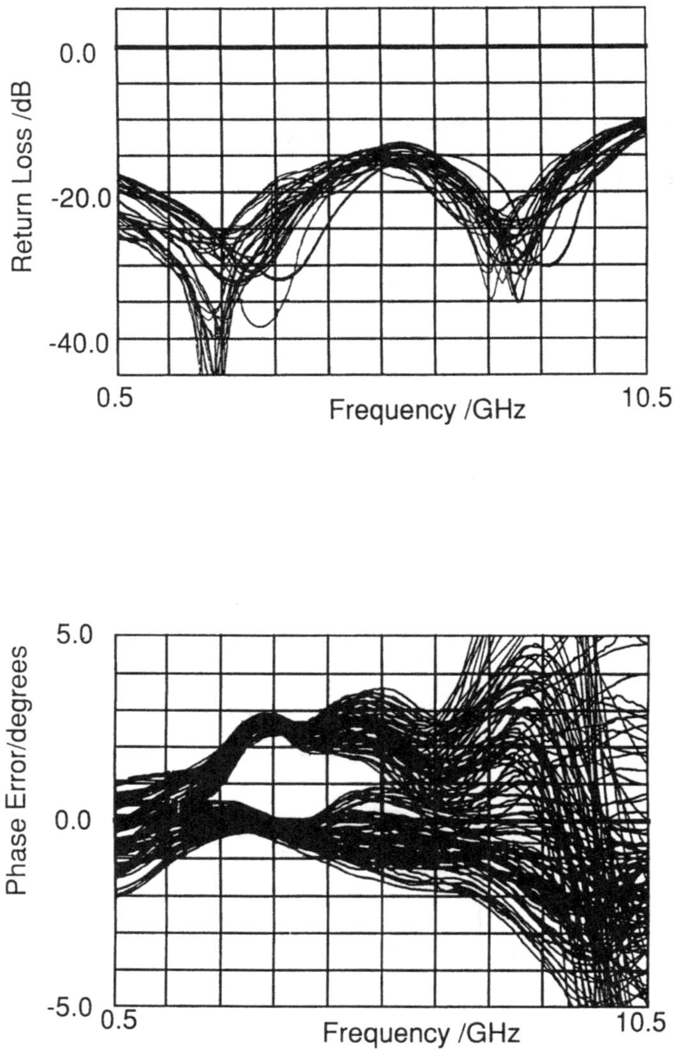

Figure 8.22 Insertion phase tracking between states

8.5 Analogue attenuators

In contrast with a purely digital implementation, analogue attenuators can provide a continuous range of amplitude control. Hence, they have the advantage that they can correct for any degradation in performance attributed to fabrication process variations, and enable calibration corrections to be performed when integrated into a subsystem. Analogue variable attenuators are often found in AGC control loops, vector modulators and adaptive beam-forming networks. In addition, they can be used as variable taps in direct implementation transversal and recursive filter architectures, and a variety of other signal processing functions.

8.5.1 Analogue reflection-type attenuator

An analogue reflection-type attenuator topology was demonstrated by Devlin and Minnis [24]. This technique, illustrated in Figure 8.23, is similar to the reflection-type phase shifter described in Chapter 9. However, a variable resistance reflection termination is used so that the signal amplitude can be controlled, with a transmission phase that does not vary (in the ideal case). For MMIC implementations, this variable resistance is most easily realised with a cold-FET. Lucyszyn [25] reported the measured performance for an X-band analogue attenuator using this technique. The attenuator had a centre frequency of 10 GHz and a bandwidth covering 8 to 12 GHz. The chip used a folded four-finger Lange coupler with variable resistors realised using cold-FETs. The GaAs MMIC was fabricated at the Marconi Caswell foundry, using their standard F20 foundry process. A microphotograph of the attenuator is shown in Figure 8.24. The chip has an active area of $2.4 \times 1.0\ mm^2$. The measured levels of attenuation, with the associated tuning characteristic, are shown in Figure 8.25. It can be seen that a good performance is achieved across a wide bandwidth.A dynamic range of 14 dB is achieved at centre frequency. The worst-case return loss decreases almost linearly from 24 dB at centre frequency to 19 dB at the band edges.

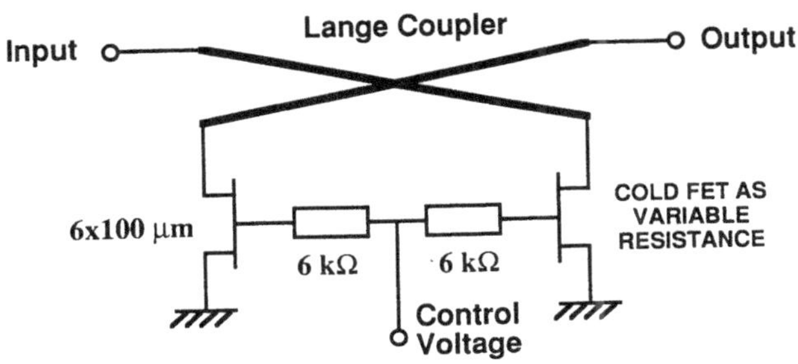

Figure 8.23 Topology of the analogue reflection-type attenuator

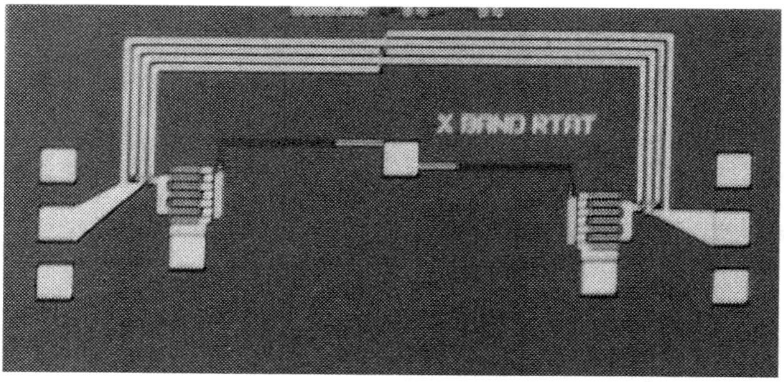

Figure 8.24 Microphotograph of the analogue reflection-type attenuator

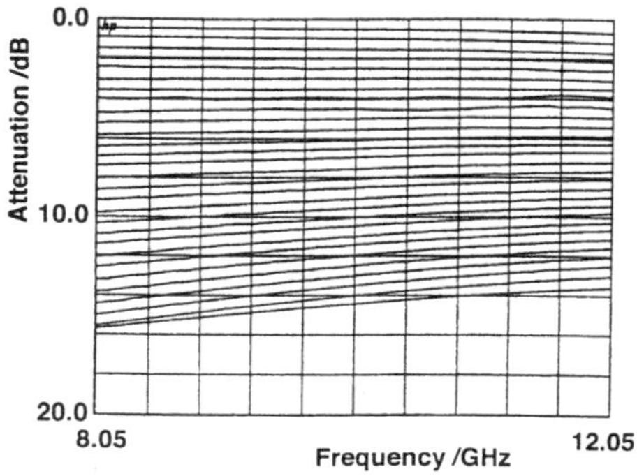

(a)

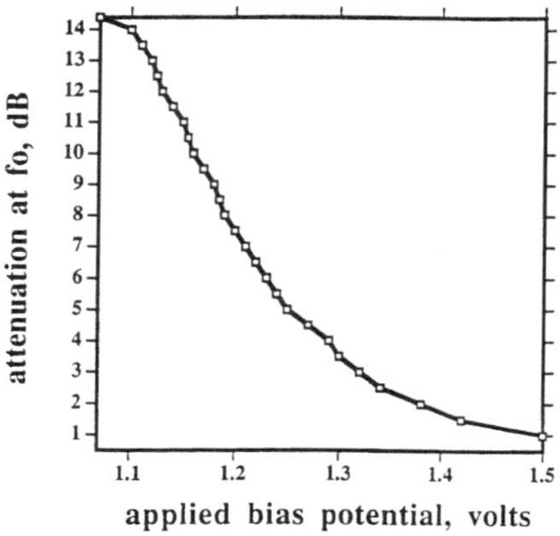

(b)

Figure 8.25 Measured performance of the attenuator: (a) attenuation against frequency and (b) attenuation against control voltage

8.5.2 Other analogue attenuators

In addition to the reflection-type analogue attenuator, diodes can be used in a balanced attenuator configuration by employing the principle of the balanced amplifier. In this type of attenuator, illustrated in Figure 8.26, the diodes are shunt-connected between a pair of Lange couplers: the variable impedance of the diode produces a continuously varying magnitude of transmission coefficient, S_{21}. The signals reflected from the diodes are absorbed in the terminations on the Lange couplers in the same way as in the balanced amplifier. This technique has been used to realise hybrid MIC attenuators with PIN diodes [26]. With special driving circuitry this design achieved an exceptional dynamic range of 120 dB in Ku-band. The same technique can be employed using Schottky barrier diodes on MMICs. The result is an octave bandwidth analogue attenuator that has an attenuation range of around 10 to 15 dB, with good input and output matches and a fairly constant transmission phase. Sweet [27] has provided a useful design procedure for this balanced type of attenuator.

Analogue attenuators can also be realised with π- and T-networks using FETs, PIN diodes or Schottky barrier diodes [28,29]. However, it is very difficult to maintain a good input and output match over the whole attenuation range with this method. Finally, rather than employing an attenuator, a variable gain amplifier can be employed for continuous gain control. The most common method is to employ a dual-gate or cascode FET arrangement, and this is described in Chapter 5. It is important to note that the dynamic range of this type of amplifier can be very restricted when operating at the limits of the gain control range.

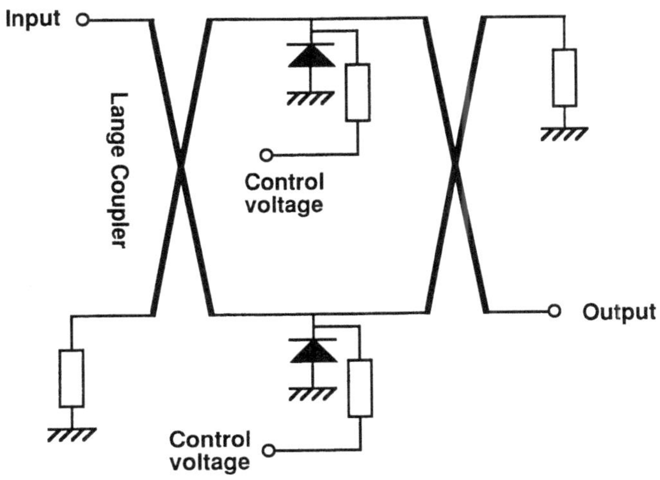

Figure 8.26 Balanced analogue attenuator employing PIN or Schottky barrier diodes

8.6 Conclusions

This chapter has reviewed the switching mechanism of GaAs FET MMIC switches and has discussed the equivalent circuit of these MMIC switches. The advantages offered by these switches over conventional PIN diode switches have been described along with their limitations. Some design examples and applications of MMIC switches have been given. The second part of the chapter discussed the requirements for MMIC attenuators. A number of techniques available for the implementation of these attenuators have been reviewed and their performance trade-offs discussed. The design example of a single chip 5-bit digital attenuator operating from 1.5 to 6.5 GHz has been discussed. The measured performance of the chip showed a good correlation with computer simulations.

8.7 References

1 AYASLI, Y.: 'Microwave switching with GaAs FETs', *Microwave Journal*, 1992, pp. 61–74

2 ERON, M.: 'Small and large signal analysis of MESFETs as switches', *Microwave Journal*, 1992, pp. 128–140

3 BUCK, C., WILLIAMS, K., and LEBLANC, R.: 'The development of MMIC switched FET models for Philips D07M foundry process', *Proceedings of IEE Colloquium on Modelling, Design and Application of MMICs*, June 1994

4 MALEVIGE, W., and SOKOLOV, V.: 'Resonated GaAs FET devices for microwave switching', *IEEE Transactions on Electron Devices*, 1981, **ED-28**, 1981, pp. 196–204

5 AYASLI, Y., VOURHAUS, J., PUCEL, R., and REYNOLDS, L.: 'Monolithic GaAs distributed FET switch circuits', *Proceedings of IEEE GaAs IC Symposium*, San Diego, 1981

6 SMUK, J., and SHIFRIN, M.: 'Monolithic GaAs multi-throw switches with integrated low-power decoder/driver logic', *Proceedings of IEEE RFIC Symposium*, 1997

7 Application Notes (Hittite Microwave)

8 IMAI, N., MINAKAWA, A., and OKAZAKI, H.: 'Novel high-isolation FET switches', *IEEE Transactions on Microwave Theory and Techniques*, 1996, **44** (5), pp.685–691

9 MIZUTANI, H., and TAKAYAMA, Y.: 'A DC-60 GHz GaAs MMIC switch using novel distributed FET', *Proceedings of IEEE Monolithic and Millimetre Wave Circuits Symposium*, 1997

10 KATZIN, P., BEDARD, B., SHIFRIN, M., and AYASLI, Y.: 'High speed, 100+ W RF switches using GaAs MMICs', *IEEE Transactions on Microwave Theory and Techniques*, 1992, **40** (11), pp. 1989–1996

11 HWANG, Y. *et al.*; 'A microwave phase and gain controller with segmented dual gate MESFETs in GaAs MMICs', *Proceedings of IEEE Microwave and Millmeter-Wave Monolithic Circuits Symposium*, 1984, pp. 1–5

12 NASTER, R, and HWANG: 'An L-band variable gain amplifier in GaAs MMIC with binary step control', *Proceedings of IEEE GaAs IC Symposium*, 1987, pp. 235–237

13 SNOW, K., KOMIAK, J., and BATES, D.: 'Wideband variable gain amplifiers in GaAs', *Proceedings of IEEE Microwave and Millimeter-Wave Monolithic Circuits Symposium*, 1988, pp. 133–137

14 SIWERIS, H.: 'A monolithic C-band variable gain amplifier with digital gain control', *Proceedings of 21st European Microwave Conference*, 1991, pp. 1475–1480

15 DEVLIN, L. *et al.*, *Proceedings ESTEC Conference on Active and Passive Beamforming networks*, Noordwijk, 1992

16 GUPTA, R. K. *et al.*: 'Modelling and CAD of an ultra-broadband monolithic 5-bit digital attenuator', Institute of Monoelectronik, 1988

17 ANDERSON, A. K., and JOSHI, J. S.: 'Wideband constant phase digital attenuators for space applications', *Microwave Engineering Europe*, 1994

18 SJOGREN, S., INGRAM, D., BIDENBENDER, M., LAI, R., ALLEN, B., and HUBBARD, K.: 'A low phase-error 44 GHz HEMT attenuator', *IEEE Microwave and Guided Wave Letters*, 1998, **8** (5), pp. 194–195

19 BAYRUNS, J., WALLACE, P., and SCHEINBERG, N.: 'A monolithic DC - 1.6 GHz digital attenuator', *IEEE MTT-S International Microwave Symposium Digest*, 1989, pp. 1295–1298

20 BEDARD, B., and MAOZ, B.: 'Fast GaAs MMIC attenuator has 5-bit resolution', *Microwaves and RF*, 1991, pp. 71–76

21 KRAFSICIK, D., ALI, F., and BISHOP, S.: 'Broadband low-loss 5- and 6-bit digital attenuators', *IEEE MTT-S International Microwave Symposium Digest*, 1995, pp. 1627–1630

22 MIYATSUJI, K., ISHIDA, H., FUKUI, T., and UEDA, D.: 'A technique for improving the distortion of GaAs variable attenuator IC using squeezed gate FET structure', *Proceedings of IEEE Microwave and Millimeter Wave Monolithics Circuits Symposium*, 1996, pp. 43–46

23 ANDERSON, A. K., and JOSHI, J. S.: 'Generic constant phase digital attenuators', *Proceedings of IEE Colloquium on the Modelling, Design and Application of MMICs*, 1994

24 DEVLIN, L. M., and MINNIS, B. J.: 'A versatile vector modulator design for MMIC', *IEEE MTT-S International Microwave Symposium Digest*, 1990, pp. 519–522

25 LUCYSZYN, S., and ROBERTSON, I. D.: 'Analog reflection topology building blocks for adaptive microwave signal processing applications', *IEEE Transactions on Microwave Theory and Techniques*, 1995, **43** (3), pp. 601–611

26 BLASER, R. J.: 'Control circuit produces linear attenuator response', 1994, *Microwaves and RF*, 1994, pp. 59–64

27 SWEET, A. A.: *MIC and MMIC amplifier and oscillator circuit design* (Artech House, Dedham, MA, 1990), pp. 168–174

28 FRANCO, D. P.: 'A miniaturised mounting for a 4-20 GHz PIN attenuator', *Microwave Engineering Europe*, 1994, pp. 31–36

29 GOLDFARB, M. E., and PLATZKER, A.: 'A wide range analog MMIC attenuator with integral 180 phase shifter', *IEEE Transactions on Microwave Theory and Techniques*, 1994, **MTT-42** (1), pp. 155–158

Chapter 9

Phase shifters

S. Lucyszyn and J. S. Joshi

9.1 Introduction

A phase shifter is a control device found in many microwave communication, radar and measurement systems. This chapter describes many of the design techniques for MMIC phase shifters, implemented under either analogue or digital control. In order to understand the subtle differences between the two main generic types of phase shifter, both the true phase shifter and the true delay line must first be defined.

A true phase shifter can be defined as a control device that has a flat group delay frequency response within its defined bandwidth of operation, the level of which does not change as the insertion phase is varied. Its two characteristic features are:

1. a flat relative phase shift frequency response, at all levels of relative phase shift
2. constant group delay, resulting in no change in the timing of an input RF pulse envelope.

The frequency characteristics of a true phase shifter are illustrated in Figure 9.1. True phase shifters can be employed in multiple space diversity receiver-combiners for aligning RF signals within a pulse envelope without changing the timing of the pulse edges. However, they should not be employed in wideband beam-forming networks for large aperture phased-array antennas, in order to avoid the effects of 'phase squinting' and 'pulse stretching'.

A time shifter can be defined as a control device that has a flat group delay frequency response, with a level that changes as its insertion phase varies, within a defined bandwidth of operation. Its two characteristic features are:

1. a linear relative phase shift frequency response, with a gradient that changes as the value of relative phase shift varies
2. a flat group delay frequency response, with a level that varies, resulting in a change in the timing of an input RF pulse envelope.

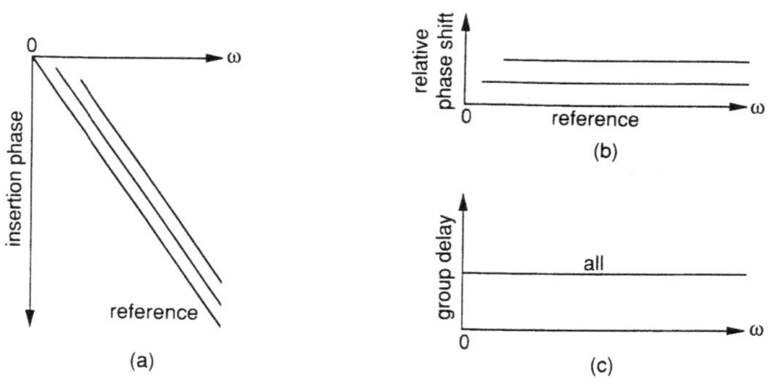

Figure 9.1 Frequency characteristics of a true phase shifter: (a) insertion phase, (b) relative phase shift and (c) group delay

A special case of a time shifter is a true delay line, which is defined by the following expression:

$$\Delta\tau = \frac{\Delta\angle S_{21}(\omega)}{\omega} \tag{9.1}$$

where $\Delta\tau$ = relative shift in group delay, $\Delta\angle S_{21}(\omega)$ = relative phase shift at ω, and ω = angular frequency.

The frequency characteristics of the true delay line are illustrated in Figure 9.2. True delay lines find many applications in general wideband microwave signal processing applications, including beam-forming networks for phased-array antennas. It must be noted that when the non-idealities of the circuit elements are included in a design, an ideal true phase shifter circuit may exhibit frequency characteristics of a true delay line, and vice versa.

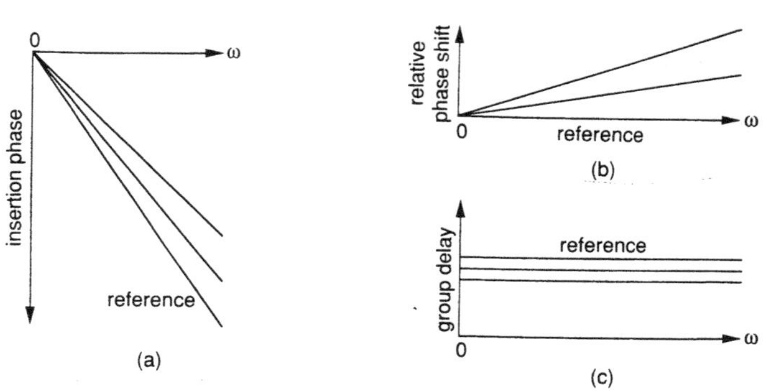

Figure 9.2 Frequency characteristics of a true delay line: (a) insertion phase, (b) relative phase shift and (c) group delay

9.2 Analogue implementations

When compared with a purely digital implementation, analogue control devices have the following advantages:

1. they may only require one control wire for an infinite number of phase states
2. no special foundry processing is required to realise high quality switches
3. they require almost no control power with a passive reflection topology incorporating varactor diodes and cold-FETs
4. if using a reflection topology with active circulators, they can make much more efficient use of limited chip space
5. they may not suffer from quantisation errors
6. since the phase variation can be continuous, any degradation in performance attributed to fabrication process variations can be corrected
7. re-calibration can be performed after integration into a subsystem
8. any degradation in performance attributed to ageing or adverse operating conditions can be corrected.

As a result of these significant advantages, analogue control devices appear ideal for large adaptive phased arrays and other high performance microwave signal processing applications, such as implementing software radio or software radar. However, it should also be pointed out that a purely analogue implementation is not inherently compatible with digital system architectures and may not provide the best overall performance across wide bandwidths

9.2.1 Introduction to analogue phase shifters

Today's spread spectrum communication, radar, EW, remote sensing, high data rate digital transmission and state-of-the-art microwave measurement systems all require very large RF bandwidths. While the microwave and RF community has addressed the problem of bandwidth for most microwave devices, high performance analogue phase shifters have only recently received such attention. The principal reason for this is the multi-dimensional nature of a suitable design. It is a relatively straightforward task to design a microwave phase shifter to meet a specified level of relative phase shift at a specific frequency. However, this task becomes more difficult if the target relative phase shift has to be maintained across a very large frequency range. The difficulty is further compounded when this wideband performance is required to be continuously tuned over a large relative phase shift range. This three-dimensional scenario has assumed that all the other device specifications, such as insertion loss and return loss, are kept within acceptable limits over the wide bandwidth and large tuning range. It will be shown that effective solutions to the three-dimensional problem can be implemented in monolithic technology as well as in conventional hybrid technology.

9.2.2 Single-stage reflection-type phase shifters

The basic topology of a single-stage reflection-type phase shifter (RTPS) is shown in Figure 9.3. For simplicity, the directional coupler is assumed to be lossless, perfectly matched and having infinite isolation. Also, there are two identical reflection terminations connected to this coupler. With reference to Figure 9.3, a proportion of the input power emerges from the direct port ④, where the resulting incident voltage wave is reflected from its terminating impedance, Z_T. A proportion of the reflected signal is transmitted back to the input port ① and a proportion is coupled into the isolated port ③. Similarly, the remainder of the input power emerges from the coupled port ②, where it is reflected and sent to both the input and isolated ports.

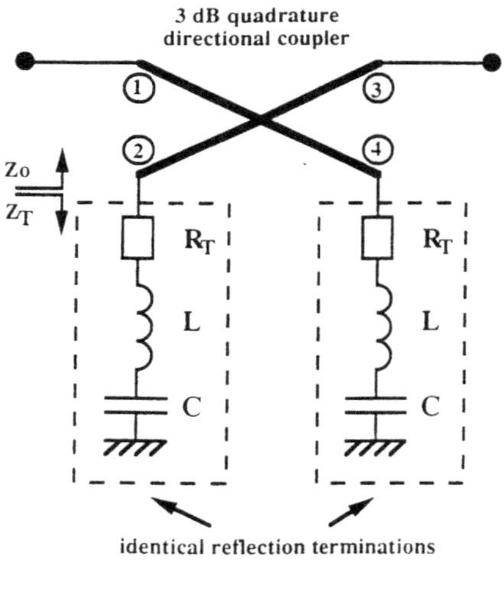

Figure 9.3 Basic topology of a single-stage reflection-type phase shifter

It can be deduced that the voltage wave emerging from the input port will have zero amplitude when equal power split and phase quadrature are maintained between the coupled and direct ports, resulting in perfect input and output impedance matching. Also, with a 3 dB directional coupler and purely reactive reflection terminations, the voltage wave emerging from the isolation port will have an amplitude equal to that entering the input port, resulting in zero insertion loss.

9.2.2.1 General design approaches

The RTPS was first introduced more than three decades ago by Hardin *et al.* [1]. Hardin employed a directional coupler and simple series-tuned circuit reflection terminations, incorporating a single varactor diode in each. The resulting frequency characteristic of the relative phase shift resembles a single hump. This hump provides a maximum fractional bandwidth of approximately 5–10%, for a maximum phase error of ±5°, within a maximum relative phase shift range of 180°. This approach was further developed by Searing [2] and Garver [3] without much improvement in the bandwidth. Attempts to widen the bandwidth were proposed by Henoch and Tamm [4] and Ulriksson [5], by synthesising a parallel resonance mode in the reflection terminations. Impedance transformers were used to effectively produce a double humped frequency response. This technique increased the bandwidth by a factor of two. Here, the component values of the reflection terminations could only be determined by intensive optimisation and the designer's judgement. In recent years, a number of variations to the basic analogue RTPS have been reported in monolithic form without significant advances in their overall performance [6–11].

9.2.2.2 *Optimum design approaches*

For a RTPS having single reactive element reflection terminations, two optimum design approaches have been identified to provide the maximum bandwidth for a defined level of maximum phase error. The first design approach gives the *minimum group delay variation* at the centre frequency. The second design approach gives the *maximum relative phase shift* at the centre frequency. With both, bandwidth is directly proportional to the level of phase error. When compared with the first, the second approach provides an increase of 25% in the level of maximum relative phase shift; however, this is at the expense of a doubling of the relative shift in group delay.

A rigorous theoretical investigation, complete design examples and the first proof-of-concept measured performance for an X-band MMIC implementation has been reported [12]. The GaAs MMIC was fabricated at Marconi Caswell, using their standard F20 foundry process. A microphotograph of the phase shifter is shown in Figure 9.4. Standard 0.5 μm MESFETs are used to realise interdigitated planar Schottky varactor diodes (IPSVDs) [13–15]. A back-to-back arrangement was adopted for the varactor diodes. This increases the maximum RF power level of the devices and reduces the errors in the resulting relative phase shift levels when operating at higher RF power levels [6]. As RF voltage increases, the capacitance of one of the capacitors increases and that of the other capacitor decreases. Therefore, the net change in capacitance, due to excess RF power, should ideally be zero. With the use of a folded Lange coupler [16], the X-band MMIC phase shifter has an active area of only 2.4×1.0 mm^2.

Figure 9.4 Microphotograph of the optimum design MMIC analogue phase shifter

With the phase shifter operating in the *maximum relative phase shift* condition, the measured relative phase shift performance, with its associated tuning characteristic, is shown in Figure 9.5*a*. Here, the RMS phase error increases almost linearly from zero at centre frequency, f_0, to ±2.5° at the band edges. The corresponding group delay performance, with its associated variation in group delay, is shown in Figure 9.5*b*. The calculated variation in group delay has also been superimposed onto the measured results, demonstrating a very close agreement. The mean level of insertion loss is 1.5 dB, with an RMS amplitude error of ±0.7 dB maintained at all levels of relative phase shift and at all frequencies. The worst-case return loss decreases almost linearly from 33 dB at the centre frequency to 18 dB at the band edges.

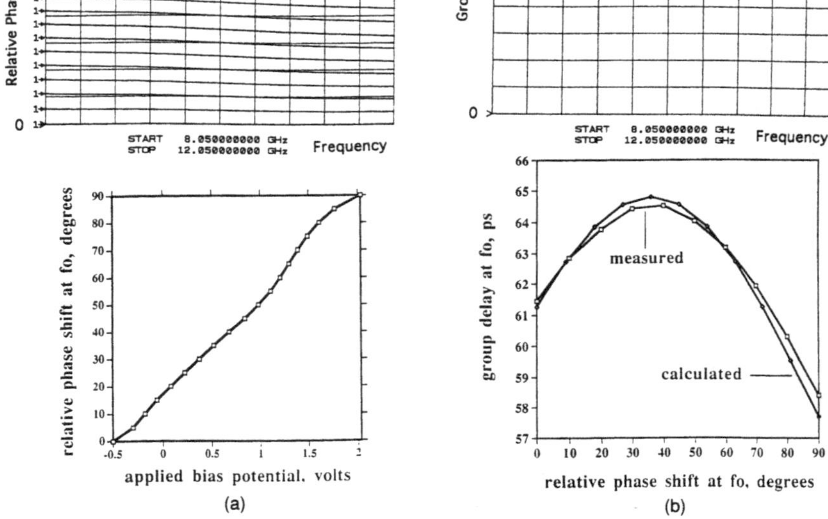

Figure 9.5 Frequency response of the optimum design MMIC phase shifter: (a) relative phase shift and (b) group delay

9.2.3 Single-stage reflection-type delay lines

Variable delay lines are required in adaptive beam-forming networks of wideband phased-array antennas. Delay lines should ideally have the following characteristics:

1. a large number of phase states, with high accuracy and precision
2. the relative shift in group delay, $\Delta\tau = (\Delta\angle S_{21})/\omega$, within the bandwidth
3. negligible PM/AM conversion
4. very low insertion loss performance
5. good return loss performance
6. low control power.

Common classes of general time shifters that are compatible with both hybrid and monolithic MIC technologies are switched-line delay lines [17] and digital reflection-type delay lines [18].

It has been found that the single-stage RTPS, which has a single reactive element reflection termination, can exhibit the time shifter frequency characteristics [19]. However, a true delay line characteristic can only be exhibited over a very narrow frequency range. In order to achieve a wideband performance, the reflection terminations must consist of a variable capacitor in series with a fixed inductor [20]. The non-idealities of the Lange coupler can also compensate for the non-linear behaviour of the group delay frequency response [12]. Simple empirical design equations for an octave bandwidth analogue delay line, employing a traditional 4-finger Lange coupler, are as follows:

$$L = \frac{4.0}{fo}\ \text{nH}, \quad C_{max} = \frac{3.0}{fo}\ \text{pF}, \quad \text{and} \quad R_T \le \frac{Zo}{10}\ \Omega$$

where f_O = centre frequency, having units of GHz

With an ideal 3 dB quadrature coupler, the relative phase shift is equal to the amount of change in the phase angle of the voltage reflection coefficient of the reflection termination. Therefore, to a good approximation, the tuning ratio, m, required by the variable capacitor can be determined from the following, for $\omega_1 \le \omega \le \omega_2$:

$$m = \frac{1}{1 - \frac{\Delta\angle S_{21}(\omega)\big|_{max}}{180^o} \cdot \left(\frac{\omega_{os}\big|_{min}}{\omega}\right)} \qquad \text{where } \omega_{os}\big|_{min} = \frac{1}{\sqrt{LC_{max}}} \tag{9.2}$$

The first proof-of-concept analogue reflection-type delay line was demonstrated using hybrid technology [21]. Here, the centre frequency was 750 MHz and a high performance was measured over a 40% fractional bandwidth. An MMIC analogue delay line has been demonstrated at X-band [12]. The GaAs MMIC was fabricated at Marconi Caswell using their F20 process. A microphotograph of the analogue delay line is shown in Figure 9.6. As with the optimally designed MMIC phase shifter, standard 0.5 μm MESFETs were used to realise the IPSVDs, which were then arranged back-to-back. With a folded Lange coupler, the X-band MMIC delay line has an active area of only 2.4 × 1.0 mm^2.

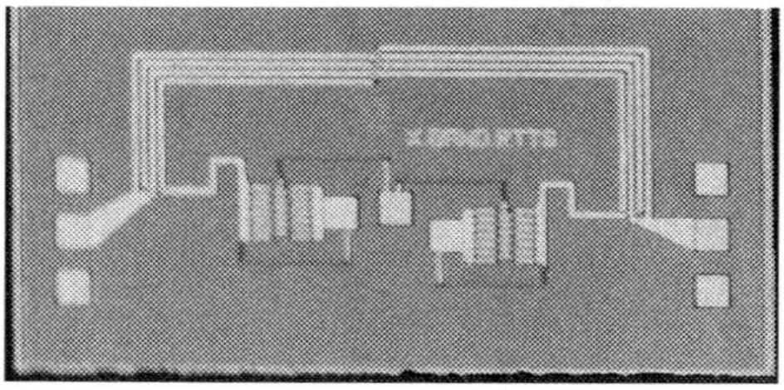

Figure 9.6 Microphotograph of the MMIC analogue delay line

The measured relative phase shift performance for the delay line, and its associated tuning characteristic, are shown in Figure 9.7*a*. The corresponding group delay performance, and its associated variation in group delay, are shown in Figure 9.7*b*. The calculated variation in group delay for an ideal variable delay line has been superimposed onto the measured results. Again, a very close agreement is demonstrated. The variation in group delay for the phase shifter operating in the *minimum group delay variation* condition has also been included for comparison. The mean level of insertion loss is 1.6 dB, with an RMS amplitude error of ±0.6 dB maintained at all levels of relative phase shift and at all frequencies. The worst-case return loss decreases almost linearly from 28 dB at centre frequency to 18 dB at the band edges.

9.2.4 Cascaded-match reflection-type phase shifters

Single-stage RTPSs have either a narrow bandwidth, with a good phase error performance at all phase states, or a wide bandwidth, with a poor phase error performance at most phase states. In addition, because single-stage RTPSs have an inherent variation in the their group delay response, they technically only approximate to being true phase shifters.

A novel method for implementing a true wideband analogue phase shifter, with a low phase error performance and constant group delay, is to combine the simplicity of Hardin's approach with the type of rippled frequency response produced by Henoch and Ulriksson. Here, the phase shifter is split into two matched cascaded stages. The topology of an ideal cascaded-match reflection-type phase shifter (CMRTPS) [22] is shown in Figure 9.8. Each stage is similar to the one proposed by Hardin, except that the resonant frequency of the series-tuned reflection termination of the second stage is always much greater than that of the first stage. With ideal 3 dB quadrature coupling maintained across the desired bandwidth, the relative phase shift would simply be the superposition of the relative phase shifts in the reflection coefficients produced by the reflection terminations of both stages. In practice, the non-ideal effects of the directional couplers also have to be corrected for.

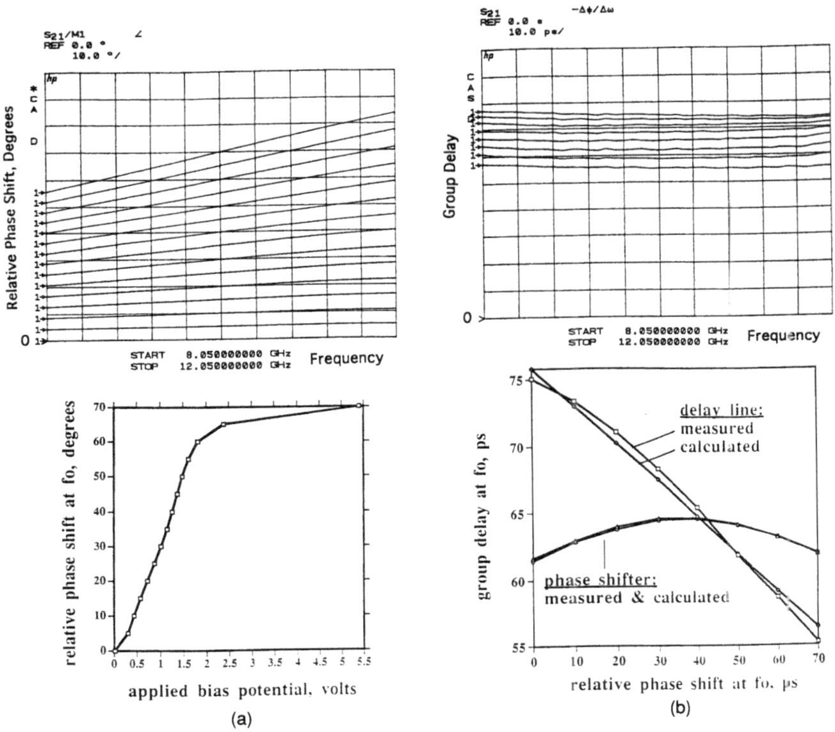

Figure 9.7 Frequency responses of the MMIC delay line: (a) relative phase shift and (b) group delay

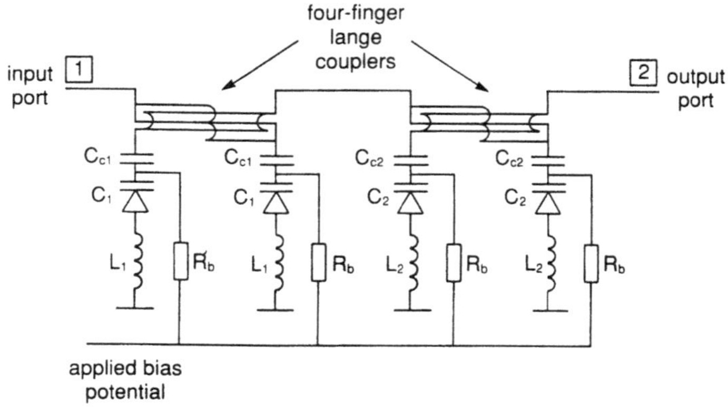

Figure 9.8 Topology of the cascaded-match reflection-type phase shifter

With the CMRTPS, the non-linear relative phase shift behaviour of one stage compensates for that of the other to produce a wideband true phase shifter. In order to maintain a low phase error over more than an octave bandwidth, at all levels of relative phase shift up to 180°, it has been found that the following conditions must be satisfied by the reflection terminations [22]:

1. the resonant frequency of the second stage reflection termination must correctly scale that of the first stage, at the reference bias potential (usually zero volts)
2. the change in the resonant frequency of the second stage reflection termination must correctly track the change in that of the first stage
3. the reflection coefficient of the second stage reflection termination must have a phase angle whose gradient, with respect to frequency, correctly scales that of the first stage, at their respective frequency of resonance.

The result is a phase shifter that has the same bandwidth performance as the directional coupler or circulator employed. Simple empirical design equations for an octave bandwidth CMRTPS, to be realised with an alumina substrate and employing traditional 4-finger Lange couplers, are given below:

$$L_1 = \frac{3.9}{f_o} \text{ nH}; \quad C_1(0) = \frac{48}{f_o} \text{ pF}; \quad C_{C1} > \frac{960}{f_o} \text{ pF};$$

$$L_2 = \frac{2.55}{f_o} \text{ nH}; \quad C_2(0) = \frac{8.55}{f_o} \text{ pF}; \quad C_{C2} = \frac{13.5}{f_o} \text{ pF};$$

$$\text{and} \quad f_C = 1.087 \cdot f_o$$

where f_C = directional couplers' frequency of maximum coupling

A rigorous theoretical investigation and a simulated design example for a Ku-band hybrid MIC phase shifter has been reported [22]. The simulated results for this CMRTPS design demonstrate that a very low maximum RMS phase error and maximum RMS amplitude error of ±1.7° and ±0.15 dB, respectively, can be achieved over a full octave bandwidth and at all levels of relative phase shift up to 180°. For an MMIC implementation, component values determined from these design equations can be used as a starting point in the modelling process.

9.2.4.1 One octave bandwidth (6.5-13 GHz) design example

Since the traditional 4-finger Lange coupler is commonplace in MMICs, it was chosen for the first proof-of-concept monolithic CMRTPS design [23]. With F20 processing specifications, Jansen Microwave's LINMIC+ CAD software was used to design the coupler (courtesy of Ron Arnold at Marconi Caswell). It was found that for equal power coupling, the finger width and separation distance should be 19 μm and 11 μm, respectively. For a frequency of maximum coupling of 10.9 GHz, the total length of the coupler is 2.7 mm. In practice, the two couplers had to be folded, with a single bend in each, in order to reduce the aspect ratio so that the design would fit into its allotted space in the wafer array. Also, the length of the couplers had to be shortened by 9.3%, again because of the limited available chip space. In the first stage reflection termination, an effective zero bias junction capacitance of 4.62 pF was implemented with a 12 × 300 μm IPSVD, resulting in a total gate width of 3.6 mm. The first stage coupling capacitor, C_{c_1}, was omitted from the design. In the second stage reflection termination, an effective zero bias junction capacitance of 0.77 pF was implemented using a 4 × 150 μm IPSVD, resulting in a total gate width of 0.6 mm. The second stage coupling capacitance of 1.5 pF was realised using 55 × 55 μm² silicon nitride MIM capacitors. All inductors were realised using 20 μm wide microstrip lines with a characteristic impedance of 100 Ω at 10 GHz. High value mesa etched resistors (~ 6 kΩ) were used to prevent RF leakage and as forward bias current limiters. A microphotograph of the monolithic CMRTPS, with its maximum dimensions of 3.0 × 2.0 mm² is shown in Figure 9.9.

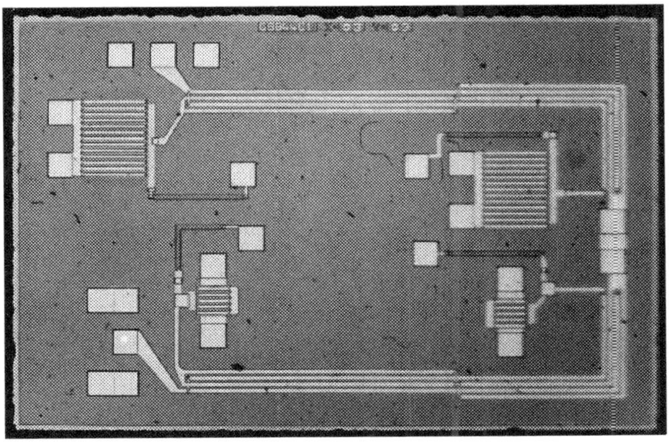

Figure 9.9 Microphotograph of the one octave bandwidth monolithic CMRTPS

Figure 9.10*a* shows the relative phase shift frequency responses. The tuning curve (also known as the phase modulation (PM) characteristic) for this MMIC, at the centre frequency, is given in Figure 9.10*b*. A maximum relative phase shift of 90° is achieved with a total capacitance ratio of $C(0)/C(-7.5\ \text{V}) = 4.4$, using the F20 process. If forward bias is applied to 0.7 V or reverse bias increased from 7.5 V to 17 V, there is only a 3.5° increase in the level of maximum relative phase shift. Therefore, the absolute maximum relative phase shift that can be introduced with this experimental phase shifter design is ~ 95°. Figure 9.10*c* shows the corresponding phase error performance, with the reference taken at the centre frequency. A maximum RMS phase error of ±2.6° is maintained across the octave bandwidth and for all levels of relative phase shift. It can be seen that if the required bandwidth and/or the maximum relative phase shift level is reduced, this phase error will be significantly reduced.

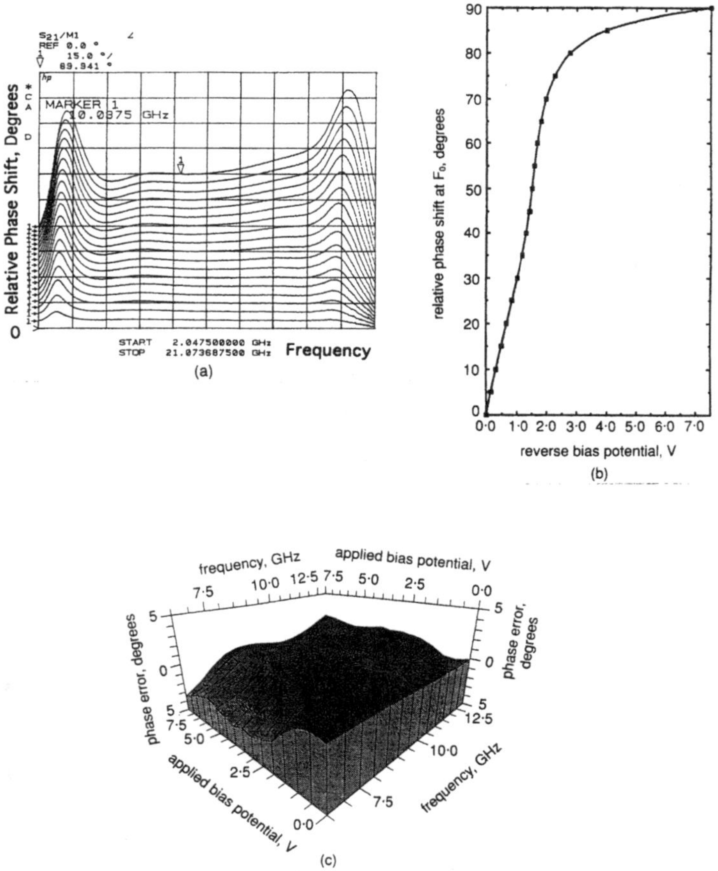

Figure 9.10 Measured relative phase shift responses: (a) frequency responses, (b) tuning curve and (c) phase error performance

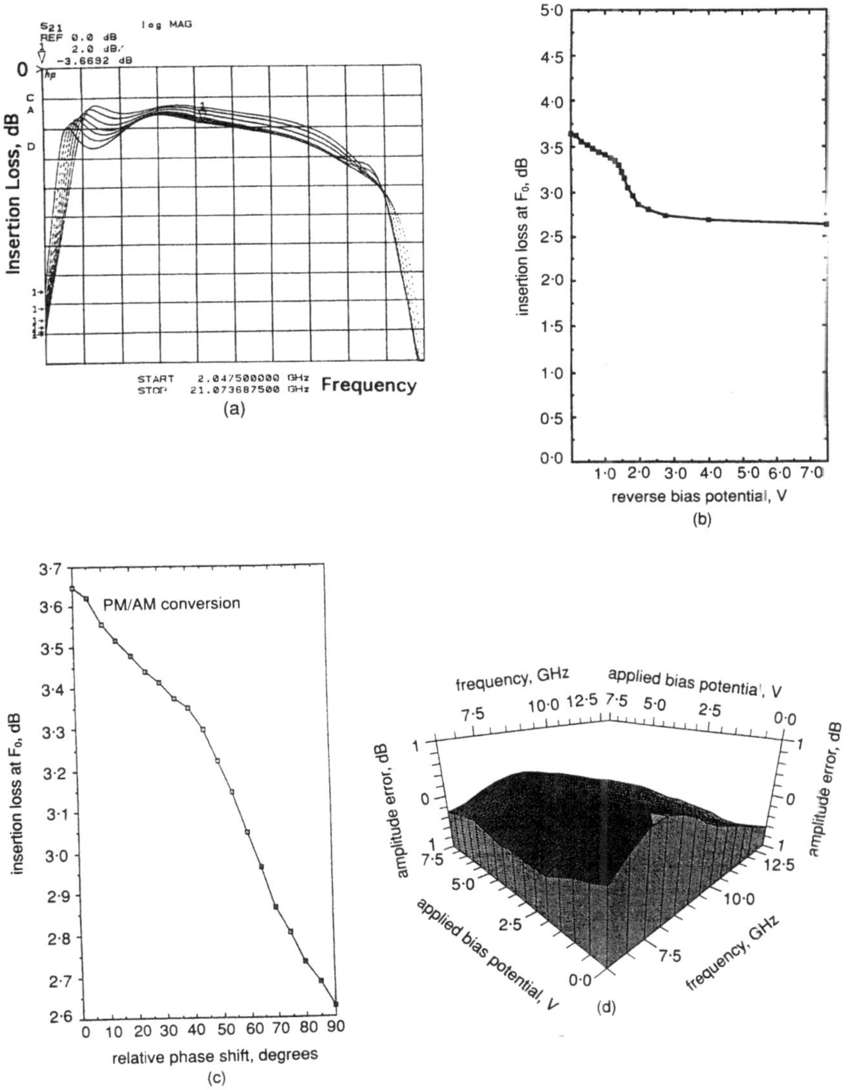

Figure 9.11 Measured insertion loss response: (a) frequency responses, (b) bias dependency, (c) PM/AM conversion and (d) amplitude error performance

Figure 9.11*a* shows the insertion loss frequency responses. The 3.14 dB level of mean insertion loss is attributed to both the couplers and IPSVDs. The fingers on the Lange coupler lie on a lossy polyimide interlayer dielectric. Therefore, a significant reduction in the insertion loss would be achieved if this lossy dielectric layer was removed. As for the IPSVDs, their effective *Q*-factors can be increased by changing their topographies [15]. The bias dependency of insertion loss (also known as the amplitude modulation (AM) characteristic) for this MMIC, at the centre frequency, is shown in Figure 9.11*b*. It can be seen that insertion loss decreases with an increase in the reverse bias potential. This is contrary to what would normally be expected, since the series resistance of F20 IPSVDs increases with an increase in the reverse bias potential [13,14]. However, at most frequencies, the overall loss of the reflection termination decreases with an increase in the reverse bias potential, due to a greater increase in the corresponding reactance of the junction capacitance; i.e. the *Q*-factor of the IPSVD increases with an increase in the reverse bias potential. Figure 9.11*c* shows the corresponding PM/AM conversion. The amplitude error performance is shown in Figure 9.11*d* (with the reference taken at the centre frequency). An RMS amplitude error of ±0.45 dB is maintained across the octave bandwidth and for all levels of relative phase shift.

The input return loss performance is shown in Figure 9.12. A good input match is achieved, with a corresponding return loss of approximately 20 dB across most of the octave bandwidth. This level of input match is very difficult to achieve with digital switched-filter phase shifters. The performance degrades towards the lower frequency end, because of the shortened length of the couplers. Direct 2-port *S*-parameter measurements on a separate test coupler indicated an equal power coupling of 3.5 dB at the frequency of maximum coupling of 12 GHz (which also indicates a 0.5 dB coupler insertion loss). Therefore, with the correct coupler length, the return loss would be 20 dB across the whole octave bandwidth. The group delay was confined within the range of 115 ± 13 ps, at all bias levels, within the octave bandwidth.

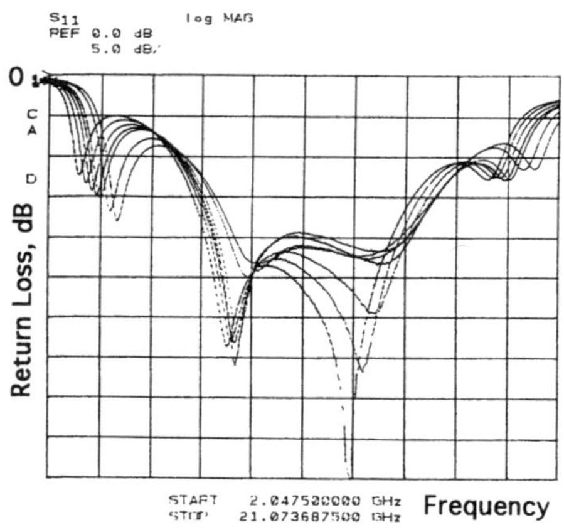

Figure 9.12 Measured input return loss frequency responses

9.2.4.2 Two octave bandwidth (4.4-16.1 GHz) design example

It can be seen from the frequency responses of the previous octave bandwidth design example that the potential bandwidth could far exceed that of one octave. With reference to Figure 9.10*a*, the relative phase shift responses have a positive gradient above 11.5 GHz. This gradient increases with an increasing reverse bias potential. This type of behaviour indicates that the resonant frequency tracking of the second stage reflection termination could be improved. With a positive gradient, the amount of coupling in the reflection termination of the second stage should be reduced with an appropriate reduction in the value of the second stage coupling capacitance. In addition, it has been found that a multilayer coupler can have a similar octave bandwidth performance to that of a 4-finger Lange coupler [24]. The multilayer coupler has the advantage of taking up less than half the chip space, while making bends much easier to implement compared with the Lange coupler. As a result, a more compact two octave bandwidth CMRTPS was designed [25] employing this experimental coupler. It was found that a 33% reduction in the second stage coupling capacitance and a 50% reduction in the values of all inductances were required in order to achieve a two-octave bandwidth performance, when compared with the first MMIC design. The F20 foundry process was used to fabricate the MMIC. A microphotograph of the MMIC, which has dimensions of 2.75 × 2.0 mm^2, is shown in Figure 9.13. It can be seen that there is ample scope for increasing the level of integration further.

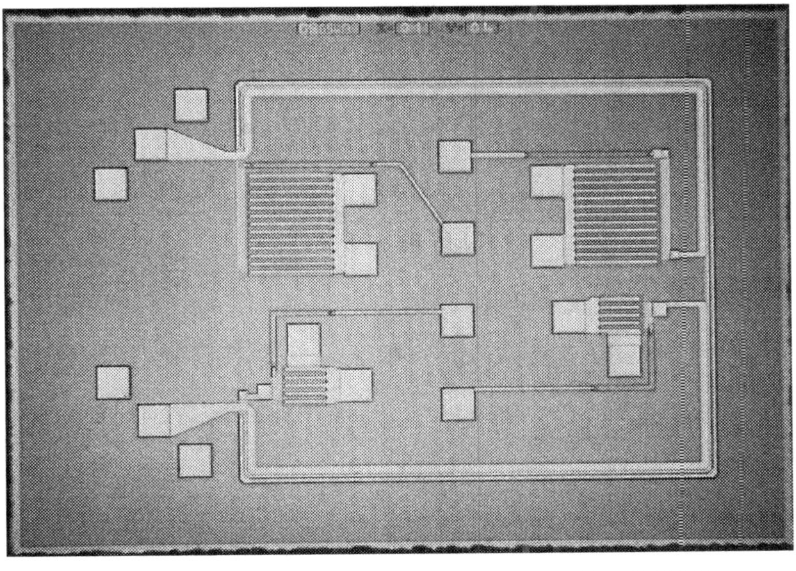

Figure 9.13 Microphotograph of the two octave bandwidth monolithic CMRTPS

The measured performance of this second MMIC phase shifter design is shown in Figure 9.14. A maximum relative phase shift of 98° was achieved with a capacitance ratio of $C(0)/C(10\ \text{V}) \sim 4.6$, using the F20 process. The maximum RMS phase error is only ±2.8° across the 4.4 to 16.1 GHz frequency range and at all bias levels. The mean insertion loss is 7 dB at the centre frequency of 10 GHz.

This is more than double that found with the first MMIC design. This is because of the higher insertion loss of the multilayer coupler, due to the coupled electric field being confined inside the lossy polyimide interlayer dielectric. The measured input return loss is better than 9 dB across the 4.8 to 16.1 GHz frequency range and at all bias levels. This modest input match performance is thought to be attributed to the poor experimental coupler designs. A very flat group delay response of 100 ± 4 ps, in the 9.8 to 16.1 GHz frequency range, was also achieved.

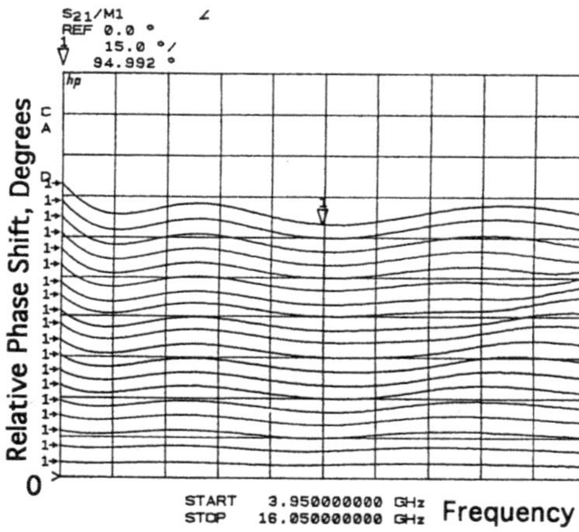

Figure 9.14 Measured relative phase shift responses of the two octave bandwidth CMRTPS

9.2.5 Ultra-wide bandwidth analogue phase shifters

9.2.5.1 Decade bandwidth (1-19 GHz) hybrid-MIC design example

The first realisation of a decade bandwidth 90° CMRTPS is described in this section [26]. Here, a 2 to 18 GHz 3 dB quadrature directional coupler was employed. As demonstrated with the previous two MMIC design examples, the potential bandwidth of the CMRTPS can increase as the series inductance in the reflection terminations decreases. Therefore, a decade bandwidth CMRTPS should have negligible series inductance. It should be noted that a reduction in the series inductance values can also reduce the maximum relative phase shift level. However, if abrupt junction IPSVDs are employed then they will be the limiting factor on the maximum relative phase shift level that can be achieved.

With a normal hybrid-MIC realisation, significant series inductance and stray parasitics are inherent. To effectively eliminate this inductance and minimise stray parasitic capacitances, monolithic technology was used to realise the reflection terminations. The MMICs sit on very low inductance chip mounts and are connected to the coupler's ports by a multitude of short bond wires. Monolithic technology also greatly simplifies the fabrication of the phase shifter.

A 6 × 150 μm IPSVD was previously characterised to a high degree of accuracy [13,14]. The resulting equivalent circuit model was used in the design of the IPSVDs for the reflection terminations of both stages. The single IPSVD in the first stage reflection termination was implemented with a 18 × 300 μm MESFET. For the second stage reflection termination, two IPSVDs were connected in series, both being implemented with 6 × 150 μm MESFETs. The coupling capacitances for the first and second stage reflection terminations are 22 pF and 5 pF, respectively. A photograph of the complete phase shifter realisation, with dimensions of 5.1 × 2.5 cm^2, is shown in Figure 9.15.

Figure 9.15 Photograph of the decade bandwidth hybrid-MIC CMRTPS

The relative phase shift performance of the decade bandwidth design is shown in Figure 9.16*a*. It can be seen that an overall flat response is achieved between 1 and 19 GHz. Since the transmission phase is highly sensitive to impedance discontinuities, the ripples present in the relative phase shift measurements can be attributed to the couplers' 'wiggly lines' (used for phase velocity equalisation) and the production grade wedge-shaped launchers (used for the coax-to-microstrip transitions); no ripples were present in the simulations. The use of a 1-18 GHz stripline 3 dB quadrature directional coupler, such as the one manufactured by Krytar (Model 1831) [27], is ideal for ultra-wideband phase shifter applications, as they have no significant impedance discontinuities. The insertion loss performance at zero bias is shown in Figure 9.16*b*. The measured input return loss is better than 10 dB over 87% of the 1 to 19 GHz frequency range. An overall flat group delay response of 2.00 ± 0.37 ns was also obtained across the decade bandwidth.

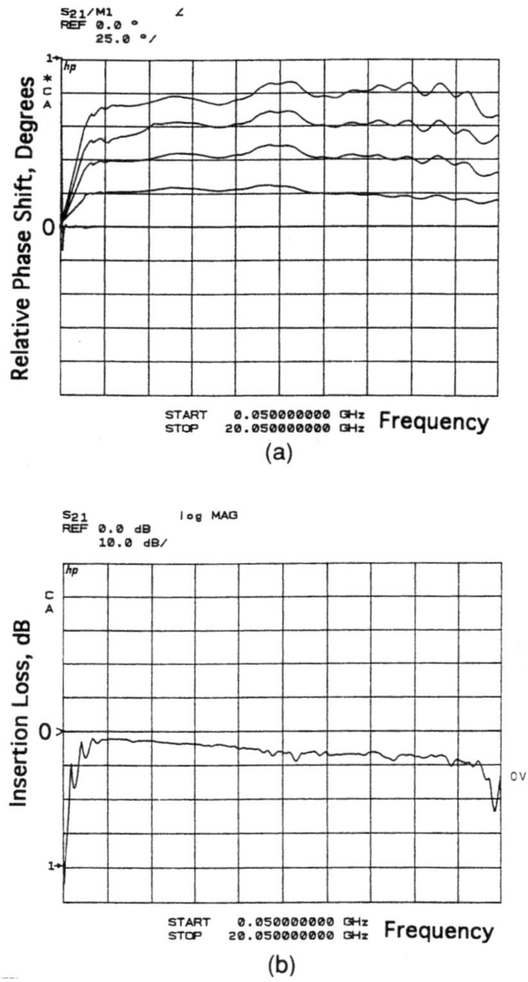

Figure 9.16 Measured performance of the bandwidth hybrid-MIC CMRTPS: (a) relative phase shift and (b) insertion loss

9.2.5.2 Decade bandwidth (0.5-5.5 GHz) MMIC design example

For the past two decades, most RTPSs have employed directional couplers. Unfortunately, no practical decade bandwidth 3 dB quadrature coupler has yet been reported that is spatially compatible with monolithic microwave technology. The design and simulated performance for a monolithic two-stage CMRTPS, which employs decade bandwidth active circulators, has been previously described [28]. A simplified representation of the decade bandwidth CMRTPS is illustrated in Figure 9.17. The design of the active circulator is described in Chapter 11. It has been found that the potential bandwidth of a high performance CMRTPS increases as the simplicity of its reflection terminations increases. Also, it has been seen with the previous decade bandwidth CMRTPS design that the series inductances must be omitted from the reflection terminations. The IPSVDs selected for the first and second stage reflection terminations are implemented with 48 × 150 μm and 6 × 150 μm MESFETs, respectively. The coupling capacitances for the first and second stage reflection terminations are 19.5 pF and 3.3 pF, respectively. From the initial mask layouts, the reflection termination of the first stage requires approximately half the area of that required by the active circulator. Therefore, one of the main advantages of using active circulators in preference to directional couplers is that only one reflection termination is required per stage, thus saving a considerable amount of expensive chip space. With careful design, the complete MMIC could fit within an active area of only 1.5 × 1.5 mm^2.

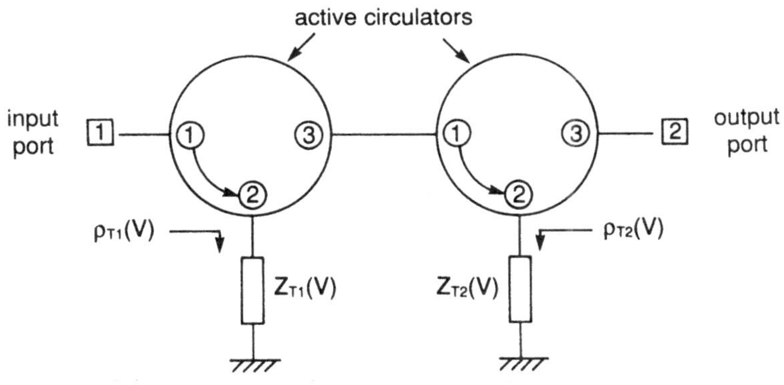

Figure 9.17 Topology of the decade bandwidth monolithic CMRTPS

Analytical closed-form expressions for this type of CMRTPS implementation have been published [28]. The relative phase shift responses for the 0.5–5.5 GHz design are shown in Figure 9.18*a*. A maximum RMS phase error of ±1.5° is achieved for all relative phase shift levels up to 85°, over a full decade bandwidth. Figure 9.18*b* shows the insertion loss responses of the phase shifter. The mean level of insertion loss is 15 dB, with an RMS amplitude error of ±1.8 dB across the full decade bandwidth. It was found that since the IPSVD of the second stage is physically much smaller than that of the first stage, the variations in the *Q*-factor of the second stage IPSVD is much greater than that of the first stage. As a result, the high level of amplitude modulation is attributed to the second stage IPSVD. The input return loss responses of the phase shifter are shown in Figure 9.18*c*. It can be clearly seen that an excellent input impedance match is achieved across the whole decade bandwidth. This performance is attributed to the high return loss and high isolation characteristics of the active circulators [29].

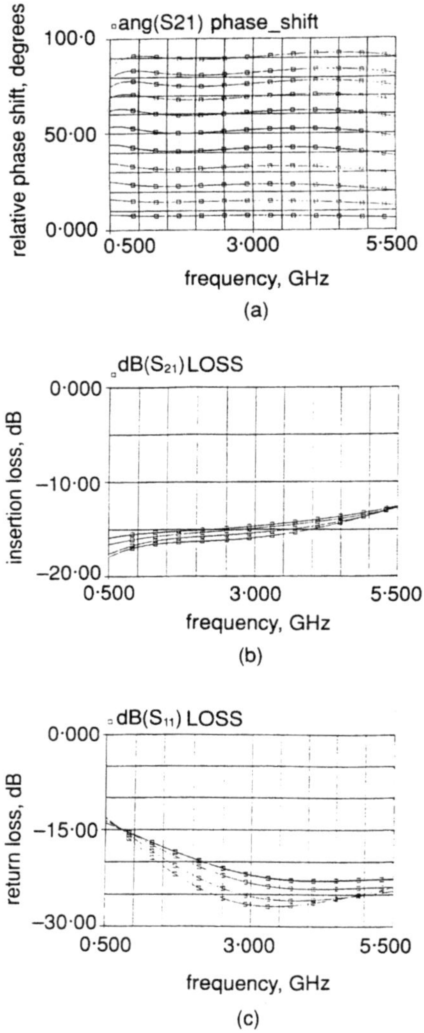

Figure 9.18 Simulated performance of the decade bandwidth monolithic CMRTPS: (a) relative phase shift, (b) insertion loss and (c) input return loss

In conclusion, when compared with directional couplers, active circulators provide the following advantages:

1. they can have non-dispersive characteristics
2. ultra-wide bandwidths can be achieved
3. only a single reflection termination is required per stage
4. small chip size.

However, the complete phase shifter design using active circulators also has the following disadvantages (which can, in nearly all cases, be overcome):

1. it is non-reciprocal, which is inherent
2. excessive total gate width of ~10 mm (reducing the yield)
3. DC power limit on the common source bias resistors (ultra-wideband active loads can be employed [30])
4. RF power limit (further investigation is required)
5. low upper frequency limit (consider using quasi-circulators [31] or unilateral circuits [32])
6. poor noise figure (replace MESFETs with HEMTs)
7. excessive insertion loss (replace MESFETs with HEMTs and/or change topology).

9.2.6 Ultra-wide bandwidth distributed analogue delay lines

The loaded-line phase shifter was first introduced by Dawirs and Swarner [33]. A transmission line is loaded with shunt capacitive reactance elements to ground. Also, the induced phase shift per section must be kept small (i.e. < 60°), in order to achieve acceptable return losses. For these reasons, a distributed approach should be used, whereby a transmission line is periodically loaded. Here, the Bragg frequency must be greater than the highest frequency of operation. Unfortunately, this type of phase shifter is considered inefficient, as considerable chip space would be required to realise a 360° design.

A low loss MMIC analogue delay line was recently reported by Nagra and York [34], based on the distributed approach. Here, Schottky junction varactors were designed to have minimal physical, and therefore, electrical sizes. The advantage of this approach over that of the reflection topology is that the useful frequency range extends down to DC. Unfortunately, such loaded-lines should ideally be implemented in a straight line, especially with guided-wave structures that are susceptible to unwanted moding at discontinuities. As a result, a CPW loaded-line may have to be as long as 20 mm in length in order to achieve a maximum relative phase shift of 360° at 20 GHz [34]. For phased array applications, which require a relative phase shift of 360° at the lowest frequency of operation, this would lead to impractical chip dimensions. Moreover, bespoke Schottky junction varactor diodes could add considerable cost when compared to FET-based varactor diodes within standard foundry processes. Finally, group delay performance may be very poor, unless designed carefully.

Another approach is to replace the Schottky junction varactor diodes with MIM capacitors that have been implemented with barium strontium titanate ($BaSrTiO_3$) insulators. This ferroelectric film has an electric field tunable dielectric constant. A analogue monolithic distributed loaded-line was realised by Erker *et al.* [35]. Here, a maximum relative phase shift of 90° (155°) was achieved at 20 GHz (30 GHz), using a 3 mm length of CPW loaded-line.

9.2.7 Millimetre-wave implementations

At millimetric frequencies, much has been reported on monolithic mixers using Schottky barrier diodes and hybrid frequency multipliers using *p-n* junction varactor diodes. However, little has been reported on millimetre-wave monolithic analogue phase shifters. One exception to this is the work by Pao *et al.* [36], demonstrating a 60 GHz analogue phase shifter realised using microstrip technology. Here, interdigitated *p-n* junction varactor diodes were employed in two single-bit branch-line phase shifters connected in cascade. The results demonstrated a worst-case RMS phase error of ±21.0°, for a maximum relative phase shift of 160°, across a 59 to 61 GHz frequency range. Also, within this frequency range the mean level of insertion loss was 6 dB with a worst-case RMS amplitude error of ±1.8 dB. No results were given for the return loss performances.

An example of a two-stage RTPS, where the reflection terminations in one stage are different from those of the other has been reported [37]. Here, a compromise is made between the optimal relative phase shift that can be achieved with two identical RTPSs in cascade and the optimal bandwidth that can be achieved with an ideal CMRTPS. Using pHEMT-based varactor diodes, from the Marconi Caswell H40 process, a 180° relative phase shift level was achieved over a frequency range of around 30 GHz to 40 GHz [37]

Precision analogue phase shifters, for which the worst-case RMS phase error is less than approximately ±1.0°, are very difficult to realise at millimetric frequencies, particularly with a monolithic implementation. This is because the insertion phase linearity is very sensitive to dispersion, unwanted modes of propagation and unwanted parasitic effects that can easily occur at these frequencies. A precision millimetre-wave monolithic analogue CMRTPS has been reported [38], which uses uniplanar technology to achieve the desired linear insertion phase responses. Here, a multilayer CPW 3 dB quadrature directional coupler is employed for this MMIC phase shifter implementation. Microstrip IPSVDs have been accurately characterised well into the millimetric frequency range [13–15]. However, shunt IPSVDs are inherently more compatible with CPW technology, since the cathode (drain/source) electrodes make direct electrical contact with the upper ground plane. Moreover, because the cathode parasitic capacitance is greatly reduced in this CPW application, the total capacitance ratio of the varactor diode increases. Therefore, the maximum relative phase shift level can increase from 90° to 120°, when compared with an equivalent microstrip implementation of a monolithic CMRTPS. The CPW varactor diodes were realised using the F20 MESFET technology. A photograph of the monolithic phase shifter, which has a total active area of 1.4 × 1.0 mm^2, is shown in Figure 9.19.

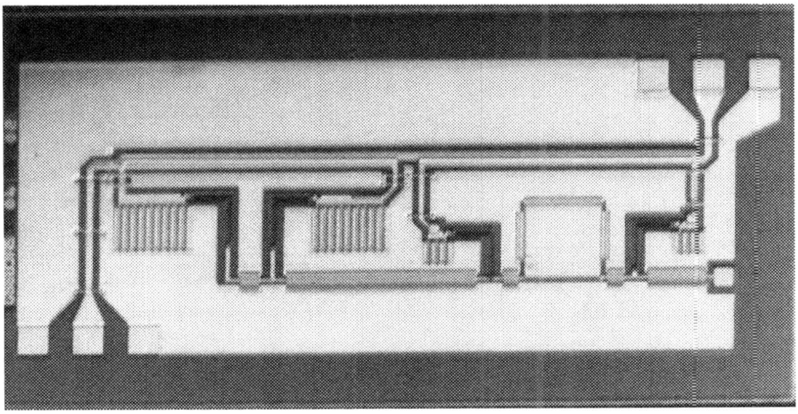

Figure 9.19 Microphotograph of a millimetre-wave analogue phase shifter

The measured relative phase shift performance and corresponding tuning curve, for a centre frequency f_0 = 31.5 GHz, are shown in Figures 9.20*a* and 9.20*b*, respectively. A worst-case RMS phase error of ±1.0° is achieved for all levels of relative phase shift up to 120°, across the 31 to 32 GHz frequency range. It can be seen that linear tuning is almost achieved. The measured insertion loss performance and corresponding PM/AM conversion characteristic are shown in Figures 9.20*c* and 9.20*d*, respectively. The mean level of insertion loss is 16 dB and the worst-case RMS amplitude error is ±1.0 dB. The worst-case input return loss for the phase shifter is 15 dB. These measured results demonstrate the potential of this novel uniplanar millimetre-wave analogue phase shifter. When compared with the results obtained by Pao *et al.*, a dramatic reduction was achieved in the levels of worst-case RMS phase and amplitude errors, across the same fractional bandwidths.

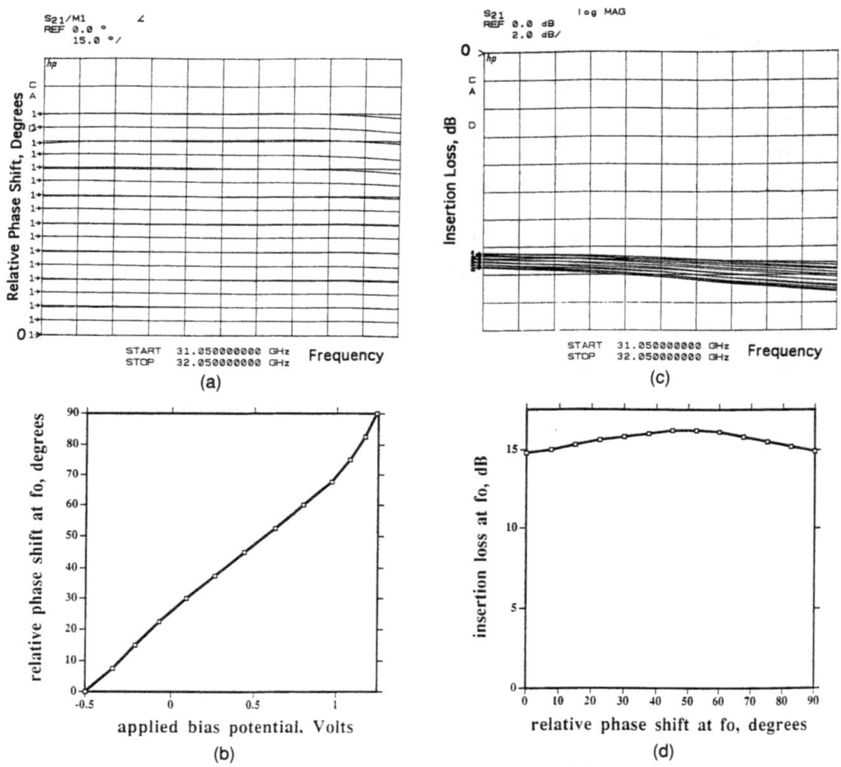

Figure 9.20 Measured performance of the millimetre-wave CMRTPS: (a) relative phase shift, (b) tuning curve, (c) insertion loss; and (d) PM/AM conversion

9.2.8 Dual-gate MESFET

A dual-gate MESFET can be used to implement an analogue phase shifter [39,40]. The input RF signal is applied to one of the gate electrodes (G1) and a shunt inductance is connected to the other gate electrode (G2). A variable gate bias is then applied to G2. This will result in a voltage-controlled tuned circuit at G2, since the inductor is in parallel with the voltage-dependent gate-source capacitance associated with G2. The output RF signal is taken from the drain. This type of active phase shifter can achieve a maximum relative phase shift of approximately 120° and has the obvious advantage of being very compact. However, it has a relatively narrow bandwidth performance and can have poor PM/AM conversion characteristics with large values of relative phase shift.

9.2.9 Phase splitter-power combiner

9.2.9.1 Passive design

Another method of implementing an analogue phase shifter is to combine a fixed phase-splitting network with a variable power-combining network. One common example of this type of phase shifter is the I–Q vector modulator, illustrated in Figure 9.21. Here, the input signal power is equally split into two orthogonal channels. A balanced modulator (defined by Garver as a device with a voltage transmission coefficient that can vary between ±1, through zero [3]) in each channel controls the amplitudes of the two orthogonal signals. Therefore, any value of magnitude or phase angle can be achieved for the overall voltage transmission coefficient of the circuit.

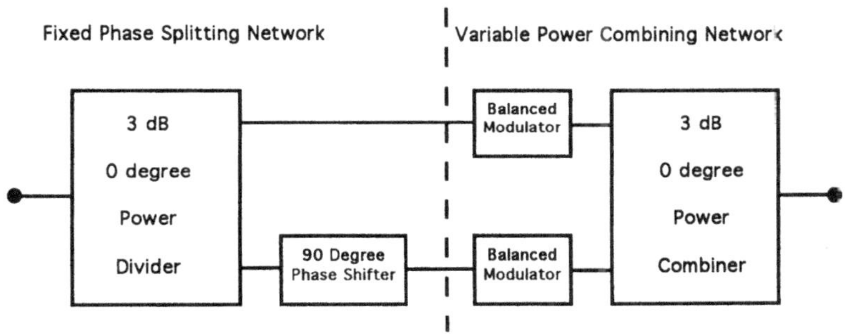

Figure 9.21 Schematic diagram of an I–Q vector modulator

A noteworthy example of a monolithic I–Q vector modulator was reported by Devlin and Minnis [41] and this subject is covered in much greater detail in Chapter 11. A Lange coupler was used at the input and a Wilkinson coupler was used at the output. The balanced modulators were implemented with a reflection-type attenuator [12].

A lumped-element phase splitter/power combiner phase shifter was reported by Frost *et al.* [42]. Here, a lumped-element Wilkinson coupler was used for both the 3 dB power divider and combiner. A 120° phase difference between the two channels was achieved by inserting a lumped-element high-pass filter in one channel (to increase its insertion phase by 60°) and an equivalent low-pass filter in the other channel (to decrease its insertion phase by 60°). A variable matched MESFET π-attenuator was used in each channel to control the amplitudes of the corresponding signals. Because variable attenuators were used instead of balanced modulators, the maximum theoretical level of relative phase shift is only 120°.

These passive designs have two drawbacks. The first is that at least two independent control lines are required. The second is that they have an inherent 6 dB minimum circular locus insertion loss. However, they can operate at higher power levels than most other types of monolithic phase shifter.

9.2.9.2 Active design

Active phase splitting/power combining phase shifter designs have also been reported [43–45]. Mazumder *et al.* [43] employed an identical 4-way power divider and combiner, consisting of three Wilkinson 0° couplers in each. The phase splitting network used four different lengths of transmission lines to produce the appropriate amount of phase offset in the four channels. A dual-gate MESFET was then assigned to each channel in the variable power combiner, to control the amplitudes of the four orthogonal signals and to overcome the inherent 12 dB minimum circular locus insertion loss. Salvage *et al.* [44] employed two all-pass networks and two differential amplifiers to realise a much smaller active 4-way orthogonal phase splitter. As with the previous example, the active variable power combiner employed dual-gate MESFETs. The concept of weighting and combining the output signals from a phase splitting network in an architectural form was proposed by Magarshack [45]. Here, an arrangement of standard building blocks are used, consisting of active phase splitters and active variable power combiners.

There are a number of drawbacks with these active designs. For example, at least four control lines are required and the circuit can be rather complicated, especially when differential amplifiers and dual-gate MESFETs are employed. Also, unwanted parasitics will limit the bandwidth and the maximum frequency of operation, and the noise figure and power handling performances may be poor.

9.3 Digital implementations

The beam-forming network of an adaptive phased-array antenna is one very important application of a variable phase shifter. Since phased arrays are often made up of many elements, it is advantageous to reduce the size, weight and cost of the electronics supporting each element. Size and weight constraints are increasingly challenging for airborne and space flight systems, especially when high performances are expected.

In the design of multi-bit digital phase shifters (where an n-bit phase shifter provides $M=2^n$ discrete phase states), the primary goal is to meet the required relative (or differential) phase shift for each discrete phase state, across the desired frequency range. Ideally, the insertion loss through the phase shifter should be invariant for all possible phase states. This invariant performance makes their application easier, as the insertion loss setting in other parts of the chain need not be adjusted. In addition, good input and output return losses are required in order to avoid any ripples in the insertion phase and insertion loss responses when they are cascaded with other components such as amplifiers or attenuators. Moreover, good impedance matches are required for each individual phase shifter stage (or bit) in order to minimise inter-stage reflections. Other requirements imposed from overall system considerations are temperature stability and power handling. Not all of these requirements can be simultaneously fulfilled, and a number of trade-offs are involved in the design of state-of-the-art MMIC phase shifters.

There are four main types of digital phase shifter: switched-line, reflection-type, loaded-line and switched-filter. In addition, there are a number of intrinsic 1-bit digital phase shifters. This section reviews the various techniques for the design of digital phase shifters and their monolithic implementation.

9.3.1 Switched-line

A simplified switched-line phase shifter is illustrated schematically in Figure 9.22. In the switched-line approach, PIN diodes or switching-MESFETs [46–49] are employed to route the input RF signal into the appropriate length of matched transmission line. Ideally, the amount of induced group delay is directly proportional to the difference in the electrical lengths of the selected line and the reference line. This approach actually yields a time shifter response in the ideal case, as discussed in Section 9.1, and is more accurately described as a switched-line variable delay line.

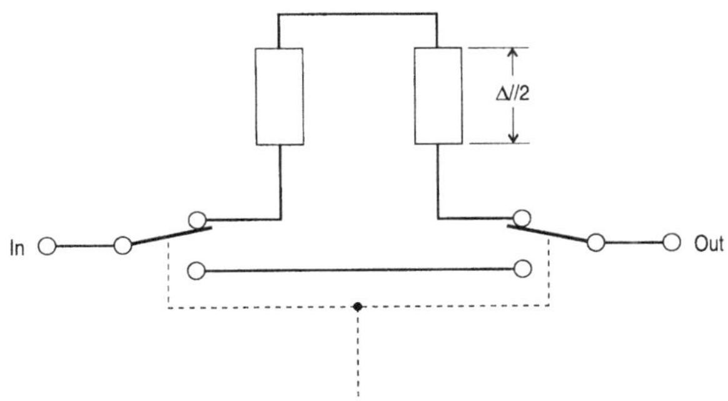

Figure 9.22 Schematic diagram of a switched-line delay line

With this purely digital implementation a wide bandwidth can be achieved, but the size of the switched-line implementation is dependent on the lowest frequency of operation and the number of bits. As a result, it may only be cost effective to have a 1 or 2-bit implementation for applications below a few gigahertz, whereas a 5- or 6-bit implementation is not uncommon for millimetre-wave applications. A number of conventional MMIC switched-line implementations have been reported over the past decade [50–55]. However, the most notable example of a state-of-the-art switched-line delay line was reported by Rogeaux *et al.* [56].

The satisfactory operation of a single-pole double-throw (SPDT) switch is essential for the performance of the variable delay line. A resonating inductor is usually placed between the drain and source terminals of the switching-FET, in order to tune-out the off-state drain-source capacitance. This technique was first demonstrated by Dunn [52]. Here, the on-state insertion loss of the FET switch was less than 2 dB and the off-state isolation was better than 18 dB, across a frequency range of 42 to 46 GHz. The 4-bit phase shifter was designed and fabricated using a 0.6 μm MESFET process on a 90 μm thick GaAs substrate. The chip size was 2.5 × 1.3 mm^2. The measured performance of all the 16 phase states showed that at the centre frequency, the worst-case insertion phase deviation was 3.3° from the desired value. The insertion loss for all states was within 1.5 dB over the 42 to 46 GHz frequency range.

Switched-line phase shifter techniques have also been used at Ka-band frequencies by Maruhashi [57] and others. They used unresonated switches based on ohmic electrode sharing technology (OEST) [57]. In the OEST scheme the parasitic capacitance between the series FET and shunt FET is reduced by using a common electrode. This results in broadband characteristic up to millimetre waves without the use of inductors. The authors demonstrated a 4-bit phase shifter from 33 to 36 GHz with 7° RMS phase deviation over the frequency band; an average insertion loss of 15 dB with 0.9 dB RMS variation was obtained for all the 16 states. The input and output VSWRs were better than 1.5 and 2.0, respectively, at 34.5 GHz.

9.3.2 Reflection-type

9.3.2.1 Delay lines

Figure 9.23 shows three variations of an ideal digital reflection-type delay line. The positions of the short circuits control the change of insertion phase through the circuit. The short circuit positions are controlled by at least one switch, as shown. The first configuration, shown in Figure 9.23*a*, employs a circulator; the input signal is circulated and routed down the reflection termination's transmission line to the short circuit end, which then reflects the signal back into the circulator. This reflected signal is then circulated to the output port. As the signal travels twice along the reflection termination's transmission line, the differential line length required for a given phase shift is only half that for an equivalent switched-line implementation. A 3 dB quadrature directional coupler is another signal separation device that can be used, as illustrated in Figure 9.23*b*, for a 1-bit case (i.e. 2 states). In this case, two identical switched transmission line reflection terminations are required. Nakahara *et al.* demonstrated this approach in monolithic form [58].

Another variation of the reflection-type delay line is illustrated in Figure 9.23*c*, for a 2-bit case (i.e. 4 states). Here, a single-stage 3 dB quadrature directional coupler is employed. Identical tapped transmission lines terminate the coupled and direct ports of the directional coupler. A number of equally spaced shunt switches can connect the transmission line to ground, but only one switch per reflection termination is in the on-state at any time. Ideally, the amount of induced group delay is directly proportional to the distance along the line from the on-state switch to the reference off-state switch. Although this provides a wide bandwidth and requires only minimal chip space, there is a practical limit on the number of possible phase states at higher microwave frequencies because of the physical size of the switches. Also, where power is at a premium, as in the case of satellite or portable applications, considerable control power would be required for large aperture arrays if monolithic PIN diode switches are employed [48]. The reflection-type delay line has been demonstrated in monolithic form by Wilson *et al.* [18]. Here, a 4-bit X-band phase shifter was designed, fabricated and tested on a single chip with dimensions of 3.7×2.3 mm^2. An important feature in their work was that Schottky diode structures were used as switching elements rather than MESFETs. These were fabricated using selective ion implantation techniques, compatible with MMIC processing.

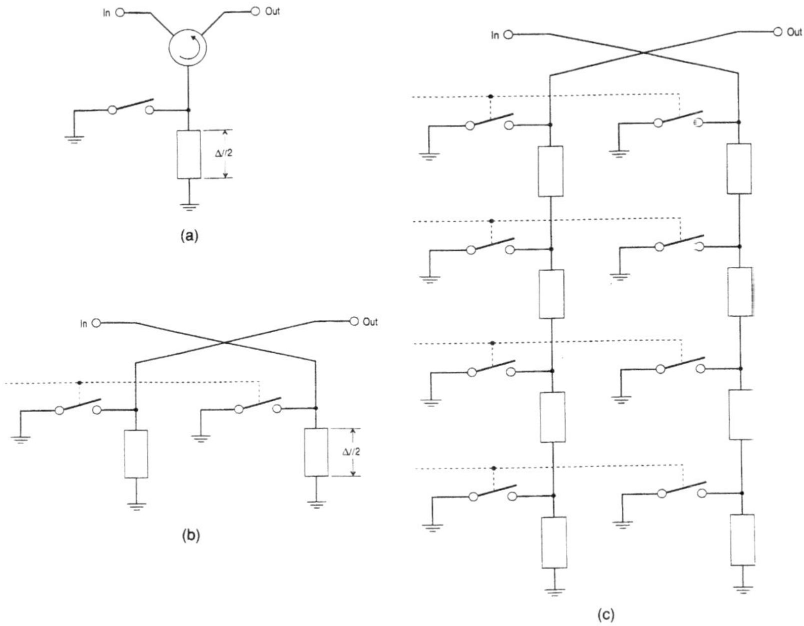

Figure 9.23 Digital reflection-type delay lines: (a) 1-bit employing an active circulator, (b) 1-bit employing a 3 dB quadrature directional coupler and (c) 5-state employing a 3 dB quadrature directional coupler

9.3.2.2 Phase shifters

Two examples of digital reflection-type phase shifters are illustrated in Figure 9.24. With reference to Figure 9.24*a*, when the switch state is changed, the ideal device can have an intrinsic relative phase shift of 180°. In practice, because of the non-idealities of the switches, an impedance-matching transformer is inserted between the switch and the directional coupler to minimise the amplitude modulation [59]. In addition, arbitrary values of relative phase shift are possible by suitable modification of these matching circuits. This principle was demonstrated in monolithic form at C-band frequencies by Andricos *et al* [60]. Here, a 6-bit phase shifter was implemented with five digital bits and one analogue bit on a single chip.

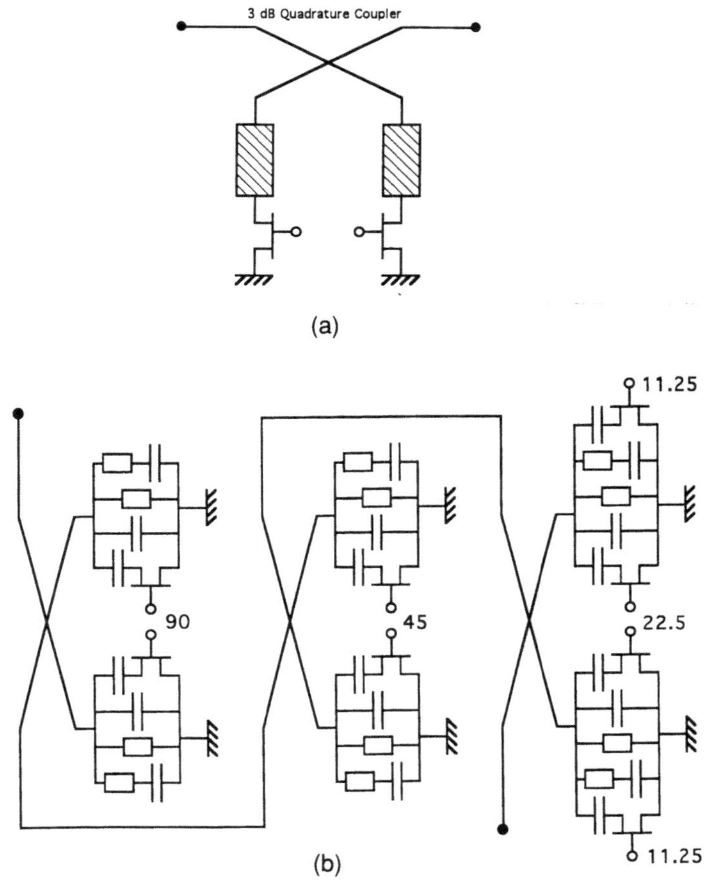

Figure 9.24 Digital reflection-type phase shifters: (a) with impedance-transformed switches and (b) with switchable load impedances

The 180° and 90° bits used the reflection-type arrangement, employing Lange couplers. Switching-FET sizes and circuit parameters were optimised for best input/output return losses and minimal insertion loss. The 90° bit had a single transmission line section between the Lange coupler and the switching-FET, while the 180° bit required additional open-circuit stubs for impedance matching. Over a 5 to 6 GHz frequency range, the 180° and 90° bits had a maximum insertion loss of 2 dB, a maximum phase error of ±5°, and an input/output return loss of 14 dB. The disadvantage of using reflection-type phase shifters at these frequencies is the excessive length of the Lange coupler, which in this case is almost 5 mm long.

Another example of a digital reflection-type phase shifter is illustrated in Figure 9.24*b*. Here, the reflection terminations consist of a switchable load impedance. A single-chip, 32-state implementation of this technique, covering the frequency range of 4 to 20 GHz, was demonstrated by Boire *et al.* [61].

9.3.3 Loaded-line

A digital version of the loaded-line approach is illustrated in Figure 9.25. A transmission line with a characteristic impedance Z_C and an electrical length θ has two identical state-switchable load admittances, $Y(i) = G(i) + jB(i)$ (where i = 1, 2 and refers to the two bias states of the switching devices), connected in shunt on either side of the line. A change between one admittance state and the other produces a change in the insertion phase of the network.

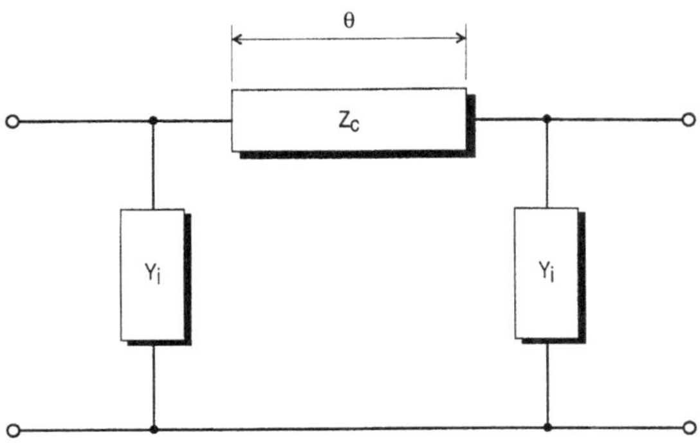

Figure 9.25 Basic loaded-line phase shifter

The design of the loaded-line phase shifter requires the appropriate choice of the six unknown parameters: θ, Z_c, $G(1)$, $G(2)$, $B(1)$ and $B(2)$. For low-loss switching devices, the ohmic losses associated with the shunt admittances are small, and $G(i)$ can be assumed to be zero as a good approximation. A common design approach, assuming an arbitrary value of transmission line length, employs CAD techniques to select a configuration that results in the required relative phase shift over a wide bandwidth with low insertion loss. Atwater [62] has shown that the choice of the electrical length is not arbitrary, but depends on the relative values of load admittance.

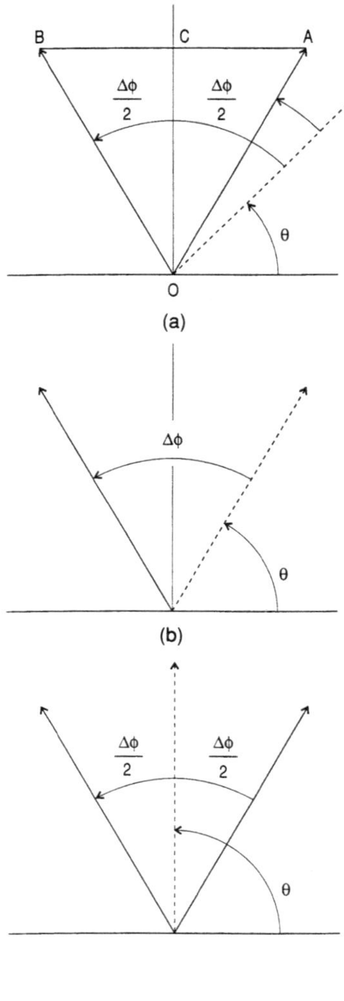

Figure 9.26 Phasor diagrams for the different loaded-line modes of operation: (a) Class I, *(b)* Class II *(unload/load) and (c)* Class III *(complex conjugate)*

A set of loading modes for the simple loaded-line phase shifter was defined by Opp and Hoffman [63] and analysed by Atwater [62]. Class I loading corresponds to the condition when $B(1)$ and $B(2)$ are non-zero and unequal. Class II corresponds to the case when $B(1)$ is zero and the insertion phase corresponds to the electrical length of the transmission line. The desired value of relative phase shift occurs when the other susceptance load, $B(2)$, is applied. This mode can also be called unload/load, as the line load is first removed and then applied. Class III loading corresponds to the condition when $B(1) = -B(2)$. In this condition, the loading admittances switch the insertion phase by half the maximum value of the differential phase shift. The loading is switched between complex conjugate admittances and a uniform output amplitude is maintained at both phase states. Figure 9.26 shows the phasor diagram for a loaded-line phase shifter operating in the different classes. Two approaches can be adopted for implementing the two states of the load admittances. In the first, ideal SPDT switches select the appropriate load susceptance, $B(1)$ or $B(2)$. In the second, a lossless impedance transformer is used to convert the terminal impedances of a non-ideal switch into the required values for $B(1)$ and $B(2)$.

The principal advantage of this loaded-line technique is that it is free from spurious resonances and it does not require a circulator or directional coupler. This technique is, however, only suitable for smaller values of relative phase shift, since it becomes more difficult to achieve a good impedance match with larger values of relative phase shift. Also, the return loss of a loaded-line phase shifter gets progressively worse as the frequency deviates from band centre, making it a narrowband device.

Because of their simplicity, loaded-line phase shifters have received a great deal of attention and have been implemented in monolithic form up to X-band frequencies [53,58–60,64,65] and at millimetre-wave frequencies by Slobodnik [66]. The C-band 6-bit phase shifter design reported by Andricos *et al.* [60] utilised the loaded-line approach for the 11.25°, 22.5° and 45° bits. Here, a quarter wavelength 50 Ω through-line with 100 Ω shunt stubs was used for the two least significant bits. The 45° bit shunt stub impedance was 56 Ω. The worst-case input/output return losses for the 11.25°, 22.5° and 45° bits were 20 dB, 15 dB and 12 dB, respectively. The maximum insertion loss was approximately 1 dB and the phase errors were ±2° over a 5 to 6 GHz frequency range. Ayasli *et al.* [65] used the loaded-line approach for the design of the 22.5° and 45° bits in an X-band 4-bit phase shifter. The switched-line approach was adopted for the 90° and 180° bits. Slobodnik *et al.* [66] also employed the loaded-line approach for the design of the 22.5° and 45° bits of a 4-bit phase shifter, operating over the 35 to 37 GHz frequency range.

9.3.4 Switched-filter

A conventional switched-filter phase shifter is illustrated in Figure 9.27. It works by switching the RF signal path between a high-pass filter and a low-pass filter, by means of two SPDT switches. For this reason, this approach is also known as a high-pass/low-pass phase shifter. This technique is similar to a switched-line phase shifter, but with the transmission lines replaced by filters. A low-pass filter consisting of series inductors and shunt capacitors provides phase lag (or delay) to signals passing through it. A high-pass filter consisting of series capacitors and shunt inductors provides phase lead (or advance) to signals passing through it. Both filters are matched for insertion loss at the operating frequency but have different insertion phases. Therefore, when the signal path is switched between the two filters, the desired change in insertion phase can be achieved without any significant change in the insertion loss. For wider bandwidths, the number of filter elements needs to be increased. However, the bandwidth of the phase shifter depends not only on the characteristics of the two filters but also on the bandwidth and isolation of the two SPDT switches. The overall chip size of a switched-filter phase shifter is considerably smaller than an equivalent switched-line, reflection-type or loaded-line implementation, because the long distributed transmission lines are replaced with much smaller lumped-elements. For this reason, it is the most popular type of digital MMIC phase shifter for low microwave frequency applications [64,67–70]. With the switched-filter phase shifter, the amount of active area used is dominated by the size of the SPDT switches.

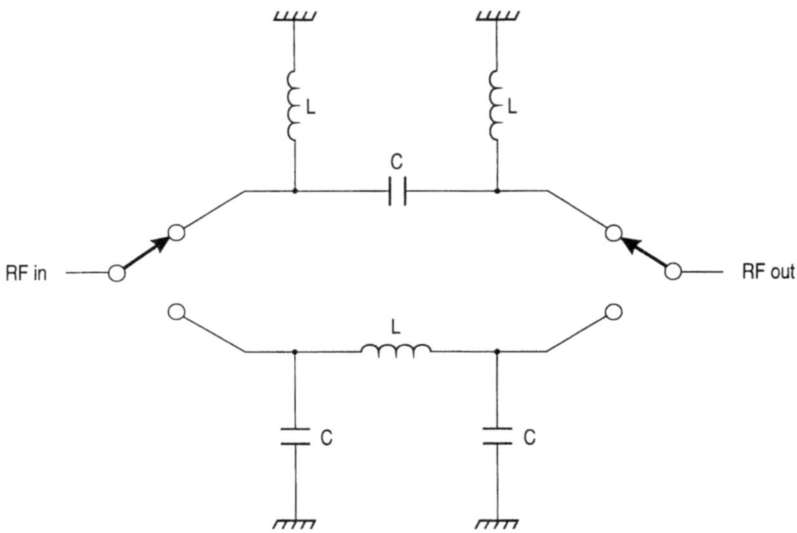

Figure 9.27 A conventional switched-filter phase shifter

With the conventional switched-filter phase shifter, the design approach can be broken down into the individual designs for the two SPDT switches and the two filters. Unlike hybrid MIC digital phase shifters, which usually employ PIN diode-based SPDT switches, MMIC implementations traditionally employ MESFET-based SPDT switches. This is because additional, and therefore very expensive, foundry processing is required to realise PIN diodes [48]. Two notable examples of a switched-filter phase shifter that do employ PIN diode switches were reported by Coats *et al.* [71] and also by Teeter *et al.* [72]. With switching-FETs, when the control voltage is zero the FET is in the on-state and can be simply represented by a small resistance between its drain and source terminations. When the reverse bias control voltage is greater than the pinch-off voltage, V_p, the FET is in the off-state and can be simply represented by a capacitance between its drain and source terminations. This capacitance can be quite large. Therefore, if the FET is in a shunt configuration, the insertion loss of the switch can be unacceptably high. If the FET is in a series configuration, the RF isolation of the switch can be unacceptably low. In order to minimise both the levels of insertion loss and RF signal leakage through the switch, the FET can be resonated with a parallel inductor. However, with this approach the bandwidth of the discrete switches, and therefore the phase shifter will be limited.

An alternative approach was proposed by Ayasli *et al.* [70] and others. Here, the MESFETs are embedded in the design of the filter to absorb the unwanted parasitic capacitances, resulting in a significant improvement in the intrinsic bandwidth of the phase shifter. The bridge topology of an embedded switched-filter phase shifter is shown in Figure 9.28*a*. It can be seen that discrete SPDT switches no longer exist in the phase shifter design. Instead, the series and shunt MESFETs that make up the SPDT switches are dispersed throughout the phase shifter topology. The complementary control signal lines, C and $\overline{C}$ to the MESFETs, have been configured so that when C is at logic '0' (i.e. zero potential) the topology of the phase shifter is equivalent to a fifth-order high-pass filter, as illustrated in Figure 9.28*b*. Any capacitor in parallel with an inductor is assumed to have a capacitive reactance that is too large to shunt the associated inductive reactance. When C is at logic '1' (i.e. > V_P potential), the topology of the phase shifter is equivalent to a five-section low-pass filter, as illustrated in Figure 9.28*c*. In both filter configurations, the MESFET's on-state resistance values are assumed to be negligible when compared with the reactances in series or parallel with them.

This technique was applied by Ayasli *et al.* in the design of a 2-bit 2 to 8 GHz phase shifter [70]. The gate periphery and impedances of the individual MESFETs were treated as parameters for optimisation. The MESFET gate periphery used in the circuit varied from 83 μm to 1333 μm. The 180° bit was designed for a 50 Ω system and the 90° bit was designed for a 25 Ω system, with an impedance transformer inserted between them. Similarly, Schindler and Miller [73] employed the bridge topology shown in Figure 9.28*a* for the design of the 180° bit in a 3-bit phase shifter operating from 18 to 40 GHz. Reber and Felde [74] designed a 6-bit phase shifter operating from 4 to 7 GHz. All bits, except the 180 bit, were implemented using the embedded MESFET technique. The 180° bit used the conventional switched-filter topology. This 6-bit phase shifter has an active GaAs area of only 3.8 × 3.0 mm^2. The insertion loss variation for all the phase states was measured to be ±2.6 dB over the 4 to 7 GHz frequency range and ±0.9 dB over the 5 to 6 GHz frequency range. The RMS phase errors were only 3.8° and 2.4°, respectively, for the two frequency bands. However, modest return losses of 7 dB and 11 dB, respectively, were measured over the two frequency bands.

9.3.5 Intrinsic phase shifters

The techniques described so far have been general, since they can be used to achieve most values of differential phase shift. Some techniques, however, provide an intrinsic value of differential phase shift. Phase shifters with intrinsic designs can have excellent performance characteristics.

9.3.5.1 Boire phase shifter

Boire *et al.* [75] demonstrated a wideband 180 bit phase shifter. The basic configuration of the phase shifter consists of two SPDT switches that change the RF signal path between a shorted 3 dB quadrature coupler and a π-network, as illustrated in Figure 9.29. The coupler has its direct and coupled ports terminated with ideal short circuits, and the π-network is constructed with a series transmission line between two shunt transmission lines. The even- and odd-mode admittances of the directional coupler, with an electrical length θ, are Y_{oe} and Y_{oo}, respectively. If the characteristic admittance of the shunt lines of the π-network is chosen to be Y_{oe}, with an electrical length θ, and the characteristic impedance of the series line is $2(Z_{oe}.Z_{oo})/(Z_{oe}-Z_{oo})$, with an electrical length θ, then its transmission parameters are equivalent to those of the shorted directional coupler preceded by an ideal phase-reversing transformer. In other words, the two networks are similar in all aspects, except that the transmission phase between the two is 180°. This relationship is independent of the electrical length of the two networks, and thus independent of frequency.

Boire *et al.* first demonstrated this technique using hybrid MIC technology, operating across the 4.5 to 18 GHz frequency range [75]. They then went on to demonstrate the principle with monolithic technology, in a 5-bit phase shifter operating over the 1.5 to 7.0 GHz frequency range [61]. This technique has also been implemented in monolithic technology by Miller and Anderson [76], in a 2-bit phase shifter operating over the 4 to 12 GHz frequency range. The 90° bit was implemented using an embedded MESFET switched-filter topology similar to that in Figure 9.28*a*. The microphotograph and measured performance of this MMIC phase shifter are shown in Figure 9.30. More recently, Slobodnik *et al.* demonstrated the Boire 180° bit phase shifter in a 4-bit monolithic design operating over the 35 to 37 GHz frequency range [66]. One drawback of the Boire 180° bit phase shifter is the large differential insertion loss between the 2-phase states.

9.3.5.2 Switched active balun

Another technique that has been adopted for implementing a 180° bit is to switch between the output ports of an active balun. This approach makes use of the inherent 180° phase difference between the source and drain terminals of a FET. Walters and Fikart used this technique in their 5-bit phase shifter design operating from 5 to 6 GHz [68]. The drawback of this technique is that unconditional stability needs to be maintained for the active balun, while simultaneously matching the gate, drain and source of the device. Moreover, the output port impedances and parasitics are unequal, thus causing significant amplitude and phase imbalances. A technique for overcoming these problems, by cross-connecting identical active baluns, has been investigated and successfully implemented in a monolithic L-band bi-phase modulator by Goldfarb *et al.* [77].

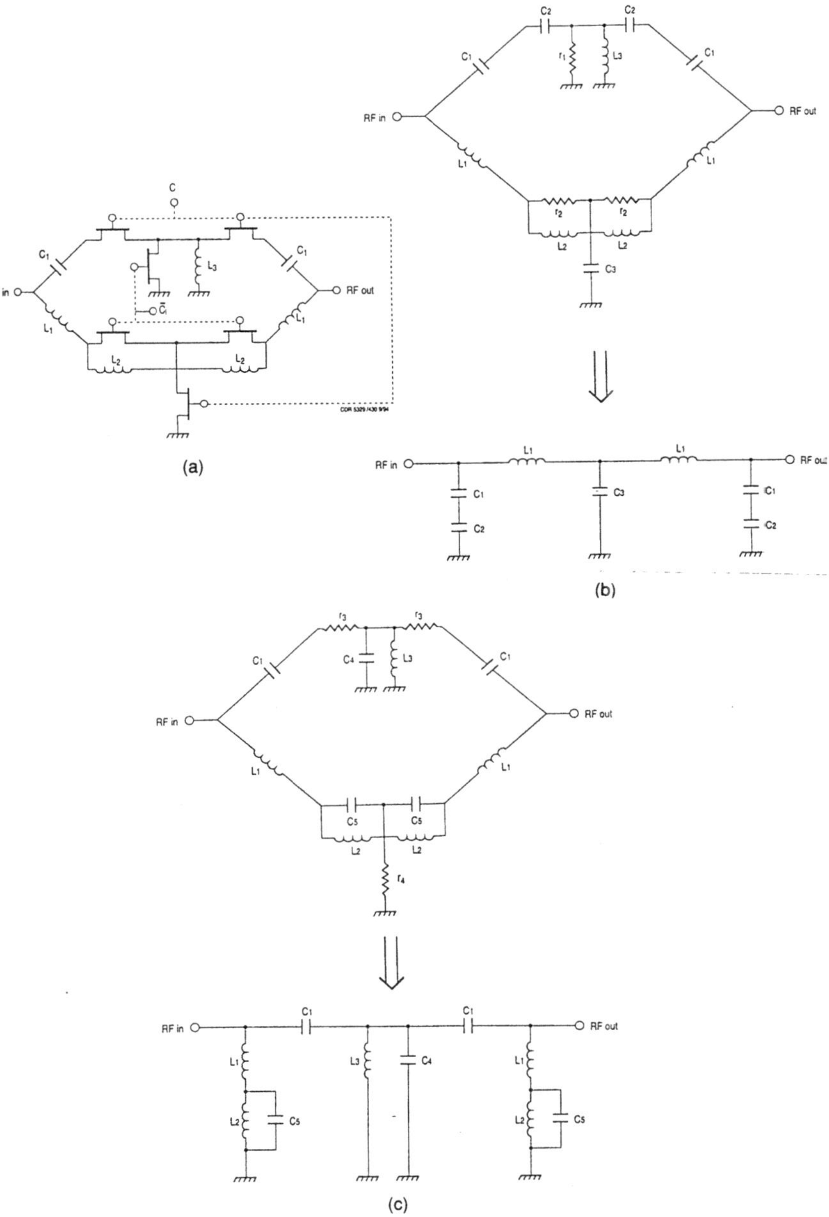

Figure 9.28 Embedded MESFET switched-filter phase shifters: (a) basic topology, (b) equivalent high-pass filter configuration and (c) equivalent low-pass filter configuration

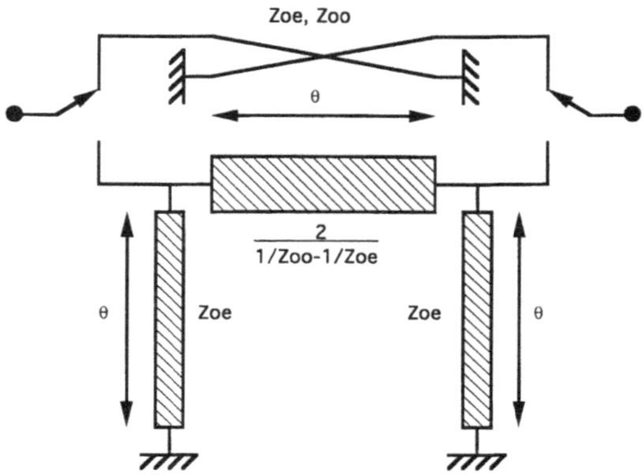

Figure 9.29 The Boire 180° bit phase shifter

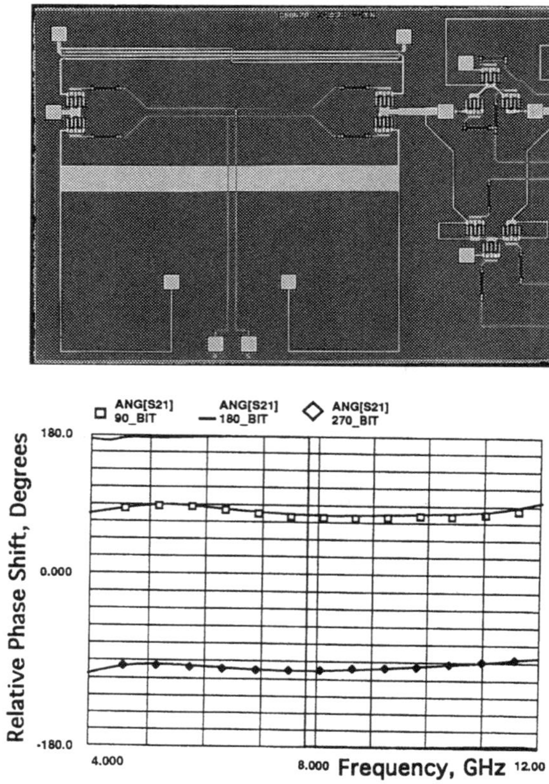

Figure 9.30 X-band 2-bit digital phase shifter: (a) microphotograph and (b) measured relative phase shift responses

9.3.5.3 Switched-coupler

A 90° bit can be implemented by switching between the direct and coupled ports of a 3 dB directional coupler. SPDT switches at the coupler's ports are used to ensure that the unused port is terminated in the coupler's impedance. This ensures that any mismatches at the ports do not degrade the performance of the phase shifter. Figure 9.31 shows the topology of a switched-coupler 90° phase shifter. This technique was demonstrated using monolithic technology by Moore and Miller [78], and Figure 9.32 shows a microphotograph of the chip. This type of phase shifter has a bandwidth that is comparable with that of the directional coupler used. Therefore, a one and a half octave bandwidth Lange coupler can, in principle, be employed to realise a one and a half octave bandwidth 90° bit phase shifter. The main disadvantage with this phase shifter is that it suffers from an inherent 3 dB insertion loss.

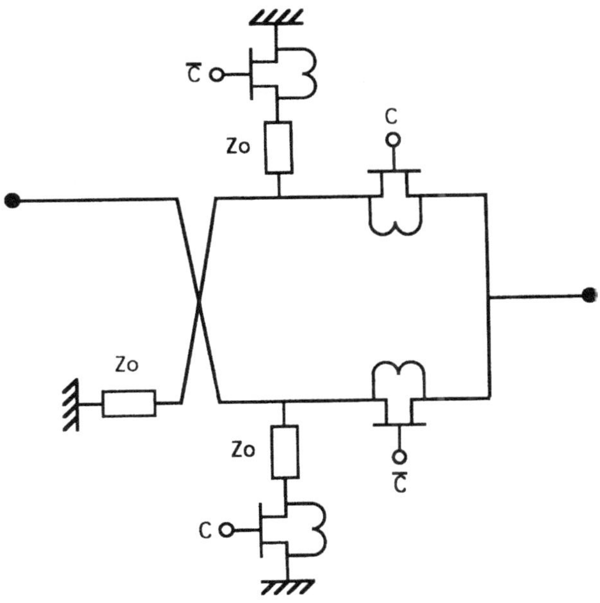

Figure 9.31 Switched-coupler 90 °bit phase shifter

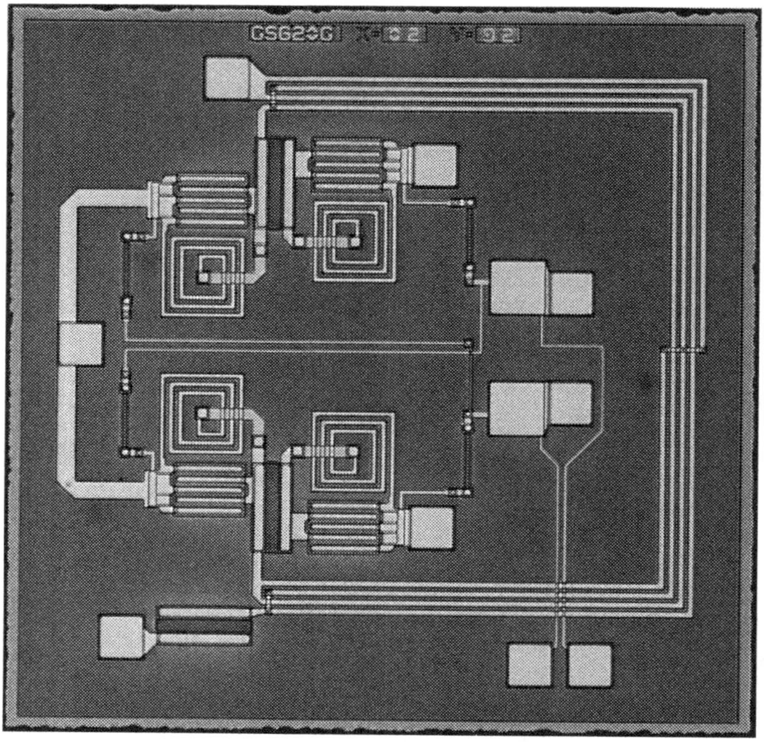

Figure 9.32 Microphotograph of a switched-coupler X-band phase shifter

9.3.6 MEMS digital delay lines

9.3.6.1 Recent work on MEMS phase shifters

The past two years have seen a flurry of papers on microelectromechanical systems (MEMS) digital delay lines. Pillians *et al.* [79] reported a 4-bit monolithic switched-line delay line with microstrip on high-resistivity silicon. Here, MEMS capacitive membrane switches were employed, with $C_{ON}/C_{OFF} \sim 100$ and an actuation voltage of 45 V. The 5% fractional bandwidth was due to the extensive use of resonant stubs; however, at 34 GHz the insertion loss was low at ~ 2.5 dB and the return loss was > 15 dB.

Malczewski *et al.* [80] also reported a two-stage 2-bit reflection-type delay line (i.e. 4-bit in total), with tapped delay line reflection terminations. The same transmission line medium, substrate and MEMS switches were used as with the switched-line example. At X-band, the measured insertion loss was around 1.5 dB, with 60% of this loss being attributed to the Lange couplers.

MEMS distributed loaded-lines have also been reported. By applying a single bias voltage to either the signal conductor of the CPW line or the MEMS bridges lying over this centre conductor, the effective distributed capacitance of the line can be changed, which in turn changes the phase velocity, and thus the associated time delay through the transmission line. Using this principle, Borgioli *et al.* [81] reported 1-bit ultra-wide bandwidth distributed digital loaded-line. Again, MEMS capacitive membrane switches were employed, with $C_{ON}/C_{OFF} \sim 7.5$ and having an actuation voltage of 75 V. Here, an 8.6 mm long CPW line on a glass substrate achieved a DC to 35 GHz bandwidth, with a relative phase shift and insertion loss of 270° and 1.7 dB, respectively, at 35 GHz. Because of the variation in the characteristic impedance of the CPW line, from 66 Ω in the off-state to 38 Ω in the on-state, the inherent impedance mismatching creates unwanted insertion loss ripples in the frequency response. The same team then went on to report a 3-bit distributed loaded-line using the same transmission line medium, substrate and MEMS switches [82]. In addition to the pioneering work from the University of California (Santa Barbara), UCSB, on novel distributed loaded lines, the University of Michigan has also been following in this area. Using the same topology and almost identical MEMS technology as UCSB, Hayden and Rebeiz reported a 1-bit X-band distributed loaded line that achieved a relative phase shift of 270° at 10 GHz [83]. The off- and on-state insertion losses were 0.48 and 0.72 dB, respectively.

9.3.6.2 Recommendations for MEMS implementations

As previously described, in recent years there has been a great deal of activity in the area of MEMS technology applied to realising low loss digital phase shifters. In principle, such phase shifters could be placed directly between antenna elements and their associated T/R module, to create a fully distributed 2-D phased array antenna system. The low losses associated with the phase shifter would relax both transmitter power amplifier and receiver low noise amplifier specifications, thus helping to dramatically reduce the cost of future front-end microwave subsystems.

For a 4-bit switched-line delay line, a number of relatively long reference transmission lines are required at microwave frequencies. The main drawback with conventional monolithic technology is that insertion loss may vary considerably between states. This limitation can be minimised by utilising low loss micromachined lines. For example, micromachined thin-film microstrip could be used and this is ideal for meandering, thus minimising the size of the MEMS structure. Here, suitable micromachining techniques need to be developed that can provide the low transmission losses and be integrated with low loss/high isolation SPDT switches. Moreover, if membrane supported CPW lines were employed, great care must be made to avoid unwanted moding, since insertion phase can be very sensitive to multi-moding.

Similar techniques recommended for the switched-line delay line can also be associated with the reflection-type delay line. The key advantage here, though, is that of the reduced area needed. However, a low loss 3dB quadrature directional coupler needs to be developed. This would be possible with a micromachined multilayer broadside coupler. To date, no reflection-type digital delay line has been realised using micromachining and MEMS switch technologies combined.

For a 4-bit implementation of the distributed loaded-line, the lengths of transmission line become excessive. Moreover, even meandered transmission lines would require considerable chip space. Also, the inherently poor return loss performance may result in unacceptable insertion loss ripples. Finally, no measured group delay performance has yet to be reported for any of the distributed loaded-line delay lines. The reason for this is that the insertion loss ripples translate to producing insertion phase ripples. The resulting ripples in the group delay would preclude their use in some practical applications, such as high performance phased array antenna subsystems. For all these reasons, this type of delay line is expected to be the least likely to meet the demanding electrical specifications of future high performance applications.

9.4 Summary

In this chapter, the distinctive characteristics of a true phase shifter and a true delay line have been defined. The main techniques for the design of both analogue and digital MMIC implementations have been reviewed and their relative advantages and drawbacks discussed. It has been demonstrated that the traditional single-stage reflection-type phase shifter topology can indeed be used to implement a variable delay line. Conversely, the general topology of the switched-line delay line can be used to implement a digital switched-line phase shifter.

The various methods used to implement the phase shifters, in both distributed and lumped-element forms, have been described, from planar to multilayer to MEMS realisation technologies. It can be seen that MMIC phase shifters and delay lines are capable of providing the high levels of performance required for future communication, radar and measurement applications.

9.5 References

1 HARDIN, R. N., DOWNEY, E. J., and MUNUSHIAN, J.: 'Electronically-variable phase shifters utilizing variable capacitance diodes', *Proceedings of the IRE*, **48**, 1960, pp. 944–945

2 SEARING, R. M.: 'Variable capacitance diodes used as phase-shift devices', *Proceedings of the IRE*, 1961, **49**, pp. 640–641

3 GARVER, R. V.: 'Broadband binary 180° diode phase modulator', *IEEE Transactions on Microwave Theory and Techniques*, 1965, **MTT-13**, pp. 32–38

4 HENOCH, B. T., and TAMM, P.: 'A 360° reflection-type diode phase modulator', *IEEE Transactions on Microwave Theory and Techniques*, **MTT-19**, pp. 103–105

5 ULRIKSSON, B.: 'Continuous varactor-diode phase shifter with optimized frequency response', *IEEE Transactions on Microwave Theory and Techniques*, 1979, **MTT-27**, pp. 650–654

6 DAWSON, D. E., CONTI, A. L., LEE, S. H., SHADE, G. F., and DICKENS, L. E.: 'An analog X-band phase shifter', *IEEE Microwave and Millimeter-Wave Monolithic Circuits Symposium Digest*, 1984, pp. 6–10

7 CHEN, C. L., COURTNEY, W. E., MAHONEY, L. J., MANFRA, L. J., CHU, A., and ATWATER, H. A.: 'A low-loss Ku-band monolithic analog phase shifter', *IEEE Transactions on Microwave Theory and Techniques*, 1987, **MTT-35**, pp. 315–320

8 KRAFCSIK, D. M., TMHOFF, S. A., DAWSON, D. E., and CONTI, A. L.: 'A dual varactor, analog phase shifter operating 6 to 18 GHz', *IEEE Microwave and Millimeter-Wave Monolithic Circuits Symposium Digest*, 1988, pp. 83–86

9 BASTIDA, E. M., DONZELLI, G. P., and SCOPELLITI, L.: 'GaAs monolithic microwave integrated circuits using broadband tunable active inductors', *Proceedings of 19th European Microwave Conference*, 1989, pp. 1282–1287

10 BIANCHI, G., PINTO, G., and GIULIANI, C.: 'MMIC 12-13 GHz voltage controlled phase shifter', *Proceedings of ESA International Workshop on Monolithic Microwave Integrated Circuits for Space Applications*, ESTEC, Noordwijk, 1990

11 ALI, F., and MYSOOR, N.: 'An analog MMIC phase modulator for X-band satellite transponder applications', *IEEE Microwave and Guided Wave Letters*, **2** (11), pp. 445–446

12 LUCYSZYN, S., and ROBERTSON, I. D.: 'Analog reflection topology building blocks for adaptive microwave signal processing applications', *IEEE Transactions on Microwave Theory and Techniques*, 1995, **MTT-43** (3)

13 LUCYSZYN, S., GREEN, G., and ROBERTSON, I. D.: 'Accurate millimeter-wave large signal modeling of planar Schottky varactor diodes', *IEEE MTT-S International Microwave Symposium Digest*, Albuquerque, pp. 259–262

14 LUCYSZYN, S., LUCK, J., GREEN, G., and ROBERTSON, I. D.: 'Enhanced modelling of interdigitated planar Schottky varactor diodes', *IEEE Asia-Pacific Microwave Conference Digest*, Adelaide, 1992, pp. 273–278

15 LUCYSZYN, S., GREEN, G., and ROBERTSON, I. D.: 'Interdigitated planar Schottky varactor diodes for tunable MMIC applications', *Proceedings of the IEEE/ESA European Gallium Arsenide Applications Symposium*, ESTEC, Noordwijk, 1992

16 LANGE, J.: 'Interdigitated stripline quadrature hybrid', *IEEE Transactions on Microwave Theory and Techniques*, 1969, **MTT-17**, pp. 1150–1151

17 RUTZ, E. M.: 'A stripline frequency translator', *IRE Transactions on Microwave Theory and Techniques*, 1961, pp. 158–161

18 WILSON, K., NICHOLAS, J. M. C., McDERMOTT, G., and BURNS, J. W.: 'A novel MMIC X-band phase shifter', *IEEE Transactions on Microwave Theory and Techniques*, 1985, **MTT-33**, pp. 1572–1578

19 LUCYSZYN, S.: 'Ultra-wideband high performance reflection-type phase shifters for MMIC applications', PhD Thesis, University of London, 1993

20 LUCYSZYN, S., ROBERTSON, I. D., and AGHVAMI, A. H.: 'A high performance analogue time shifter for wideband phased-array antennas', *IEE Colloquium Digest on Recent Advances in Microwave Sub-systems for Space and Satellite Applications*, London, 1993, pp. 4/1–4

21 LUCYSZYN, S., ROBERTSON, I. D., and AGHVAMI, A. H.: 'High performance wideband analogue time shifter', *Electronics Letters*, 1993, **29** (10), pp. 885–887

22 LUCYSZYN, S., and ROBERTSON, I. D.: 'Synthesis techniques for high performance octave bandwidth 180° analog phase shifters', *IEEE Transactions on Microwave Theory and Techniques*, 1992, **MTT-40** (4), pp. 731–739
23 LUCYSZYN, S., and ROBERTSON, I. D.: 'High performance octave bandwidth MMIC analogue phase shifter', *Proceedings of 22nd European Microwave Conference*, Espoo, 1992, pp. 221–224
24 LUCYSZYN, S., and ROBERTSON, I. D.: 'An improved multi-layer quadrature coupler for MMICs', *Proceedings 21st European Microwave Conference*, Stuttgart, 1991, pp. 1154–1158
25 LUCYSZYN, S., and ROBERTSON, I. D.: 'Two octave bandwidth monolithic analog phase shifter', *IEEE Microwave and Guided Wave Letters*, 1992, **2** (8), pp. 343–345
26 LUCYSZYN, S., and ROBERTSON, I. D.: 'Decade bandwidth hybrid analogue phase shifter using MMIC reflection terminations', *Electronics Letters*, 1992, **28** (11), pp. 1064–1065
27 KRYTAR: 'Broadband, 3 dB 90° hybrids covering 1 to 18 GHz', *Microwave Journal*, 1985
28 LUCYSZYN, S., and ROBERTSON, I. D.: 'Decade bandwidth MMIC analogue phase shifter', *IEE Colloquium Digest on Multi-octave Microwave Circuits*, London, 1991, pp. 2/1–6
29 DOUGHERTY, R.: 'Circulate signals with active devices on monolithic chips', *Microwaves and RF*, 1989, pp. 85–89
30 ROBERTSON, I. D., and AGHVAMI, A. H.: 'Ultrawideband biasing of MMIC distributed amplifiers using improved active load', *Electronics Letters*, 1991, **27** (21), pp. 1907–1909
31 KOTHER, D., HOPF, B., SPORKMANN, T., and WOLFF, I.: 'Active CPW MMIC circulator for the 40 GHz band', *Proceedings of 24th European Microwave Conference,* Cannes, 1994, pp. 542–547
32 ROBERTSON, I. D., and AGHVAMI, A. H.: 'Novel monolithic ultra-wideband 4-port junction using distributed amplification techniques', *IEEE MTT-S International Microwave Symposium Digest*, Albuquerque, 1992, pp. 1051–1054
33 DAWIRS, H. N., and SWARNER, W. G.: 'A very fast, voltage-controlled, microwave phase shifter', *Microwave Journal*, 1962, pp. 99–107
34 NAGRA, A. S., and YORK, R. A.: 'Distributed analog phase shifters with low insertion loss', *IEEE Transactions on Microwave Theory and Techniques*, 1999, **47** (9), 1999, pp. 1705–1711
35 ERKER, E. G., NAGRA, A. S., LIU, Y., PERIASWAMY, P., TAYLOR, T. R., SPECK, J., and YORK, R. A.: 'Monolithic Ka-band phase shifter using voltage tunable $BaSrTiO_3$ parallel plate capacitors', *IEEE Microwave and Guided Wave Letters*, 2000, **10** (1), pp. 10–12
36 PAO, C. K., CHEN, J. C., LAN,) G. L., WANG, D. C., WONG, W. S., and HERMAN, M. I.: 'V-band monolithic phase shifters', *10th Annual IEEE GaAs IC Symposium Technical Digest*, 1988, pp. 269–272
37 ASHTIANI, A. E., NAM, S, LUCYSZYN, S., and ROBERTSON, I. D.: 'Monolithic Ka-band 180-degree analog phase shifter employing HEMT-based varactor diodes', *Proceedings of IEE Colloquium on Microwave and Millimetre-Wave Oscillators and Mixers*, London, 1998, pp. 7/1–6

38 LUCYSZYN, S., ROBERTSON, I. D., and AGHVAMI, A. H.: 'Monolithic analogue phase shifter and cascode FET amplifier using uniplanar techniques', *Proceedings of 24th European Microwave Conference*, Cannes, 1994, pp. 554–559

39 TSIRONIS, C., and HARROP, P.: 'Dual gate GaAs M.E.S.F.E.T. phase shifter with gain at 12 GHz', *Electronics Letters*, 1980, **16** (14), pp. 553–554

40 PENGELLY, R. S.: 'GaAs monolithic microwave circuits for phased array applications', *IEE Proceedings F*, 1980, **127** (4), pp. 301–311

41 DEVLIN, L. M., and MINNIS, B. J.: 'A versatile vector modulator design for MMIC', *IEEE MTT-S International Microwave Symposium Digest*, 1990, pp. 519–522

42 FROST, R. D., FISHER, D. A, and PECK, D. E: 'A GaAs MMIC voltage-controlled phase shifter', *Microwave Journal*, 1991, pp. 87–94

43 MAZUMDER, S. R., TSAI, T. L., and TSAI, W. C.: 'A frequency translator using dual-gate GaAs FETs', *IEEE MTT-S International Microwave Symposium Digest*, 1983, pp. 346–348

44 SALVAGE, S. T., HASH, R. J., and PETTED, B. E.: 'An octave band GaAs analog phase shifter', *IEEE MTT-S International Microwave Symposium Digest*, 1989, pp. 1051-1054

45 MAGARSHACK, J.: 'A new digital phase shifter architecture suitable for MMIC's', *IEEE Transactions on Microwave Theory and Techniques*, 1994, **MTT-42** (1) pp. 154–156

46 SHARMA, A. K.: 'Solid-state control devices: state of the art', *Microwave Journal*, 1989, pp. 95–112

47 CAVERLY, R. H.: 'Distortion in GaAs MESFET switch circuits', *Microwave Journal*, 1994, pp. 106–114

48 KOBAYASHI, K. W., TRAN, L., OKI, A. K., and STREIT, D. C.: 'A 50 MHz-30 GHz broadband co-planar waveguide SPDT PIN diode switch with 45-dB isolation' *IEEE Microwave and Guided Wave Letters*, 1995, **5** (2), pp. 56–58

49 MIZUTANI, H., FUNABASHI, M. and KUZUHARA, K.: 'DC-50 GHz small size HJFET MMIC switch for high power applications', *Proceedings of European Microwave Conference*, Prague, 1996, pp. 949–952

50 GUPTA, A., KAELIN, G., STEIN, R., HUSTON, S., LP, K., and PETERSEN, W.: 'A 20 GHz 5-bit phase shift transmit module with 16 dB gain', *IEEE GaAs IC Symposium Digest*, 1984, pp. 197–200

51 BAUHAHN, P., BUTTER, C., SOKOLOV, V., and CONTOLATIS, A.: '30 GHz multibit monolithic phase shifters', *IEEE Microwave and Millimeter-Wave Monolithic Circuits Symposium Digest*, 1985, pp. 4–7

52 DUNN, V. E., HODGES, N. E., SY, O. A., and ALYASSINI, W.: 'MMIC phase shifters and amplifiers for millimeter-wavelength active arrays', *IEEE MTT-S International Microwave Symposium Digest*, 1989, pp. 127–130

53 OAKI, A., INOUE, M., TSUDA, Y., ORIME, N., and UCHINO, S.: 'A small X-band phase shifter with small phase increments', *Proceedings of 3rd Asia-Pacific Microwave Conference*, Tokyo, 1990, pp. 873–876

54 LAU, C. L., FENG, M., HWANG, T., LEPKOWSKI, T., ITO, C., DUNN, V., and HODGES, N.: 'Millimeter wave monolithic ICs using direct ion implantation into GaAs LEC substrates', *IEEE GaAs IC Symposium Digest*, 1990, pp. 73–76

55 AUST, M., WANG, H., CARANDANG, R., TAN, K., CHEN, C. H., TRINH, T., ESFANDIAR, R., and YEN, H. C.: 'GaAs monolithic components development for Q-band phased array application', *IEEE MTT-S International Microwave Symposium Digest*, 1992, pp. 703–706
56 ROGEAUX, E., GROGNET, V., CAZAUX, J. L., COPPEL, F., GEEN, M., and GATTI, G.: 'Standard MMICs for space application: A broadband high resolution phase-shifter', *Proceedings of 24th European Microwave Conference*, Cannes, 1994, pp. 1483–1488
57 MARUHASHI, K. *et al.*: 'A Ka-band 4-bit monolithic phase shifter using unresonated FET switches', *IEEE MTT-S International Microwave Symposium Digest*, 1998, pp. 51–54
58 NAKAHARA, K., CHAKI, S., ANDOH, N., MATSUOKA, H., TANINO, N., MITSUI, Y., OTSUBO, M., and MITSUI, S.: 'A novel three phase-state phase shifter', *IEEE MTT-S International Microwave Symposium Digest*, 1993, pp. 369–372
59 MILLER, P., and JOSHI, J. S.: 'MMIC phase shifters for space applications', *Proceedings of ESA International Workshop on Monolithic Microwave Integrated Circuits for Space Applications*, ESTEC, Noordwijk, 1990
60 ANDRICOS, C., BAHL, I., and GRIFFIN, E.: 'C-band 6-bit GaAs monolithic phase shifter', *IEEE Transactions on Microwave Theory and Techniques*, 1985, **MTT-33**, pp. 1591–1596
61 BOIRE, D. C., ONGE, G. ST., BARRATT, C., NORRIS, G. B., and MOYSENKO, A.: '4:1 bandwidth digital five bit MMIC phase shifters', *Microwave and Millimeter-Wave Monolithic Circuits Symposiium Digest*, 1988, pp. 69–73
62 ATWATER, H. A.: 'Circuit design of the loaded-line phase shifter', *IEEE Transactions on Microwave Theory and Techniques*, 1985, **MTT-33** (7), pp. 626–634
63 OPP, F. L., and HOFFMAN, W. F.: 'Design of digital loaded-line phase shift networks for microwave thin-film applications', *IEEE Transactions on Microwave Theory and Techniques*, 1968, **MTT-16** (7), pp. 462–468
64 OZAKI, J., ASANO, T., WATANABE, S., TATEMATSU, M., and KAMIHASHI, S.: 'X-band MMIC phase shifter with small loss variation', *Proceedings of 3rd Asia-Pacific Microwave Conference*, Tokyo, 1990, pp. 869–872
65 AYASLI, Y., PLATZKER, A., VORHAUS, J., and REYNOLDS, L. D.: 'A monolithic single chip X-band four-bit phase shifter', *IEEE Transactions on Microwave Theory and Techniques*, 1982, **MTT-30** (12), 1982, pp. 2201–2206
66 SLOBODNIK, A. J., WEBSTER, R. T., and ROBERTS, G. A.: 'A monolithic GaAs 36 GHz four-bit phase shifter', *Microwave Journal*, 1993, pp. 106–111
67 DEVLIN, L. M.: 'Digitally controlled, 6 bit, MMIC phase shifter for SAR applications', *22nd European Microwave Conference Digest*, Espoo, 1992, pp. 225–230
68 WALTERS, P. C., and FIKART, J. L.: 'A fully integrated 5-bit phase shifter for phased array applications', *IEEE International Symposium Digest on MMICs in Communications Systems*, London, 1992
69 GOLDFARB, M. E., and PLATZKER, A.: 'A wide range analog MMIC attenuator with integral 180 phase shifter', *IEEE Transactions on Microwave Theory and Techniques*, 1994, **MTT-42** (1), pp. 156–158

70 AYASLI, Y., MILLER, S. W., MOZZI, R., and HANES, L. K.: 'Wideband monolithic phase shifter', *IEEE Transactions on Microwave Theory and Techniques*, 1984, **MTT-32** (12), pp. 1710–1714

71 COATS, R., KLEIN, J., PRITCHETT, S. D., and ZIMMERMANN, D.: 'A low loss monolithic five-bit PIN diode phase shifter', *IEEE MTT-S International Microwave Symposium Digest*, 1990, pp. 915–918

72 TEETER, D., WOHLERT, R., COLE, B., JACKSON, G., TONG, E., SALEDAS, P., ADLERSTEIN, M., SCHINDLER, M., and SHANFIELD, S.: 'Ka-band GaAs HBT PIN diode switches and phase shifters', *IEEE MTT-S International Microwave Symposium Digest*, 1994, pp. 451–454

73 SCHINDLER, M. J., and MILLER, M. E.: 'A 3 bit K/Ka band MMIC phase shifter', *IEEE MTT-S International Microwave Symposium Digest*, 1988, pp. 95–98

74 REBER, R. H., and FELDE, H.-P.: 'A monolithic single-chip C-band 6-bit phase shifter', *Proceedings of 21st European Microwave Conference*, Stuttgart, 1991, pp. 479–482

75 BOIRE, D, Internatio. C., DEGENFORD, J. E., and COHN, M.: 'A 4.5 to 18 GHz phase shifter', *IEEE MTT-S International Microwave Symposium Digest*, 1985, pp. 601–604

76 MILLER, P., and ANDERSON, A. K.: 'Development of gain/phase control blocks for space applications', *IEEE ESA Proceedings of the European Gallium Arsenide Applications Symposium*, ESTEC, Noordwijk, 1992

77 GOLDFARB, M. E., COLE, P. J. B., and PLATZKER, A.: 'A novel MMIC biphase modulator with variable gain using enhancement-mode FETs suitable for 3 V wireless applications', *IEEE Microwave and Millimeter-Wave Monolithic Circuits Symposium Digest*, 1994, pp. 99–102

78 MOORE, M. T., and MILLER, P.: 'Microwave IC control components for phased-array antennas', *IEE Electronics and Communication Engineering Journal*, 1992, pp. 123–130

79 PILLANS, B., ESHELMAN, S., MALCZEWSKI, A., EHMKE, J., and GOLDSMITH, C.: 'Ka-band RF MEMS phase shifters', *IEEE Microwave and Guided Wave Letters*, 1999, **9** (12), pp. 520–522

80 MALCZEWSKI, A., ESHELMAN, S., PILLANS, B., EHMKE, J., and GOLDSMITH, C. L.: 'X-band RF MEMS phase shifters for phased array applications', *IEEE Microwave and Guided Wave Letters*, 1999, **9** (12), pp. 517–519

81 BORGIOLI, A., LIU, Y., NAGRA, A. S., and YORK, R. A.: 'Low-loss distributed MEMS phase shifter', *IEEE Microwave and Guided Wave Letters*, 2000, **10** (1), pp. 7–9

82 LIU, Y., BORGIOLI, A., NAGRA, A. S., and YORK, R. A.: 'K-band 3-bit low-loss distributed MEMS phase shifter', *IEEE Microwave and Guided Wave Letters*, 2000, **10** (10), pp. 415–417

83 HAYDEN, J. S., and REBEIZ, G. M.: 'Low-loss cascadable MEMS distributed X-band phase shifters', *IEEE Microwave and Guided Wave Letters*, 2000, **10** (4), pp. 142–144

Chapter 10

Integrated antennas

V. F. Fusco

10.1 Introduction

This chapter describes the technology choices and circuit topologies associated with monolithic integrated antennas. MMICs are well known for their applications in transmit/receive modules, although in recent years, there has been a great deal of research undertaken to extend the level of integration to include the antenna subsystem. This area is a difficult one for traditional MMIC designers who are largely unfamiliar with antenna techniques, and vice versa. Fortunately, there is synergy between classical MMIC and antenna technologies, although relatively few engineers have a thorough appreciation of both disciplines. It is hoped that this chapter will go some of the way towards bringing the two technologies together.

Recent trends in mobile wireless communications systems indicate the need for unrestricted mobility associated with improved bandwidth availability. This has led to developments in mobile broadband systems technology [1]. For example, business communications in a wireless office, where the potentially high data rates available through broadband wireless could lead to systems that could improve video conferencing quality [2]. Present indicators show that integration of millimetre-wave technology, using MMIC techniques, can be made economically viable [3–5], although assembly and integration of the MMIC components with the antenna appear to be largely unresolved issues at present. Frequencies into the terahertz range are also under consideration for use in remote sensing, radio astronomy, radar and covert communications [6]. Much has been written about how MMIC integration of front-ends, minus the antenna, can be achieved and multifunctional and hybrid realisations have been reported [7–9]. In the front-end, a key critical component is the antenna. Many perceived applications will require planar, miniature antenna arrays, and in a number of important cases (e.g. in mobile communications, automotive sensors, and non-geostationary satellites), antennas that can be continuously aimed at the target will be required. These requirements insist on added functionality at the antenna. This can be obtained by using integrated active antenna techniques [10] or the assembly of separate integrated passive antennas and integrated circuit devices.

10.2 Basic integrated antenna requirements

As with all volume electronics, the driving factor for the antenna element is a lowering of cost, mass and volume in a way that gives unimpaired electrical specification. Cost is directly related to the antenna complexity, the manufacturing process and the non-recurrent engineering time required for right-first-time designs. Some of the fundamental requirements for integrated antenna development will now be discussed.

Integrated antennas are invariably printed or etched into some type of dielectric or ferromagnetic substrate material. The first problem that must be considered are losses due to dissipation in the substrate dielectric and metallisation. If thin low loss substrates are used, then radiating elements tend to exhibit a high Q-factor and consequently narrow bandwidth. The bandwidth, B, for an impedance-matched antenna operated away from resonance is related to the input VSWR of the antenna and its loaded quality factor, Q, by:

$$Q = \frac{\mathrm{VSWR} - 1}{B\sqrt{\mathrm{VSWR}}} \tag{10.1}$$

Here, B is the band of frequencies over which the antenna's VSWR (or return loss) is less that some specified maximum value, usually taken as 2:1 (or -10 dB). The use of a thicker and/or lower permittivity substrate reduces Q, and therefore, increases B. Thus, the volume of the antenna is an important issue in the design of the element in order for it to comply with a given electrical specification. With integrated circuit antennas, the use of micromachining techniques potentially allows control of the antenna dimensions in the z-direction. This aspect will be discussed in Section 10.8.

The next significant factor is the generation of surface-waves and their effect on the radiation characteristics of the antenna [11]. Surface-waves can modify the far-field radiation patterns, especially with respect to antenna cross-polarisation and sidelobe response. For a wide microstrip line of the type that may be used in a patch antenna design, constructed on a dielectric substrate (having a relative permittivity ε_r), the cut-off frequency at which the dominant leakage mechanism of the TM_0 mode becomes relevant is given by:

$$f_c = \frac{75}{h\sqrt{\varepsilon_r - 1}} \tag{10.2}$$

where h is the substrate thickness in mm and the cut-off frequency f_c is in GHz. Thus, for a microstrip patch antenna constructed on GaAs (ε_r = 12.9) and operating at 77 GHz, the substrate should be no thicker than 280 μm. In practice, wafer thickness is of major concern in the design of integrated circuit antennas, since additional manufacturing processes involving wafer thinning are generally required in order to alleviate surface-wave effects. In addition, a thinned wafer is more prone to damage during handling in the fabrication process. Alternatively, the use of the z-dimension in an integrated circuit antenna allows the possibility, by micromachining, of introducing air-spacers into the dielectric in order to reduce surface-wave effects.

From these considerations, it can be seen that the use of thin, high dielectric constant substrates will lead to inferior bandwidth performance when compared to an antenna constructed on thick, low dielectric constant substrates, at least up to the point where surface-wave effects start to impair the performance of the antenna. Thus, for integrated circuit antennas, the use of semiconductor substrate materials such as silicon (ε_r = 11.7) or GaAs could potentially lead to difficulties in broadband systems, unless special precautions are adopted.

10.2.1 Additional Requirements

The designer of integrated antenna structures is always faced with a variety of conflicting requirements with respect to the performance of the antenna; some of these are listed in Table 10.1

Table 10.1 Antenna element basic specifications

Power capacity	Cross polarisation levels
Radiation efficiency	Impedance matching
Beamwidth	Sidelobe response
Polarisation type	Physical shape

In addition to the parameters listed in Table 10.1, the use of integrated circuit techniques adds new considerations. First, the antenna or antenna array can be non-integrated with respect to the circuit function. That is to say, the antenna and the circuit are designed separately and function via a common interface impedance standard in such a way that their separate functions are distinct (e.g. a standard front-end receiver topology). Next, the antenna or array can be partially integrated. Here, the antenna and the circuit can interact with each other and circuit pulling can occur (c.f. a self oscillating mixer [12]). Finally, a fully integrated system where the circuit and the antenna interact both through the circuit and the near-field and/or surface-wave coupling (e.g. a dielectric slot power combiner [13]). When these are taken into consideration, additional parameters must be included, such as those in Table 10.2.

Table 10.2 Integrated antenna extended specifications

Transmit	Receive
Frequency stability	Noise figure
Spectral content	Intermodulation distortion
Linearity	
AM, FM & PM noise	

These additional constraints place a huge demand on CAD software that can quantitively analyse the performance of such systems [14,15].

10.3 Reported applications of integrated antennas

Examination of the literature shows that integration has been applied to a wide variety of antenna structures that are suitable for photolithographic production and manufacture by monolithic processing techniques. A subset of some of these applications is listed in Table 10.3.

Table 10.3 Integrated antenna applications

Integrated passive antennas and arrays	Integrated active antennas
Dual band mobile communications [16]	Doppler sensor [12]
Dual band satellite communications [17]	Military phased array [24]
Dual polarization mobile communications [18]	Mobile wireless [25]
Cellular communications [19]	Satellite communications [26]
	Retrodirective vehicular tracking [27]
Active antennas	Vehicular identification [28]
	Road condition monitor [29]
VHF receive amplifier [20]	Tagging [30]
VHF transmitter [21]	BPSK modulator [31]
Phased array beam steering [22]	ISM transducer [32]
TV receiver [23]	Imaging [33]

10.4 Integrated antenna retrodirective array example

An example of how integrated antennas can be coupled with MMIC components, to build minimum complexity antenna arrays with added functionality, is now discussed. For reduced RF power operation, mobile wireless communication/transponder systems require the accurate pointing of high gain antennas. In a mobile wireless situation, a retrodirective array (RDA, also known as a self-phased array), performs beam steering action automatically, by virtue of the array architecture used. Here, phasing information is derived directly from the signals each array element receives. Thus, the antenna array continuously adapts its phase response to track an incoming signal, without prior knowledge of the spatial position of this incoming signal. Van Atta described an array, shown in Figure 10.1*a*, which could perform this function [34].

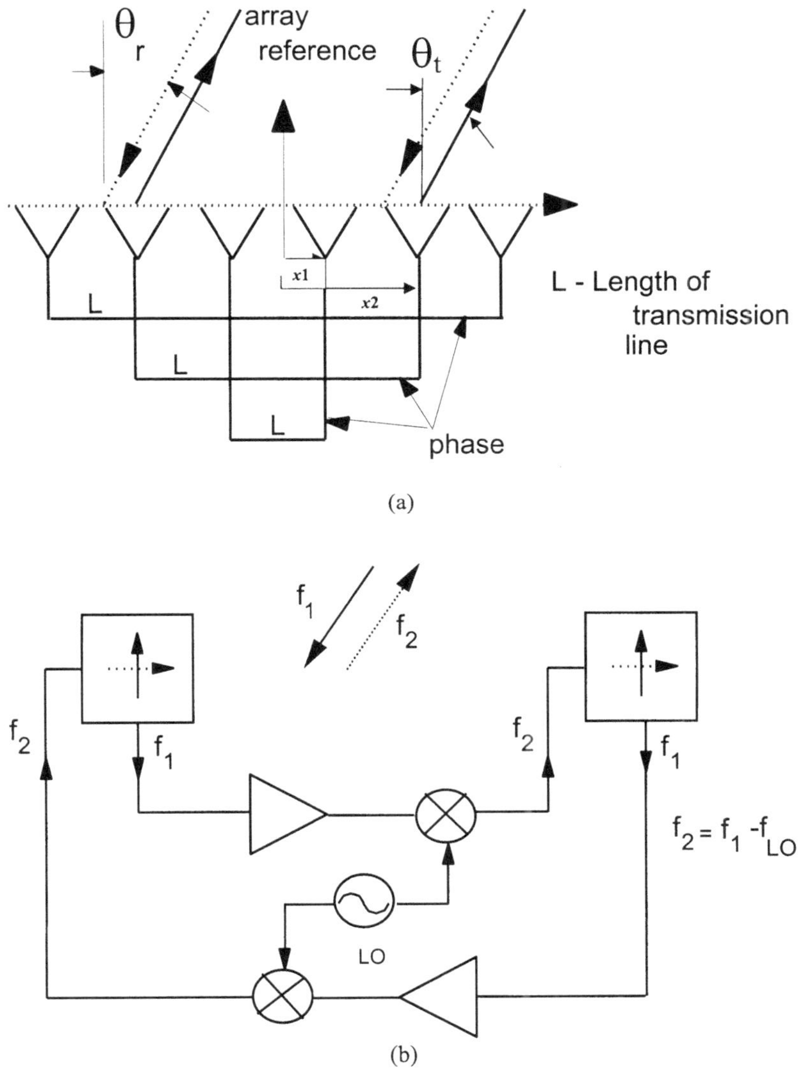

Figure 10.1 Retrodirective array: (a) Van Atta Array and (b) frequency-offset Van Atta array

Here, the incoming wavefront samples, obtained by the antenna array elements to the right of the centre of the array, are retransmitted from their mirror image element on the left side of the array. When $\theta_t = \theta_r$, retrodirective action occurs and the retransmitted wavefront arrives in-phase back at its point of origin. By using dual, linearly polarised, integrated antennas, the original transmitted and the copy re-transmitted signals have increased isolation from each other.

This reduces the mutual interference between the received and transmitted signals [35]. In addition, a frequency-offset, f_{LO}, is introduced between transit and receive signals, in order to allow discrimination between them, as shown in Figure 10.1*b*.

The performance of a four-element retrodirective array, constructed using GaAs MMICs (for the mixer and amplifier) and a soft-board (for the array and interconnects), is presented in Figure 10.2. This is a monostatic measurement and was carried out by moving co-located transmit/receive antenna around the retrodirective antenna. All patterns are normalised to 0 dB at $\theta = 0°$. The enhanced 3 dB response for the retrodirective array, with respect to a passive array, indicates that wider azimuthal coverage is obtained with the retrodirective array, since the beam from this array is self-tracking. It should be noted that the RDA response is actually a plot of the locus of the individual angular responses of each of the far-field maximum of the individual return patterns, formed for each angle of incident signal exciting the array.

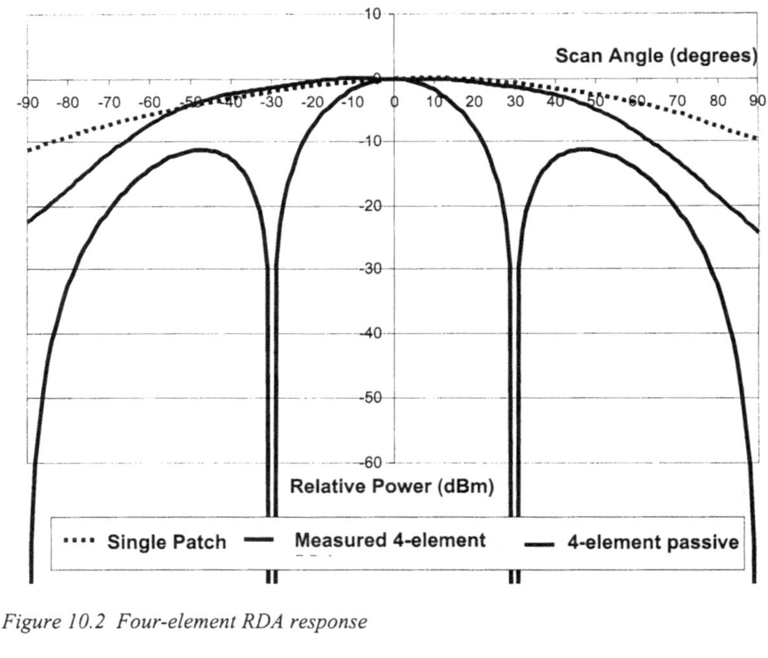

Figure 10.2 Four-element RDA response

Next, the design and behaviour of a compact self-phased integrated mixer active antenna receive array are shown. The array performs automatic beam steering at low IF frequencies by the use of a pilot carrier inserted in-channel with the information signal. Phase compensation is achieved by an integrated silicon diode, using an in-band pilot carrier as a phase reference. The pilot carrier is used as a remote local oscillator to down-convert the incoming signal at the diode. The method of operation is described in [36] and illustrated in Figure 10.3*a*.

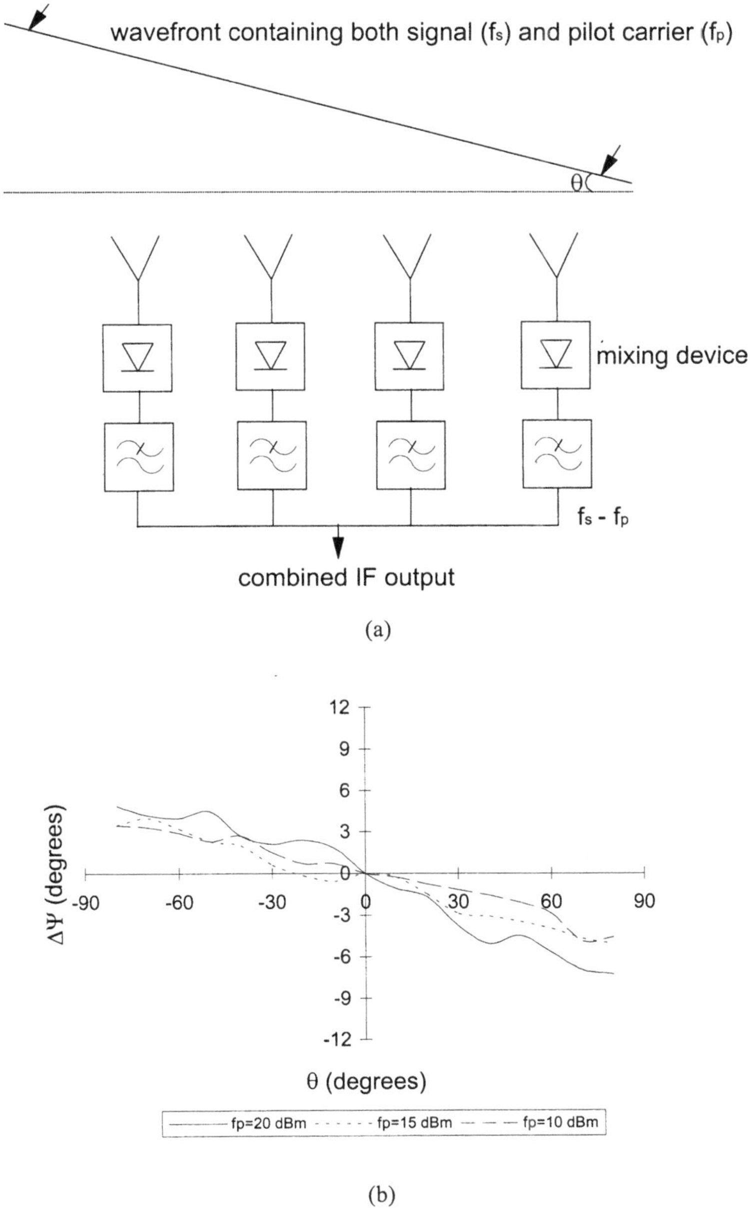

Figure 10.3 Self-tracking integrated mixer receiver

To verify the theory, a two-element array was constructed; the signal frequency, f_s, and pilot frequency, f_p, were 5.2 GHz and 5.18 GHz, respectively, selected to represent a HIPERLAN application. Figure 10.3*b* shows the phase difference between the two IF outputs. Here, a maximum phase deviation between IF outputs of $\Delta\Psi = 5^\circ$ occurs for incident signals of up to 60°, thus indicating that self-steering is occurring on receive. The circuit is robust to remote LO power level. An all integrated version on high resistivity silicon (HRS), for use as a low-cost tracking receiver, has been constructed [37,38].

10.5 Integrated antenna selection

Planar integrated antennas are attractive due to their compatibility with microstrip and CPW transmission line types and for their structural compatibility with monolithic technology. Many different types of planar antenna exist and each has different input impedance and far-field radiation characteristics. Patch and resonant slot antennas exhibit low gain, and hence, broad beamwidth radiation patterns, and consequently, they are useful in mobile communications transceivers or as elements in beam forming arrays. These types of elements can be made to exhibit linear, dual-linear or circular polarisation [39]. Such antennas are resonant and are therefore generally narrowband in performance. In planar integrated form, microstrip patch antennas are most naturally fed using a microstrip feed-line. On the other hand, slot antennas comprising of a narrow slit in a ground-plane can be fed by microstrip, slotline or CPW arrangements. By using inset-feed techniques microstrip patch antennas can be readily matched to 50 Ω [40], and a half-wavelength slot has an impedance of greater than 300 Ω. This can be a problem for integration of the antenna with monolithic technology, due to the ultra-fine line geometry needed to obtain 300 Ω interconnection, and for impedance matching to 50 Ω on high permittivity and mechanically thin substrates. By folding the slot or by feeding it off-centre, its input impedance can be reduced [41], and CPW folded slotline variants can be readily matched. For example, a 60% fractional bandwidth was demonstrated for an input return loss of less than -10 dB [42].

In order to enhance bandwidth performance, designers have developed planar travelling-wave antennas that are, by definition, non-resonant structures. They tend to be larger than resonant antennas, and hence, have narrower beamwidth and higher gain. Two types of travelling-wave antennas exist: leaky-wave and surface-wave antennas. With the former, as wave propagation occurs so does radiation [43], while with the later, radiation only occurs wherever the surface-wave gets disturbed by deliberately placed discontinuities [44].

The choice of an appropriate antenna for a particular application is difficult. Table 10.4 summarises some of the main properties for the most widely cited passive planar integrated antennas, as extracted from [45].

Table 10.4 Planar antenna types

Type	Directivity	Far-field pattern	Polarisation	Bandwidth
Patch	Medium	Broadside	Linear/Circular	Narrow
Slot	Medium/Low	Broadside	Linear	Medium
Ring	Medium	Broadside	Linear/Circular	Narrow
Spiral	Medium	Broadside	Linear/Circular	Wide
Bow-tie	Medium	Broadside	Linear	Wide
Vivaldi	Medium/High	End-fired	Linear	Wide
Leaky-wave	High	Scannable	Linear	Medium

Many of the applications listed in Section 10.6 operate at millimetre-wave frequencies and above. Millimetre-wave frequencies (and corresponding free-space wavelength, λ) are defined here as meaning 30 GHz (λ = 10 mm) to 300 GHz (λ = 1 mm). This range of frequencies encompasses many of the major planned commercial applications [1], such as the microwave video distribution system (MVDS) at 42 GHz, mobile broadband system (MBS) at 51 GHz, automatic cruise control (ACC), radar at 76.5 GHz, and so on. The short wavelengths associated with these frequencies mean that physically small antennas can be electrically large, while maintaining low physical mass and profile. Propagation of these ultra-short wavelength signals is significantly affected by absorption and transmission bands, due to local atmospheric conditions.

The basic approximate relationships between antenna aperture, D, beamwidth, φ, and power gain, G, are given as [46]:

$$G = \eta\left(\frac{\pi D}{\lambda}\right)^2 \tag{10.3}$$

$$\varphi = \left(\lambda/D\right) \text{ [rad]} \tag{10.4}$$

Here, η is the antenna efficiency. These equations show that it is very easy to obtain very narrow beamwidths (i.e. high angular resolution) and high gain at very high frequencies with physically small antennas. The short wavelengths associated with these antennas means that the dimensional tolerances for the antenna must be very fine, typically better than $\lambda/30$. These tolerances can be easily realised in monolithic antennas due to the precision lithography developed for use in conventional microelectronics fabrication, where < 1 μm line widths are routine.

There are three basic types of radiating structures that are currently deployed for use as radiating elements [46]: single resonant elements (or arrays of resonant elements), travelling-wave antennas and aperture antennas. Aperture antennas are further subdivided into horns, reflectors and lenses.

Monolithic reflectors are difficult to produce, even when micromachining techniques are used, and as a consequence will not be discussed further here. Lens antennas have been demonstrated using quartz and silicon as the dielectric material [47]. To be useful, the dielectric needs to be homogeneous and exhibit low RF loss. In practice, they are used mostly to couple the radiation from slot antennas [48]. Horn antennas are traditionally used as feeding structures for reflectors and lenses or as array elements. Integrated horn antennas can be formed by micromachining [49].

Planar antennas are obviously amongst the more attractive for use in monolithic technology. Of these, travelling-wave antennas are attractive at millimetre-wave frequencies, since they radiate energy from wave propagation along an open transmission line structure [43]. With this class of antenna, the angle of radiation and amount is controlled, respectively, by the guiding wave medium and its attenuation coefficient. The resulting antennas can be used to form beam-scanning arrays when fed with a variable frequency source. One example, the relatively wide bandwidth travelling-wave Vivaldi antenna, has been extensively studied in monolithic technology [50].

To control the flow of energy to or from an antenna, a wave-guiding transmission medium is required. This has to be of low loss and exhibit minimal coupling and leakage effects as higher order modes begin to propagate or radiate from discontinuities. Often, at millimetre-wave frequencies and above, the topology of the integrated antenna and/or its system integration makes the provision of such transmission line interconnects difficult. In these cases millimetre-wave energy is guided through free-space. This technique is very important since functionality enhancement can be incorporated into the antenna structure by placing elements into the beam path [51–53]. Importantly, for integrated antenna techniques, a number of circuits developed for use at very high millimetre-wave frequencies have been re-engineered for use at much lower frequencies. One example is the slot-ring mixer [54,55]. Other examples include a broad variety of transponders, non-contact tagging devices (such as those described in [30]), spatial power combiners [55], and phase shifter-less beam-steered arrays [56], which use arrays of active integrated antennas. These methods are generically known as 'quasi-optical' techniques.

All the antenna types mentioned above have been used in both passive and active integrated antennas, specific examples of which will be discussed in Section 10.6.

10.5.1 Substrate choice

The decision as to which substrate will be used to fabricate the integrated antenna, and possibly other circuit functions, will depend on many considerations. Factors such as cost, electrical, mechanical and thermal properties all have to be considered. When monolithic circuitry is to be included at microwave and millimetre-wave frequencies then the substrate choice is normally limited to GaAs, InP or HRS. A particular problem associated with the integration of the antenna onto a high dielectric constant substrate is compatibility with normal foundry processes. Mostly, the concerns here relate to the thickness of the wafer used by the foundry and its relationship to surface-wave effects. Secondly, when constructed on a semiconductor substrate of permittivity ε, the power radiated into the substrate is a factor of $\varepsilon^{3/2}$ larger than the power radiated into free-space [57]. Therefore, unless special precautions are taken, antennas constructed on GaAs or silicon radiate only 2 or 3% energy into free-space, the rest being developed as surface-waves. Consequently, the substrate thickness should typically be less than $0.05\lambda/\sqrt{\varepsilon_r}$, thereby keeping the radiation efficiency of the antenna as high as possible, without the need for extra process steps for realising micromachined trenches or wells (see Section 10.8). The relatively large overall chip area required by the antenna and final yield has very important cost implications, especially with GaAs and InP substrates.

The basic properties of several of the main types of substrates presently used in integration antenna applications are listed in Table 10.5

Table 10.5 Typical substrate RF properties

Material	Approx. ε_r	$10^4 \tan\delta$ (10 GHz)	RMS surface roughness (μm)	Thermal conductivity (W/cm²/°C)	Dielectric strength (MV/cm)
99.5% alumina	10	1-2	2-8	0.3	4
Sapphire (single crystal)	ε_{rx} = 9.4, ε_{ry} = 11.6	1	1	0.4	4
Glass	5	20	1	0.01	-
Quartz (fused)	3.8	1	1	0.01	10
GaAs	12.9	6	1	0.3	0.35
HRS	11.9	10-100	1	0.9	0.3
Air	1	0	-	2.4×10^{-4}	0.03

Alumina is used for the production of planar antennas up to 40 GHz, while sapphire and quartz are used up to 300 GHz. Substrates can be used with thin or thick film manufacturing processes for metal deposition. Various metals are used, including gold, aluminium and copper. Some problems associated with integrated circuit realisation of integrated antennas onto the semiconductor substrates used for MMIC processes are related to the thickness and type of metallisation used.

Typically, in GaAs processes the metallisation is gold, evaporated or sputtered, over a very thin seed layer used to enhance adhesion to the substrate material. The electrical conductivity properties of the deposited conductors are generally not as good as the material in its bulk form so RF conductor losses increase. Also, the metallisation layers tend to be fairly thin (e.g. 1 to 3 μm); much less than several skin depths, as normally required for loss reduction in microwave transmission media. Conductor losses, when included with the relatively high dielectric substrate losses, increase antenna loss, and thus reduce radiation efficiency. As a consequence, it may be necessary to selectively electroplate additional bulk conductor metal onto the original antenna metallisation. This of course represents an additional process step in the normal MMIC fabrication procedure. In the case of silicon, a resistivity > 3 kΩcm must be used in order to keep the substrate dielectric losses to a minimum. This choice of resistivity puts the use of HRS material in conflict with the normal silicon substrates used in VLSI circuitry, as they are generally fabricated on much lower resistivity material (e.g. 0.5 to 4 Ωcm).

Measurements on the attenuation of microstrip lines fabricated on 100 Ωcm resistivity silicon [58] yielded values of 3.6 dB/cm at 10 GHz. Recently, much work has been done to find the best HRS variant for microwave and millimetre-wave applications. When used at millimetre-wave frequencies, as the substrate on which silicon MMICs are formed, the term 'silicon millimetre-wave integrated circuit' (SIMMWIC) has been coined [59,60]. Attenuation values of about 0.6 dB/cm were measured on 5 kΩcm material at 18 GHz [60] and 0.6 dB/cm on high purity 10 kΩcm material at 90 GHz [61].

Also demonstrated was that thermal oxidisation at 1000°C could be tolerated for several hours, without a degradation in HRS substrate resistivity [62]. This opened the way for HRS compatibility with CMOS processing techniques. A demonstration of a HRS receiver, consisting of a 94.6 GHz series microstrip dipole antenna on a 100 μm thick substrate with associated CMOS electronics, was given in [63], as shown in Figure 10.4.

Figure 10.4 RF/CMOS rectifying antennas [63]
(© 1998 IEEE)

When developing planar integrated antennas for use on HRS, some precautions must be employed to maximise performance. For example, direct deposition of aluminium metallisation (normally the metallisation of choice in VLSI technology, which has been shown to give low line losses in a CPW configuration, e.g. 2.5 dB/cm on 10 kΩcm silicon at 30 GHz) onto a HRS substrate can lead to a Schottky barrier junction being formed. This effect has been used to produce a silicon microstrip patch antenna that can be directly modulated. By applying DC bias control to the patch metallisation, direct amplitude modulation of an RF carrier can be achieved [64].

The integral modulator/patch antenna demonstrated was constructed on *p*-type <100> orientation 10 kΩcm HRS wafer, having a doping concentration of 3×10^{12} cm^{-3}. The wafer thickness and the dimensions of the patch are shown in Figure 10.5. Here, a 1.3 μm thick aluminium layer was evaporated onto both sides of the wafer and the topside patterned in order to create a microstrip patch antenna.

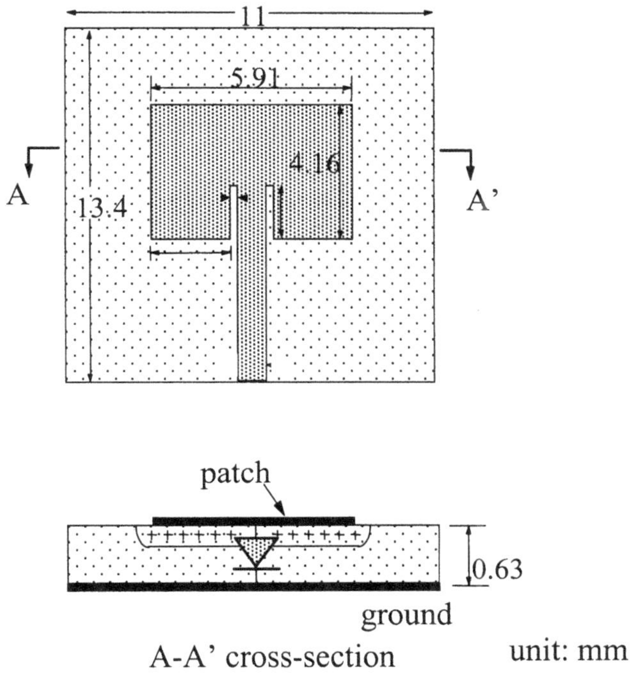

Figure 10.5 Direct modulated silicon patch antenna

There is no SiO_2 insulating layer between the upper conductor and silicon wafer. In addition, the patch pattern was produced on the polished side of the wafer, with the ground-plane on the unpolished side. As a consequence, the small area of intimate contact between the patch metal and polished silicon enables a metal-semiconductor barrier to be formed, while the larger area, having less intimate contact between the ground metal and the rougher silicon, is equivalent to many metal-semiconductor barriers in a shunt connection, resulting in an ohmic contact. The overall effect is to form a distributed rectifying diode between the patch and the ground-plane, as illustrated in Figure 10.5.

All primary antenna characteristics, such as working frequency, bandwidth and radiation power, can be controlled by means of a DC bias voltage applied to the patch metallisation. The transmitted signal's spectral response for the antenna (when operated in transmit mode) and direct signal detection by the antenna (when operated in receive mode) showed that direct amplitude modulation to the patch antenna was realised by direct encoding of the baseband modulation signal to the patch. Far-field measurements showed that DC bias control of the patch antenna acts as an electronic attenuator, thereby reducing the radiation level without significantly changing the radiation patterns associated with the patch antenna. In other words, direct amplitude modulation of an RF carrier was possible.

However, if additional circuitry requiring DC biasing is to be included in the antenna in order to modify its functionality, then this effect can be undesirable. To avoid this, a ~ 0.1 μm thick SiO_2 layer can be added to passivate the substrate. However, it was shown [65] that unless carefully controlled, this radically increases CPW line loss to 18 dB/cm at 30 GHz. This is due to the formation of induced charge at the silicon surface. The effect of the induced surface charge layer can be removed by introducing a thin layer (e.g. 0.6 μm) of polycrystalline silicon between the HRS substrate and the SiO_2 layer. This has the effect of reducing CPW line losses to about 1 dB/cm at 30 GHz [65]. For best integrated antenna or interconnect performance on HRS, the strategy outlined above appears to be the most appropriate in returning minimum line losses without recourse to micromachining techniques [66].

Throughout all of the discussion on silicon so far, in this section the emphasis has been on HRS, due to its low substrate losses. Recent work on 50 Ω CPW lines has shown that a low resistivity (e.g. 8 to 10 Ωcm) boron-doped silicon substrate, which has had a 20 μm thick porous silicon layer (PSL) formed on its surface, can exhibit losses of only 1 dB/cm at 4 GHz [67]. This is a very promising development, since it may eventually lead to a complete systems solution involving integrated antennas and VLSI cores. Here, largely conventional microelectronics techniques can be used with selective PSL formation to reduce dielectric loss in critical RF regions of the circuit.

10.5.2 Measurement issues

Of particular difficulty is the measurement of a planar integrated antenna's input return loss and far-field pattern. The difficulty arises due to the fact that the integrated antenna can only be connected to measurement equipment through an on-wafer probe. Integrated slot antennas present particular difficulties, since these radiate into both half spaces, and as a consequence the circuit cannot be directly mounted onto the metal chuck of the probe station for return loss verification. In view of this, the wafer chuck and substrate on which the antenna is formed have to be separated. In [68], a Styrofoam spacer ($\varepsilon_r \approx 1.0$) was used between the chip and a piece of RF absorber. The thickness of the Styrofoam spacer is critical, since it acts to transform the absorber impedance to that of free-space in classical radome fashion. Thickness values of 0.6 and 0.7 wavelengths were found to work best with 20 to 30 GHz slot antennas fabricated on 90 μm thick GaAs substrates. Once mounted in this way, on-wafer calibration and measurement can then take place in the usual fashion.

Conventional on-wafer probing has the chip lying parallel to the horizontal plane. However, this is incompatible with standard vertical antenna mounting configurations used in far-field test ranges. In [68], a portable test-fixture was fabricated using a specially constructed dielectric (Delrin, $\varepsilon_r = 2.1$), complete with a vertical Styrofoam chuck and dielectric positioners for the wafer probes. The wafer, containing the antenna under test, is held vertically in place by two small dielectric clamps attached to the Styrofoam chuck. The effect of this arrangement, with respect to its influence on the EM behaviour of the antenna under test, was assessed using numerical EM field simulation techniques. The test jig was optimised in order to reduce electromagnetic interaction between the probe and the integrated antenna under test.

10.5.3 Packaging

Environmental protection and the integration of monolithic antennas are important issues which, to date, have been the subject of relatively little work. In [69], an RF module was proposed that used a hermetically sealed package in order to house both MMIC devices and a planar antenna. The package used Kovar with a Borosilicate glass window (which has the same expansion coefficient as Kovar) for the antenna, and housed a complete 59.5 GHz transmitter. The transmitter consisted of an MMIC balanced mixer, band-pass filter, two 180° couplers, a 4-stage power amplifier and a microstrip patch antenna. These were all wire-bonded together, as shown in Figure 10.6.

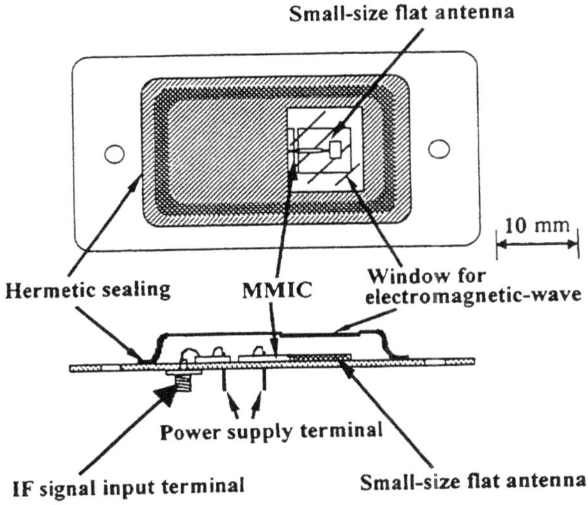

Figure 10.6 Millimetre-wave wireless module [69]
(with permission, Miller Freeman UK Ltd)

A rather different approach was adopted in [70], where a 34 GHz demonstrator was given. Two multilayer schemes were developed, that were aimed at the construction of phased array radar modules. Heat removal from the MMIC devices and self shielding of these devices from the integrated antennas also featured in the discussion, as did package resonance considerations.

A very interesting integration strategy was presented in [71]. The structure did not contain an integrated antenna, but is nevertheless worthwhile reporting here. A 2.5 to 3.5 GHz receive module for a phased array was constructed using three GaAs MMIC chips (low noise amplifier, attenuator, phase shifter). These components were integrated onto a silicon substrate using low loss temperature expansion matched (to silicon and GaAs) inorganic film of an unspecified type.

This was plasma etched to form holes, into which the GaAs devices were inserted. Plasma etching was also employed to produce via-holes for direct metallisation of the multilayer interconnects, used to connect the devices together.

The example cited in [72] demonstrated a 51 GHz receiver system, where a monolithic LNA was connected to an integrated patch antenna using conventional wire bonding and flip-chip techniques. It was shown that at 70 GHz more than 0.5 dB improvement on insertion loss can be achieved by using the flip-chip technique. The cumulative insertion loss for each method, for all of the interconnects needed to form the system, was 0.15 dB (with flip-chip) and 2.6 dB (with wire-bond).

In [73], a packaged front-end consisting of planar circuits including an antenna and waveguide components, was demonstrated. The package carrier was constructed using a multilayer low temperature co-fired ceramic, with moulded inserts to allow inclusion of copper–molybdenum heat sinks. A metallised plastic housing covered the structure and had integral waveguide-to-microstrip transitions, RF chokes and antenna included in its design.

An entirely hybrid assembly of a millimetre-wave front-end, consisting of five sequentially rotated silicon integrated rectifying antennas (rectennas) in a standard IC package, to form a polarisation insensitive multi-beam Doppler radar sensor (having an aluminium heat sink with polystyrene lens), was described in [74].

10.6 Integrated antenna examples

It has been shown that GaAs patch and silicon antenna elements can normally be designed to exhibit a return loss of around 10 dB at 65 GHz, while yielding a maximum gain per element of around 2.5 dBi [75]. Slot antennas have inherently wider bandwidth than patch antennas. They also have the advantage that coplanar MMIC devices can be easily integrated across the slot without the need for through-hole vias. Also, when mounted on a cavity, they can be made to have a radiation efficiency of up to 50% [10,41].

An intrinsically matched CPW-fed slot-ring antenna operating at 10 GHz was designed, fabricated and tested on a 10 kΩcm HRS wafer [76]. The antenna had a finite ground-plane (the size of this ground-plane was selected to be greater than 10 times the slot width, in order to minimise any edge effects), and the square slot-ring was selected to have $\lambda_g/2$ sides, thereby giving it a total loop perimeter of $2\lambda_g$. The measured return loss was -24 dB, and far-field radiation patterns were in good agreement with simulated results. The operating bandwidth (defined for a return loss of -10 dB) was 9%, the calculated beamwidth was 120° and the gain was 5 dBi.

The slot-ring antenna was fabricated by evaporating a 650 μm thick HRS wafer on one side with a 1.5 μm thick aluminium film. The CPW-fed slot-ring antenna provides the facility for direct on-wafer RF probe testing of the structure. The planar silicon slot-ring structure is easily adaptable for use with devices such as mixer diodes, FETs, etc., without the need to penetrate the substrate or for the back-plane metallisation (necessary for microstrip patch antennas).

Since a $2\lambda_g$ square wire-loop dipole antenna has a real input impedance at resonance of approximately 200 Ω, and then due to the slotline/dipole duality [77,78], the input impedance of the designed slot-ring antenna is estimated to be 200/4 = 50 Ω. Thus, the input impedance of this antenna is matched to 50 Ω without the need for extra matching circuitry.

A design methodology for the integration of silicon rectennas with CMOS amplifiers was given in [79]. The arrangement consisted of a 61 GHz microstrip patch antenna with integral Schottky diodes constructed on HRS. The objective was to define precisely the diode bias and CMOS audio amplifier configuration that maximises the minimum detectable signal level sensitivity of the arrangement. It was shown that the frequency selectivity of the rectenna was mainly governed by the bandwidth of the match between diode and antenna. For a bandwidth of ±1.5 GHz, at 61 GHz, the amplifiers must have an input impedance of between 10 Ω and 1 kΩ with a Schottky diode that typically generates 20 nV/√Hz of noise voltage at frequencies above 10 kHz. Below this frequency, the $1/f$ noise of the Schottky diode dominates noise behaviour.

An all silicon transmitter was demonstrated in [80]. In this paper, a coplanar slot resonator formed a slot antenna at 89 GHz and a double Read IMPATT diode provided the power source. The effect of surface-wave losses on antenna radiation efficiency was noted. The energy leaked by this mechanism coupled from the rear of the chip to a *W*-band measurement system. About 4.4 mW of power was radiated and the oscillation exhibited -82 dBc phase noise at 100 kHz from the carrier. The arrangement was stabilised for temperature and its spectral purity enhanced in [81] by mounting onto an integrated resonant cavity.

The techniques developed in [80] were extended in [82] to produce, on a HRS substrate, a planar integrated microstrip patch radiator operating at 147 GHz, using a silicon double Read IMPATT diode. This device generated 100 mW of power with 4.5% DC-to-RF efficiency, and had an oscillator purity of -75 dBc/Hz at 100 kHz from the carrier. To avoid surface-wave effects, the wafer was only 70 μm thick. The IMPATT diode was beam-lead mounted in a via-hole that was produced using micromachining techniques (Section 10.8). Figure 10.7 shows the layout of the planar *D*-band oscillator with the beam-lead configuration adopted.

A circularly polarised rectenna was developed in [83] using a novel arrangement of patch antennas, fabricated on <100> 100 μm thick HRS. The antennas are nearly square, such that the phase difference between the TM_{01} and TM_{10} modes is 90°, in order to create circular polarisation. Two antennas are then used in a dual-antenna arrangement and a silicon detector diode flip-chip mounted between their output connections. Better than 14 dB cross-polarisation discrimination was obtained with this configuration. The intended applications for antennas were for its use in either a polarimetric radar system or a detector in a wireless communications link.

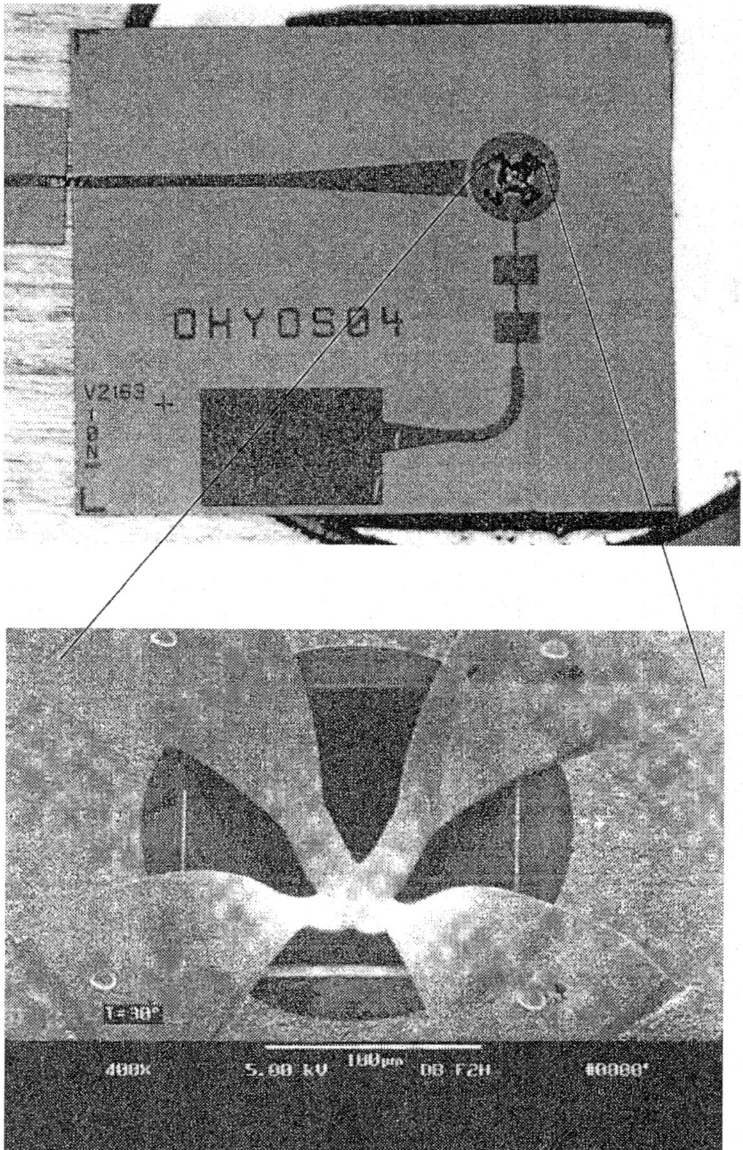

(b)

Figure 10.7 Planar D-band oscillator with beam-lead diode: (a) circuit and (b) beam-lead diode [82] (© 1997 IEEE)

By using an all silicon rectenna and self-oscillating detector (SOD), coupled using normally undesirable surface-wave leakage, a 67 GHz homodyne sensor was constructed. The SOD was provided with an LO feed that is applied to the rectenna array with a 90° phase lag. Again, an IMPATT diode was used as the SOD's active device. With this configuration, a direction sensitive short range CW Doppler can be created using homodyne I/Q signal detection. The use of surface-wave coupling to the rectenna allows direction sensitive Doppler extraction and improves the minimum detectable signal level of the system by 10 dB above that which is obtained by using the noisy IMPATT SOD alone [84].

The combination of a double slot-ring antenna detector and an extended hemispherical silicon lens was reported in [85]. Operating over 86–90 GHz, the structure reported had approximately 7 dB double sideband conversion loss. By using a CPW-fed network the antenna was uniplanar, i.e. it did not require any through-hole vias and had a performance that was within 3 dB of the best waveguide mixers available at that time. In the realisation reported, the diode was discrete and attached using silver epoxy glue in series between the slot antennas, to achieve a sum-mode pattern. The use of the silicon lens increased the directivity of the antenna to more than 22 dB with a Gaussian coupling efficiency of 93%.

A broadly similar concept, this time using a hemispherical lens to produce a semi-infinite dielectric, was proposed in [86]. Here, a 100 μm GaAs substrate was used to produce a 94 GHz slot-ring antenna with 1.8 kΩ radiation resistance. This was placed on a dielectric-filled circular metal waveguide. It was then shown, using an integral equation solution method, that the arrangement behaves in a similar fashion to a slot-ring placed on a semi-infinite substrate. The waveguide was then matched to a radiating aperture consisting of a solid metal ground-plane. The resulting configuration was termed a bath-plug antenna and exhibited a radiation resistance value of around 100 Ω.

By using CPW techniques (on 2.5 kΩcm silicon) integrated with an anti-parallel diode pair and a slot-ring antenna, the work above was extended so that a dual polarised sub-harmonically pumped receiver could be developed. The circuit operated between 86 and 92 GHz, with better than 25 dB isolation between antenna ports, 8 to 9 dB conversion loss and 1800 K double-sideband noise temperature [87].

The CPW configuration used for the slot-ring antenna, when connected to the dielectric lens with the same dielectric constant, creates an apparent dielectric half-space which is incapable of supporting surface-waves. Hence, a relatively thick silicon wafer can be used. This means that the wafer is easier to handle during manufacturing. A second advantage is that since the antenna radiates preferentially into the substrate, the presence of the lens will lead to narrow band, high gain, far-field radiation patterns without the need for a larger array of elements.

The use of a silicon lens with a GaAs substrate, patterned with a log-periodic antenna and integrated with a 0.8 μm planar Schottky diode, was given in [88]. The device had 15 dB conversion loss and 8900 K noise temperature, at 761 GHz, when measured at room temperature. This can be seen in Figure 10.8.

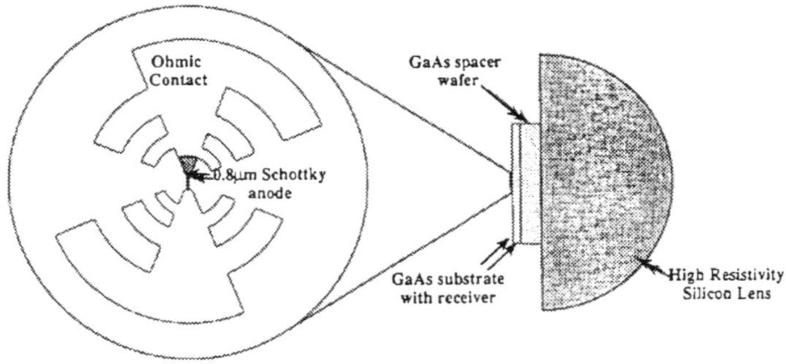

(a)

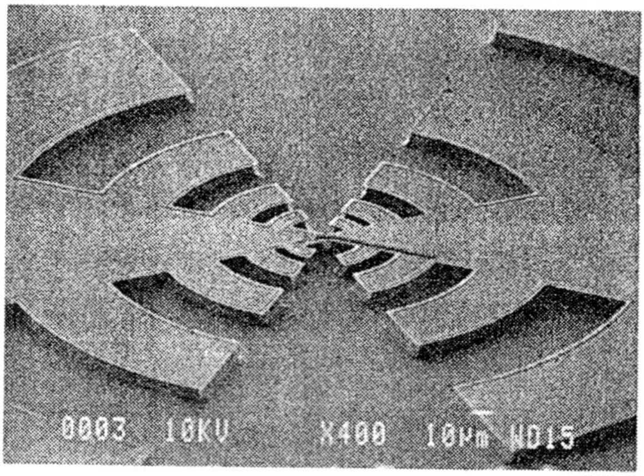

(b)

Figure 10.8 Planar integrated Schottky receiver: (a) receiver and (b) rectenna [88] (© 1993 IEEE)

A completely integrated GaAs transmit antenna was demonstrated in [89], as illustrated in Figure 10.9. In this configuration, a non-resonant tapered slot (or notch) antenna was loosely coupled to a CPW resonator, to form the resonant circuit for a Gunn diode with integral copper heat sink. Unfortunately, the radiation properties of the structure were not reported.

The idea of utilising the back-side metallisation of a GaAs MMIC wafer, normally reserved for the lower ground-plane, for realising an antenna element was shown in [90]. Here, polyimide and silicon nitride dielectric layers were used to separate a CPW corporate feed structure from a patch. Then a direct corporate interconnection between corporate feed and antenna is made using the through-hole via made available in the GaAs process used. At 35 GHz, approximately 35 dB of return loss was obtained at the input to the feed network. In [91], a low noise HEMT amplifier was integrated using indirect aperture coupling to a slot antenna etched on the rear face of a MMIC.

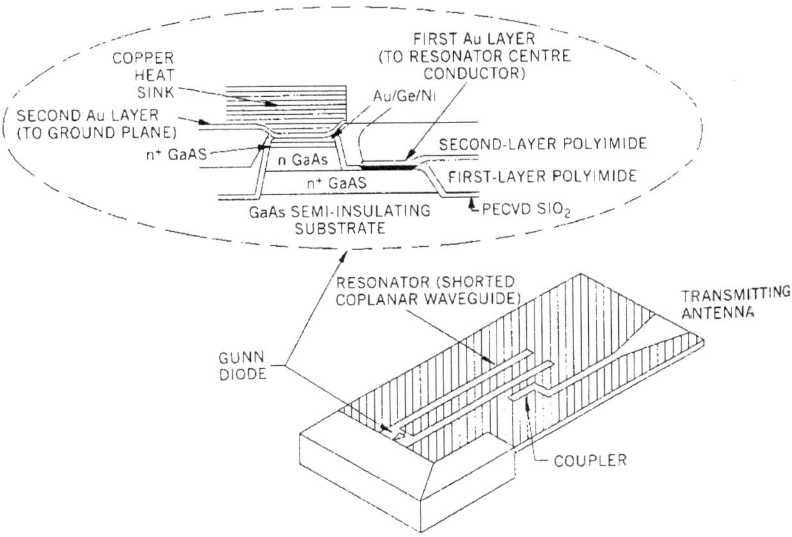

Figure 10.9 Monolithic Gunn integrated notch antenna [89]
(Reproduced from Navarro/integrated active antennas and spatial power combining copyright © 1984 IEE, reprinted by permission of John Wiley and Sons Inc.)

The idea of exploiting the various layers available in a standard semi-insulating GaAs process to realise an integrated antenna has been developed beyond the examples given above. As discussed previously, the bandwidth and efficiency of an integrated antenna are determined by the height of the substrate. Normally, the bandwidth of the elements (if placed directly on the substrate) will be less than a few per cent. Therefore, it would be very useful if their bandwidth performance could be improved. It is known that by stacking planar microstrip patches bandwidth improvements can be made [92].

Consider a typical GaAs process consisting of five layers. The first, third and fifth layers are metal (1 μm thick), while the second and fourth layers are 2 μm thick polyimide dielectric (ε_r = 3.4). These layers sandwich the semi-insulating GaAs substrate to give the total structure a nominal thickness of 400 μm. As in [90], interconnection between layers is made using through-hole vias made to pass through windows placed in the polyimide layers. In [93], a stacked dual-band microstrip patch element and a single dual band integrated spur line antenna were demonstrated. The fifth and third metal layers in the MMIC were used to fabricate co-centred rectangular patch elements. The resonant frequency of the upper patch is slightly higher than the lower patch, and the patches are electromagnetically coupled through the polyimide dielectric layer to increase bandwidth.

For single elements, separately constructed on each layer, the fractional bandwidth (defined by a -16dB return loss) was between 4% and 6%, at 35.5 GHz and 35 GHz, respectively. When stacked, a dual-resonant mode was formed at 35.65 GHz and 38.9 GHz, with a -16dB defined fractional bandwidth of 3.5% and 1.67%, respectively. Although not demonstrated, an iterated design of the top patch (to reduce its operating frequency) may have led to an overlap in these separate dual mode responses. Hence, a single broader band response could have been created. A spur-line antenna, consisting of a notched microstrip patch antenna placed on metal layer five, also showed narrowband dual-frequency responses at 33.7 GHz and 36.75 GHz. Here, the element required 50% less area than a stacked multilayer integrated GaAs MMIC antenna operating at the same frequencies.

A very different and novel strategy for a beam steered antenna was presented in [94]. This paper demonstrated a radical approach to solving the problem of beam steering antennas. The operation of this antenna is different to all other configurations presented thus far, as it uses the direct interaction between millimetre-wave energy and solid-state plasma formed in a silicon waveguide. The resulting structure was termed an MMIC millimetre-wave scanning plasma-grating antenna. The antenna consists of a 1D array of 200 PIN diodes, each with a separate (p^+) anode and a common (n^+) grounded cathode. The upper electrodes are individually connected to a current source. When switched on, the diode forms a hot pixel, creating an electron–hole plasma. These carriers prohibit the flow of millimetre-waves through the silicon slab waveguide. The hot pixels are switched in groupings, which are arranged with a period Λ. This presents the millimetre-wave signal with an amplitude diffraction grating. The arrangement is excited by a dielectric rod waveguide at an angle greater than the critical angle of the silicon slab air interface, so that a leaky-wave cannot be formed. However, when the grating is activated (by turning the PIN diode grating on), a leaky-wave will result at an angle determined by the period of the PIN array grating. Since the period of the grating can be selectively altered (by switching PIN diode groups), beam scanning can be made to occur. The array was tested at 94.3 GHz and beam steering over a 30° angle was obtained.

10.7 Photonic bandgap antennas

As pointed out in Section 10.2, planar antennas constructed on high dielectric constant materials (e.g. silicon or GaAs) are compatible with normal microelectronic fabrication processing procedures, but are subject to narrow-bandwidth, inefficient radiation, high cross-polarisation, etc. due to surface-wave effects. It would be ideal if high dielectric constant materials could be used to fabricate antennas without these detrimental effects. Recently, one possibility to overcome this has become available by using specialised techniques, whereby an artificial dielectric substrate material is synthesised [95]. This technique utilises a photonic bandgap (PBG) substrate. This is a periodic structure that exhibits band-stop transmission characteristics (i.e. high reflection coefficient) over a range of frequencies defined by the periodicity of the structure. This region is called the bandgap [96] and gives an additional degree of flexibility in the design of electromagnetic structures, where control can be exercised over normally unwanted wave propagation effects (e.g. surface-wave leakage).

A series of photonic bandgap lattices have been proposed recently, which, due to their planar nature, lend themselves to application in planar integrated antennas [97,98]. A 2D PBG lattice has been designed, which enhances the radiation characteristics of microstrip patch antennas [99]. The properties of the lattice are designed in such a way that the dominant microstrip loss mechanism (i.e. the transverse magnetic TM_0 surface-wave mode) is inhibited over a wide frequency range. The concept is shown in Figure 10.10 for a surface-wave suppressing PBG substrate surrounding a microstrip patch antenna [99].

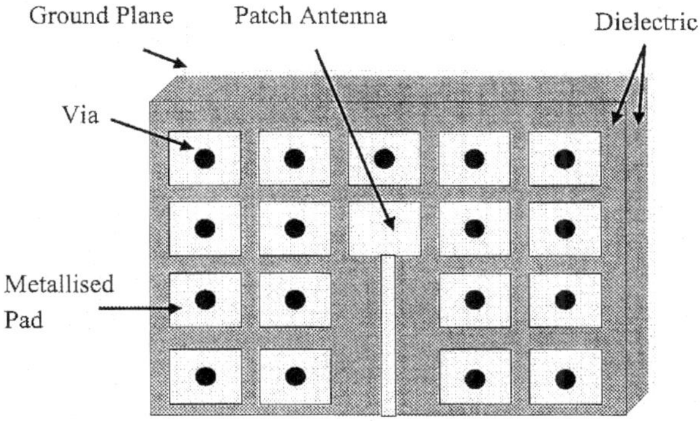

Figure 10.10 Patch antenna surrounded by 2D PBG lattice

In [99], a 635 μm thick dielectric substrate ($\varepsilon_r = 10.2$) has a square lattice of pads. Each pad had dimensions of 240 × 240 μm^2, grounded at their centres by metal vias, with 200 μm gaps between them. With this geometry, the structure was shown to exhibit band-stop properties from 8.6 to 12.66 GHz and 23.44 to 27.36 GHz, irrespective of polarisation and wave direction. A microstrip patch antenna was designed to operate at 14 GHz, so that any TM_0 surface-waves would be prevented from propagating (due to the pair of band-stop responses exhibited by the structure). The patch dimensions are 3.0 × 4.3 mm^2 and an inset microstrip feeding arrangement was used.

The resulting antenna showed a 5.4% fractional bandwidth, over which the return loss was better than -16 dB, which was about 3.4 times greater than that from a conventional microstrip patch antenna designed to operate at the same frequency on a homogeneous dielectric substrate. The photonic bandgap antenna exhibited less TM_0 leaked power into the substrate than the reference antenna, since there were reduced ripples in the backlobe radiation in far-field patterns. In addition, the antenna was about 85% efficient and exhibited about 6.8 dBi (i.e. around 1.6 dB higher gain than its conventional counterpart).

In [99], it was suggested that the suppression characteristic of the PBG engineered substrate could be used to reduce mutual coupling effects between adjacent microstrip patch antennas, which could, in-turn, help alleviate scan blindness in planar integrated phased arrays.

In [100], it was shown that a dipole antenna printed over a 3D PBG substrate (consisting of a periodic grid of square cylinders) could have its gain enhanced significantly. Up to 15 dBi gain in the *H*-plane at 27° beam angle was reported at 28 GHz. In the same paper, the use of the PBG as a superstrate placed above a conventional printed dipole was also reported to give enhanced directivity.

The radiation patterns of integrated dipole antennas on other types of 3D PBG material, such as a stacked multilayer rectangular lattice formed using dielectric rods, have been demonstrated and an optimum geometry for best boresight radiation pattern found in [101]. This work also showed that within the stop-band the PBG structure behaved as a dissipationless reflector. The concept of using a 3D stacked dielectric rod arrangement as a reflector was further explored in [102], for use with a monopole element at 12.7 GHz, and with a 2D PBG lattice in [103]. A similar woodpile arrangement was shown to improve the gain of a slot antenna by 2 to 3 dB at 13 GHz [104].

10.8 Micromachined antennas

It has already been shown that the low leakage loss at high frequencies of HRS makes it a feasible foundry material for microwave and millimetre-wave circuits. However, when used as a microstrip or slotline antenna substrate, silicon results in small antenna size, narrow antenna bandwidths and low radiation efficiency. At millimetre-wave frequencies, any antennas developed using this material will have small dimensions (e.g. $1.0 \times 1.6\ \text{mm}^2$ for a rectangular microstrip patch or $0.8 \times 0.8\ \text{mm}^2$ for a square slot-ring, both at 35 GHz). Several methods were reported recently to overcome these problems. These include cavity and photonic bandgap methods.

10.8.1 Trench etching

10.8.1.1 Patch antennas

Micromachining technology can employ wet-etching of the substrate to create edges that taper the antenna away from the patch or slot edges. As a result, an overhanging conductor remains so that EM fields can radiate directly into free-space [105–106]. Figure 10.11 shows the arrangement for a trenched microstrip patch antenna.

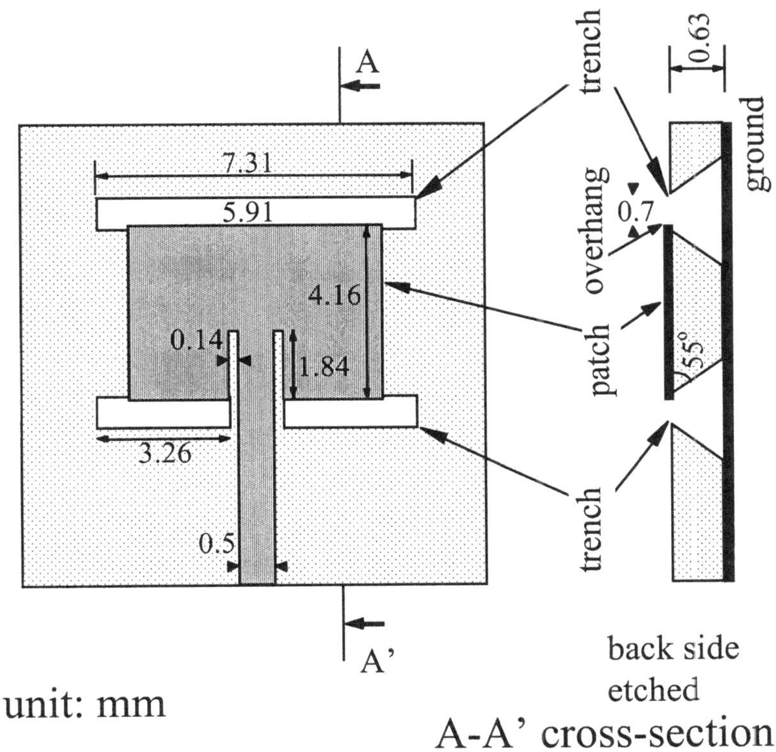

Figure 10.11 Trenched microstrip patch antenna

It can be seen from Figure 10.12 that with the back-side etching approach, the localised fields of a microstrip line or a slotline are less bound to the silicon substrate than they would be if: (i) no trench was used, or (ii) if the etching was done from the front-side of the wafer. This technique effectively increases the size of the microstrip/slot antenna for a given operating frequency, while simultaneously improving radiation patterns and enhancing radiation efficiency.

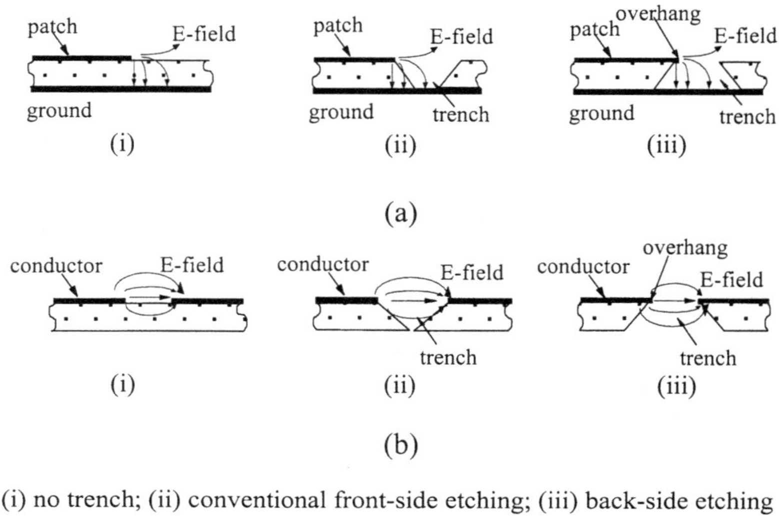

Figure 10.12 Comparison of edge field radiation of trenched: (a) microstrip patch and (b) slotline antenna

The micromachined antennas were built on a HRS wafer of <100> crystal orientation and 10 kΩcm resistivity. To achieve the required front-side conductor overhang, the wafer was first through-trenched from the back-side using an anisotropic etchant, potassium hydroxide (KOH). Due to the <100> orientation, 55° tapered edges are produced. The polished front-side of the wafer was metallised with aluminium and then patterned, so that the microstrip patch was optically aligned with the sharp edges of the two trench sides. The next step was to generate the overhang cantilever effect. For this, the wafer was dipped into an isotropic etchant, CP4 (a mixture of nitric acid, hydrofluoric acid and water), for 10 seconds. Since the CP4 etching speed for silicon is much greater than that of aluminium, in 10 seconds the sharp trench edges on the silicon will be etched away by 2 μm and the aluminium is only etched by 0.2 μm. In this way, the aluminium conductor overhang and a reverse taper silicon edge effect are created, as shown in Figure 10.13.

Figure 10.14 shows trenched microstrip patch arrays on a HRS wafer of <100> orientation and 10 kΩcm resistivity. Here, trenches are micromachined at patch radiating edge pairs, either from the back-side or the front-side of the wafer. In the front-side etching cases the trench depth is 400 μm and a 1.3 μm, aluminium metallisation at the back-side of the wafer is produced as a ground-plane. The wafer is 630 μm thick so the trench etching does not penetrate the wafer. While in the back-side etching cases, the 630 μm thick wafer is through-trenched and a copper foil (coated with a highly conductive pressure sensitive adhesive) is attached firmly to the wafer's back-side for use as a ground-plane. The aluminium conductor overhang effect is created for each back-side etched case.

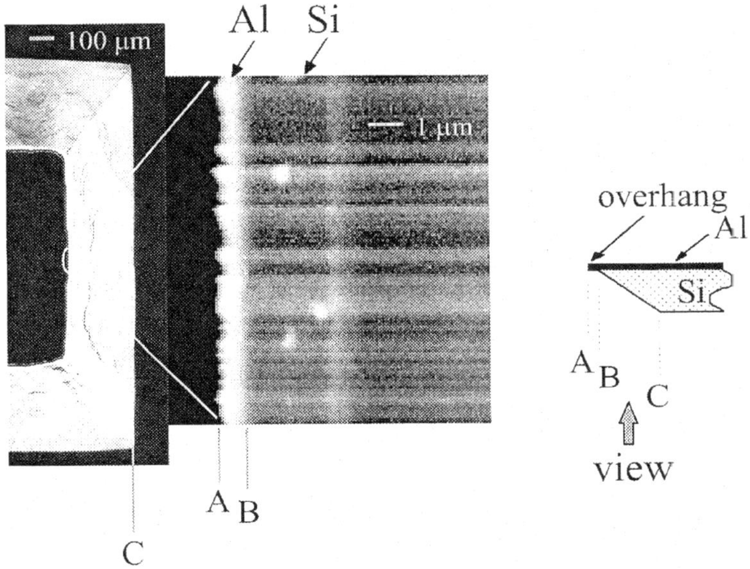

Figure 10.13 Reverse side wet-etched silicon wafer with cantilevered aluminium conductor

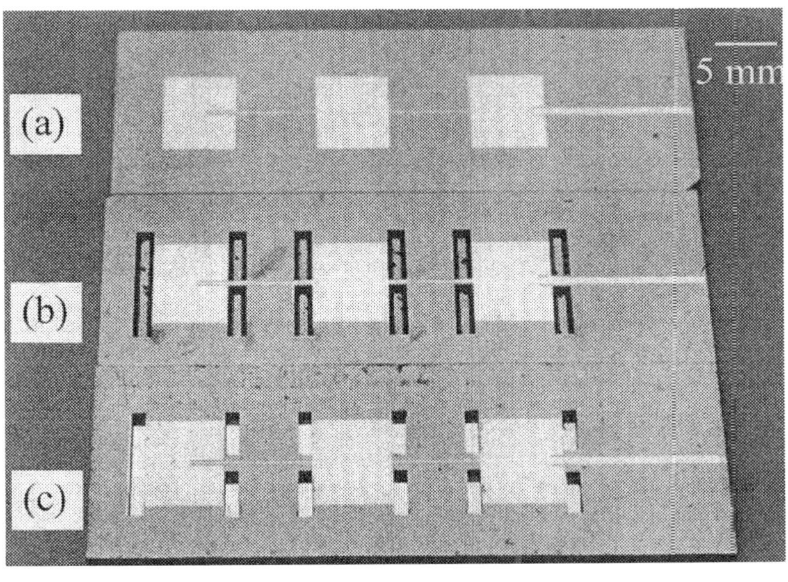

Figure 10.14 Photograph of micromachined patch antenna arrays: (a) untrenched, (b) front-side trenched and (c) back-side trenched

It was found that trenches placed at the non-radiating edges of a front etched patch have less effect on the antenna resonant frequency than those placed at the radiating edges. When etched from the rear, the trenches have considerably more effect on the resonant frequency. This is to be expected, since the slot radiation pair of the microstrip antenna is now operating mostly into air and not into dielectric, as was the untrenched case. Radiation-edge placed slots dominate the resonant response of the antenna. It was also seen that the combined technique of back-side etching and patch conductor overhanging trench edge effect increases the resonant frequency (from 10.5 to 13.78 GHz); this is much more than that of the normal front-side etching method (from 10.06 to 10.56 GHz). This means that for a fixed resonant frequency, a back-side trenched patch antenna would have a larger size than an untrenched one. This feature is an advantage for a millimetre-wave patch antenna, as the dimensional increase eases the difficulty of fabrication tolerance requirements.

The trenched patch has a wider operating bandwidth of 720 MHz, for less than -10 dB return loss, as compared to 515 MHz for the untrenched case. In addition, the back-side trenched patch with cantilevered conductor has more uniform radiation patterns than those of the untrenched case. Also, the radiation power levels of a trenched patch are approximately 1.2 dB higher than an un-trenched patch. These results suggest that a higher radiation efficiency has been achieved in the trenched patch. This is because the absence of silicon at the radiating edges of the patch makes far more efficient radiation of the EM fields, as they are mostly in air.

10.8.1.2 Slotline antennas

Slotline antennas have the advantage of easy integration with two or three-terminal active devices when compared with microstrip based planar antennas since via-holes are not required. A simple and compact FET active slot-ring antenna was proposed in [107]. This antenna was constructed on silicon using the techniques described above [106], as shown in Figure 10.15. In order to physically support the substrate and conductor inside the slot-ring, silicon bridges were required at each corner of the square loop (i.e. to protect the substrate from being etched away).

Two types of slot-ring antenna were fabricated: one with trenches and silicon bridges along the slot-ring and another without. An ATF13336 packaged GaAs FET was attached (using conductive epoxy glue) onto each of the two antennas. It was shown that the trenches increased the oscillating frequency of the active slot-ring antenna. This means that for a fixed resonant frequency, a trenched slot-ring antenna would have a larger size than an untrenched one. Again, this is helpful in relaxing dimensional restrictions for millimetre and sub-millimetre-wave applications. Simulation showed that the trenched silicon substrate is equivalent to an untrenched one with the same thickness substrate ($\varepsilon_r = 8.9$). Similar to the microstrip patch antenna case, the *E*-plane and *H*-plane radiation power levels of the trenched slot-ring antenna are approximately 1 to 2 dB higher than those of the untrenched case. This indicates that the trenched active slot-ring antenna can achieve higher radiation efficiency.

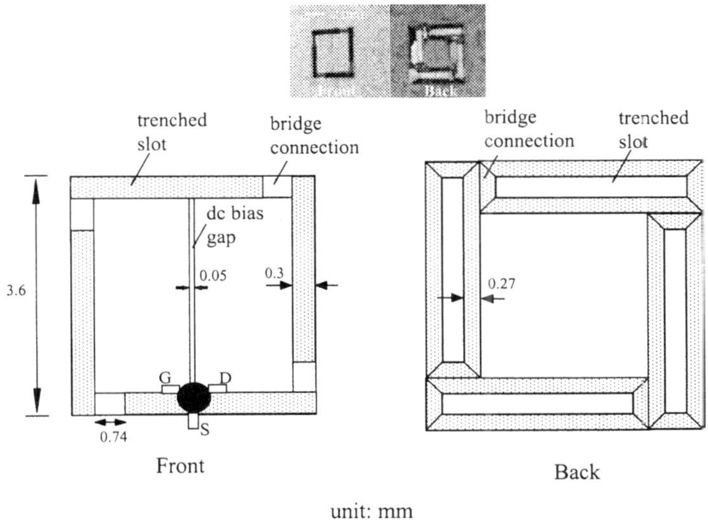

Figure 10.15 Trenched active slot-ring antenna on silicon

10.8.2 Cavity etching

Semiconductor cavities can be formed by wet-etching holes in the dielectric material [108,109] and then stacking the resulting laminations, as shown in Figure 10.16. At 35 GHz, a minimum of four such laminations (each formed from 625 μm through-etched silicon blanks) are required to obtain reasonable unloaded-Q values. For wet-etching of a cavity resonator [109], anisotropic crystallographic wet-etching produced through-holes with 56° sloping edges. This procedure was calibrated so that blind-holes (i.e. top and bottom caps) could also be created. However, the relatively slow etch rate of 2 μm/min resulted in the masking material also being attacked. As a result, a rough floor was left on the bottom of the hole forming the cavity walls. To overcome this, a faster wet-etch of CP4 (with a rate of 11 μm/min) was used to produce a smoother floor surface.

Cavities of various depths were created by forming laminations of blind-hole and through-hole silicon laminations [109,110]. For cost reduction, the spacer and bottom cap laminations are formed using suitably metallised low resistivity silicon material. Here, a variety of techniques and metals were used in order to establish a procedure that would result in metallisation of the near vertical cavity walls. Evaporation and sputtering of gold, aluminium and copper was attempted. Evaporated 1.8 μm thick aluminium produced excellent results, giving 2.7 μΩcm resistivity and good adhesion to the substrate. A copper evaporation process was also developed which yielded 6.5 μm thick layers having 1.7 μΩcm resistivity. One concern for the design of the silicon cavity is that in order to obtain reasonable unloaded Q-factors at millimetre-wave frequencies (e.g. 35 GHz), the depth of the cavity needs to be at least 2.0 mm. As the standard silicon substrate thickness is only 625 μm, several laminations are required.

Figure 10.16 Photograph of stacked lamination silicon micromachined cavity

By gluing pre-metallised laminations together using an alignment jig, characteristics that followed almost exactly classical rectangular metal cavity theory were obtained. It was proven by this approach that three or four laminations provided acceptable Q-factors. When closed cavity resonators were tested, loaded Q-factors of 340 for copper and 24 for aluminium were measured. Thus, the copper metallisation was found to be superior for realising higher-Q resonant structures [111].

A novel antenna that deliberately used an electrically thin cavity has also been developed. As the metallisation was less than the skin depth, almost unimpeded field penetration through the silicon material could be made to occur [111]. In this way it was demonstrated that the cavity could be used as an antenna in its own right. The approach could also be used to obtain simplified coupling in arrays of cavity-backed active antenna oscillators, removing the need for internal iris coupling that requires micromachining of the cavity walls.

As an alternative to wet-etching, dry sand abrasion can be used by stacking two silicon laminations. A wideband microstrip patch antenna was produced in [112], which had a performance close to that obtained from antennas on a lower dielectric constant substrate. Here, a 254 μm thick <111> orientated HRS wafer was patterned with a microstrip patch antenna. This was then glued on top of a second silicon wafer, this time having a lower resistivity. Since the dimensions of the spacer are not critical, the through-hole was made using abrasive etching in-conjunction with a metal mask (chemical etching could also have been used). For the antenna, operated at 16 GHz, its equivalent effective dielectric constant was estimated to be 2.8. The bandwidth of the antenna (for better than -10 dB return loss) was 12.5%, compared to 2% for a conventional patch antenna constructed on a substrate (with ε_r = 10 and of the same thickness). The gain of the antenna was measured at 6.6 dBi. These elements had excellent radiation pattern characteristics, which were similar to those obtained from antennas on a conventional low permittivity microwave substrate material.

In [113], a microstrip patch antenna was demonstrated using this technique on HRS. This yielded efficiency and bandwidth increases of 28% and 64%, respectively, at 21 GHz. The overall thickness of the substrate used was 635 μm and the thinned region was 305μm. Adhesive copper foil was used to form the ground-plane. Here, a selective etching procedure was used to form a blind hole, in order to thin the wafer in the region of the integrated antenna, while giving the advantage of an air-filled, cavity-backed resonator and structural support elsewhere.

Several other concepts have been developed that exploit the vertical dimension of a silicon wafer in order to produce reduced dielectric losses. One major example is the suspended membrane technique [114]. An example of this type of line is shown in Figure 10.17.

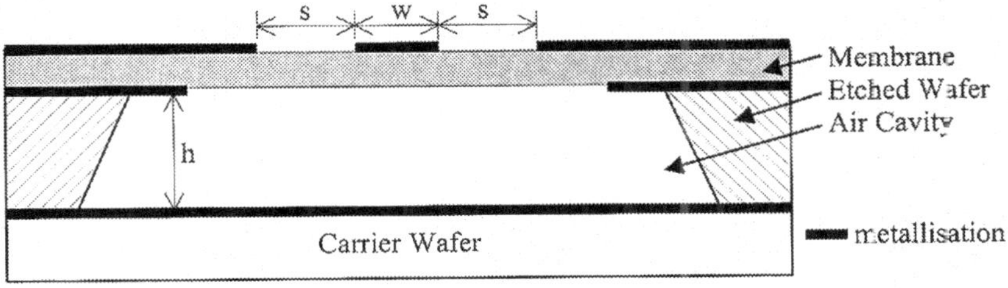

Figure 10.17 microshield line cross-section

This type of line uses micromachining techniques to etch a silicon wafer, which has a 1.5 μm thick SiO_2-Si_3N_4-SiO_2 membrane grown on it. The membrane acts as an etchant back-stop and structural support for the antenna metallisation. The antenna, or other circuit pattern, is deposited using standard lithographic methods. The cavity etching methods employed are similar to those used for the etched trenched antennas. The resulting structure has very low loss and the additional volume provided by the cavity can be exploited to give wider bandwidth patch antennas (before the onset of surface-wave effects) than can be produced on thin substrates. The resulting structure, called a microshield line, has almost ideal CPW lines with losses as low as 0.3 dB/cm at 35 GHz [114]. An additional benefit of this type of circuit is that it is self-packaging and can, therefore, be used to reduce cost for systems integration.

A planar millimetre-wave antenna was produced using this technique in conjunction with a GaAs substrate [115] and a cavity-backed end-fired slotline antenna was constructed [116]. The technique has also been used to fabricate micromachined horn antennas and arrays of horn antennas for imaging applications [117], as illustrated in Figure 10.18.

Here, dipole probes were patterned on a 1 μm thick dielectric membrane with a pyramidal cavity etched in silicon. The cavity acts as a horn reflector that focuses incident energy onto the dipole elements. The horn array is composed of stacked <100> orientated silicon wafers, and the pyramidal holes forming the horn are defined by the substrates <111> crystal planes to yield a flare angle of 70 6°. Antenna arrays operating at 94 GHz were fabricated with beam-lead diodes attracted to the dipole elements. The resulting structures were very robust to temperature and vibration induced shock. Imaging on a 16 × 16 element array, composed using the same principle, was reported at 802 GHz on an array measuring only 8 × 8 mm^2 [117]. In addition, a 92 GHz 5 × 5 array of dual-polarized horn antennas (having better than 23 dB polarisation isolation) was also reported for use in polarimetric receivers [118]. Since the flare angle of a horn antenna created by this technique is restricted to 70.6°, flare apertures of greater than 1.6λ are not possible with aperture phase aberration effects occurring.

To combat this, a step-profiled approach was adopted in [119], which reduced the flare angle between 30° and 40° and yielded a gain of up to 20 dBi.

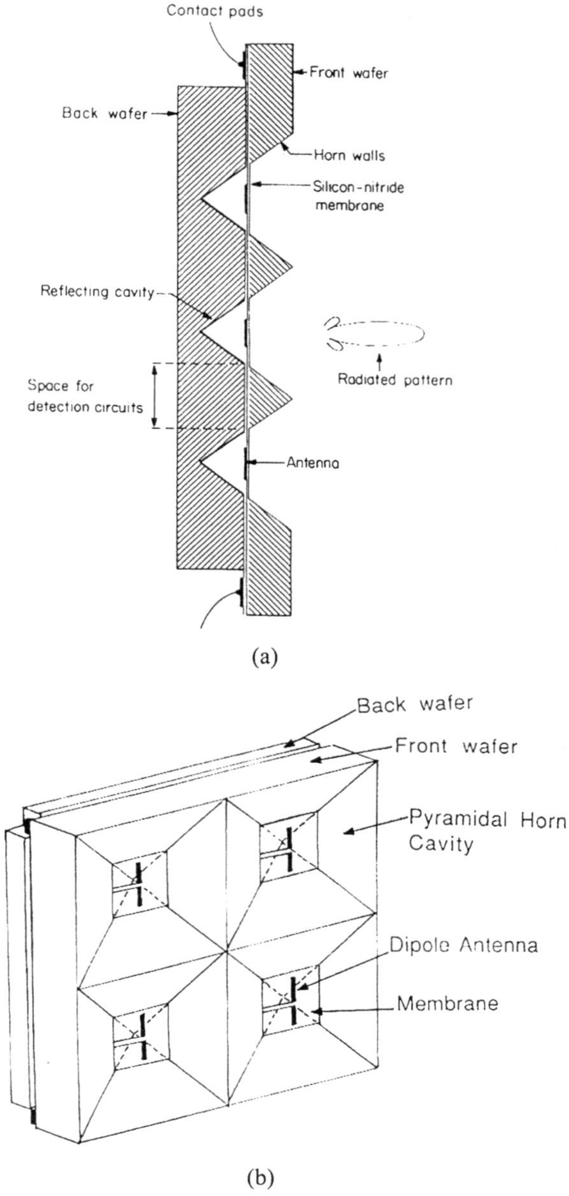

Figure 10.18 micromachined horn antenna [107] (©1992 IEEE)

The use of deep X-ray lithography and metal plating procedures (associated with the LIGA micromachining process) [120] can be used to form tall metal structures, up to 250 μm high, with near vertical sidewalls. This feature was used in [121] to produce an end-fired slotline tapered antenna. Here, due to the metallisation thickness that can be achieved with this process, it was possible to create the antenna as a self-supporting structure. This was obtained by depositing a 150 μm thick layer of nickel, which had enough mechanical rigidity to support itself without the need for an underlying substrate. When the GaAs substrate, onto which the metal had been deposited, was etched away, the 74 GHz antenna extended unsupported for 7 mm. To this end, the losses associated with the dielectric material were removed.

10.9 Microelectromechanical systems antennas

A relatively recent advancement of silicon micromachining, called microelectromechanical systems (MEMS), can be used to implement novel antenna circuits with enhanced functionality [122], such as switches [123] and tuneable capacitors [124], which could be used as control elements in phased array systems. Antennas with mechanically reconfigurable characteristics have also been developed.

MEMS allow the variable positioning of suspended membranes and cantilevers in a way that can be integrated with MMIC devices. With MEMS structures, mechanical movement can be induced by using electrostatic forces to produce physical displacement of elements. The ability to produce miniaturised moveable parts (i.e. not anchored to the substrate) is established either by exploiting differential etch rates between the crystallographic directions, due to the orientation of the silicon crystal planes, or by undercutting etch resistant features created by selective heavy boron doping [66].

An extremely interesting application of this technology to an integrated antenna was given in [125]. Here, a mechanically reconfigurable vee-antenna was reported. The planar antenna can have its far-field radiation pattern electrically altered by electrostatically-controlled microactuators. The 17.5 GHz antenna is illustrated in Figure 10.19. In this circuit, a 3-layer polysilicon surface micromachining process was used. Here, the arms of the vee-antenna used to form the radiating aperture are moved through actuators. The ends of these arms are located by hinges anchored to the substrate material. Lateral movement of the actuators (20 nm per 70 V to 120 V biasing pulse) is translated to rotational movement of the arms by the use of moveable hinges that are not anchored to the substrate. Each antenna arm is capable of independent movement, giving the possibility of far-field radiation beam-steering and beam-shaping. The directivity for the antenna was estimated (from measured 3 dB beamwidths) to be about 38, while cross-polarisation levels of below 20 dB were also reported.

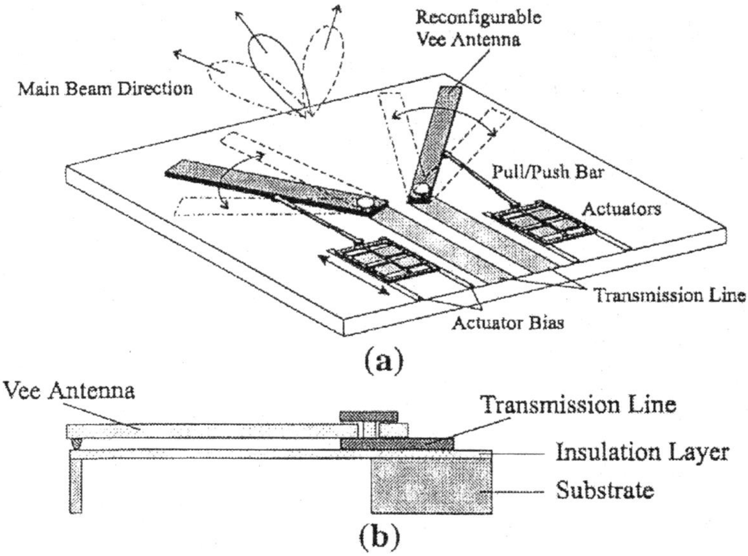

Figure 10.19 MEMS reconfigurable vee-antenna [125] (© 1999 IEEE)

With a fixed vee angle of 75°, the arms were rotated to 30° and 45° off-boresight, demonstrating beam steering, as shown in Figure 10.20. The design was further enhanced by the use of a MEMS sliding back-short, which acted as an impedance tuning unit.

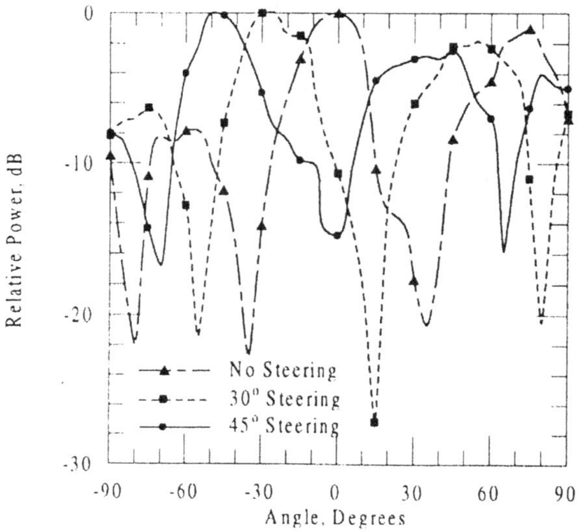

Figure 10.20 MEMS Reconfigurable V-antenna beam-steered E-plane patterns [125] (© 1999 IEEE)

10.10 Conclusions

It has been shown that all of the basic requirements normally given to an antenna designer as a specification (e.g. input return loss, power gain, etc.) must be augmented with additional considerations. This augmented design scenario impacts the designer in several diverse ways. Firstly, to assist the designer in evaluating which integrated circuit antenna is appropriate for a cited application, a wide variety of systems applications were reported. Secondly, the designer needs to be aware of the particular properties of planar antenna types that can be readily integrated in MMIC form. With this in mind, a taxonomy of different radiating structures was provided. A detailed discussion of the impact on design using popular semiconductor substrate materials (e.g. GaAs or silicon) was given. This was followed by discussions of major issues related to antenna behaviour characterisation and packaging for systems insertion.

A guided tour of various reported GaAs and silicon MMIC antennas and their functionality and operational attributes was also given, as a focus for realistic expectation of performance and as a stimulus for ideas for prospective designers of next generation wireless systems. Special techniques for high performance monolithic antennas involving micromachining techniques were discussed. Even though these techniques involve additional MMIC processing, they offer the prospect for considerable design flexibility and extend the opportunity for enhanced electrical performance (e.g. radiation efficiency) for integrated antennas. Finally, this chapter offers some indication as to where integrated circuit antennas development is going. These developments include antennas fabricated on artificial dielectric materials (e.g. the so-called photonic bandgap antennas) and mechanically reconfigurable antennas that exploit MEMS technology.

It is hoped that the reader of this chapter will reflect on the significant progress that has been made in this topic over the past ten years. In addition, the reader is encouraged to consider the development of completely integrated wireless systems, which these developments could exploit, as well as providing a basis on which to extend the ideas suggested in this chapter.

10.11 References

1 'Mobile broadband systems (MBS)', *European Radiocommunications Office*, 1997

2 SYLVIE, S.: 'Market analysis for mobile broadband system', *RACE Mobile Telecommunications Summit*, Cascais, Portugal, 1995, pp. 548–552

3 LI, D., LUO, S., and KNOX, R.: 'Millimeter-wave FMCW radar transceiver/antenna for automotive applications', *Applied Microwave and Wireless*, 1997, pp. 58–68

4 CACHIER, G.: Microwave and millimeter wave communications: status, market, and future trends', *Proceedings of 28th European Microwave Conference*, Amsterdam, 1998, pp. 122–125

5 MACDONALD, P., and MATLOUBIAN, M.: 'Millimeter wave technology for commercial applications', *Microwaves and RF*, 1996, pp. 114–117

6 REBEIZ, G.: 'Millimeter wave and terahertz integrated circuit antennas', *Proceedings of the IEEE*, 1992, **80** (11), pp. 1748–1770

7 MEINEL, H.: 'Recent advances on millimeter wave PCN system development in Europe', *IEEE MTT-S International Microwave Symposium Digest*, 1995, pp. 401–404

8 VIAUD, J.P. *et al.*: Modulator millimeter wave transceiver design for todays' digital radio links', *Microwave Engineering Europe*, 1999, pp. 49–54

9 TELLIEZ, I. *et al.*: 'Millimeter wave integrated front end for mobile broadband systems', *RACE Mobile Telecommunications Summit*, Cascais, Portugal, 1995, pp. 122–125

10 NAVARRO, J., and CHANG, K.: 'Integrated active antennas and spatial power combining' (Wiley Interscience, 1996)

11 JAMES, J.R., and HENDERSON, A.: 'High-frequency behaviour of microstirp open circuit terminations', *IEE Journal on Microwaves, Optics and Acoustics*, 1979, **3** (5), pp. 205–218

12 FUSCO, V.F.: 'Self-detection performance of active microstrip antennas', *Electronics Letters*, 1992, **28** (14), pp. 1362–1363

13 PERKONS, A.R., QIAN, Y., and ITOH, T.: 'TM surface-wave power combining by a planar active-lens amplifier', *IEEE Transactions on Microwave Theory and Techniques*, 1998, **40** (6), pp. 775–783

14 FUSCO, V.F., DREW, S., and McDOWALL, D.S.: 'Synthesis and performance of an active microstrip antenna', *International Journal of Microwave and Millimeter Wave Computer Aided Engineering*, 1994, **4** (1), pp. 110–118

15 FUSCO, V.F., HALL, P.S., and CRYAN, M.: 'Analysis and design of integrated circuit antenna modules', in GUPTA, K. C., and HALL, P. S. (Eds): *Circuit Simulation Methods* (John Wiley & Sons, 1999)

16 LIU, Z.D., and HALL, P.S.: 'Dual band antenna for hand held portable telephones', *Electronics Letters*, 1996, **32** (7), pp. 609–610

17 CHEBOLU, S., DEY, S., MITTRA, R., and ITOH, M.: 'Dual band stacked microstrip antenna array for mobile satellite applications, *Proceedings of IEEE AP-S International Symposium*, Newport Beach, 1995, pp. 598–601

18 YAMUZAKI, M., RAHARDJO, E.T., and HANEISH, M.: 'Construction of a slot coupled planar antenna for dual polarisation', *Electronics Letters*, 1994, **30** (22), pp. 1814–1815

19 DOUGLAS, M.G., and JOHNSTON, R.H.: 'A compact two way diversity microstrip U patch antenna', *Proceedings of IEEE AP-S International Symposium*, Newport Beach, 1995, pp. 978–981

20 MEINKE, H., and LANDSTORFER, P: 'Noise and bandwidth limitations with transistorised antennas', *Proceedings of IEEE Antennas and Propagation Symposium*, 1968, pp. 272–275

21 RAMSDALE, P.A. and MACLEAN, T.S.M.: 'Active loop-dipole aerials, *IEE Proceedings*, 1971, **118** (12), pp. 1698–1710

22 AL-ANI, A.H., CULLEN, A.L., and FORREST, J.R.: 'A phase-locking method for beam steering in active array antennas', *IEEE Transactions on Microwave Theory and Techniques*, 1974, **22** (6), pp. 698–703

23 TAGUCHI, M., FUJIMOTO, T., and TASAKA, K.: 'CPW fed active dipole antenna for Television Receivers', *Electronics Letters*, 1994, **30** (22), pp. 1815–1816

24 STEYSKAL, H., MAILLOUX, and R.J., TURTLES, J.P.: 'EHF active phased array development in the USA', *Proceedings of 23rd European Microwave Conference*, 1994, pp. 80–83
25 SANCHEZ-HERNANDEZ, D., and ROBERTSON, I.: '60 GHz-band active patch antenna for spatial power combining arrays in European mobile communications systems', *Proceedings of European Microwave Conference*, 1995, pp. 1773–1778
26 ROEDERER, A.G.: 'Semi-active satellite antennas', *Proceedings of International Antennas Conference*, Nice, 1992, **12-14**, pp. 569–577
27 FUSCO, V.F., and KARODE, S.L.: 'Self-phasing antenna array techniques for mobile communications applications', *IEE Journal of Communications Engineering*, 1999, pp. 274–287
28 MONTIEL, C.M., FAU, L., and CHANG, K.: 'A Self-mixing active antenna for communication and vehicle identification applications, *IEEE MTT-S International Microwave Symposium Digest*, 1996, pp. 333–336
29 THIEME, M.O. *et al.*: 'A novel circularly polarised W-band direct detection receiver for six-port polarimetric radar system', *IEEE MTT-S International Microwave Symposium Digest*, 1997, pp. 1269–1272
30 POBENZ, C.W., and ITOH, T.: 'A microwave non-contract identification transponder using sub-harmonic interrogation', *IEEE Transactions on Microwave Theory and Techniques*, 1995, **43**, pp. 1673–1679
31 PASSIOPOULOS, G. *et al.*: 'V-Band single chip direct carrier BPSK modulation transmitter with integrated patch antenna', *IEEE MTT-S International Microwave Symposium Digest*, 1998, pp. 305–308
32 LAI, A.K.Y., and NG, J.H.: 'A low cost active transceiving antenna', *IEEE MTT-S International Microwave Symposium Digest*, 1998, pp. 1047–1050
33 GOLDSMITH, P.F. *et al.*: 'Focal plane imaging systems for millimeter wavelengths', *IEEE Transactions on Microwave Theory and Techniques*, 1993, **41**, pp. 1664–1675
34 VAN ATTA, L. C.: 'Electromagnetic reflector', US Patent Office, no. 2908002, 1959
35 KARODE, S.L., and FUSCO, V.F.: 'Self-tracking duplex communication link using planar retrodirective antennas', *Transactions on IEEE Antennas and Propagation*, 1999, **47** (6), pp. 993–1000
36 GUPTA, S., and FUSCO, V.F.: 'Self-phased re-transmitting integrated mixer antenna array', *International Journal of Electronics*, 1998, **86**, pp. 207–215
37 WU, Y., FUSCO, V.F., and CHEN, Q.: 'Integrated self–steered silicon detector array', *IEE Proceedings of Microwaves, Antennas and Propagation*, 2000, **147**, pp. 106–110.
38 FUSCO, V.F., CHEN, Q., GAMBLE, H.S., and ARMSTRONG, B.M.: 'Self-steered silicon receiver', *Proceedings of 2nd Topical Meeting on Silicon Monolithic Integrated Circuits in RF Systems*, Garmish, Germany, 2000, pp. 67–70
39 POZAR, D.M., and SCHAUBERT, D.H.: 'Microstrip antennas: the analysis and design of microstrip antennas and arrays' (IEEE Press, 1995)
40 SCHAUBERT, D.H.: 'A review of some microstrip antenna characteristics', *ibid*, pp. 59–67
41 BOOKER, H.G.: 'Slot aerials and their relation to complementary wire aerials', *Journal of the IEE*, part IIIA, 1946, **93**, pp. 620–626

42 WELLER, T.M., KATEHI, P.B., and REBEIZ, G.M.: 'Single and double folded slot antennas on semi-infinite substrates', *IEEE Transactions on Antennas and Propagation*, 1995, **53** (12), pp. 1423–1428

43 KRAUS, J.D.: 'Antennas' (McGraw-Hill, 1988, 2nd edn.)

44 COLLIN, R.E.: 'Field theory of guided waves' (IEEE Press, 1990, 2nd edn.)

45 DEAL, W.R., QIAN, Y., and ITOH, T.: 'Planar integrated antenna Technology', *Microwave Journal*, 1999, pp. 128–144

46 OLVER, A.D.: 'Basic Principles of Antennas', in *The Handbook of Antenna Design* (Peter Peregrinus, London, 1982), Chap. 1

47 KASILINGHAM, D., and RUTLEDGE, D.: 'Focussing properties of small lenses', *International Journal of Infrared and Millimeter Waves*, 1986, **7** (10), pp. 1631–1647

48 GELSTROPHE, R.V. *et al.*: 'High performance millimeter wave antennae employing broad wall slot arrays', *Proceedings of 4th International Conference on Antennas and Propagation*, UK, 1985, pp. 580–584

49 REBEIZ, G.M. *et al.*: 'Monolithic millimeter-wave two-dimensional horn imaging arrays', *IEEE Transactions on Antenna and Propagation*, 1990, **28**, pp. 1473–1482

50 YNGVESSON, K.S. *et al.*: 'The tapered slot antenna - a new integrated element for millimeter-wave applications', *IEEE Transactions on Microwave Theory and Techniques*, 1989, **37**, pp. 365–374

51 SAAVEDRA, C.E., VAUGHAN, M.J., and COMPTON, R.C.: 'An M-PSK modulator for quasi-optical wireless applications', *IEEE MTT-S International Microwave Symposium Digest*, 1966, pp. 1243–1246

52 MADER, T., BUNDY, S., and POPOVIC, Z.B.: 'Quasi-optical VCO's', *IEEE Transactions on Microwave Theory and Techniques*, 1993, **41** (10), pp. 1775–1781

53 WEIKLE, R.M. *et al.*: 'Transistor oscillator and amplifier grids', *Proceedings of the IEEE*, 1992, **80** (11), pp. 1800–1809

54 STEPHAN, K.D. *et al.*: 'A Quasi-optical polarization-duplexed balanced mixer for millimeter-wave applications', *IEEE Transactions on Microwave Theory and Techniques*, 1983

55 POBANZ, C.W., and ITOH, T.: 'Quasi-optical microwave circuits for wireless applications', *Microwave Journal*, 1995, pp. 64–82

56 STEPHAN, K.D.: 'Inter-injection-locked oscillator for power combining and phased arrays, *IEEE Transactions on Microwave Theory and Techniques*, 1986, **34** (10), pp. 1017–1025

57 RUTLEDGE, D.B., NEIKIRK, D.P., and KASILINGAM, D.P.: 'Infrared and millimeter waves' (Academic Press, Orlando, 1983), vol. 10, pp. 1–20

58 HYLTIN, T.M.: 'Microstrip transmission on semiconductor dielectrics', *IEEE Transactions on Microwave Theory and Techniques*, 1965, **13**, pp. 777–781

59 ROSEN, A. *et al.*: 'Silicon as a millimeter-wave monolithically integrated substrate - a new look', *RCA Review*, 1981, **42**, pp. 633–660

60 STABILE, P.J., and ROSEN, A.: 'A silicon technology for millimeter-wave monolithic circuits', *RCA Review*, 1984, **45**, pp. 587–605

61 PONCHAK, G.E.: 'RF transmission lines on silicon substrates', *Proceedings of 29th European Microwave Conference*, Munich, 1999, pp. 158–161

62 STROHM, K.M., LUY, J.F. *et al.*: 'Silicon technology for monolithic millimeter-wave integrated circuits', *Mikowellen & HF Magazin*, 1988, **14** (8), pp. 750–760

63 STROHM, K.M., BUECHLER, J., and KASPER, E.: 'SIMMWIC rectennas on high resistivity silicon and CMOS compatibility', *IEEE Transactions on Microwave Theory and Techniques*, 1998, **46**, pp. 669–676

64 FUSCO, V.F., and CHEN, Q.: 'Direct signal modulation using a silicon microstrip patch antenna', *IEEE Transactions on Antennas and Propagation.*, 1999, **47** (6), pp. 1025–1028

65 GAMBLE, H.S. *et al.*: 'Low-loss CPW lines on surface stabilized high-resistivity Silicon', *IEEE Microwave and Guided Wave Letters*, 1999, **9** (10), pp. 395-–397

66 MOORE, D.F., and SYMS, R.R.A.: 'Recent developments in micromachined silicon', *Electronics and Communication Engineering Journal*, 1999, **11** (6), pp. 261–270

67 NAM, C., and KWON, Y.: 'Coplanar waveguides on silicon substrate with thick oxidised porous silicon (OPS) layer', *IEEE Microwave and Guided Wave Letters*, 1998, **8** (11), pp. 369–371

68 ROY, L., LABONTE, S., and SIMONS, N.R.S.: 'Measurement techniques for integrated-circuit slot antennas', *IEEE Transactions on Instrumentation and Measurement*, 1997, **46** (4), pp. 1000–1004

69 HIRACHI, Y., NAKANO, H., and KATO, A.: 'A cost-effective RF-module with built-in patch antenna for millimeter-wave wireless systems', *Proceedings of 29th European Microwave Conference*, Munich, 1999, **3**, pp. 347–350

70 GRIFFIN, D.W., and PARFITT, A.J.: 'Electromagnetic design aspects of packages for monolithic microwave integrated circuit-based arrays with integrated antenna elements', *IEEE Transactions on Antennas and Propagation*, 1995, **43** (9), pp. 927–931

71 DEBORGIES, F., LEMOINE, T., MAY, R., and OUDART, M.: 'Low-cost highly integrated S-band receiver', *Proceedings of 29th European Microwave Conference*, Munich, 1999, pp. 299–300

72 BAUMANN, G. *et al.*: '51 GHz front end with flip chip and wire bond interconnections from GaAs MMICs to a planar patch antenna', *IEEE MTT-S International Microwave Symposium Digest*, 1995, pp. 1639–1642

73 MENZEL, W.: 'Packaging and interconnect techniques for complex millimeter-wave front-ends', *Proceedings of 28th European Microwave Conference*, 1998, pp. 497–502

74 RUSSHOFER, R.H., and BIEBL, E.M.: 'SIMMWIC system application design: design tradeoffs and system performance', *Proceedings of Topical Meeting on Silicon Monolithic Integrated Circuits in RF Systems*, 1998, pp. 209–213

75 FUSCO, V.F., CHEN, Q., SALAMEH, D., and BRABETZ, T.: 'Silicon and soft-board millimeter wave antennas for broadband mobile wireless', *Proceedings of 2nd Topical Meeting on Silicon Monolithic Integrated Circuits in RF Systems*, 2000

76 YANG, S., and FUSCO, V.F.: 'Intrinsically matched high resistivity silicon slot-loop antenna', *Microwave and Optical Technology Letters*, 1999, **20** (1), pp. 26–28

77 STUTZMAN, W.L., and THIELE, G.A.: 'Antenna theory and design' (John Wiley & Sons, 1998), pp. 206–210

78 JORDAN, E.C. *et al.*: 'Electromagnetic waves and radiating system' (Prentice Hall, 1968), pp. 513–519

79 RASSHOFER, R.H., HERB, A., and BIEBL, E.M.: 'Optimal design of detectors and video amplifiers in silicon monolithic integrated systems', *Proceedings of Topical Meeting on Silicon Monolithic Integrated Circuits in RF Systems*, 1998, pp. 205–208

80 STROHM, K.M., BUECHLER, J., LUY, J.F., and SCHAFFLER, F.: 'A silicon technology for active high frequency circuits', *Microelectronic Engineering*, 1992, **19**, pp. 717–720

81 PRESTING, H., BUECHLER, J. *et al.*: 'Silicon monolithic mm-wave integrated circuit (SIMMWIC) devices mounted up-side-down on a copper heat sink integral with cavity resonator', *IEEE Transactions on Microwave Theory and Techniques*, 1994, pp. 1750–1755

82 WOLLITZER, M., BUECHLER, J., and LUY, J.F.: 'High efficiency planar oscillator with RF power of 100 mW near 140 GHz', *IEEE MTT-S International Microwave Symposium Digest*, 1997, pp. 1205–1208

83 RASSHOFER, R.H., THIEME, M.O., and BIEBL, E.M.: 'Circularly polarized millimeter-wave rectenna on silicon substrate', *IEEE Transactions on Microwave Theory and Techniques*, 1998, **46** (5), pp, 715–718

84 RASSHOFER, R.H., and BIEBL, E.M.: 'Surface-wave coupling of active antennas for homodyne sensor systems', *IEEE MTT-S International Microwave Symposium Digest*, 1998, pp. 1891–1894

85 GAUGHIER, G. *et al.*: 'A low-noise 86-90 GHz uniplanar Schottky-receiver, *IEEE MTT-S International Microwave Symposium Digest*, 1993, pp. 325–327

86 KOT, J. S., NIKOLIC, N., and BIRD, T.S.: 'Integrated-circuit antenna for high-permittivity substrates', *IEE Proceedings Microwave, Antennas and Propagation*, 1997, **144** (1), pp. 47–51

87 RAMAN, S., and REBEIZ, G.M.: 'Single and dual polarized slot-ring sub-harmonic receivers', *IEEE MTT-S International Microwave Symposium Digest*, 1997, pp. 565–568

88 GEARHART, S.S. *et al.*: 'A wide-band 760 GHz planar integrated Schottky receiver', *IEEE Microwave and Guided Wave Letters*, 1993, **3**, pp. 205–207

89 WONG, N., SCHWARZ, S.E., and HIERL, T.: 'Monolithically integrated Gunn oscillator at 35 GHz', *Electronics Letters*, 1984, **20**, pp. 603–604

90 ROBERTSON, I.D.: 'A monolithic 35 GHz back-face patch antenna using multi-layer techniques', *Proceedings of 23rd European Microwave Conference*, 1993, pp. 348–350

91 ROY, L.: '30 GHz monolithic low noise amplifier antenna', *IEEE MTT-S International Microwave Symposium Digest*, 1997, **2**, pp. 967–970

92 DANIEL, J.P., DUBOST, G., TERRAT, C., CITERNE, J., and DRISSI, M.: 'Research on planar antennas and arrays: "Structures Rayonnantes"', *IEEE Antennas and Propopagation Magazine*, 1993, **35** (1), pp. 14–38

93 SANCHEZ-HERNANDEZ, D., WONG, Q.H., REZAZADEH, A.A., and ROBERTSON, I.D.: 'Millimeter-wave dual-band microstrip patch antennas using multi-layer GaAs technology', *IEEE Transactions on Microwave Theory and Techniques*, 1996, **44** (9), pp. 1590–1592

94 MANASSON, V.A., and SADOUNIK, L.S.: 'Monolithic electronically controlled millimeter-wave beam steering antenna', *Proceedings of Topical Meeting on Silicon Monolithic Integrated Circuits in RF Systems*, 1998, pp. 215–217
95 GAUTHIER, G.P., COURTAY, A., and REBEIZ, G.M.: 'Microstrip antennas on synthesised low dielectric constant substrates', *IEEE Transactions on Antennas and Propagation*, 1997, **45**, pp. 1310–1315
96 JOAMOPOULOS, J.D., MEADE, R.D., and WINN, J.N.: 'Photonic crystals' (Princeton University Press, 1995)
97 JASPER, L.J., and TRAN, G.T.: 'Photonic band gap (PBG) technology for antennas', *Proceedings of the SPIE*, 1996, **2843**, pp. 80–82
98 QIAN, Y. *et al.*: 'A novel approach for gain and bandwidth enhancement of patch antennas', *Proceedings of IEEE Radio and Wireless Conference*, 1998, pp. 221–223
99 QIAN, Y. *et al.*: 'A microstrip antenna using novel photonic bandgap structures', *Microwave Journal*, 1999, pp. 66–76
100 YANG, D.H., ALEXOPOULOS, G., and YABLONOUITCH, J.I.: 'Photonic bandgap materials for high-gain printed circuit antennas', *IEEE Transactions on Antennas and Propagation*, 1997, **45** (1), pp. 185–187
101 SIGALAS, M.M. *et al.*: 'Dipole antennas on photonic bandgap crystals – experiment and simulation', *Microwave and Optical Technology Letters*, 1997, **15** (3), pp. 153–158
102 SMITH, G.S., KESLER, M.P., and MALONEY, J.G.: 'Dipole antennas used with all-dielectric woodpile photonic-bandgap reflectors: gain, field patterns and input impedance', *Microwave and Optical Technology Letters*, 1999, **21**, (3), pp. 191–196
103 KESLER, M.P., MALONEY, J.G., and SHIRLEY, B.L.: 'Antenna design with the use of photonic bandgap materials as all-dielectric planar reflectors', *Microwave and Optical Technology Letters*, 1996, **11** (4), pp. 169–174
104 LEUNG, W.Y. *et al.*: 'Slot antennas on photonic bandgap crystals', *IEEE Transactions on Antennas and Propagation*, 1997, **45** (10), pp. 1569–1570
105 CHEN, Q., FUSCO, V.F., ZHENG, M., and HALL, P.S.: 'Trenched silicon microstrip antenna arrays with ground pane effects', *Proceedings of 29th European Microwave Conference*, Munich, 1999, pp. 263–266
106 CHEN, Q., FUSCO, V.F., RAZA, S.H., ZHENG, M., HALL, P.S., and GAMBLE, H.S.: 'Trenched silicon microstrip patch antenna with cantilevered conductor', *Proceedings of European Wireless Conference*, 1998, pp. 87–93
107 HO, C. H., LU, F., and CHANG, K.: 'New FET active slot ring antenna', *Electronics Letters*, 1993, **19** (6), pp. 521–522
108 WOOD, D.: 'The fabrication of silicon microsystems', *Electronics and Communication Engineering Journal*, 1999, **11** (6), pp. 253–266
109 FUSCO, V.F.: 'Micromachined techniques for silicon antennas', *Microwaves and RF*, 1998, pp. 190–194
110 FUSCO, V.F., CHEN, Q., ZHENG, M., and HALL, P.S.: 'Micromachined silicon antenna', *Proceedings of IEEE ICMMT'98 International Conference on Microwave and Millimeter Wave Technology*, Beijing, 1998, pp. 289–292

111 WU, Y., CHEN, Q., FUSCO, V.F., ZHENG, M., and Hall, P.S.: 'Radiation leakage from an under-metallised silicon cavity', *IEEE MTT-S International Microwave Symposium Digest*, 1999, pp. 1331–1334
112 ZHENG, M., CHEN, Q., HALL, P.S., and FUSCO, V.F.: 'Broadband microstrip patch antenna on micromachined silicon substrates', *Electronics Letters*, 1998, **34** (1), pp. 3–4
113 PAPAPOLYMEROU, I., DRAYTON, R.F., and KATEHI, L.P.B.: 'Micromachined patch antennas', *IEE Transactions on Antennas and Propagation*, 1998, **46** (2), pp. 275–283
114 WELLER, T.M., KATEHI, L.P.B., and REBEIZ, G.M.: 'High performance microshield transmission line components', *IEEE Transactions on Microwave Theory and Techniques*, 1995, **43** (3), pp. 534–543
115 STOTZ, M., GOTTWALD, G., HASPEKLO, H., and WENGER, J.: 'Planar millimeter-wave antennas using Si N_x – membranes on GaAs', *IEEE Transactions on Microwave Theory and Techniques*, 1996, **44** (9), pp. 1593-1595
116 EKSTROM, H. *et al.*: '348 GHz endfire slotline antennas on thin dielectric membranes', *IEEE Microwave and Guided Wave Letters*, 1992, pp. 357–358
117 REBEIZ, G.M., KATHEI, L.P.B. *et al.*: 'Integrated horn antennas for millimeter wave application', *IEEE Antennas and Propagation Magazine*, 1992, **34** (1), pp. 7–16
118 ALI-AHMAD, W.Y., and REBEIZ, G.M.: '92 GHz dual-polarized integrated horn antennas', *IEEE Transactions on Antennas and Propagation*, **39**, pp. 820–825
119 ELEFTHERIADES, G.V., ALI-AHMAD, W.Y., and REBEIZ, G.M.: 'A 20 dB quasi integrated horn antenna', *IEEE Microwave and Guided Wave Letters*, 1992, **2**, 1992, pp. 73–75
120 MUNCHMEYER, D., and LANGEN, J.: 'Manufacture of three-dimensional microdevices using synchrotron radiation', *Review of Scientific Instruments*, 1992, **63** (1), pp. 713–721
121 GEARHART, S.S., WILLKE, T., and ONGGOSAMUSI, E.: 'Integrated micromachined transmission lines and endfire slotline antennas', *Proceedings of the SPIE*, 1997, **3225**, pp. 109–118
122 CHIAO, J-C., FU, Y., CHOUDHURY, D., and LIN, L-Y.: 'MEMS millimeterwave components', *IEEE MTT-S International Microwave Symposium Digest*, 1999, pp. 463–466
123 FENG, Z. *et al.*: 'Design and modelling of RF MEMS tunable capacitors using electro-thermal actuators', *IEEE MTT-S International Microwave Symposium Digest*, 1999, pp. 1507–1510
124 MULDAVIN, J.B., and REBEIZ, G.M.: '30 GHz tuned MEMS switches', *IEEE MTT-S International Microwave Symposium Digest*, 1999, pp. 1511–1514
125 CHIAO, J. *et al.*: 'MEMS reconfigurable Vee antenna', *IEEE MTT-S International Microwave Symposium Digest*, 1999, pp. 1515–1518

Chapter 11

Transceivers

I. D. Robertson and S. Lucyszyn

11.1 Introduction

RFICs and MMICs are finding more and more applications in the commercial sector, but fierce competition between manufacturers and between different technologies has led to immense pressure on manufacturing costs: single-chip solutions with minimum chip size are demanded. In order to produce competitive products, it is necessary to be familiar with more than just the basic techniques of MMIC design. One notable example of a commercial success is Marconi's fully integrated transceiver for wireless LAN applications [1], illustrated in Figure 11.1. This design serves to demonstrate that extremely high levels of RF front-end integration are possible, as long as the designers are armed with suitable CAD packages and a full range of advanced circuit techniques.

Nevertheless, technology moves on and there is now immense demand for the 'single chip radio', and a transceiver design must increasingly integrate RF, mixed signal and digital functions. In fact, single chip radios have existed for some time: examples are paging receivers and RFID transponders. These are relatively low power, low data rate applications, however. The goal of a single chip mobile phone presents much greater challenges because of the high level of signal processing required, the high transmitter power, low receiver noise figure, the need for immunity from interference and the demand for low DC power consumption. As yet, these requirements cannot all be best met by any single technology. For example, whilst BiCMOS can be used for probably 90% of the functionality of a mobile phone, there remain key functions such as the PA, filters and TX/RX switch that can be realised more successfully in other technologies. As always, it is the cost and performance of the final product that matters, and not the technical wizardry of the chosen solution. So, there are trade-offs throughout the design process and improvements tend to be evolutionary rather than revolutionary.

This chapter is intended to give an insight into some of these more advanced techniques for the design of integrated transmitters and receivers. In many cases, new architectures have been developed as a means of replacing expensive components, and this is especially true of filters.

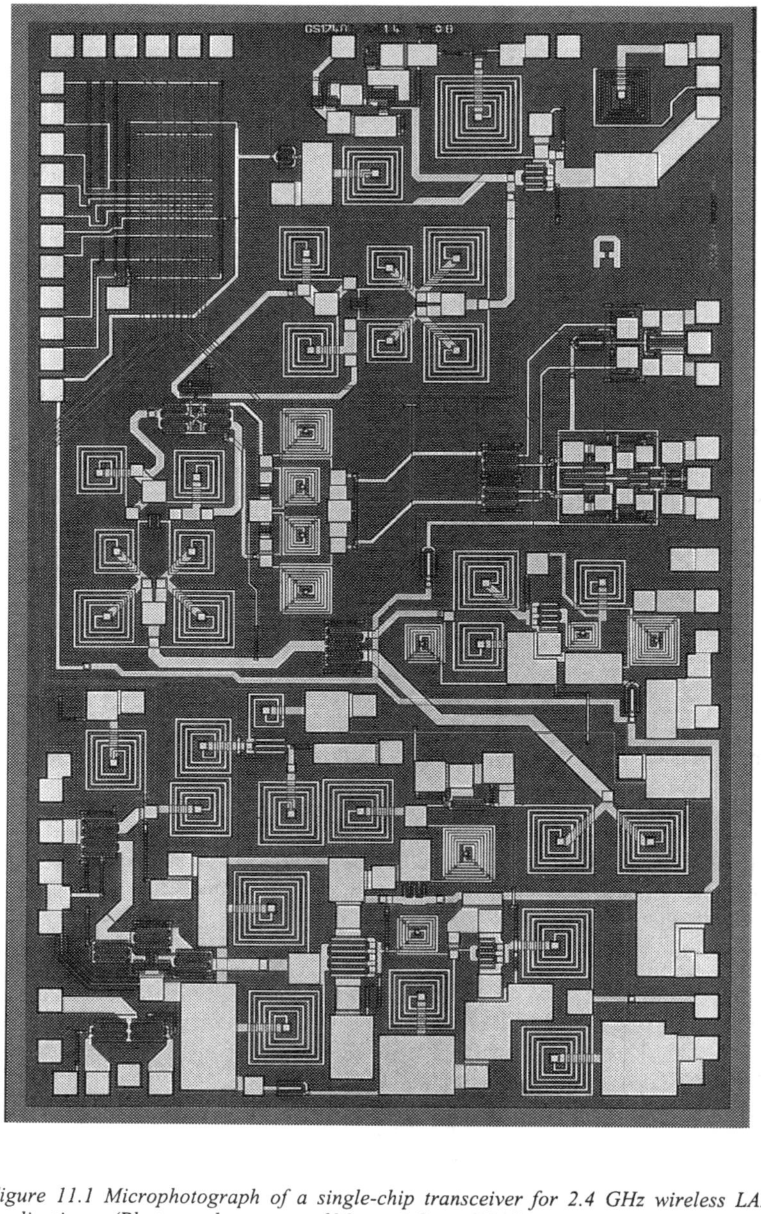

Figure 11.1 Microphotograph of a single-chip transceiver for 2.4 GHz wireless LAN applications (Photograph courtesy of Marconi Caswell Ltd)

11.2 Conventional up/down-conversion architectures

Integrated transmitters and receivers inevitably suffer from compromises when compared with traditional hybrid approaches: components such as isolators and filters cannot simply be inserted in order to improve port isolations or reduce spurious signal levels, and the tight packing of multiple circuit functions leads to unwanted coupling and unexpected feedback paths. Hence, there are many difficulties in designing integrated transmitters and receivers, including:

1. integration of low phase noise oscillators
2. integrating high-selectivity filters
3. achieving high transmitter efficiency
4. integrating the diplexing function
5. minimising coupling and leakage
6. implementing RF and baseband functions on a single chip
7. ensuring adequate decoupling for the numerous DC bias lines.

If the frequency is low enough, RF CMOS or Silicon and SiGe BiCMOS technology will enable RF and baseband functions to be on a single chip. Otherwise, a multi-chip module approach will probably be necessary, since the natural partitioning of functions between different device technologies is difficult to overcome. Broadly speaking, GaAs technology is excellent for realising RF circuit function (PA, TX/RX switch, etc.), whereas silicon technology is far better for realising digital and mixed-signal circuitry (DSP, etc.). It should be emphasised that because there are so many possible design approaches and combinations of technologies, there is rarely a standard solution for any given application.

Achieving high power efficiency is discussed in detail in Section 11.6. In a communications transceiver the diplexing function is performed either by filters and/or a circulator (for full duplex operation) or a TX/RX switch (for time-division duplexed systems). Active circulators are discussed later, and switches are described in detail in Chapter 8.

11.2.1 Filtering solutions

The problem of filtering can be addressed in many ways. A simple solution is to rely on off-chip filters. However, this method inconveniences the end-user or customer, since it makes it harder to build a complete system around the chip. As time-to-market and manufacturing cost are of increasing importance, the added design complexity and increased component count are major drawbacks. Active filters and direct conversion techniques are described in further sections. Both these methods lead to some performance compromises, but they find widespread application.

Unfortunately, with a conventional super-heterodyne receiver it is very difficult to do without high selectivity SAW-type filters, which are needed at the IF. However, it is relatively easy to take the signal off-and-onto the chip at these low frequencies without complicating the packaging or board-level design. In contrast, if off-chip RF filters are needed then there can be challenging signal routing problems. A common approach is to use image-rejection mixers [2,3] and single-sideband modulators, described in Chapter 7. Figure 11.2 shows the block diagram of a typical silicon bipolar IC for the 900MHz band, using an image rejection approach at RF and off-chip SAW filters at the IF. At frequencies where quadrature couplers are too large, a number of other techniques can be employed to realise the important 90° power splitter: either RC-filter based methods [4], or methods using multilayer passive components [5]. For high image rejection levels it is possible to use an adaptive system [6], continually fine-tuned electronically for minimum amplitude and phase error.

Whilst phasing-type image-rejection techniques may not, by themselves, achieve sufficient rejection of unwanted signals, they may ease the task so that external filters can be realised as part of the complete module — for example, microstrip filters can easily be integrated in MCM technology, but by themselves may not offer enough selectivity. MCM technology, or at least a cohesive chip/package design strategy, will also ease the problem of DC bias decoupling, since high value off-chip capacitors can be placed close to the chip.

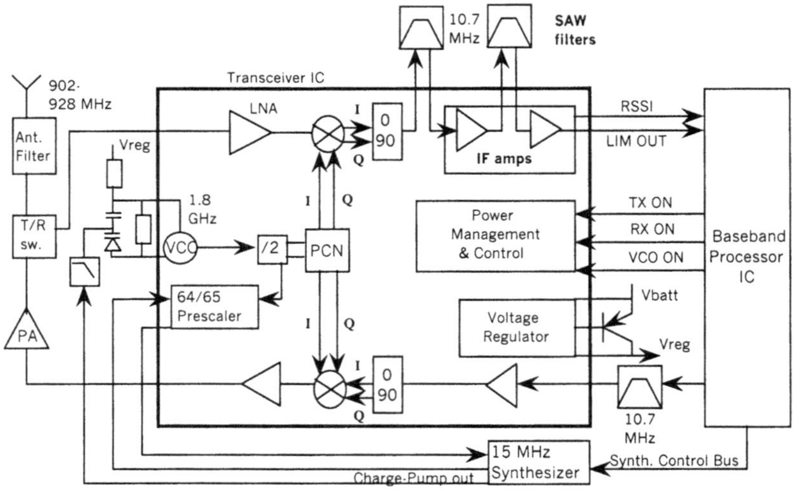

Figure 11.2 Image reject transceiver IC block diagram [3]

11.2.2 Frequency synthesisers

Synthesisers can be readily integrated at frequencies up to several GHz. Relevant techniques include direct digital synthesis (DDS), divide-by-N synthesis and fractional-N synthesis. Divide-by-N synthesis is the classic phase-locked loop (PLL) technique, where the output signal of a VCO is divided down to the frequency of a reference signal (such as 10 MHz). Basic variations of this theme are illustrated in Figure 11.3. The loop characteristics of this method must be designed very carefully to achieve fast frequency switching. The major limitations of the technique are that the frequency is limited to integer multiples of the reference frequency, and fast switching times are difficult to achieve. Direct digital synthesis uses a look-up table of values for the sinusoidal signal, and a reference clock is used to cycle through them, with a D-to-A converter followed by a filter used to create the analogue sinusoid. This technique has proved quite revolutionary, providing a synthesiser which is very amenable to chip integration. Very fast phase-continuous frequency switching is possible, and the frequency step size can be very small. However, the maximum frequency of operation is relatively low and the DC power consumption and chip area can be quite high.

The fractional-*N* technique can offer major advantages in principle: the technique allows non-integer values of the reference frequency to be generated and offers fast switching speeds. However, the technique is relatively complex and has higher spurious output and phase noise. Hybrid techniques, such as combinations of DDS with fractional-*N* synthesis can be employed to combat these drawbacks. Advanced techniques such as dithering and noise shaping can also be applied. The reader is referred to one of the many good texts on frequency synthesis for more information on these different techniques [7]. Finally, the anti-jitter circuit (AJC) has recently been shown to be a powerful new approach to improving the performance of synthesisers [8]. It has been proposed that this could be integrated into a fractional-*N* architecture in order to clean-up the spurious signals.

At frequencies above a few GHz it is extremely difficult to design a fully-integrated local oscillator (LO) subsystem. Some approaches that have been used are [9–11]:

1. frequency multipliers [12,13]
2. analogue frequency dividers [14]
3. a harmonic mixer to down-convert the oscillator output signal
4. sub-harmonic injection-locking of a fundamental oscillator.

Passive multipliers with diodes or resistive-mode FETs and active multipliers with transistors can be used. The balanced approach gives the advantage of input signal isolation, so that the filtering requirement at the output is relaxed, which is vital for monolithic solutions where filters are too large and lossy. In the frequency multiplication approach, however, considerable additional gain will be required to maintain a sufficient LO power level, and this increases chip area.

The use of a regenerative analogue frequency-divider is described in Chapter 6. For a given transistor f_t , the analogue divider can operate up to higher frequencies than a digital divider. Furthermore, the HEMT device, which is the dominant device for mm-wave circuits, is highly unsuited to the design of digital dividers. Hence, the analogue divider is extremely important for mm-wave PLL systems. The most compact alternative is to employ sub-harmonic injection locking, in which an oscillator circuit operating directly at the LO frequency is phase-locked by the injection of a much lower frequency reference signal.

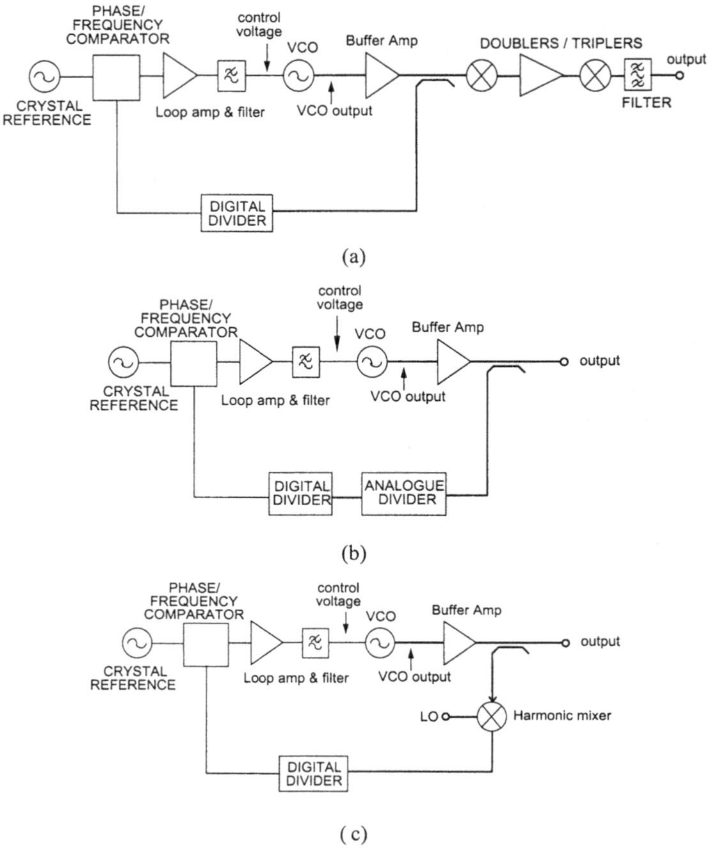

Figure 11.3 Divide-by-N high frequency synthesisers: (a) frequency multiplication, (b) analogue frequency division and (c) down-conversion by harmonic mixing

11.2.3 Mixers

Harmonic mixer techniques are becoming increasingly more common in millimetre-wave applications, and offer the double benefit of reduced LO input frequency and increased LO-to-RF isolation [15]. Resistive mixers, discussed in Chapter 6, are also widely used for millimetre-wave transceivers [16,17] since they have simple DC biasing, good linearity and require modest LO power. Figures 11.4 and 11.5 show a 38 GHz transmitter and receiver, which employ resistive mixers. Image rejection mixers and single-sideband upconverters are widely used in all frequency ranges as they ease filtering requirements at the RF input and output. Combinations of the various techniques, such as an even harmonic image rejection mixer [18] or a subharmonically-pumped SSB mixer [19], often provide a number of advantages, providing that the additional complexity can be tolerated.

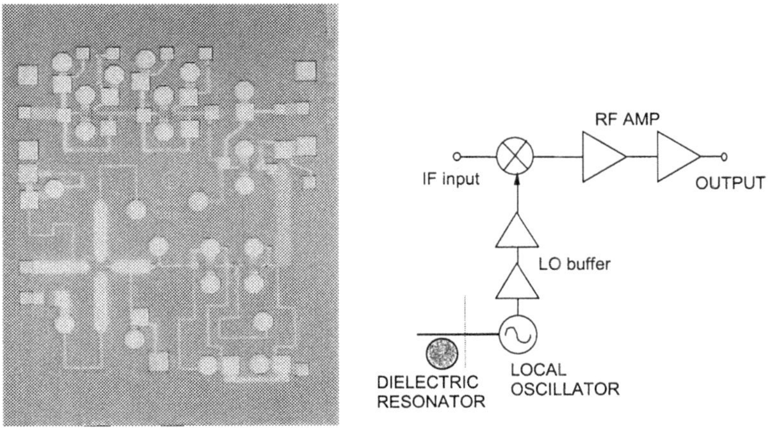

Figure 11.4 Microphotograph and block diagram of a pHEMT transmitter chip designed for operation at 38 GHz

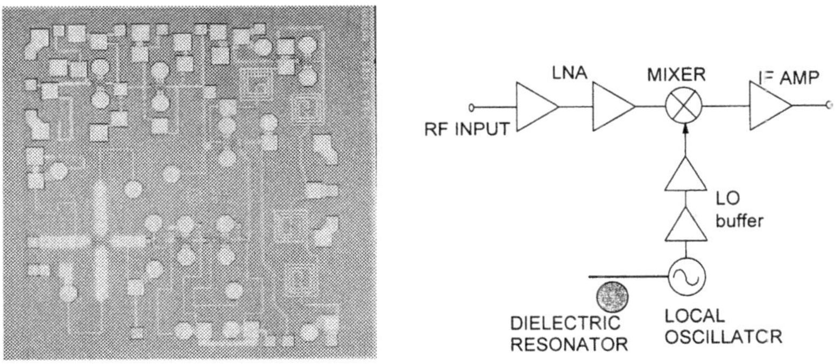

Figure 11.5 Microphotograph and block diagram of a pHEMT receiver chip designed for operation at 38 GHz

11.2.4 Active antenna transceivers

At millimetre-wave frequencies, it is straightforward to integrate directional antennas onto the chip, as described in Chapter 10. For many short-range sensing and communications applications it is thus possible to realise a complete active antenna transceiver, as illustrated in Figure 11.6. A number of innovative approaches to minimising circuit complexity have been demonstrated, including self-mixing oscillators [20], self-oscillating mixers and novel methods of making use of a single LO circuit for both transmitter and receiver [21,22]. As the technology pushes further into the millimetre-wave frequency range, quasi-optical power combining and integrated phased-array antennas have become practical possibilities [23].

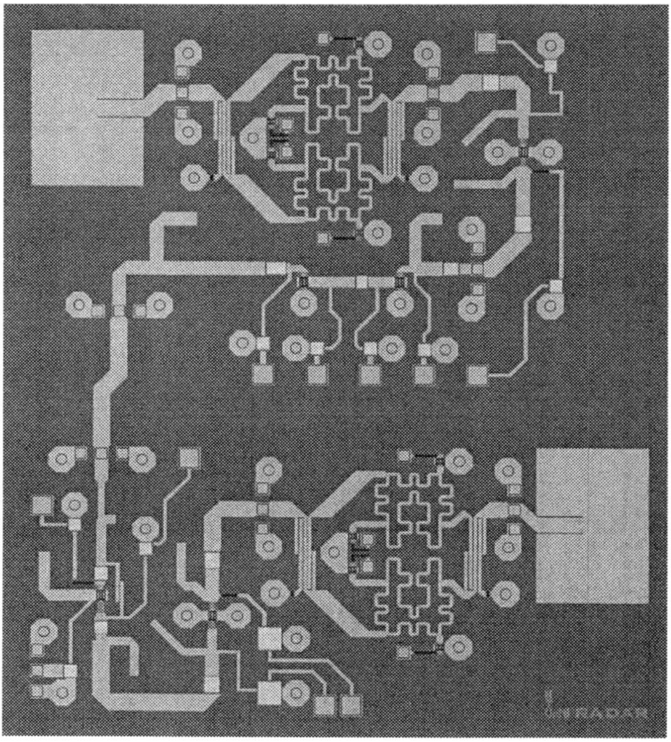

Figure 11.6 Layout of an integrated 77.6 GHz FMCW radar chip with integrated antennas (designed at the University of Surrey for fabrication at NTU in Singapore)

11.3 Direct conversion architectures

Direct conversion architectures reduce the complexity of the RF subsystems by making use of digital signal processing techniques, either applied directly to baseband signals or applied to a sampled IF signal [24,25]. The advantages include lower component count, leading to lower cost, and increased flexibility: a transceiver employing DSP techniques can easily be adapted to different data rates, modulation schemes and services. In contrast, the conventional architectures have many critical parameters fixed by, for example, the bandwidth of channel filters. This adaptive nature is key to 'software radio' and indeed 'software radar' concepts: systems in which DSP performs most of the functions traditionally associated with RF electronics (filtering, carrier recovery, etc.), and which are then largely defined by software. The goal of a radio in which an A-to-D converter is placed directly at the antenna is a dream shared by many. Realistically, the compromise solution of a simple up/down converter followed by D-to-A and A-to-D converters operating at IF or baseband will offer much higher dynamic range and a much less demanding A/D conversion specification.

11.3.1 Direct modulation transmitters

Direct modulation of the microwave carrier signal has been found to be a very attractive means of reducing hardware complexity in communications applications. In traditional transmitter designs the modulator operates at an IF frequency (such as 70 MHz), and a complex chain of mixers, filters and amplifiers is required to upconvert the modulated signal to the transmission frequency. In contrast, if the modulator is designed to operate directly at the microwave carrier frequency, the RF hardware requirement is minimal. This approach has been used for low-cost VSAT terminals, and it has been shown that the output spectrum can be band-limited by pre-filtering the baseband data [26,27]. Simple modulators for schemes such as ASK, BPSK and FSK can be designed using a wide range of techniques: ASK modulators could use switches and BPSK modulators could use analogue or digital 180° phase shifters. FSK can be achieved with a switched VCO or by using one of many frequency translation techniques.

For quadrature modulation schemes, such as QPSK and *m*-QAM, the I–Q vector modulator, shown in Figure 11.7, is used widely. The I and Q baseband signals can be careful processed in software and generated by D-to-A conversion. In this way, many functions such as spectral shaping and predistortion can be implemented in software, and the RF circuits are very simple.

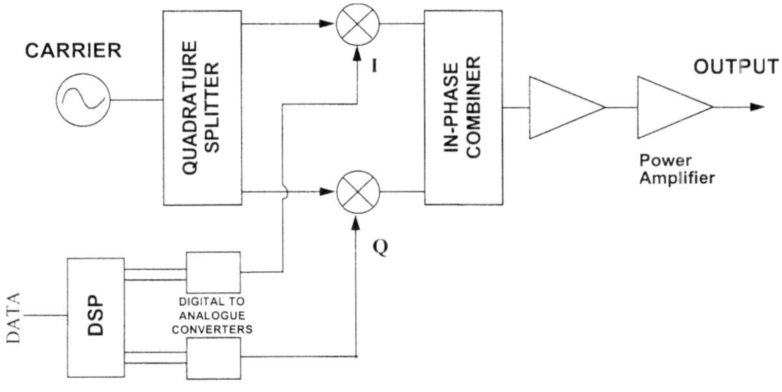

Figure 11.7 Generic direct conversion transmitter

11.3.2 Direct conversion receivers

Direct detection of the microwave signal is possible by converting the incoming modulated carrier straight to two orthogonal baseband components. The block diagram of a direct conversion receiver is shown in Figure 11.8. The main drawbacks of this architecture arise because the LO is exactly at the RF frequency; this makes it difficult to prevent LO leakage out of the antenna, and the LO self-mixing causes a problem with DC offsets. The dynamic range is also limited compared with a superheterodyne receiver, a consequence of using hardly any real filters. Nevertheless, comparing this architecture with Figure 11.2, it is clear that this is far more amenable to single-chip integration, and it has been widely used as a result. It has been widely reported that the problems of LO leakage and DC offsets can be solved by using subharmonically-pumped mixers [28–30]. Alternatively, it is possible to employ a six-port direct digital demodulator technique [31]. This technique makes use of the six-port network concept, widely studied for use in reflectometers, and applies DSP techniques to recover the data directly from the signal.

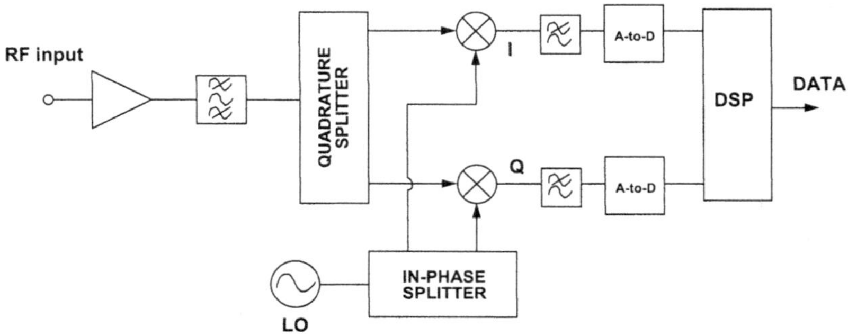

Figure 11.8 Generic direct conversion receiver

11.3.3 Direct IF sampling receivers

If the signal is downconverted to a low IF instead of directly to baseband, many of the problems associated with the direct conversion technique are avoided. Figure 11.9 shows the block diagram of a direct IF sampling receiver. The low IF frequency is directly sampled using a high frequency, high dynamic range A-to-D converter. These do exist, but are relatively expensive and power hungry. As a result, the IF sampling technique has been widely applied to base stations, but less so to mobile terminals. The advantage of the technique for a base station is reduced complexity and cost, and the fact that the receiver can be reprogrammed for different data rates, channel spacings, etc.

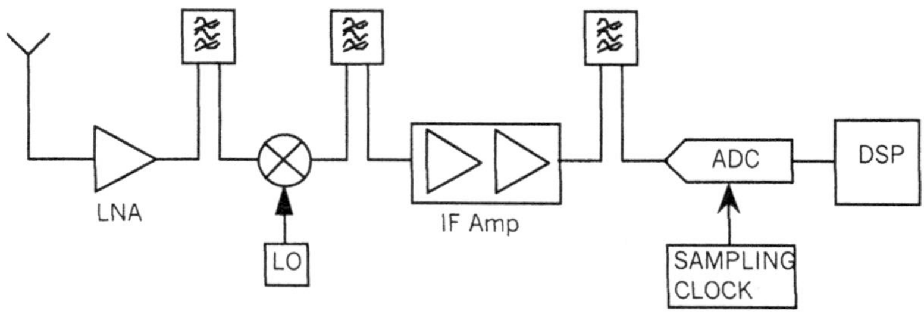

Figure 11.9 Direct IF sampling receiver

11.4 Modulators, demodulators and frequency translators

This section describes the design of modulators, demodulators and frequency translators. Modulators and demodulators operating directly at the microwave (or millimetre-wave) carrier frequency are an attractive means of simplifying the hardware requirements in a communication system. Frequency translators take many forms, the most familiar of which is the RF-to-IF downconverter used in heterodyne receivers in the form of mixers. However, other types are often used in which the input and output signals are both at microwave frequencies. The difference in frequency may be quite large (many GHz) or very small (a few kHz). Rather different approaches are necessary in these two extremes.

Mixers are a fundamental element for the design of modulators, demodulators, transmitters and receivers, frequency translators and phase-locked loops. Double-balanced diode mixers are a ubiquitous component since they have isolation between all three ports and excellent rejection of unwanted mixing products. For applications up to a few GHz they are available very cheaply as packaged modular components from companies such as Mini-Circuits. The double-balanced diode mixer is a very adaptable component which has many different uses, including upconverters, downconverters, frequency translators, modulators, demodulators and phase detectors. The distinction between the use of the mixer in these applications can be very subtle. Double-balanced diode mixers have this multifunctionality because of the DC-coupled IF port, the complete port-to-port isolation and the high level of rejection of unwanted mixing products. However, for individual applications, only some of these parameters may be important.

On MMICs, the traditional double-balanced diode mixer may require a large chip area, but the alternatives, such as FET mixers, generally do not have the same multifunctionality. For example, the Gilbert cell multiplier is a good replacement in many applications, but the inherent unilateral nature of an active mixer means that, for example, a mixer designed for a downconverter cannot then be used in reverse for an upconverter. So, for modulators, demodulators and general frequency translator applications, the question must be first addressed as to what exactly is required from the mixer and which parameters are important. Furthermore, in some cases it may be preferable to use a completely different approach: for example, rather than using a mixer, an m-ary ASK modulator could use a variable attenuator, and an m-ary PSK modulator could use a variable phase shifter.

11.4.1 Vector modulators

There are two generic types of microwave vector modulator. The first is the phase-splitting power-combining vector modulator. The second type of vector modulator simply consists of an attenuator (α) in cascade with a phase shifter (ϕ). The phase-splitting power-combining vector modulator traditionally comes in either a four-channel or a two-channel version. With the four-channel version, a four-way orthogonal 6 dB power divider is used to create the four orthogonal channels. An individual uniphase amplitude modulator is assigned to each channel. The output signals from these amplitude modulators are then combined using a four-way in-phase power combiner. The four-channel vector modulator has an inherent 12 dB transmission loss penalty (for a circular locus), and therefore, usually employs active amplitude modulators that can provide enough gain to overcome this loss. The advantage of the four-channel topology is that quite simple amplitude control circuits can be used — for example, dual-gate FETs or other variable gain amplifier circuit [32,33].

With the two-channel version (the I–Q vector modulator [34,35]), a quadrature 3 dB power divider is used to create the two orthogonal channels. An individual bi-phase amplitude modulator is assigned to each channel. The output signals from these amplitude modulators are then combined using an in-phase 3 dB power combiner. This is the topology shown in the direct modulation transmitter of Figure 11.7. The two-channel vector modulator has an inherent 6 dB transmission loss penalty for a circular locus (or 3 dB for a square constellation). Mixers can be used for the individual bi-phase amplitude modulators. Up to 5 GHz or so, the most common approach is to use a Gilbert cell design [36,37]. At higher frequencies, lumped [38,39] or distributed techniques can be used in diode mixer circuits, resistive mixers, or active mixers, as described in Chapter 7. The 'IF' port must be DC coupled, and LO isolation is necessary, and this will determine which of the mixer topologies are suitable. Mixers are not the only means of realising bi-phase amplitude modulators, however: for example, a simple reflection-type topology can give excellent performance [40,41].

The $\alpha-\phi$ vector modulator [42] can be considered as a one-channel vector modulator, with no transmission loss penalty. It can be implemented with either a uniphase amplitude modulator and a 360° variable phase shifter, as shown in Figure 11.10, or a bi-phase amplitude modulator and a 180° variable phase shifter. The levels of AM/PM conversion and PM/AM conversion of the amplitude modulator and phase shifter, respectively, can be calibrated out by each other.

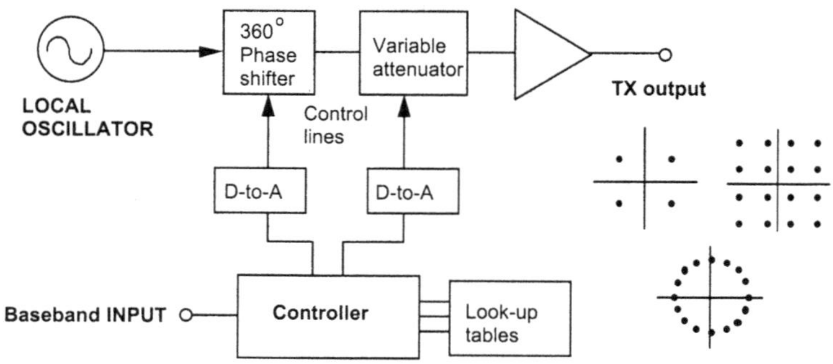

Figure 11.10 Block diagram of an $\alpha-\phi$ type vector modulator using an analogue phase shifter and attenuator

Since the vector modulator can realise any combination of signal amplitude and phase it is a very powerful control component with many applications. It can be used in phased array antenna beam-forming, as well as for modulation of a microwave carrier with virtually any digital modulation scheme (e.g. ASK, FSK, BSPK, QPSK and *m*-QAM). By placing the vector modulator under computer or

DSP control, the unit can switch between the different types of modulation with no change in the RF hardware. One limitation of the approach is that of carrier signal leakage to the output, but this can be reduced by using a harmonic modulator [43].

For millimetre-wave applications, the push-pull vector modulator, based on analogue reflection-type attenuators, has been shown to be a simple yet effective solution, and a wide range of signal processing functions have been demonstrated [44–46]. The schematic diagram of the push-pull vector modulator of this type is shown in Figure 11.11. The circuit uses the transistors as passive variable-resistance coupler terminations and the push-pull topology compensates for the unwanted reactance component, enabling circuits to be realised at very high frequencies. Figure 11.12 shows a microphotograph of a vector modulator operating at 110 GHz (chip size 1.4 × 1.1 mm^2).

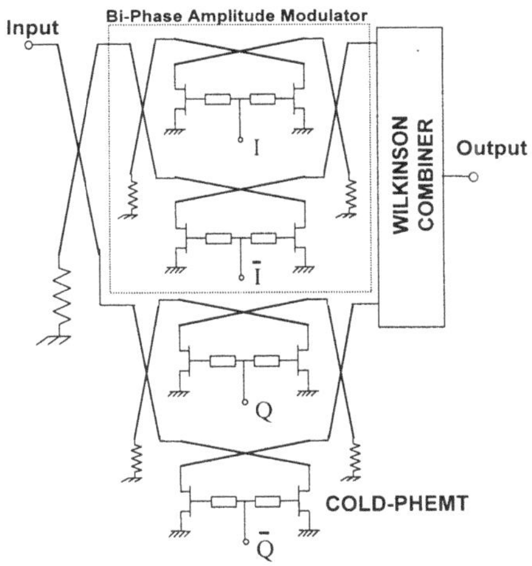

Figure 11.11 I–Q Vector modulator topology using push-pull reflection-type bi-phase amplitude modulators

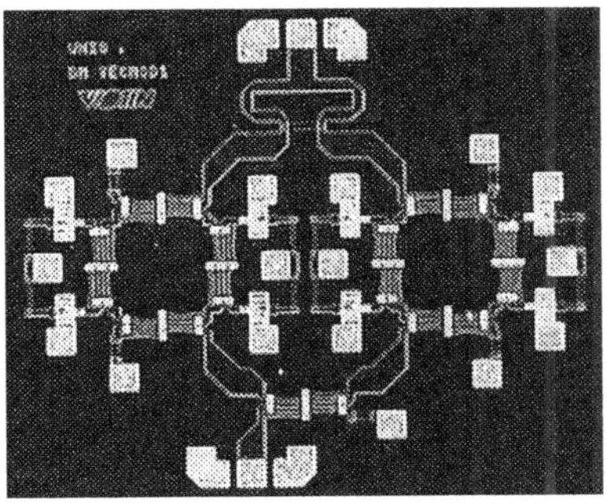

Figure 11.12 Microphotograph of a 110 GHz push-pull I–Q vector modulator (designed at the University of Surrey and fabricated by EONCOM Ltd in South Korea)

11.4.2 Frequency translators

Frequency translators take many forms, and it is useful to separate out the various types into three classifications: large-shift, small-shift and ultra-small-shift. This classification is based on the fractional frequency shift (FFS) that is to be performed:

$$\mathrm{FFS} = \frac{\text{translation frequency}}{\text{input centre frequency}} \tag{11.1}$$

Large-shift frequency translators (with FFS $\geq$ 50%) are used as up-converters in transmitters and down-converters in heterodyne receivers, and normally use filters to remove unwanted mixer outputs. Small-shift frequency translators (where FFS < 50%) are employed in non-regenerative transponders (or repeaters) used in terrestrial relay stations and satellite communication systems, where frequency separation is required to avoid the signal transmitted from the output high power amplifier from interfering with the low noise amplifier at the input. For example, in the latter application, the higher frequency uplink may be at Ku-band, while the lower frequency downlink may be at X-band. Ultra-small-shift frequency translators (FFS < 1%) are employed in numerous applications, including:

1. novel signal processing for communications and radar systems [45–47]
2. calibration of Doppler radars by emulating Doppler return signals
3. smart radar jammers, by generating false Doppler return signals for providing velocity deception and for creating ghost targets [48]
4. homodyne vector network analysers [49]
5. frequency scanning antennas [50].

An important application of an ultra-small-shift frequency translator is direct baseband modulation of an RF carrier. Here, the input signal is the unmodulated RF carrier, the translating input is the baseband signal, and the output is a modulated RF carrier.

Figure 11.13 shows three spectra representing the corresponding FFS classifications: (*a*) large-shift, (*b*) small-shift and (*c*) ultra-small-shift. The large-shift case is illustrated as a downconverter which might be used in a receiver. The terminology is very familiar: the local oscillator is the pump signal, the RF is the input signal, the IF is the low frequency output signal (the difference frequency, RF-LO) and the image (2LO-RF) is the mirror-image of the RF on the other side of the LO. The sum frequency, RF+LO, is easily filtered out.

With reference to Figure 11.13*b*, in the case of small-shift frequency translators there can be considerable confusion in terminology because the local oscillator (the modulating or translating signal) is at a lower frequency than the 'IF' output signal. To further compound this confusion, the IF port of a double-balanced mixer may often be used for this low frequency pump signal. As a result, is it less confusing if the RF is labelled RF_{in}, the local oscillator is still labelled LO and the 'IF' signal is labelled RF_{out}. The principle outputs of the mixer are still the sum (RF_{in}+LO) and difference frequencies (RF_{in}-LO). However, the difficulty now encountered is that because of the small LO frequency, the sum and difference frequencies are now very close, being equally spaced either side of RF_{in}; they are termed the upper sideband (USB) and lower sideband (LSB), respectively. Since they are very close, they are difficult to separate with filters. Furthermore, as the translation frequency is reduced, a point is reached where the translation frequency is equal to half the bandwidth of the input RF signal and the LSB and USB meet. At this point only a theoretical brick-wall filter could be considered. Beyond this point, there would be inherent distortion due to spectral overlap. As a result, small-shift frequency translators generally employ phasing techniques (using a pair of mixers or modulators fed with quadrature signals) to cancel out the unwanted sideband at the output. The mixer topology that achieves this cancellation of the undesired sideband is the single sideband (SSB) modulator, [51,52] described in Chapter 7. Unlike large-shift frequency translators, the LO signal power can be almost negligible with the small-shift frequency translator, in particular when Schottky devices are employed in reflection topologies.

The case of the ultra-small frequency shift, illustrated in Figure 11.13*c*, is shown as the modulation of the RF carrier with a baseband signal. For ultra-small frequency translations there are alternative methods of frequency conversion, using sinusoidal and serrodyne modulators, which tend to employ multipliers rather than mixers. With the former, direct carrier modulation is a good example of ultra-small sift frequency translation.

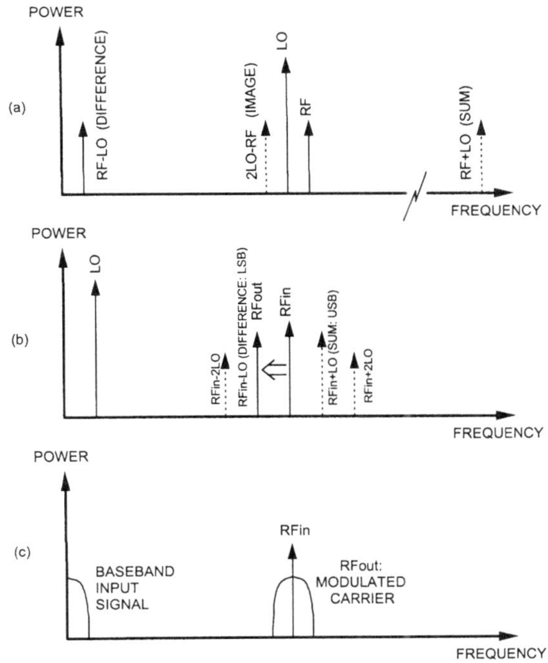

Figure 11.13 Spectra for frequency translators: (a) large shift, (b) small shift and (c) ultra-small shift

11.4.3 Serrodyne modulators

In practice, ultra-small-shift frequency translations can conveniently be performed using either a non-mixing sinusoidal modulator or a serrodyne modulator. With both techniques, the generation of significant unwanted sidebands can be avoided, improving RF power efficiency (allowing a conversion loss of 0 dB) and removing the need for an unrealisable filter.

The sinusoidal modulator simply introduces a continuous linear change, with time, in its insertion phase. An increasing change causes a positive frequency translation, by an amount which is proportional to the rate of change. Conversely, a decreasing change causes a negative frequency translation. Non-mixing sinusoidal modulators have been realised using a mechanically rotating dielectric slab [53], rotating magnetic field [54], electronic phase shifters and a vector modulator. The first two methods are not compatible with monolithic technology. With a vector modulator, the I and Q signals are equal in amplitude, but in phase quadrature. This leads to a circular rotation in the constellation — i.e. the phase varies linearly with time. It can be seen that in this application the I–Q vector modulator is identical to the single-sideband modulator.

The serrodyne (or sawtooth) modulator is a derivative of the sinusoidal modulator. Here, a klystron valve, travelling wave tube or phase shifter is modulated with a sawtooth waveform, such that one period of the sawtooth results in an induced phase shift which linearly sweeps through an integer multiple of 360°. With a 360° phase shifter, the amount of frequency translation is equal to the repetition rate of the sawtooth waveform, as illustrated in Figure 11.14. The direction of frequency translation can be changed simply by inverting the sawtooth profile. The serrodyne modulator employing a 360° phase shifter is highly suited to monolithic implementations.

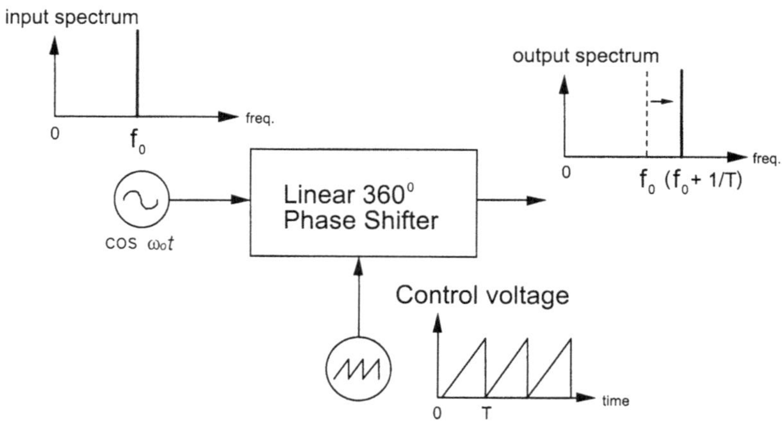

Figure 11.14 Serrodyne modulator principle

If a digital phase shifter is used, unwanted sidebands are generated as a result of quantisation errors and any phase errors and amplitude-modulation in the phase shifter response. With an analogue phase shifter, the level of suppression of unwanted sidebands is very sensitive to tuning non-linearity and amplitude modulation. In the case of an analogue phase shifter, the tuning non-linearity can be compensated for by predistorting the sawtooth using coefficients stored in a look-up table [55], or by deliberate under-modulation [56]. Figure 11.15 shows a microphotograph of a 24 GHz serrodyne modulator realised with a uniplanar 360° phase shifter and using under-modulation for reduced sidelobe levels [57]. More recently, a serrodyne modulator has been demonstrated in HBT technology [58].

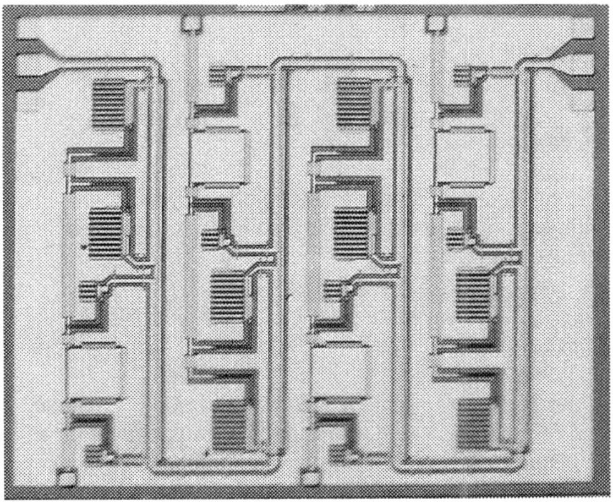

Figure 11.15 Microphotograph of a 24 GHz uniplanar serrodyne modulator

11.5 Active filters

In an integrated communications transceiver, filters and diplexers are so large and expensive that the full benefits of MMIC technology are not readily retained when the complete RF subsystem is considered. There are many applications, however, where a low cost and/or low mass integrated circuit based module is an essential part of the enabling technology. For example, advanced satellite payloads using phased arrays may have hundreds of transceiver modules, and it is desirable to integrate the filters into the MMIC modules. Conventional passive filters are not a practical solution, however, because of the size limitations of MMICs and the low Q-factor of MMIC lumped elements. It is only at frequencies well over 20 GHz that microstrip or CPW coupled-line filters might be acceptably small. As a result, there is a great deal of activity worldwide in the area of MMIC active filters. The most important techniques that have been reported in the literature for microwave active filters are active inductor based filters, filters employing actively-coupled passive resonators, techniques using negative resistance elements and transversal/recursive filters using FETs.

The advantages of active filters are:

1. small size and mass
2. low cost in mass production
3. high selectivity
4. easy integration with amplifiers, mixers, oscillators, etc.
5. potential for electronic tuning.

However, the drawbacks associated with active techniques are:

1. poor noise figure
2. non-linearity
3. danger of oscillation
4. complex bias circuitry and significant DC power
5. sensitivity to fabrication tolerances
6. environmental sensitivity.

Clearly, for their small size, these techniques are very attractive, but at the same time they are unlikely to be used for applications such as diplexers, where high power and low noise are important.

11.5.1 Active inductor based filters

With MMIC technology, large inductances are traditionally realised using planar spiral or stacked spiral inductors. Unfortunately, these inductors can suffer from excess series resistance and undesirable high frequency resonances. Thus, spiral inductors can place a fundamental limit on the performance of a circuit. As a result, a considerable amount of interest has been shown in the use of active inductors. Active inductors can be created using a number of simple transistor circuits, or with a capacitively-terminated gyrator or negative impedance converter.

The first high-Q active inductors were reported over two decades ago [59–62]. Here, the single inverted-transistor topology did not facilitate independent electronic tuning of both the inductance and series resistance. The use of these inductors to realise L–C resonator type active filters was presented in two particularly notable papers [60,61]. More recently, the active inductor demonstrated by ATR in Japan [63] has created considerable interest in active inductors for MMIC applications. Here, Hara *et al.* demonstrated the implementation of a relatively lossy broadband monolithic active inductor using a common-source cascode-FET with resistive feedback topology. Two of these active inductors were successfully employed in a broadband actively matched amplifier. Later, a significant reduction in the series resistance of the active inductor was measured with the use of a common-gate FET feedback arrangement, and a completely lossless active inductor was simulated using a common-gate cascode-FET feedback element [64].

Using the simple common-source cascode-FET with resistive feedback topology, Bastida *et al.* demonstrated the inductance tuning ability of the lossy active inductor [65]. Here, a floating cold-FET was used to implement a voltage controlled feedback resistor. The application of these tuneable active inductors

(TAIs) was demonstrated in a hybrid 180° analogue reflection-type phase shifter with MMIC TAI reflection terminations, and in a monolithic band-tuneable amplifier. A TAI in which both the inductance and series resistance can be controlled electronically was first demonstrated in a monolithic active filter in 1994 [66].

11.5.2 Actively-coupled passive resonators

The well known cascaded cells approach to active filter design involves breaking the required filter transfer function down into its constituent factors, and then realising each factor with the appropriate passive filter section. The complete filter transfer function is then achieved by cascading these sections together via amplifiers, which prevent the resonators interacting. Sussman-Fort presented a second-order bandpass filter using this technique [67] and Bonetti (Comsat Labs) applied the technique to the design of a 4 to 8 GHz bandpass MMIC active filter [68].

11.5.3 Techniques using negative resistance elements and active loops

In these techniques, the Q-factor of a resonator is increased by amplifying the signal within the resonator, either with an active loop or by coupling negative resistance into the resonator. The active loop technique has been applied to a microstrip resonator by the University of Limoges. Negative resistance elements formed from transistors have been used by Itoh [69] to achieve a 200 MHz wide bandpass filter centred on 10.5 GHz, with approximately 20 dB rejection 200 MHz from the band edges. For a single pole filter, a 1 dB compression level of -13 dBm was reported, measured at the input. This shows that the power handling of active filters can be expected to be considerably worse than an amplifier using the same device, and this is due to the high peak values of voltage or current that are inevitable in a high-Q resonator.

The use of a simple FET negative resistance circuit has been demonstrated for loss compensation in microstrip dual-mode ring resonator filters [70,71] and in lumped-element monolithic filters [72,73]. The topology of the lumped-element monolithic filter using this negative resistance technique is shown in Figure 11.16. This filter is far less complicated than the active inductor filter and has lower DC power consumption.

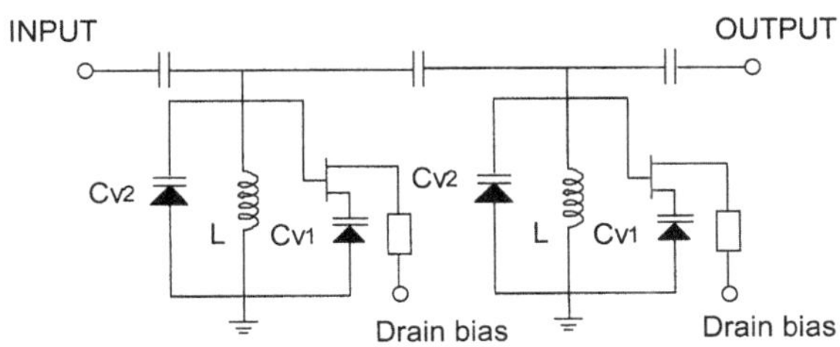

Figure 11.16 Active filter topology employing FET negative resistance elements

11.5.4 Transversal and recursive filters

Transversal and recursive filter techniques have been taken from the digital filter domain and applied to microwave filters by a number of researchers [74,75]. The basic transversal filter uses a transmission line with taps taken off at regular intervals, with the outputs from the taps then summed together. The result is that the output waveform at a particular instant is made up of a series of weighted samples of the past, present and 'future' instantaneous values of the input signal. It should be noted, however, that for a true transversal filter as many as 200 taps might typically be required. The recursive filter is similar, but the important difference is that it employs a second series of taps, this time fed back from the output. This additional flexibility has the important result that an arbitrary Q-factor can be achieved as positive feedback is employed.

The well known MMIC active filter work by Schindler and Tajima [76] uses a combination of conventional filtering with transversal elements to 'trim' the response. Their filter is based on the distributed amplifier, but uses bandpass structures directly integrated into the gate and drain lines. They achieved a 9.7 to 11.4 GHz passband with 3 dB insertion loss, and with 40 dB rejection approximately 1 GHz from the band edges. Work at the University of Limoges has involved using Lange couplers in transversal and recursive filters [77]. The Lange splits the signal into two paths, each of which has a certain weighting factor, and then the two paths are recombined in a second Lange coupler. The measured result of this simple filter shows a gentle bandpass response as the number of taps is so low. It is shown that the true transversal filter has a cyclic frequency response. This feature was not apparent in Rauscher's or Schindler's filters because they used a 'hybrid' approach. Generally, it appears that transversal and recursive techniques are mainly applicable to broadband filters with octave bandwidths.

11.5.5 Practical applications

MMIC active filters have tremendous performance capability and can be directly integrated with other circuitry. Although they have a poor noise figure and limited power handling, they are expected to be used first in applications where these performance criteria are not so important, such as in IF chains. Process variations make some form of electronic tuning essential, and varactor tuning is expected to play a vital role in narrowband active MMIC filters. Varactor diodes provide well controlled tuning characteristics and draw no current. Hence, a multi-pole filter could be tuned up by employing varactor bias potentiometers, which become analogous to the tuning screws in waveguide cavity filters. Such an approach cannot easily be avoided if brick-wall responses are required. Other tuning methods could involve control of DC bias, laser-trimming of resistors, switched spiral inductors and the use of conveniently removable air-bridges in the circuit. Recently, attention has turned to finding ways in which an active filter can be tuned adaptively with complex control loops [78]. With more development, it is certain that active filters will be used extensively at ever-greater frequencies.

11.6 Power amplifier linearisation

With standard foundry processes, achieving only a few watts is generally very challenging because of thermal limitations and the lack of true power devices. For applications requiring more power, special MMIC processing is necessary. MMICs can then, in principle, be used for power amplifiers (PAs) operating up to around 10–20 W. Beyond this, hybrid-MIC power amplifiers, or even travelling-wave tube (TWT) amplifiers, are necessary in order to achieve the required output power and efficiency. For power-limited applications such as mobile and satellite communications, it is necessary to operate the amplifier as close to saturation as possible. However, when operating close to saturation, signals are distorted by the non-linearities of the PA. This non-linear behaviour is quantified by the AM-AM (amplitude modulation to amplitude modulation) and AM-PM (amplitude modulation to phase modulation) conversion characteristics of the amplifier. The AM-AM and AM-PM conversion effects have the following deteriorative effects on the performance of a communication system:

1. Degradation of the bit error rate (BER) of the system. This is partly due to distorted amplitude and phase of the signalling elements in the transmitted signal constellation, and partly due to intersymbol interference.

2. Spectral spreading of the transmitted signal, which increases undesirable interference to adjacent channels.

3. In frequency-multiplexed (FDM/FDMA) systems the different carrier frequencies mix together, generating intermodulation products which cause adjacent channel interference.

These effects are pronounced in any transmission system where the PA must be operated close to saturation for high efficiency. For modulation schemes with little or no envelope variation, such as GMSK, solid-state PAs can operate fairly close to saturation (e.g. 0.1–0.5 dB backoff) for high efficiency. However, for spectrally efficient modulation schemes like 16-QAM, which result in large carrier envelope variations, the amplifier may have to operate several decibels below its maximum power in order to avoid problems caused by AM-AM and AM-PM conversion. This results in a very large reduction in efficiency. There are two distinct approaches to improving this situation. One method is to adaptively control the quiescent DC bias point of the transistors [79], as shown in Figure 11.17. This is particularly applicable to class-A amplifiers and signals with envelope variations. With a direct conversion transmitter, the envelope can be directly generated from the I and Q signals. The second method is to employ linearisation techniques. These techniques are based on compensation for the AM-AM and AM-PM conversion effects, so that the overall characteristics of the linearised PA approach those of a linear amplifier.

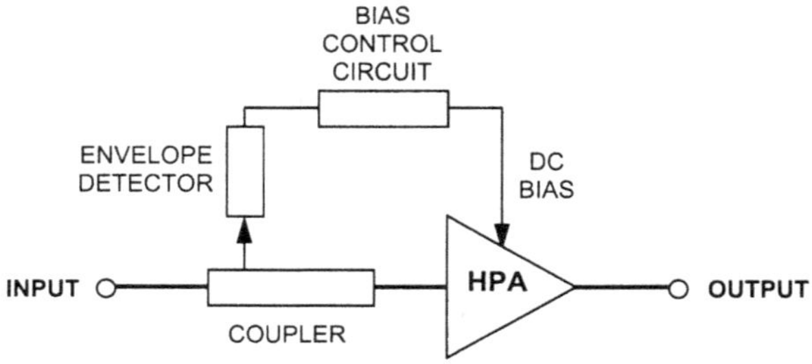

Figure 11.17 GaAs FET power amplifier with adaptive gate bias control

11.6.1 Linearisation techniques

The three main techniques that have been used can be classed as feed-forward, feedback and predistortion linearisers. Feedforward techniques employ an error amplifier which injects a distortion signal with equal amplitude but opposite phase at the PA output. The error signal is usually created by an error-determining subtraction loop around the PA, as shown in Figure 11.18. The drawback of this approach is that the error amplifier consumes considerable DC power and there are considerable losses at the PA output. As a result, the efficiency is low, but for multicarrier situations the reduction in intermodulation products is very desirable. Feedforward linearisation has also been applied to mixers and direct modulation transmitters [80,81].

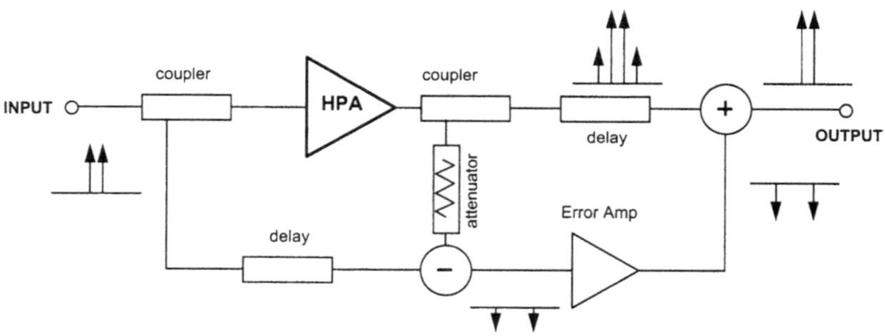

Figure 11.18 Block diagram of a linearised amplifier using feedforward

Predistortion techniques can be implemented at RF, IF or baseband. Baseband predistortion is carried out before modulation of the carrier, and involves introducing deliberate distortion into the constellation such that the distortion introduced by the PA restores an ideal constellation. State-of-the-art techniques use DSP and employ a look-up table method that incorporates spectral shaped filters and baseband predistortion. It has been shown that this technique substantially improves the performance of a digital communication system for practical CNR values [82].

RF or IF predistorters are circuits that use the non-linearity of elements such as transistors or diodes to create non-linear AM-AM and AM-PM characteristics which are the inverse of those of the HPA, so that the overall characteristics approach those of an ideal linear amplifier. For example, the lineariser may have an insertion gain which increases with increasing input power (gain expansion) in order to combat the effect of gain compression in the power amplifier. Although the maximum saturated output power cannot be increased, the power amplifier can be operated closer to saturation in order to improve its efficiency. At best, the characteristics of a soft-limiter can be achieved. Systems with these types of characteristics have been found to give a considerable improvement in performance. RF predistorters have been demonstrated both in hybrid MIC and MMIC technologies [83–87].

The final class of linearisation uses negative feedback. In microwave amplifiers, however, there is often too little gain for this to be used directly. Instead, the transmitted signal is tapped off and a low frequency component is extracted from it for feedback purposes; this could be the signal envelope or some intermodulation products, or the signal can be demodulated to recover the baseband signal itself. For quadrature modulation schemes, such as QPSK and 16-QAM, one technique involves demodulating the amplifier output signal and using the actual transmitted baseband values of I and Q as feedback signals. The demodulated I and Q signals are fed back to the modulator for adaptive predistortion of the signal constellation. This is known as 'Cartesian feedback' or 'adaptive baseband predistortion', and has been demonstrated successfully using analogue feedback loops [88,89] and digital techniques [90–94]. A highly-integrated approach using DSP and look-up tables techniques is shown in Figure 11.19. This technique has become very popular for RFIC and MMIC transceivers because it offers such an elegant solution.

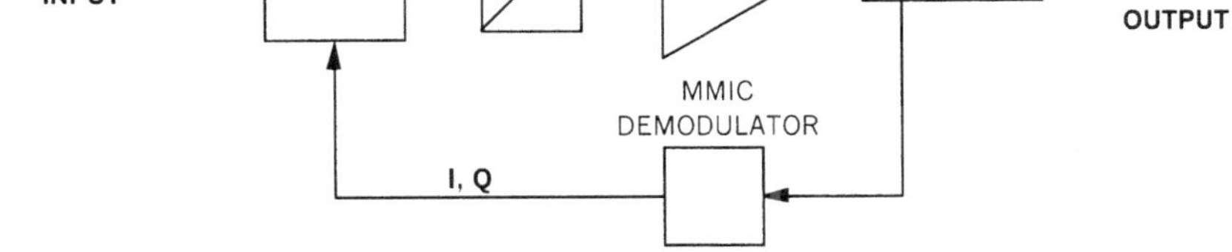

Figure 11.19 Block diagram of a linearised amplifier using adaptive baseband predistortion with a DSP-based modulator

11.7 Active isolators and circulators

11.7.1 Active isolators

Isolators provide a unilateral signal path. This may be required in order to absorb unwanted reflected signal power or to avoid interaction within various parts of a circuit or system. Isolators are sometimes required when the ports of devices or circuits have to be protected from unwanted impedance mismatches. For instance, when the load impedance presented to an oscillator changes, the conditions necessary to maintain satisfactory operation may not be met. This may cause the amplitude and/or frequency of oscillation to change, or even a cessation of oscillation altogether. With MMIC technology, active isolators are used to buffer the output port of an oscillator. The characteristics of an ideal active isolator include at least 20 dB of reverse isolation, constant forward insertion loss and input and output return losses better than 15 dB, all maintained across the desired bandwidth of operation. A simple method of implementing an active isolator is to use a cascode-FET amplifier. This circuit can provide a large gain and high reverse isolation.

With classical distributed amplifiers, a constant insertion gain can be maintained across multi-octave bandwidths. However, the isolation performance degrades with increasing frequency due to coupling through the gate-drain capacitance. A novel alternative to the distributed amplifier for implementing a multi-octave MMIC active isolator was proposed by Pyndiah and Bogaart [95]. The circuit consists of a simple feedback amplifier with the input and output ports swapped over. A single MESFET with an active load makes this a very compact design. The measured results demonstrated a minimum isolation of 20 dB, constant insertion loss of 9 dB and input and output return losses of -20 dB and -18 dB, respectively, all maintained within a 0.5 to 20 GHz frequency range.

11.7.2 Active circulators

As its name suggests, a circulator is a multiple-port device that can route a signal unilaterally from one RF port to another. Figure 11.20 shows three different types of circulator. Common applications for a circulator are illustrated in Figure 11.21. All of these can be implemented using any one of the types of circulator shown in Figure 11.20. The circulator is used to separate two signals propagating in opposite directions. In a diplexer, these are the transmitted and received signals at port 2. In a reflection-type phase shifter, parametric amplifier or reflectometer, they are the incident and reflected waves corresponding to a terminating impedance, Z_T, at port 2. In an isolator, any reflections entering port 2 are circulated into the matched load at port 3.

A general approach to realising a classical circulator consists of a symmetrical arrangement of three directional couplers and three isolators. For MMIC realisation, Bahl [96] proposed the use of either Wilkinson 0° couplers or Lange couplers with active isolators. Unfortunately, this approach may require a prohibitive amount of active area for applications below X-band. A highly economical solution would be to remove the directional couplers altogether and to employ the type of active isolator proposed by Pyndiah and Bogaart [95].

This approach was adopted three decades ago by Tanaka *et al.* [97], using bipolar junction transistor technology in a very low frequency active circulator.

Until recently, there has been little interest in the active circulator. This was mainly due to its narrowband performance and the problems associated with a hybrid realisation. Nowadays, more interest is being shown in active circulators, since their bandwidths have increased considerably as a result of the advances in transistor technology. Also, active circulators are totally compatible with monolithic technology [98–103].

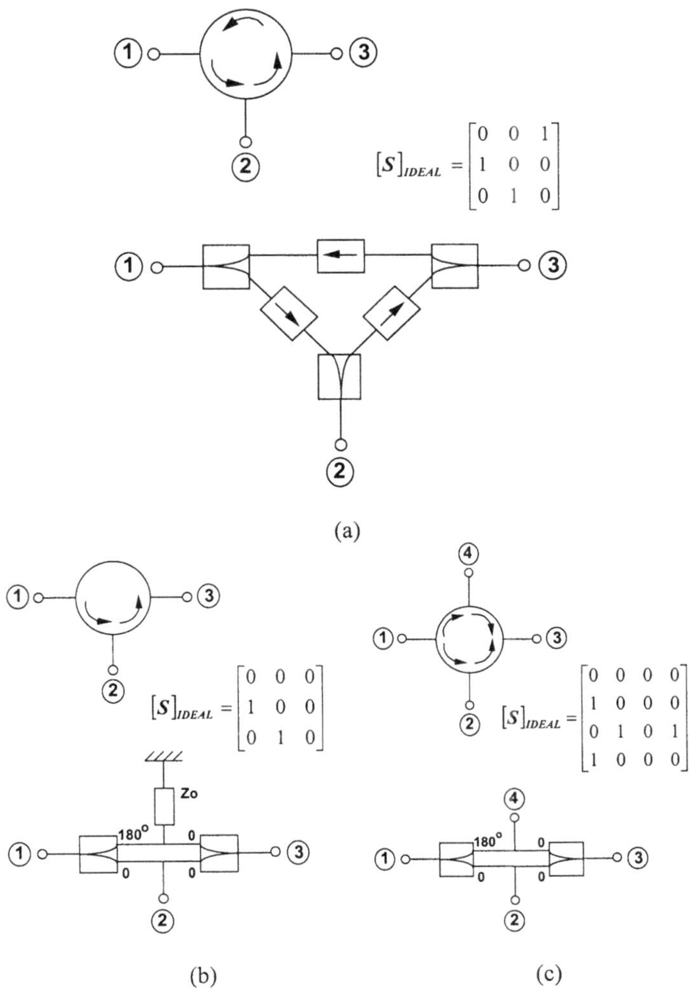

Figure 11.20 Types of circulator: (a) classical circulator, (b) quasi-circulator and (c) unilateral four-port junction

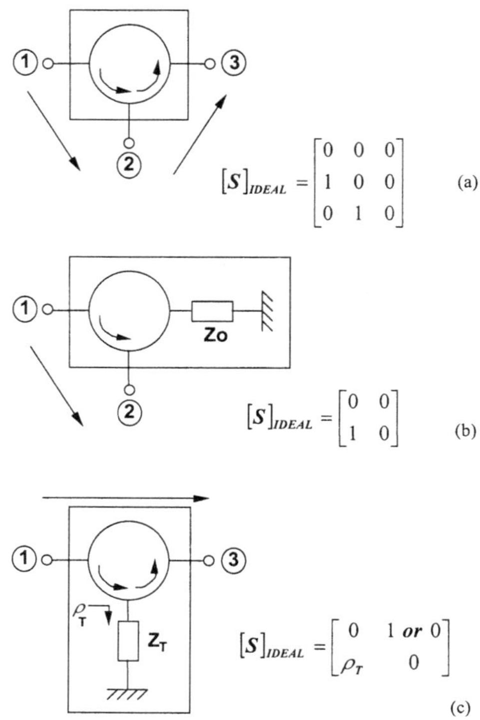

Figure 11.21 Circulator applications: (a) diplexer, (b) isolator and (c) reflection-type phase shifter/attenuator, parametric amplifier or reflectometer

A MESFET equivalent of the classical circulator demonstrated by Tanaka *et al.* was reported by Dougherty [100]. A modified version of the topology, employing standard F20 4 × 75 μm MESFETs, is shown in Figure 11.22. With the MESFETs operating at half I_{dss}, the simulated response for this circulator shows that 20 dB of isolation and a constant insertion loss of 3.7 dB are possible across more than a decade bandwidth (0.5 to 5.5 GHz). From an initial mask layout design, this circulator requires an active area of less than 1.0 mm^2. The model of this active circulator was used to simulate a high performance, decade bandwidth, analogue phase shifter [104].

A general approach to realising a quasi-circulator is to connect the outputs from a balun to the inputs of a power combiner: one of the links between the balun and combiner is terminated with a matched load, while the other link provides the main circulating port. An active realisation of a monolithic quasi-circulator was reported by Hara *et al.* [98]. The balun was implemented with a single MESFET power divider and the power combiner was implemented with two common-gate MESFETs. This very compact quasi-circulator achieved a good general performance across a 0.1 to 15 GHz frequency range. A classical circulator can be realised by simply connecting three identical quasi-circulators in a ring. The simulated performance of a 0.1 to 10 GHz active circulator implemented using this technique was reported by Hara *et al.* [98]. A HEMT equivalent of the monolithic quasi-circulator demonstrated by Hara *et al.* has been recently reported by Kother *et al.* [99]. Here, pHEMTs were used in a 40 GHz realisation, while InP-HEMTs were used in 55 GHz and 80 GHz realisations.

The topology of a unilateral four-port junction is the same as for the quasi-circulator, but with the matched load removed. Therefore, both links can provide main circulating ports. An active realisation of a monolithic-based unilateral four-port junction has been reported [102] and used in a novel reflectometer [105]. The balun was implemented with a distributed power divider and the power combiner was implemented with a distributed power combiner.

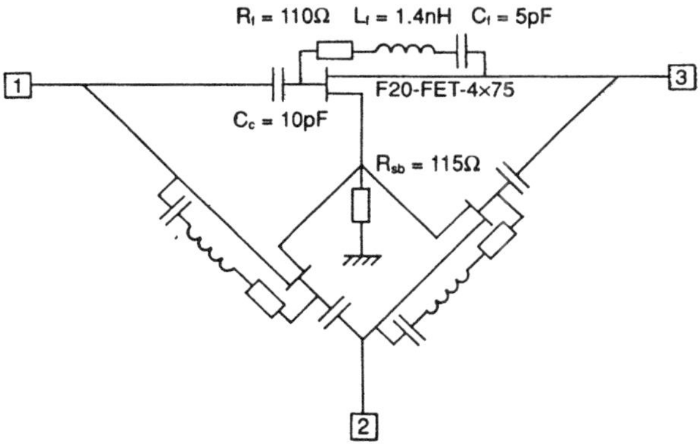

Figure 11.22 Active circulator circuit diagram

11.8 Optoelectronic integrated circuits (OEICs)

11.8.1 Applications

Optoelectronics and optical fibres have an increasingly important role in microwave and millimetre-wave applications. Examples of this are optically fed phased arrays [106,107] and the use of picosecond optical pulses for ultra-broadband MMIC measurements, which is described in Chapter 12. At the same time, microwave techniques have an increasingly important role in optoelectronic systems, as data rates have gone beyond 40 Gbit/s and as ultra-short pulses are used in many applications. Examples of this are optical communications [108,109] and optical fibre sensing [10]. Other applications exist which use a combination of optical and microwave techniques. Examples include fixed telecommunications networks having a wireless local loop and microcellular mobile communications systems employing a fibre-optic backbone (including 'radio-over-fibre' systems [111,112]).

The optically-fed phased array antenna employs conventional MMIC-based transmit-receive modules, but signals are distributed to the modules using optical fibres rather than coaxial cable or hollow waveguides. This approach is feasible for active phased arrays because the optical fibres need only carry low power signals such as IF or baseband signals, control and monitoring signals and a local oscillator reference signal. The advantages of the optical feed are lower transmission loss, a reduction in mass and the possibility of using optical techniques for signal processing and beam-forming. The disadvantage is, of course, the extra complexity because of the need for electrical/optical and optical/electrical interface components. This problem would be overcome nicely if the MMICs had optical inputs and outputs, or if direct optical control could be applied to the circuits in some way. Because GaAs and InP are direct band-gap semiconductors, III-V technology has a major advantage for integrating microwave and optoelectronic devices.

For long-haul communications the 1300 nm low dispersion window and 1550 nm low loss window wavelengths are preferred, and InP offers the requisite band-gap. As a result, InP or metamorphic (InP on GaAs substrates) devices are the best candidates for OEICs. A conceptual InP-based OEIC transceiver is illustrated in Figure 11.23. Wireless local loop and microcellular mobile systems employ a fibre-optic backbone, with the final communication link performed by a short microwave or millimetre-wave radio link. Since a large number of base stations are required, a number of schemes have been proposed for making the radio base stations low cost by using IF or RF subcarriers [112] and frequency division multiplexing. This approach means that the base station hardware is simpler, and does not have to cope with the +40 Gbit/s signal that would result from a time-division approach.

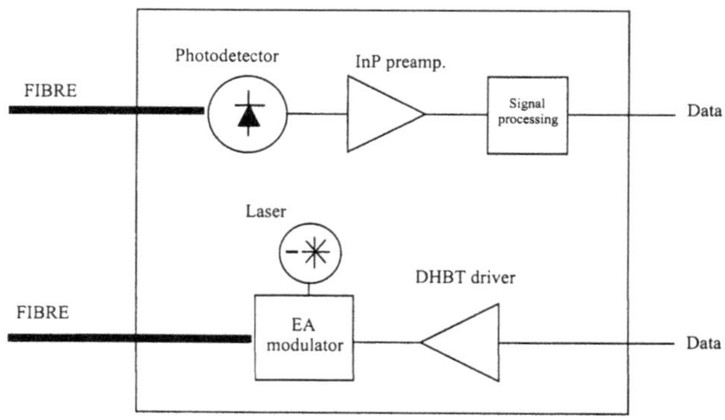

Figure 11.23 InP OEIC communications transceiver

11.8.2 Optically-controlled circuits

The optical control of circuits allows ultra-fast switching to be performed, with a near ideal RF-to-baseband signal isolation and immunity to electromagnetic interference [113–115]. Optical control has been applied to phase shifters, amplifiers, attenuators, switches and oscillators. There are two distinct approaches that can be used. In direct control methods, the component performing the circuit function (e.g. transistor or diode) is directly illuminated to control its electrical characteristics. With indirect methods, the component that is illuminated (e.g. a photodiode) is separate from the circuit function but provides it with a control voltage or current. The advantage of the latter approach is that the photodetector component can be optimised for optical sensitivity and does not have to perform a circuit function. The photodetector-circuit interface can be optimised to improve the efficiency and linearity of the control characteristic.

11.8.2.1 Phase shifters

An optically-controlled millimetre-wave bulk semiconductor phase shifter has been developed for insertion between rectangular waveguides [116]. When the absorbed energy within the semiconductor is equal to or greater than the band-gap energy level, free charge carriers (electron–hole pairs) are produced, forming a surface plasma layer in the semiconductor. As the optical illumination intensifies, the number of electron–hole pairs increases. Since the dielectric constant of the resulting electron–hole plasma region is higher than that of the rest of the guide, a sudden increase in the induced phase shift occurs, until saturation is reached.

More recently, transmission-line phase shifters implemented with a CPW line on a lossless/lossy/lossless (AlGaAs/GaAs/AlGaAs) layered semiconductor substrate have been reported [117,118]. Here, the regions of semiconductor between the signal line and ground planes are exposed to the optical illumination, generating the electron–hole plasma in the buried GaAs layer. The conductivity of this lossy layer is increased by increasing the intensity of the optical illumination. At a fixed frequency, as the conductivity of the lossy layer is increased, the characteristic propagating mode changes from a lossy dielectric mode to a slow wave mode, due to the increase in the effective capacitance of the structure. As the conductivity is further increased, the slow wave mode eventually changes to a skin effect mode.

11.8.2.2 Switches

In the past decade, the optical control of GaAs MESFETs and HEMTs as switches has been demonstrated [115]. Plasma produced within the active layer of the semiconductor causes a photovoltaic effect across the depletion region and a photoconductive effect in the undepleted region. With a reverse gate-source bias voltage equal to the pinch-off voltage, the MESFET is in the 'off' state. By illuminating the MESFET with a sufficiently high intensity, a voltage can be generated at the Schottky junction that is enough to overcome the pinch-off voltage and switch the MESFET to the 'on' state.

11.8.2.3 Amplifiers and attenuators

FET and HEMT amplifiers can be optically controlled directly by illuminating the active region exposed in the gate-source and gate-drain gap. Both photovoltaic and photoconductive effects alter the transistor's intrinsic elements such as g_m, C_{gs} and the channel resistance. It has been shown [113] that a change in g_m can be made the dominant mechanism, which leads to a variation in the amplifier gain. The largest gain variation is achieved with the device operated close to pinch-off, and by optimising the photovoltaic effect by employing a large gate bias resistor (> 50 kΩ). Gain variation of 20 dB has been achieved in this way, but the amplifier's dynamic range is rather limited at this bias point. As an alternative, indirect methods can be used, in which a photodetector converts the optical signal into an electrical signal, which then controls the amplifier's DC bias. This approach has been demonstrated for the optical control of a monolithic distributed amplifier [119].

11.8.2.4 Oscillators

Oscillators can be switched, frequency-tuned or injection-locked using optical control [113]. The key reason for controlling oscillators with optical signals is to enable large numbers of oscillators to be phase-locked to a common reference without requiring a bulky electrical distribution network. Applications include phased array antennas and active antennas employing spatial power combining [120]. Optical control can be performed directly on the oscillating device (MESFET, HEMT, HBT, Gunn or IMPATT diode) or it can be performed indirectly by illuminating an ancillary device within the resonant circuit. Subharmonic optical injection-locking of 21.5 GHz FET oscillators has been demonstrated [121].

11.8.2.5 Mixers

The separation of the RF, LO and IF signals in a mixer is a problem. In an optically-pumped mixer the LO is introduced into the mixing device by illuminating it with a laser beam that has been modulated by the LO signal. Alternatively, the LO can be directly applied as an electrical signal and the RF can be modulated onto the laser beam. The advantage is that high isolation of the signals can be achieved.

In optical mixing, two lasers with a 'small' frequency separation can be used to create microwave or millimetre-wave signals from the beat frequency [122]. Alternatively, this approach may be used with an optical phase-locked loop (OPPL) to generate microwave or millimetre-wave intensity modulated light [107].

11.8.3 Circuits for optoelectronic applications

A common application of MMICs in optoelectronic applications is the use of ultra-broadband amplifiers in high-speed optical communications systems [123]. MMICs find use in the receiver side, amplifying the signal from a photodiode, and in the transmitter side, as medium power drivers for optical modulators. Links operating at 10 Gbit/s have become commonplace, and systems operating at > 40 Gbit/s per carrier have already being developed. With dense wavelength division multiplexing (DWDM), aggregate data rates of over 3 Tbit/s have been demonstrated. Two amplifier topologies are particularly suited to providing the required bandwidth: the distributed amplifier and the Darlington-pair feedback amplifier. Distributed amplifiers (with MESFETs or HEMTs) have the disadvantage that they cannot easily be DC-coupled, and that it is hard to maintain good gain flatness down to very low frequencies. As a result, the Darlington-pair approach has become increasingly popular, since with HBT technology this can operate to frequencies well beyond 20 GHz — indeed, nearly 100 GHz bandwidth has been achieved with InP HBTs. The advantages of employing HBT technology are that DC coupling and >> 20 GHz bandwidth can be achieved and that transimpedance amplifiers can be designed, these being best suited for amplifying the current from p-i-n photodiode receivers. MMICs have also been used for pre-detection pulse-shaping: This has been demonstrated both with passive lumped-element equalising filters [124] and using transversal-style filtering embedded directly within the distributed amplifier structure [125]. InP HBT technology has become established as the most promising candidate for OEICs as in principle all the components for a high data rate transceiver can be integrated onto a single chip, as illustrated in Figure 11.23.

In applications where optical control of MMICs is attractive, such as large phased-arrays, the concept of MMICs with optical inputs and outputs has been proposed as a means of miniaturising the overall system [106]. The III-V compounds are highly suited to this because the active devices can be used as input ports, through direct optical control and on-chip external electro-optic modulators can provide optical output ports. However, it should not be forgotten that the advanced device engineering made possible by MBE and MOCVD means that silicon–germanium technology is by no means excluded from the optical domain.

11.9 References

1 DEVLIN, L. M., BUCK, B. J., CLIFTON, J. C., DEARN, A. W., and LONG, A. P.: 'A 2.4 GHz single chip transceiver', *IEEE Microwave and Millimeter-Wave Monolithic Circuits Symposium Digest*, 1993, pp. 23–26

2 HANKE, A., HADJIZADA, K., HEINEN, S., PUMA, G.L., and GEPPERT, W.: 'A 2.7 V image reject receiver for DECT'. *Proceedings of Radio Frequency Integrated Circuits (RFIC) Symposium*, 1998, pp. 93–96

3 KATZIN, P., BROKAW, A.P., DAWE, G., GILBERT, B., LYNN, L., and MOURANT, J.M.: 'A 900 MHz image-reject transceiver Si bipolar IC'. *Proceedings of Radio Frequency Integrated Circuits (RFIC) Symposium*, 1998, pp. 97–100

4 MAHFOUDI, M., and ALONSO J.I.: 'A simple technique for the design of MMIC 90° phase-difference networks', *IEEE Transactions on Microwave Theory and Techniques*, 1996, I, pp. 1694–1702

5 TOYODA, I., NISHIKAWA, K., KAMOGAWA, K. *et al.*: 'X-band Si bipolar transistor single-chip transceiver using three-dimensional MMIC technology (1998 [RFIC])'. *Proceedings of Radio Frequency Integrated Circuits (RFIC) Symposium*, 1998, pp. 289–292

6 LOVALACE, D., and DUREC, J.: 'A self calibrating quadrature generator with wide frequency range', *Proceedings of Radio Frequency Integrated Circuits (RFIC) Symposium*, 1997, 147–151

7 CRAWFORD, J. A.: 'Frequency Synthesizer Handbook' (Artech House, Norwood, MA, 1994)

8 UNDERHILL, M.J.: 'The adiabatic anti-jitter circuit', *Proceedings of Joint Meeting of the European Frequency and Time Forum and IEEE International Frequency Control Symposium*, 1999, **2**, pp. 611–614

9 SUWAKI, H., NAKAGAWA, T., and OHIRA, T.: 'An MMIC local oscillator for 16-QAM digital microwave radio systems', *Proceedings of 23rd European Microwave Conference*, 1993, pp.158–160

10 PEREZ, J., DORTA, P., and SIERRA, F.: 'A comparison of the performance of three different phase locked oscillators fabricated at 21 GHz', *IEEE MTT-S International Microwave Symposium Digest*, pp.305–308

11 NAKAGAWA, T., HIROTA, T., and OHIRA, T.: 'A novel MMIC approach to sampling phase/frequency detection for microwave quartz-lock oscillators', *IEEE MTT-S International Microwave Symposium Digest*, pp.1301-1304

12 BARNES, A.R., BANNISTER, D.C., and MOORE, M.T.: 'A 44 GHz low noise block downconvertor MMIC suitable for EHF satellite communication applications', *Proceedings of Radio Frequency Integrated Circuits (RFIC) Symposium*, 1998, pp. 223–226

13 CHEN, S.W. *et al.*: 'A high-performance 94-GHz MMIC doubler', *IEEE Microwave and Guided Wave Letters*, 1993, pp.167–169

14 NUCHTER, P., and MENZEL, W.: 'A MM-wave frequency divider', *IEEE MTT-S International Microwave Symposium Digest*, 1992, pp.695–697

15 KOK, YON-LIN, AHMADI, M., WANG HUEI, ALLEN, B.R., and LIN TZUEN: 'A Ka-band monolithic single-chip transceiver using sub-harmonic mixer', *Proceedings of Radio Frequency Integrated Circuits (RFIC) Symposium*, 1998, pp. 235–237

16 VERWEYEN, L., SIWERIS, H.J., NEUMANN, M. *et al.*: 'Coplanar transceive MMIC for 77 GHz automotive applications based on a nonlinear design approach'. *Proceedings of Radio Frequency Integrated Circuits (RFIC) Symposium*, 1998, pp. 33–36

17 NAM, S., GOKDEMIR, T., BAREE, A. H. *et al.*: 'Fully integrated Q-band MMIC transmitter and receiver chips using resistive PHEMT mixers', *IEEE MTT-S International Microwave Symposium Digest*, 1998, pp. 223–226

18 KAWAKAMI, K., SHIMOZAWA, M., IKEMATSU, H., ITOH, K., ISOTA, Y., and ISHIDA, O.: 'A millimeter-wave broadband monolithic even harmonic image rejection mixer'. *IEEE MTT-S International Microwave Symposium Digest*, 1998, **III**, pp. 1443–1446

19 OKAZAKI, H., and YAMAGUCHI, Y.: 'Wide-band SSB subharmonically pumped mixer MMIC', *IEEE Transactions on Microwave Theory and Techniques*, 1997, **II**, pp. 2375–2379

20 MONTIEL, C.M., FAN L., and CHANG K.: 'An X-band self-mixing oscillator antenna for transceiver and spatial power-combining applications', *IEEE Transactions on Microwave Theory and Techniques*, 1998, pp. 1546–1551

21 LAI, A.K.Y., and NG, J.H.H.: 'A low-cost active transceiving antenna', *IEEE MTT-S International Microwave Symposium Digest*, 1998, **II**, pp. 1047–1050

22 RASSHOFER, R.H., and BIEBL, E.M.: 'Surface-wave coupling of active antennas for homodyne sensor systems', *IEEE Transactions on Microwave Theory and Techniques* , 1998, **II**, pp. 2457–2462

23 QIAN, Y., and ITOH, T.: 'Progress in active integrated antennas and their applications', *IEEE Transactions on Microwave Theory and Techniques*, 1998, pp. 1891–1900

24 KERMARREC, C., DAWE, G., TEWKSBURY, T., and BROWN, T.: 'The future of RFIC transceiver technology', *Proceedings of Radio Frequency Integrated Circuits (RFIC) Symposium*, 1997, pp. 65–68

25 ATKINSON, S., and STRANGE, J.: 'Integration trends with RF silicon technologies in mobile radio applications', *Proceedings of Radio Frequency Integrated Circuits (RFIC) Symposium*, 1998, pp. 83–86
26 CANNISTRARO, V., LIOU, J.-C., and McCARTER, S.: 'Direct modulation lowers VSAT equipment costs', *Microwaves & RF*, 1990, pp. 99–101
27 KUMAR, S.: 'Directly modulated VSAT transmitters', *Microwave Journal*, 1990, pp. 255–264
28 KAWAKAMI, K., TAJIMA, K., SHIMOZAWA, M., ITOH, K., KASAI, N., and IIDA, A.: 'Fully monolithic integrated even harmonic quadrature ring mixer with an active matched 90 degree power divider for direct conversion receivers', *IEEE MTT-S International Microwave Symposium Digest*, 1997, **II**, pp. 657–660
29 SHIMOZAWA, M., KAWAKAMI, K., ITOH, K., IIDA, A., and ISHIDA, O.: 'A novel sub-harmonic pumping direct conversion receiver with high instantaneous dynamic range', *IEEE MTT-S International Microwave Symposium Digest*, 1996, **II**, pp. 819–822
30 SHIMOZAWA, M., KAWAKAMI, K., IKEMATSU, H. *et al.*: 'A monolithic even harmonic quadrature mixer using a balance type 90 degree phase shifter for direct conversion receivers', *Proceedings of Radio Frequency Integrated Circuits (RFIC) Symposium*, 1998, pp. 209–212
31 LI, J., BOSISIO, R.G., and WU, K.: 'A new direct digital receiver performing coherent PSK reception', *IEEE MTT-S International Microwave Symposium Digest*, 1995, **III**, pp. 1007–1010
32 McPHERSON, D., STUBBS, M. G., WIGHT, J. S., and LUCYSZYN, S.: 'A 28GHz HBT vector modulator and its application to an LMCS feedforward power amplifier', *Proceedings of 28th European Microwave Conference*, 1998, pp. 523–538
33 MAZUMDER, S.R., TSAI, T.L., and TSAI, W.C.: 'A frequency translator using dual-gate GaAs FETs', *IEEE MTT-S International Microwave Symposium Digest*, 1983, pp. 346–348
34 TRUITT, A., CERNEY, J., and MASON, J.S.: 'A 0.3 to 3 GHz monolithic vector modulator for adaptive array systems', *IEEE MTT-S International Microwave Symposium Digest*, 1989, pp. 421–422
35 PYNDIAH, R., JEAN, P., LEBLANC, R., and MEUNIER, J.-C.: 'GaAs monolithic direct (1-2.8) GHz Q.P.S.K. modulator', *Proceedings of 19th European Microwave Conference*, 1989, pp. 597–602
36 ITOH, J., NAKATSUKA, T., SATO, K. *et al.*: 'A low distortion GaAs quadrature modulator IC', *Proceedings of Radio Frequency Integrated Circuits (RFIC) Symposium*, 1998, pp. 55–58
37 COLOMINES, S., AMAUD, T., PLANA, R., PARRA, T., and GRAFFEUIL, J.: 'Design of high performances Gilbert-cell mixers for GSM/DCS front-ends', *Proceedings of Radio Frequency Integrated Circuits (RFIC) Symposium*, 1998, pp. 143–146
38 TELLIEZ, I. *et al.*: 'A compact monolithic microwave demodulator-modulator for 64-QAM digital radio links', *IEEE Transactions on Microwave Theory and Techniques*, 1991, **MTT-39** (12), pp. 1947–1954
39 POLLMAN, M.D., TRANTANELLA, C., SHIFRIN, M., APARIN, V., and UPTON, D.: 'A low-cost packaged MMIC chip set for 5.8 GHz ISM band applications'. *Proceedings of Radio Frequency Integrated Circuits (RF IC) Symposium*, 1997, pp. 33–36

40 DEVLIN, L.M., and MINNIS, B.J.: 'A versatile vector modulator design for MMIC', *IEEE MTT-S International Microwave Symposium Digest*, 1990, pp. 519–522

41 LUCYSZYN, S., and ROBERTSON, I.D.: 'Analog reflection topology building blocks for adaptive microwave signal processing applications', *IEEE Transactions on Microwave Theory and Techniques*, 1995, **MTT-43**, pp. 2617–2622

42 NORRIS, G.B., BOIRE, D.C., ST. ONGE, G. *et al.*: 'A fully monolithic 4-18 GHz digital vector modulator', *IEEE MTT-S International Microwave Symposium Digest*, 1990, pp. 789–792

43 ITOH, K., SHIMOZAWA, M., KAWAKAMI, K., IIDA, A., and ISHIDA, O.: 'Even harmonic quadrature modulator with low vector modulation error and low distortion for microwave digital radio'. *IEEE MTT-S International Microwave Symposium Digest*, 1996, **II**, pp. 967–970

44 ASHTIANI, A., NAM, S., d'ESPONA, A., LUCYSZYN, S., and ROBERTSON, I. D.: 'Direct multilevel carrier modulation using millimeter-wave balanced vector modulators', *IEEE Transactions on Microwave Theory and Techniques*, 1998, **MTT-45** (12), pp. 2611–2619

45 CHONGCHEAWCHAMNAN, M., ANG, K. S., KPOGLA, D., NAM, S., LUCYSZYN, S., and ROBERTSON, I. D.: 'Low-cost millimeter-wave transmitter using software radio techniques', *IEEE MTT-S International Microwave Symposium Digest*, 2000, pp. 1949–1952

46 McPHERSON, D.S. , ROBERTSON, I. D., and LUCYSZYN, S.: 'A W-band vector modulator and its application to software radar for automotive collision avoidance', *IEEE MTT-S International Microwave Symposium Digest*, 2000, pp. 1423–1426

47 LUCYSZYN, S., ROBERTSON, I.D., and AGHVAMI, A.H.: 'Novel applications in microwave communication systems for small-shift frequency translators', *Proceedings of IEEE International Conference on Telecommunications*, Dubai, 1994

48 MADNI, A.M., and WAN, L.A.: 'Solid-state preamplifier/frequency translator finds velocity deception applications', *Microwave Systems News*, 1983, pp. 71–86

49 SCHIEK, B., and EUL, H.-J.: 'Network analysis with a single sideband generator of a high image suppression', *Proceedings of 21st European Microwave Conference*, 1991, pp. 503–508

50 SKOLNIK, M.I. (Ed.): 'Radar handbook' (McGraw-Hill, NY, 1970), Chap. 13: 1970

51 ARISI, S.J.: 'Monolithic, lumped element, single sideband modulator', *IEEE MTT-S International Microwave Symposium Digest*, 1992, pp. 1047–1050

52 BAREE, A.H., ROBERTSON, I.D., and BHARJ, J.S.: 'A compact wideband MMIC image-rejection mixer chip', *Proceedings of IEEE Asia-Pacific Microwave Conference*, 1994, pp. 241–244

53 FOX, A.G.: 'An adjustable waveguide phase shifter', *Proceedings of the IRE*, 1947, **35**, pp. 1489–1498

54 BOYD, C.R.: 'Analogue rotary-field ferrite phase shifters', *Microwave Journal*, 1977, pp. 41–43

55 LUCYSZYN, S., PILCHEN, Y., ROBERTSON, I.D., and AGHVAMI, A.H.: 'Ku-band serrodyne frequency translator using wideband MMIC analogue phase shifters', *Proceedings of 23rd European Microwave Conference*, Madrid, 1993, pp. 819–822

56 GARVER, R. V.: '360° varactor linear phase modulator', *IEEE Transactions on Microwave Theory and Techniques*, 1969, pp. 137–147

57 LUCYSZYN, S., ROBERTSON, I.D., and AGHVAMI, A.H.: '24 GHz serrodyne frequency translator using a 360° analog CPW MMIC phase shifter', *IEEE Microwave and Guided Wave Letters*, 1994, **4** (3), pp. 71–73

58 KUSHNER, L. *et al.*: 'An 800 MHz monolithic GaAs HBT serrodyne modulator', *IEEE GaAs IC Symposium Digest*, 1994

59 CRYSTAL, E.G., HO, R.Y.C., ADAMS, D.K., and COHN, S.B.: 'Microwave synthesis techniques', *Stanford Research Institute annual report*, Section 5, Contract DAAB07-68-0088, 1968

60 ADAMS, D.K., and HO, R.Y.C.: 'Active filters for UHF and microwave frequencies'. *IEEE Transactions on Microwave Theory and Techniques*, 1969, **MTT-17** (9), pp. 662–670

61 SNYDER, R.V., and BOZARTH, D.L.: 'Analysis and design of a microwave transistor active filter', *IEEE Transactions on Microwave Theory and Techniques*, 1970, **MTT-18** (1), pp. 2–9

62 FLIEGLER, E.: 'Operating criteria for active microwave inductors', *IEEE Transactions on Microwave Theory and Techniques*, 1971, **MTT-19** (1), pp. 89–91

63 HARA, S., TOKUMITSU, T., TANAKA, T., and AIKAWA, M.: 'Broad-band monolithic microwave active inductor and its application to miniaturized wide-band amplifiers', *IEEE Transactions on Microwave Theory and Techniques*, 1988, **MTT-36** (12), pp. 1920–1924

64 HARA, S., TOKUMITSU, T., and AIKAWA, M.: 'Lossless, broadband monolithic microwave active inductors', *IEEE MTT-S International Microwave Symposium Digest*, 1989, pp. 955–958

65 BASTIDA, E.M., DONZELLI, G.P., and SCOPELLITI, L.: 'GaAs monolithic microwave integrated circuits using broadband tunable active inductors', *Proceedings of 19th European Microwave Conference*, 1989, pp. 1282–1287

66 LUCYSZYN, S., and ROBERTSON, I.D.: 'Monolithic narrow band filter using ultra high-Q tunable active inductors', *IEEE Transactions on Microwave Theory and Techniques*, 1994, **MTT-42** (12), pp. 2617–2622

67 SUSSMAN-FORT, S.: 'A realisation of a GaAs microwave active filter', *IEEE Transactions on Microwave Theory and Techniques*, 1990, **MTT-38** (10), pp. 1524–1526

68 BONETTI, R.R. *et al.*: 'An MMIC active filter with 60-dB rejection', *IEEE MTT-S International Microwave Symposium Digest*, 1992, pp. 1195–1198

69 CHANG, C., and ITOH, T.: 'Microwave active filters based on coupled negative resistance method', *IEEE Transactions on Microwave Theory and Techniques*, 1990, **MTT-38** (12), pp. 1879–1884

70 KARACAOGLU, U., ROBERTSON, I.D., and GUGLIELMI, M.: 'A dual-mode microstrip ring resonator filter with active devices for loss compensation', *IEEE MTT-S International Microwave Symposium Digest*, 1993, pp. 189–192

71 KARACAOGLU, U., ROBERTSON, I.D., and GUGLIELMI, M.: 'Microstrip bandpass filters with MMIC negative resistance circuits for loss compensation', *IEEE MTT-S International Microwave Symposium Digest*, 1994

72 HOPF, B.P., WOLFF, I., and GUGLIELMI, M.: 'Coplanar MMIC active bandpass filters using negative resistance circuits', *IEEE Transactions on Microwave Theory and Techniques*, 1994, **MTT-42**, pp. 2598–2602

73 KARACAOGLU, U., and ROBERTSON, I.D.: 'High selectivity varactor-tuned MMIC bandpass filter using lossless active resonators', *IEEE Microwave and Millimeter-Wave Monolithic Circuits Symposium Digest*, 1994, pp. 237–240

74 JUTZI, W.: 'Microwave bandwidth active transversal filter concept with MESFETs', *IEEE Transactions on Microwave Theory and Techniques*, 1971, **MTT-19** (9), pp. 760–767

75 RAUSCHER, C.: 'Microwave active filters based on transversal and recursive principles', *IEEE Transactions on Microwave Theory and Techniques*, 1985, **MTT-33** (12), pp. 1350—1360

76 SCHINDLER, M., and TAJIMA, Y.: 'A novel active filter with lumped and transversal elements', *IEEE Transactions on Microwave Theory and Techniques*, 1989, **MTT-37** (12), pp. 2148–2153

77 BILLONET, L. *et al.*: 'Design concept for microwave recursive and transversal filters using lange couplers, *IEEE MTT-S International Microwave Symposium Digest,* 1992, pp. 925–928

78 KUHN, W.B., YANDURU, N.K., and WYSZYNSKI, A.S.: 'Q-enhanced LC bandpass filters for integrated wireless applications', *IEEE Transactions on Microwave Theory and Techniques*, 1998, **II**, pp. 2577–2586

79 SALEH, A.A.M., and COX, D.C.: 'Improving the power-added efficieny of FET amplifiers operating with varying envelope signals', *IEEE Transactions on Microwave Theory and Techniques*, 1983, **MTT-31**, pp. 51–56

80 ELLIS, T.J.: 'A modified feed-forward technique for mixer linearization', *IEEE MTT-S International Microwave Symposium Digest*', 1998, **III**, pp. 1423–1426

81 CHONGCHEAWCHAMNAN, M., BLEWETT, M. J., and ROBERTSON, I. D.: 'Feedforward linearisation applied to a direct carrier modulation transmitter', *Proceedings of IEEE International Symposium on Circuits and Systems*, 2000, **1**, pp. 655–658

82 AGHVAMI, A.H., and GEMIKONAKLI, O.: '16-ary QAM system for low data rate satellite services', *Electronics Letters*, 1989, **25** (16), pp. 1055–1057

83 INADA, R. *et al.*: 'A compact 4 GHz linearizer for space use', *IEEE MTT-S International Microwave Symposium Digest*, 1986, pp. 323–326

84 MOOCHALLA, S. *et al.*: 'An integrated Ku-band linearizer driver amplifier for TWTAs with high gain and wide bandwidth', *Proceedings of 14th AIAA International Communications Satellite Systems Conference*, 1992, pp. 167–174

85 KUMAR, S.: 'Power amplifier linearization using MMICs', *Microwave Journal*, 1992, pp. 96–104

86 KUMAR, M., WHARTENBY, J.C., and WOLKSTEIN, H.J.: 'Predistortion linearizer using GaAs dual-gate MESFET for TWTA and SSPA used in satellite transponders', *IEEE Transactions on Microwave Theory and Techniques*, 1985, **MTT-33** (12), pp. 1479–1488
87 ZHANG, W.-M., and YUEN, C.: 'A broadband linearizer for Ka-band satellite communication'. *IEEE MTT-S International Microwave Symposium Digest*, 1998, **III**, pp. 1203–1206
88 PETROVIC, V.: 'Reduction of spurious emission from radio transmitters by means of modulation feedback', *Proceedings of IEE Conference on Radio Spectrum Conservation Techniques*, 1983, pp. 44–49
89 BATEMAN, A., HAINES, D.M., and WILKINSON, R.J.: 'Linear transceiver architectures', *Proceedings of IEEE Vehicular Technology Conference*, 1988, pp. 478–484
90 CAVERS, J.K.: 'Amplifier linearisation using a dual predistorter with fast adaptation and low memory requirements', *IEEE Transactions on Vehicular Technology*, 1991, **39** (4), pp. 374–382
91 WRIGHT, A.S., and DURTHER, W.G.: 'Experimental performance of an adaptive digital linearized power amplifier', *IEEE MTT-S International Microwave Symposium Digest*, 1992, pp. 1105–1108
92 JECKELN, E.G., GHANNOUCHI, F.M., and SAWAN, M.A.: 'An L band adaptive digital predistorter for power amplifiers using direct I-Q modem', *IEEE MTT-S International Microwave Symposium Digest*, 1998, **II**, pp. 719–722.
93 QIMING, R, and WOLFF, I.: 'Improvement of digital mapping predistorters for linearising transmitters'. *IEEE MTT-S International Microwave Symposium Digest*, 1997, **III**, pp. 1691–1694
94 JECKELN, E.G., GHANNOUCHI, F.M., and SAWAN, M.: 'Adaptive digital predistorter for power amplifiers with real time, modeling of memoryless complex gains', *IEEE MTT-S International Microwave Symposium Digest*, **II**, pp. 835–838
95 PYNDIAH, R., and BOGAART, F. V. D.: 'Novel multioctave MMIC active isolator (1–20 GHz)', *Electronics Letters*, 1989, **25** (21), pp. 1420–1422
96 BAHL, I. J.: 'On the design of an active circulator', *Microwave and Optical Technology Letters*, 1988, **1** (1), pp. 18–20
97 TANAKA, S., SHIMOMURA, N., and OHTAKE, K.: 'Active circulators—the realization of circulators using transistors', *Proceedings of the IEEE*, 1965, pp. 260–267
98 HARA, S., TOKUMITSU, T., and AIKAWA, M.: 'Novel unilateral circuits for MMIC circulators', *IEEE Transactions on Microwave Theory and Techniques*, 1990, **MTT-38**, pp. 1399–1406
99 KOTHER, D., HOPF, B., SPORKMANN, T., WOLFF, I., and KOBLOWSKI, S.: 'Active CPW MMIC circulator for the 40 GHz band', *Proceedings of 24th European Microwave Conference*, Cannes, 1994, pp. 542–547
100 DOUGHERTY, R.: 'Circulate signals with active devices on monolithic chips', *Microwaves & RF*, 1989, pp. 85–86, 89
101 POLACEK, G. A.: 'Stable bias yields active MMIC circulator', *Microwaves & RF*, 1990, pp. 132–137

102 ROBERTSON, I.D., and AGHVAMI, A.H: 'Novel monolithic ultra-wideband unilateral 4-port junction using distributed amplification techniques', *IEEE MTT-S International Microwave Symposium Digest*, Albuquerque, 1992, pp. 1051–1054

103 KATZIN, P., AYASLI, Y., REYNOLDS, L., and BEDARD, B.: '6 to 18 GHz MMIC circulator', *Microwave Journal*, 1992, pp. 248–256

104 LUCYSZYN, S., and ROBERTSON, I.D.: 'Decade bandwidth MMIC analogue phase shifter', *IEE Colloquium Digest on Multi-octave Microwave Circuits*, London, 1991, pp. 2/1–6

105 ROBERTSON, I.D., and AGHVAMI, A.H: 'A novel reflectometer using a wideband monolithic active unilateral 4-port junction', *Proceedings of 23rd European Microwave Conference*, Madrid, 1993, pp. 296–298

106 HERCZFELD, P.R.: 'Hybrid photonic-microwave systems and devices', *IEICE Transactions on Electronics*, 1993, **E76-C** (2), pp. 191–197

107 SEEDS, A.: 'Optical technologies for phased array antennas', *IEICE Transactions on Electronics*, 1993, **E76-C** (2), pp. 198–206

108 MIDWINTER, J.E.: 'The threat of optical communications', *IEE Electronics & Communication Engineering Journal*, 1994, pp. 33–38

109 COCHRANE, P., HEATLEY, D.J.T., SMYTH, P.P., and PEARSON, I.D.: 'Optical telecommunications–future prospects', *IEE Electronics & Communication Engineering Journal*, 1993, pp. 221–232

110 UTTAMCHANDANI, D.: 'Fibre-optic sensors and smart structures: developments and prospects', *IEE Electronics & Communication Engineering Journal*, 1994, **6** (5), pp. 237–246

111 SHIBUTANI, M. *et al.*: 'Feasibility studies on an optical fiber feeder system for microcellular mobile communications systems', *Proceedings of IEEE International Conference on Communications*, 1991, pp. 1176–1181

112 OGAWA, H., POLIFKO, D., and BANBA, S.: Millimeter-wave fiber optics systems for personal radio communications', *IEEE Transactions on Microwave Theory and Techniques*, 1992, **MTT-40** (12), pp. 2285–2292

113 SEEDS, A.J., and DE SALLES, A.A.: 'Optical control of microwave semiconductor devices', *IEEE Transactions on Microwave Theory and Techniques*, 1990, **MTT-38** (5), pp. 577–585

114 ROSSEK, S.J., and FREE, C.E.: 'Optical control of microwave signals using GaAs FETs', *IEE Electronics & Communication Engineering Journal*, 1994, pp. 21–30

115 LUCYSZYN, S., and ROBERTSON, I. D.: 'Optically induced measurement anomalies with voltage-tunable analog-control MMICs', *IEEE Transactions on Microwave Theory and Techniques*, 1998, **MTT-46**, pp. 1105–1114

116 VAUCHER, A.M., STRIFFLER, C.D., and LEE, C.H.: 'Theory of optically controlled millimeter wave phase shifters', *IEEE Transactions on Microwave Theory and Techniques*, 1983, **MTT-31**, pp. 209–216

117 KAISER, M. *et al.*: 'Optically controlled phase shift of Schottky coplanar lines', *Electronics Letters*, 1989, **25** (23), p. 1575–1577

118 CHEUNG, P., NEIKIRK, D.P., and ITOH, T.: 'Optically controlled coplanar waveguide phase shifters', *IEEE Transactions on Microwave Theory and Techniques*, 1990, **MTT-38** (5), 586–595

119 PAOLELLA, A., and HERCZFELD, P.R.: 'Optical gain control of a GaAs MMIC distributed amplifier', *Microwave and Optical Technology Letters*, 1988 **1** (1), pp. 13–16

120 LIN, J., and ITOH, T.: 'Active integrated antennas', *IEEE Transactions on Microwave Theory and Techniques*, 1994, **MTT-42** (12), pp. 2186–2194

121 DARYOUSH, A.S.: 'Optical synchronization of millimeter-wave oscillators for distributed architectures', *IEEE Transactions on Microwave Theory and Techniques*, 1990, **MTT-38** (5), pp. 467–476

122 NI, D.C., FETTERMAN, H.R., and CHEW, W.: 'Millimeter-wave generation and characterization of a GaAs FET by optical mixing', *IEEE Transactions on Microwave Theory and Techniques*, 1990, **MTT-38** (5), pp. 608–614

123 IMAI, Y., SANO, E., and KAZUYOSHI, A.: 'Design and performance of wideband GaAs MMICs for high-speed optical communication systems', *IEEE Transactions on Microwave Theory and Techniques*, 1992, **MTT-40** (2), pp. 185–190

124 DARWAZEH, I., LANE, P.M., MARNANE, W.P. *et al.*: 'GaAs MMIC optical receiver with embedded signal processing', *IEE Proceedings, G*, 1992, **139** (2), pp. 241–243

125 MOREIRA, P., DARWAZEH, I.Z., and O'REILLY, J.J.: 'Novel optical receiver design using distributed amplifier pulse shaping network', *Proceedings of IEE Colloquium on Optical Detectors and Receivers*, Savoy Place, 1993, Digest no. 1993/173, pp. 8/1–8/4

Chapter 12

Measurement techniques

S. Lucyszyn

12.1 Introduction

All electronic sub-systems are made up of devices and networks. In order to simulate the overall performance of a sub-system under development, all the components that make up the sub-system must be accurately characterised. To this end, precision measurement techniques must be employed at component level. Not only do precision measurements enable a manufacturer to check whether devices are within their target specifications, and to monitor variations in parameter tolerances due to process variations, they also allow more accurate empirical models to be extracted from the measurements and help new modelling techniques to be validated. Also, the operation and performance of some experimental devices can often only be understood from accurate measurements and subsequent modelling. Conversely, poor measurements could result in the needless, and therefore, expensive redesign of high performance components or sub-systems.

Devices and networks are traditionally characterised using z, y or h-parameters. In order to measure these parameters directly, ideal open and short circuit terminations are required. These impedances can be easily realised at low frequencies. However, at microwave frequencies such impedances can only be achieved over narrow bandwidths (when tuned circuits are employed) and can also result in circuits that are conditionally stable (when embedded within a 'matched load' reference impedance environment) becoming unstable. Fortunately, scattering- (or S-) parameters can be determined at any frequency. To perform such measurements, the device under test (DUT) is terminated with matched loads. This enables extremely wideband measurements to be made and also greatly reduces the risk of instability; but only when the DUT is terminated with near ideal matched loads (this is irrespective of whether the measurement system is calibrated or not). S-parameter measurements also offer the following advantages:

1. any movement in a measurement reference plane along an ideal transmission line will vary the phase angle only

2. for a linear device or network, voltage or current and measured power are related through the measurement reference impedance (normally 75 or 50 Ω for coaxial lines and 1 Ω for rectangular waveguides)
3. with some passive and reciprocal structures, ideal *S*-parameters can be deduced from spatial considerations, enabling the measurements of the structure to be checked intuitively.

By applying a known incident wave to the DUT and then measuring the reflected and transmitted wave amplitudes, *S*-parameters can be calculated from the resulting wave amplitude ratios. The equipment most commonly used to perform this measurement is called a vector network analyser (VNA) [1]. The DUT can now be characterised using complete *S*-parameter measurements (along with DC measurements). The element values associated with the small-signal equivalent circuit model of the DUT can be determined using direct calculations, iterative optimisation and intuitive tuning. This process is referred to as parameter extraction.

With a monolithic microwave integrated circuit (MMIC), either a test fixture or probe station is employed to secure the MMIC in place and to provide a stable means of electrically connecting the MMIC to the measurement system [2]. In this chapter, the use of test fixtures and probe stations at ambient room temperature is reviewed and their role at thermal and cryogenic temperatures is discussed. Finally, with the increasing need for performing non-invasive (or non-contacting) measurements, experimental field probing technologies are introduced.

12.2 Test fixture measurements

Although probe stations result in much more accurate and reproducible measurements, test fixtures are still widely used. The principal reasons are that they are very much cheaper than probe stations and they offer a greater degree of flexibility, such as facilitating larger numbers of RF ports and enabling DC bias circuitry and any off-chip resonators to be located next to the chip. Also, the heat dissipation required when testing monolithic power amplifiers can be easily provided with test fixtures. In addition, test fixtures are ideally suited when RF measurements are required during temperature cycling and when cryogenic device characterisation is required [3–5].

An illustration of a basic 2-port test fixture is shown in Figure 12.1. Most test fixtures are, in principle, based on this generic design, typically consisting of four different components: a detachable metal chip carrier with high permittivity substrate, a rigid metal housing, connector/launchers and bond wires. The MMIC is permanently attached to the chip carrier with either conductive epoxy glue or solder. The metal housing is employed to hold the chip carrier and the connector/launchers in place. The launcher is basically an extension of the coaxial connector's centre conductor, which passes through the housing wall to make electrical contact with the associated chip carrier's microstrip transmission line. Bond wires or straps are used to connect the other end of the microstrip line to the MMIC under test. The parasitic element values associated with a test fixture are typically an order of magnitude greater than those of the MMIC under test.

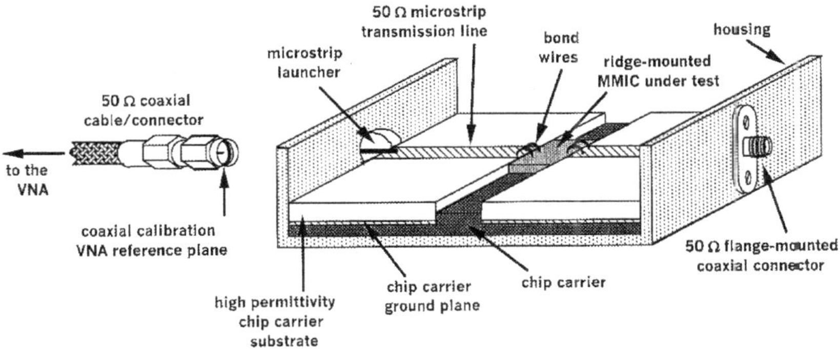

Figure 12.1 Generic design of a 2-port test fixture

Before any accurate measurements can be performed, measurement systems must first be calibrated, in order to correct for the systematic errors resulting from the numerous reflection and transmission losses within the measurement system. A calibration kit is required to perform this calibration procedure. This 'cal. kit' has a number of electrical reference standards and software that must be downloaded into the VNA's non-volatile memory or associated PC/workstation controller. For a 2-port measurement system, the calibration standards must:

1. define the primary reference planes
2. remove any phase ambiguity using open circuit and/or short circuit reflection standard(s)
3. define the reference impedance, using either delay line, matched load or attenuator impedance standard(s).

Some of the various combinations of different standards that can be employed in a 2-port calibration procedure are listed in Table 12.1. The software should contain accurate models for the associated standards and the algorithms required to implement the chosen calibration method. The accuracy of subsequent measurements ultimately depends on how well all the standards remain characterised. Any deviation in the electrical parameters of the standards will degrade the magnitudes of the effective directivity and source match for the measurement system. As a result, great care must be taken to look after these calibration standards.

Table 12.1 Common calibration methods

Method	Calibration Standard					
	Through		Reflect	Reference Impedance		
	L=0	L≠0	$\rho_1 = \rho_2$	Line	Match	Atten.
TRL	○		○	○		
LRL		○	○	○		
TRM	○		○		○	
LRM		○	○		○	
TRA	○		○			○
LRA		○	○			○
TSD	○		ρ=-1	○		

The non-idealities of a measurement system are characterised using mathematical error correction models, represented by flow diagrams (also known as error adapters or boxes). The function of the calibration procedure is to solve for the error coefficients in these models by applying the raw, uncorrected, *S*-parameter measurements of the standards to a set of independent linear equations. The basic 2-port calibration procedures have an 8-term error model (4-terms associated with each port) and require only three standards. These error terms should correspond directly to the raw hardware performance, including the directivity, source match and frequency tracking. A more accurate 12-term error model, as used in 2-port coaxial calibration, takes cross-talk and the effects of impedance mismatches at the RF switches within the VNA's test set into account. Once the calibration procedure has been performed, it can be verified by measuring separate verification standards.

12.2.1 Two-tier calibration

One method of calibrating the measurement system is to split the process into two tiers [6]. Initially, a coaxial calibration is performed, where the VNA reference planes are located at the end of its cable connectors. Historically, the VNA was calibrated using short-open-load-through (SOLT) standards. These lumped-element standards can give high quality coaxial calibrations across an ultra-broad bandwidth (e.g. DC to 50 GHz), so long as all the standards remain accurately characterised across the entire bandwidth. Since test fixtures are far from ideal, a second process is required to shift the initial VNA reference planes to the MMIC under test, in order to eliminate the effect of the test fixture. This second process is known as de-embedding or de-convolution [7]. To perform de-embedding it is necessary to accurately characterise the test fixture [8].

Another reason why you may need to characterise a test fixture is when multiple RF port MMICs are to be measured using a 2-port VNA [9]. Here, power reflected from impedance-mismatched loads on the auxiliary ports of the MMIC can result in significant measurement errors. These errors will increase as the mismatch losses increase and/or the number of RF ports increases. As a result, MMICs that have more RF ports than the VNA require all the loads to be individually characterised, and a further process of matrix renormalisation [9–13] in order to remove the effects of the mismatched loads on the auxiliary ports. Three methods that can, in principle, be employed to characterise a test fixture are: *time-domain gating*, *in-fixture calibration* and *equivalent circuit modelling*.

12.2.1.1 Time-domain gating

Some VNAs can be upgraded with a synthetic-pulse time-domain reflectometry (TDR) option [14–20]. Here, the discrete form of the inverse Fourier transform (IFT) is applied to a real sequence of harmonically related frequency-domain (F-D) measurements; in our case, of the MMIC embedded within its test fixture. This is directly equivalent to mathematically generating synthetic unity-amplitude impulses (or unity-amplitude steps), which are then 'applied' to the embedded MMIC. The resulting time-domain (T-D) reflection and transmission responses can then be analysed to provide information about the MMIC and test fixture discontinuities. In reflection measurements, it is possible to remove the effects of unwanted impedance mismatches or else isolate and view the response of an individual feature. With a multiple port test-fixture, transmission measurements can give the propagation delay and insertion loss of signals travelling through a particular path by removing the responses from the unwanted paths.

With an MMIC fed with transmission lines that only support a pure TEM mode of propagation, time and actual physical distance are simply related:

$$\textit{Physical Distance} = \begin{cases} c\,\Delta t\,\zeta/2 & \text{with reflection measurements} \\ c\,\Delta t\,\zeta & \text{with transmission measurements} \end{cases}$$

where,

$$c = \text{speed of light in free space}$$

$$\Delta t = \text{time difference, relative to a reference (e.g. t = 0)}$$

$$\zeta = \frac{1}{\sqrt{\varepsilon_r}} = \text{velocity factor}$$

$$\text{Also, frequency-domain nulls in } |S_{11}| \text{ are at frequency harmonics of } \frac{1}{\Delta t}$$

$$\text{where, } \Delta t = \text{time difference, between two reflected impulses}$$

If the feed lines are non-TEM, and therefore dispersive, impulse spreading will occur, which could significantly distort the impulse shape (in time and amplitude). If the dispersive nature is known, the frequency sweep can be pre-warped [20].

With either a banded VNA (which may cover just one of the main waveguide bands), or a broadband VNA, the *band-pass* T-D mode can be selected, where only synthetic impulses are generated. This is useful for band-limited guided-wave structures (e.g. rectangular waveguides). However, in this mode, only the magnitudes of the individual reflection and transmission coefficients are available. As a result, the exact nature of any discontinuity (e.g. resistive, inductive or capacitive) cannot be identified.

With a broadband VNA, a *low-pass* T-D mode is also available; where both synthetic impulses and synthetic steps can be generated. The *low-pass* mode is used to emulate a real-pulse TDR measurement system. This allows the user to identify the nature of any discontinuity. F-D measurements are taken from the *start frequency*, f_1, to the *stop frequency*, f_2. When compared to *band-pass*, for the same bandwidth (i.e. *frequency-span*) $B = f_2 - f_1$, the *low-pass* mode offers twice the *response resolution* in the T-D. However, with the *low-pass* mode, the F-D measurements must be harmonically related, from DC to f_2, such that $f_2 = n_{fd} f_1$, where n_{fd} is the number of points in the F-D (e.g. 51, 101, 201, 401 or 801). The DC data point is extrapolated from the f_1 measurement. However, if the measurement at f_1 is noisy, the T-D trace will be unstable and difficult to interpret.

In TDR, the *width* of a band-limited unit impulse (or window function) is defined as the interval between its two half-amplitude (i.e. -6 dB power) points. The corresponding *response resolution* is defined as the interval between two impulses that are just distinguishable from each other as separate peaks. With equal amplitude impulses, the *response resolution* is equal to the 6 dB *impulse width*. With no window function applied to the F-D measurements:

$$6\ dB\ Impulse\ Width = \begin{cases} \dfrac{1.2}{B} & \text{for } band\text{-}pass \\ \dfrac{0.6}{B} & \text{for } low\text{-}pass \end{cases}$$

$$Main\ Lobe's\ Null\text{-}to\text{-}Null\ Width = \begin{cases} \dfrac{2}{B} & \text{for } band\text{-}pass \\ \dfrac{1}{B} & \text{for } low\text{-}pass \end{cases}$$

The time *range* is the length of time that measurements can be made without encountering a repetition of the same response. The *range* must be set longer than the furthest discontinuity, otherwise aliasing will occur, where out-of-range discontinuities will fold-over and appear in-range at (*2 range - target position*):

$$Range = \frac{1}{\Delta f} \quad where \quad \Delta f = \frac{B}{n_{fd} - 1}$$

If a feature lies exactly midway between two T-D points then the energy associated with the discontinuity will be distributed between the two points, resulting in the displayed amplitude being reduced by almost 4 dB [20].

Therefore, care must be taken to ensure there is sufficient *range resolution* (or *point spacing*) in the T-D:

$$Range\ Resolution = \frac{Range}{n_{td}} \quad where \quad n_{td} = \text{number of points in the T-D}$$

The *point spacing* can be reduced to any desired level, at the expense of processing time, by using a chirp-Z fast Fourier transform algorithm. This allows *range* to be replaced by an arbitrary display *time-span* in the above *range resolution* equation.

De-embedding using synthetics-pulse TDR is not de-embedding in the true sense. It is specifically T-D gating, which can isolate a time feature and emphasise its frequency response. With time-gating, a mathematical window (called a gate or time filter) is used to isolate the embedded MMIC, so that only the MMIC's frequency response can be emphasised. When the gate is switched on, all reflections outside the gate are set to zero. This is equivalent to terminating the MMIC with the complex conjugate of its respective port impedance(s).

The synthetic-pulse TDR option can be a very useful tool, although it can suffer from a number of sources of errors [15,19,20]; some of these are listed below:

(1) Noise errors [15]

(a) **Sweep mode**: The VNA's synthesised source can be operated in either the *ramp-sweep* or *step-sweep* mode. With the former, small non-linearities and phase discontinuities generate low-level noise sidebands on the T-D impulse and step stimuli. However, with the *step-sweep* mode, the improved source stability eliminates these noise sidebands and improves the T-D's dynamic range by as much as 30 dB. Moreover, to reduce the noise floor of the T-D measurements further, the *step-sweep* mode enables more averaging of the F-D measurements, compared to the *ramp-sweep* mode, without greatly increasing the sweep time.

(b) **Bandwidth**: The noise floor in the T-D response is directly related to noise in the F-D data. Therefore, the number of F-D data points taken at, or below, the system's noise floor can be minimised by reducing the *frequency-span* to the bandwidth of the MMIC.

(c) **Test-set**: If the test-set does not have a flat response down to the *start frequency* then the reduction in the F-D's dynamic range towards f_1 will cause an increase in the T-D's noise floor, The resulting trace bounce, in the *low-pass* mode, can be improved by turning on T-D trace averaging.

(3) Frequency-domain window errors

There is usually a choice of F-D window functions (e.g. Kaiser–Bessel) that can be applied prior to the IFT, for example: *minimum (0th order)*, *normal (6th order)* and *maximum (13th order)*. The *minimum* window has a rectangular function that produces the sin *x*/*x* impulse shape, having the minimum *6 dB impulse width* but also the maximum sidelobe levels (with a minimum sidelobe suppression of only 13 dB in its power response). The other two window functions reduce the sidelobe levels (with a minimum suppression of 44 dB and 98 dB, respectively) at the expense of a wider impulse (by a factor of 1.6 and 2.4, respectively). It will be seen that a trade-off has to be made when choosing the F-D's windowing function, between the desired resolution and dynamic range in the T-D. Note that this windowing function does not affect the displayed F-D response.

(a) **Time resolution errors:** With narrow bandwidth VNAs, the impulse may be too wide. As a result, it may be difficult to resolve the MMIC and the connector/launchers features, down to the baseline, when the associated discontinuities are too close to one another. In practice, the MMIC should be separated by at least two *6 dB impulse widths* from the connector/launcher.

(b) **Dynamic range errors:** Impulse sidelobes limit the dynamic range of the T-D responses, since the sidelobes from a large impulse can hide a small adjacent target impulse.

(c) **Moding errors:** If the bandwidth of the VNA is too high, such that overmoding in transmission lines or box mode resonances occur in the test jig, the T-D responses become un-interpretable.

(d) **Out-of-band response**: The amplitude of the impulses represents the average value over the entire *frequency-span*. Therefore, the displayed amplitude of an impulse can be different from the expected value if the *frequency-span* includes an MMIC with a non-flat frequency response; for example, having highly abrupt out-of-band characteristics.

(3) <u>Discontinuity errors</u>

(a) **Masking errors:** If the target discontinuity is preceded by other discontinuities that either reflect or absorb energy, then these other discontinuities may remove some of the energy travelling to and emanating from the target discontinuity. Also, the trailing edge of earlier features can also obscure the target feature.

(b) **Multi-reflection aliasing errors:** Multiple reflections between discontinuities can cause aliasing errors. For example, if a 2-port MMIC is positioned mid-way between two connector/launchers, reflection from the furthest connector/launcher will be corrupted by multiple reflections between the MMIC and the nearest connector/ launcher.

(4) <u>Time-domain window errors</u>

In practice, a time filter having a non-rectangular response is used for gating, otherwise the sin x/x weighting would be conveyed to the F-D. There is usually a choice of T-D window functions that can be applied before the Fourier transform: *minimum, normal, wide* and *maximum*. The *minimum* window has the fastest roll-off and largest sidelobes, while the *maximum* window has the slowest roll-off and smallest sidelobes.

(a) **Baseline errors:** The *gate-start* and *gate-stop* times, which define the *-6 dB gate-span* of the filter, must be set at the baseline if low frequency distortion in the F-D is to be minimised [14].
(b) **Truncation errors:** A limited gate width may truncate lengthy target features. To minimise truncation error, the wider gates are preferred.
(c) **Sidelobe errors:** The time filter sidelobes may 'see' earlier or subsequent features. This could significantly corrupt the F-D response of the target feature. To minimise sidelobe errors, the wider gates are preferred.
(d) **Gate offset errors:** F-D distortion can occur if the *gate-centre* is offset from the centre of the target feature(s). This is because a near-symmetrical target response may lose its symmetry when applied to a time-offset gate that has significant in-gate attenuation. To minimise gate offset errors, the wider gate shapes are preferred.
(e) **Minimum gate-span errors:** The gate-span must be set wider than the minimum value, otherwise the gate will have no passband and may have high sidelobe levels.
(f) **Attenuation errors:** For a fixed gate span, the level and duration of in-gate attenuation may be excessive with wider gates.
(g) **Reflection/transmission switching errors:** If gating is performed on a voltage reflection coefficient response then the associated return loss F-D measurement is valid. If the same gating times are applied to the voltage transmission coefficient response(s) then this may not be appropriate. For example, when a 2-port MMIC is not placed midway between connectors, the transmission pulse may not be fully enclosed within the reflection response's gate. The resulting insertion loss F-D measurement will not represent accurate de-embedding.

As an example, a GaAs MMIC with a 2.9 mm length of 55 Ω microstrip through-line was placed at the centre of a 25.4 mm alumina chip carrier. An Agilent Technologies 8510B VNA was calibrated with a 20 GHz bandwidth and *401* frequency points. The *band-pass* mode was selected with a *minimum* F-D windowing function. This combination provides a minimum *response resolution* and maximum *range* values of 60 ps and 20 ns, respectively. The F-D power responses are shown in Figure 12.2*a*. The corresponding T-D response of the input port's voltage reflection coefficient is shown in Figure 12.2*b*. Here, the first and last peaks correspond to the impedance mismatches associated with the coaxial-to-microstrip transitions of the input and output ports, respectively. The two centre peaks correspond to the reflections associated with the microstrip-to-MMIC transitions. It will be apparent from Figure 12.2*b*, that accurate de-embedding would not be possible using T-D gating. This is because the unwanted reflections cannot be resolved down to the baseline. If de-embedding were attempted in the above example then the ripples in the F-D responses would be smoothed out, as one would expect, although this would not constitute accurate de-embedded measurements. In order to achieve accurate de-embedded measurements, a VNA with more bandwidth, or alternatively, a real-pulse TDR system having ultra-short impulses, can be used.

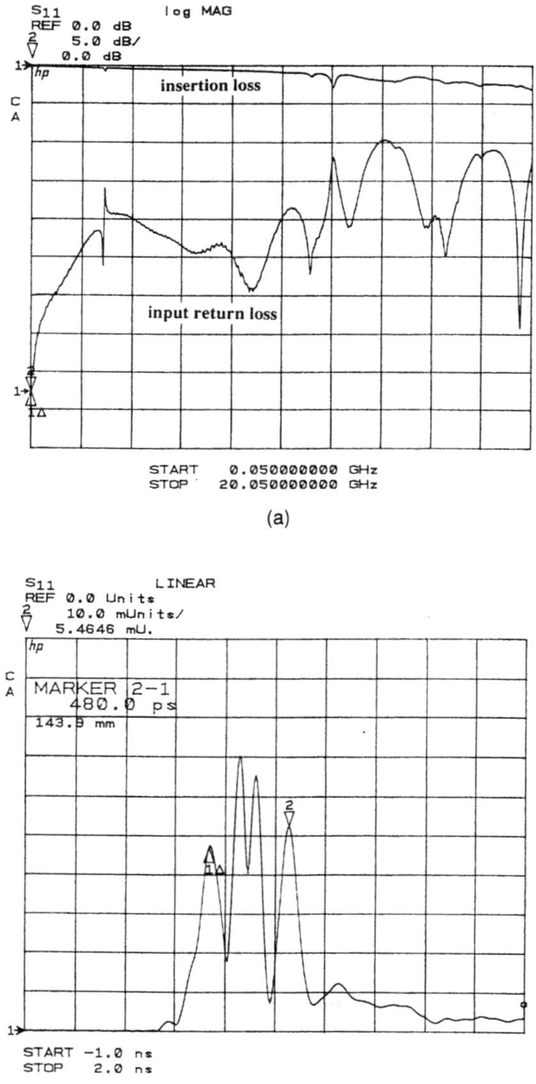

Figure 12.2 Embedded 55 Ω MMIC through-line: (a) frequency-domain power responses and (b) corresponding time-domain response for the input voltage reflection coefficient

12.2.1.2 In-fixture calibration

In general, a quality test fixture is much cheaper to buy than a probe station. Suitably designed quality test fixtures can be accurately characterised using in-fixture calibration techniques. As with coaxial calibration, the most appropriate algorithms use a combination of through, reflection and delay line standards, with common methods being TRL, TSD and LRL. The main reason for employing these types of calibration is that only one discrete impedance standard is required, such as an open or short, which is relatively easy to implement. The matched load is avoided; this is advantageous, as it is more difficult to fabricate non-planar 50 Ω loads to the same level of accuracy that can be achieved with low dispersion transmission lines. However, there are still significant disadvantages with in-fixture calibration:

1. Multiple delay lines may be required for wideband calibration (any one line must introduce between about 20° and 160° of electrical delay to avoid phase ambiguity, limiting the bandwidth contribution of each line to an 8:1 frequency range).
2. The use of multiple lines can add uncertainty to the measurements, since the launchers are continually being disturbed during calibration. Although freely available software (called *MultiCal*™) can eliminate the effects of non-repeatability, by measuring either the same line a number of times or different lengths of line, in order to reduce the uncertainty [21].
3. A frequency-invariant measurement reference impedance must be taken from the characteristic impedance, Zo, of the delay lines, however, frequency dispersion in microstrip lines may not always be corrected for. In practice, the Zo of the lines can be determined using TRL calibration [22–23] and then subsequent measurements can be renormalized to any measurement reference impedance.
4. The high level of accuracy is immediately lost with test fixtures that employ poor quality components and/or non-precision assembly.
5. The calibration substrates dictate and, therefore, restrict the location of the RF ports.
6. For devices with more than two ports the calibration procedure must be significantly extended and all the results from this routine must be easily stored and retrieved.
7. The microstrip-to-MMIC transition is not taken into account.

12.2.1.3 Equivalent circuit modelling

Test fixtures made in-house tend to be simple in design, such as the type shown in Figure 12.1, and cost only a small fraction of the price of a good quality commercial test fixture. Unfortunately, these non-ideal test fixtures suffer from unwanted resonances [24], poor grounding [25] and poor measurement repeatability. The problem of unwanted resonances can be clearly seen in the frequency-domain responses of Figure 12.2*a*. Here, the resonances at 3 GHz and 12 GHz are attributed to the production grade coaxial connectors used in the test fixture. Because of poor repeatability, employing elaborate and expensive calibration techniques to characterise such fixtures would appear unjustified, because significant measurement degradation is inherent. As an alternative, equivalent circuit models (ECMs) can provide a crude but effective means of de-embedding. This 'stripping' process results in about the same level of degradation as would be found if in-fixture calibration was used with a non-ideal test fixture, but with minimal expense and greater flexibility. Also, ECMs based on the physical structure of the fixture have demonstrated a wide bandwidth performance. The ECMs can be easily incorporated into conventional frequency-domain simulation software packages. They can also be employed to simulate packaged MMICs. An example of an ECM for a test fixture similar to the one in Figure 12.1 is shown in Figure 12.3. This model has demonstrated a sufficient degree of accuracy from DC to 19 GHz for the popular Omni-Spectra SMA connector/wedge-shaped launcher [9], which is similar to the more popular SMA printed circuit board socket.

The exact nature of the ECM, the element values and the microstrip parameter data are extracted from through-line measurements of the test fixture. Both a direct microstrip through-line and an MMIC through-line should be used in order to provide more information for the parameter extraction process, and to make it possible to model the microstrip-to-MMIC transition accurately. De-embedding can be carried out with most frequency-domain CAD packages by converting the ECM into a series of negative elements connected onto the ports of the measured data. Some CAD packages provide a 'negation' function that allows the ECM sub-circuit to be directly stripped from the measured data. With either method, the order of the node numbers is critical, and the de-embedding routine should be verified.

In addition to those already mentioned, de-embedding using equivalent circuit modelling has the following advantages:

1. dispersion in the microstrip lines does not have to be corrected for in the VNA's calibration
2. there is no restriction by the calibration procedure on the location of the RF ports
3. systematic errors resulting from variations in the characteristic impedance of the chip carrier's microstrip lines, due to relaxed fabrication tolerances, can easily be corrected for
4. bond wires [26] and the microstrip-to-MMIC transition can be modelled [27]
5. resonant mode coupling between circuit components, due to a package resonance, can also be modelled [24]. Better still, package resonances can, in some instances, be removed altogether [28–29].

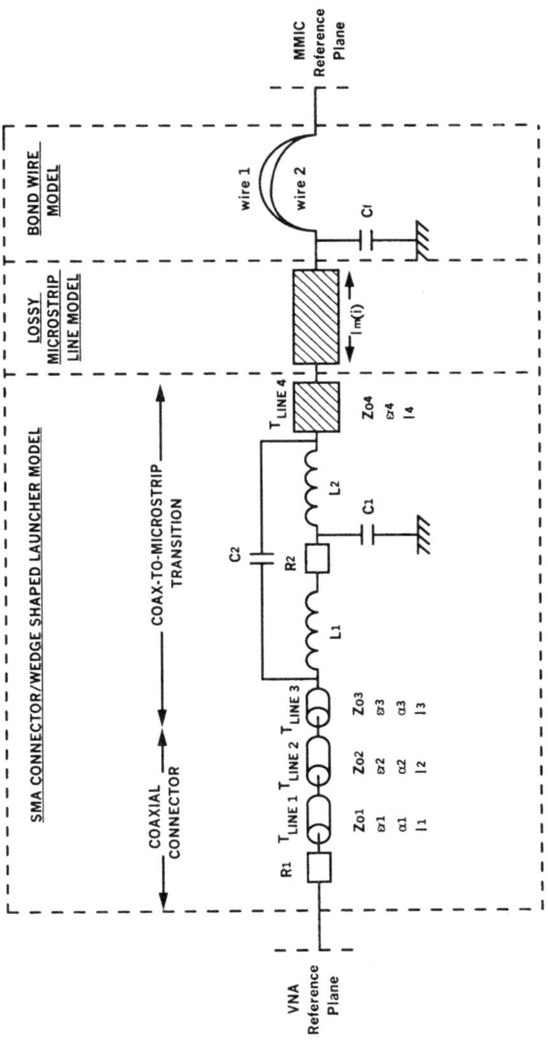

Figure 12.3 Equivalent circuit model of a microstrip test fixture

12.2.2 One-tier calibration

Improved contact repeatability and prolonged contact lifetime are two considerations that favour the two-tier process [6], as they are only assembled once with time-domain gating and ECMs. In practice, however, to achieve the best performance, in-fixture TRL or line-network-network [30] calibration is applied directly to a quality test fixture, without the need for the two tier coaxial calibration/de-embedding process. This one tier calibration procedure gives more accurate measurements than the two tier method, since de-embedding is inherently prone to errors, and the propagation of measurement errors is reduced [6]. Using this approach, the Anritsu 3680V universal test fixture can perform repeatable measurements up to 60 GHz. At the time of writing, a number of other companies produce test fixtures for accurate in-fixture calibration, including Agilent Technologies, Intercontinental Microwave, Argumens and Design Techniques. They are either split-block fixtures, with a removable centre section, or they use launchers attached to sliding carriages.

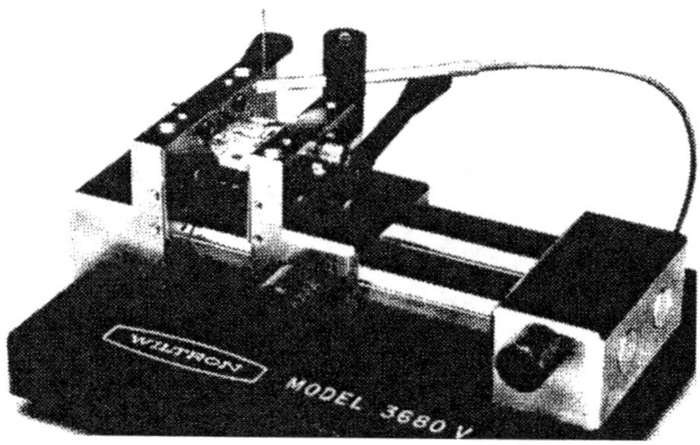

Figure 12.4 Photograph of the Anritsu 3680V universal test fixture (Courtesy of Anritsu)

With the high levels of accuracy that can be achieved using quality test fixtures, the poor characterisation of bond wires, due to the poor repeatability of conventional manually operated wire-bonding machines, becomes significant. Improvements in the modelling accuracy and physical repeatability of the microstrip-to-MMIC transition when using automatic wire-bonding assembly techniques have been reported [27]. In addition, flip-chip technology (also known as solder-bump technology) is now well established [31–39]. Here, a tiny bead of solder is placed on all the MMIC bond pads and the MMIC is placed upside down directly onto the chip carrier. When heated to the appropriate temperature, the solder flows evenly and a near perfect connection is made between the MMIC pad and its associated chip carrier pad. The advantages of this technology over bond wire technology, for the purposes of measurements, are its ultra-broad bandwidth, superior contact repeatability and high characterisation accuracy of the carrier's transmission line-to-MMIC transition.

12.2.3 Test fixture design considerations

The following guidelines are useful when selecting, designing or using a test fixture:

1. Split-block test fixtures [5,40] are ideal for 2-port in-fixture TRL calibration, since they can provide good repeatability. Here, a short circuit standard is preferred, since significant energy may be radiated with an open circuit standard.
2. Side walls can form a waveguide or resonant cavity. The size of the waveguide/cavity should be made small enough so that the dominant mode resonant frequency is well above the maximum measurement frequency. Carefully placed tuning screws and/or multiple RF absorbing pads can eliminate or suppress unwanted modes [28,29].
3. Poor grounding, due to excessively long ground paths and ground path discontinuities, must be avoided.
4. Avoid thick chip carrier substrates, wide transmission lines (sometimes used for off-chip RF decoupling) and discontinuities, in order to minimise the effects of surface wave propagation and transverse resonances at millimetric frequencies. Transverse currents can be suppressed by introducing narrow longitudinal slits into the low impedance lines.
5. Use substrates with a high dielectric constant to avoid excessive radiation losses and to minimise unwanted RF coupling effects.
6. New precision connector/launchers should be used whenever possible, and measurements should be performed below the connector's dominant TEM mode cut-off frequency.
7. Launchers should be separated from the DUT by at least 3 or 4 times the substrate thickness, so that any higher order evanescent modes, generated by the non-ideal coax-to-microstrip transition, are sufficiently attenuated at the DUT.

12.3 Probe station measurements

Until relatively recently, the electrical performance of an MMIC was almost always measured using test fixtures. Nowadays, extremely accurate MMIC measurements can be achieved using probe stations. Such techniques were first suggested for use at microwave frequencies in 1980 [41], demonstrated experimentally in 1982 [42], and introduced commercially by Cascade Microtech in 1983. During the past decade there have been rapid developments in probe station measurement techniques.

Today, the channel partnership between Cascade Microtech and Agilent Technologies provide a total solution for on-wafer probing, which can perform repeatable frequency-domain measurements at frequencies as high as 110 GHz [43], although single-sweep measurements from 45 MHz to 120 GHz are routinely undertaken at the University of Surrey. When compared with test fixtures, commercial probe station measurements have the following advantages:

1. they are available in a single-sweep system from DC to 120 GHz
2. they are more accurate and much more repeatable, since they introduce much smaller systematic errors
3. they have a simpler calibration procedure, which can be automated with on-wafer calibration and verification standards [12,44]
4. they enable the VNA measurement reference planes to be located at the probe tips or at some distance along the MMIC's transmission line; in the latter case, transition effects can be removed altogether
5. they provide a fast, non-destructive means of testing the MMIC, thus allowing chip selection prior to dicing and packaging.

Overall, the microwave probe station can provide the most cost effective way of measuring MMICs when all costs are taken into account.

12.3.1 Passive microwave probe design

At frequencies greater than a few hundred megahertz, DC probe needles suffer from parasitic reactance components, due to the excessive series inductance of long thin needles and shunt fringing capacitances. If the needles are replaced by ordinary coaxial probes that are sufficiently grounded, measurements up to a few gigahertz can be achieved. The upper frequency is ultimately limited by the poor coax-to-MMIC transition. A tapered coplanar waveguide (CPW) probe provides a smooth transition with low cross-talk. Cascade Microtech have developed tapered CPW probes that enable measurement to be made from DC to 110 GHz with a single coaxial input, and between 50 GHz and 75 GHz (for V-band) or 75 GHz and 110 GHz (W-band) with waveguide inputs [43]. The maximum frequency limit for coaxial-input probes is imposed by the onset of higher order modes propagating in the conventional coaxial cables and connectors. For W-band operation, Agilent Technologies developed a coaxial cable and connector that has an outer screening conductor diameter of only 1 mm. An illustration of the use of this 1 mm coaxial technology to give state-of-the-art performance up to 110 GHz, with a Cascade Microtech Summit 10000 probe station, is shown in Figure 12.5. A corresponding photograph of the first single-sweep, 45 MHz to 120 GHz, Agilent Technologies 8510XF vector network analyser to be commissioned in the UK, with a Cascade Microtech Summit 9000 probe station, is shown in Figure 12.6.

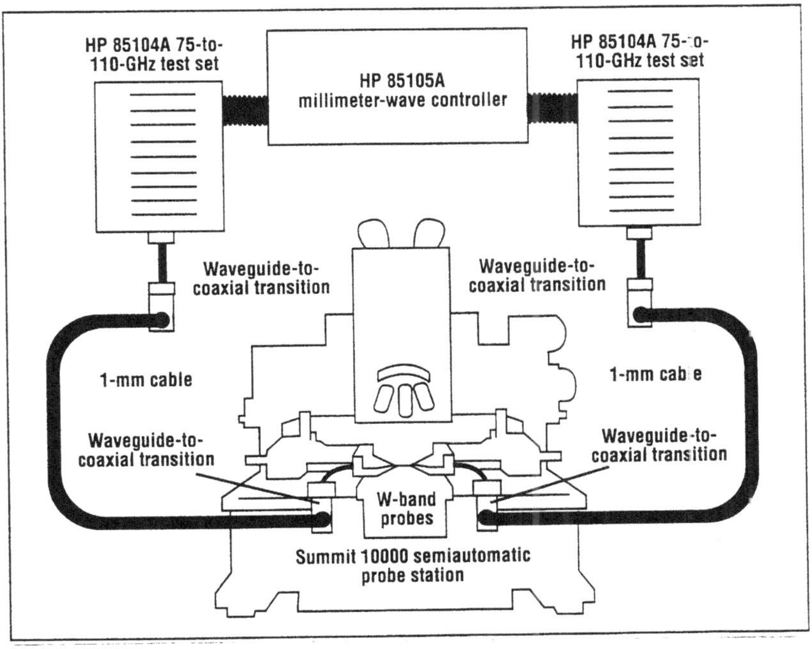

Figure 12.5 Illustration of a banded, 75 to 110 GHz, on-wafer probing system (Courtesy of Cascade Microtech Europe)

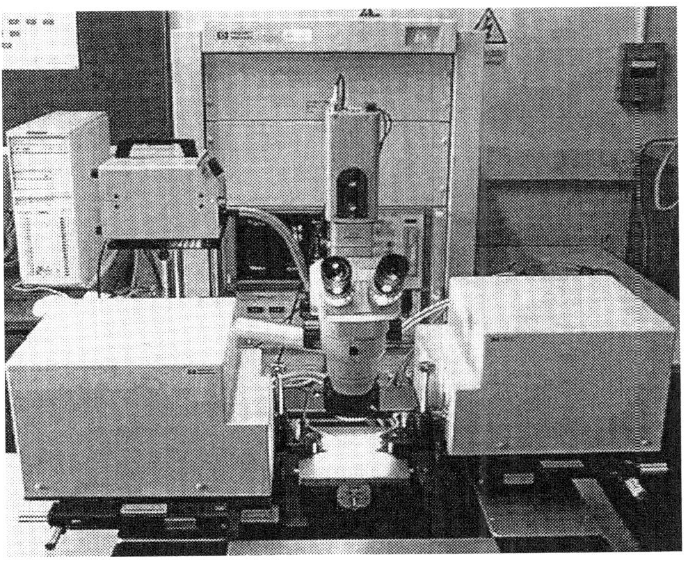

Figure 12.6 Photograph of the single-sweep, 45 MHz to 120 GHz, on-wafer probing system at the University of Surrey

Initially, the tapered coplanar waveguide probe was made from an alumina substrate or an ultra-low-loss quartz substrate. The probe tips that make the electrical contacts consist of hard metal bumps that are electroplated over small cushions of metal, allowing individual compliance for each contact. As the probes are over-travelled (in the vertical plane) the probe contacts wipe or 'skate' the MMICs' probe pads (in the horizontal plane). One of the major limitations of these tapered CPW probes is their short lifetime, since the substrate has limited compliance and the probe contacts can wear down quite quickly. As a result, the more the probe is used the more over-travel has to be applied to them. Eventually, either the probe substrate begins to crack or the probe tips fall apart. For this reason, GGB Industries developed the *Picoprobe*™. This coaxial probe is more compliant and can achieve operation between DC and 120 GHz with a coaxial input, and between 75 GHz and 120 GHz with a waveguide input [45]. From DC to 40 GHz, this probe has demonstrated an insertion loss of less than 1.0 dB and a return loss better than 18 dB. However, one potential disadvantage of coaxial probes is that the isolation between probes may be limited when operating above V-band. For even better compliancy, durability, ruggedness and flexibility, Cascade Microtech developed the *Air Coplanar*™ tipped coaxial probe [46]. This probe has demonstrated an insertion loss of less than 1.0 dB from DC to 110 GHz and can operate at temperatures from -65 °C to +200 °C. When selecting the type of microwave probe required, it is necessary to supply the vendor with the following specifications:

1. Footprint: ground-signal-ground (GSG) is the most common for MMICs, although ground-signal (GS) probes are used below 10 GHz.
2. Probe tip contact pitch (i.e. distance between the mid-points of adjacent contacts): for microwave applications, 200 μm is very common, although probes are commercially available with pitches ranging from 50 μm to 1250 μm. Smaller pads result in smaller extrinsic launcher parasitics. 100 μm pitch is commonly used from applications in the 40 to 120 GHz frequency range, while 75 μm is used above 120 GHz.
3. Probe tip contact width: 40 μm and 25 μm are typical for operation up to 65 GHz and 110 GHz, respectively:
4. Probe tip contact metal-plating: BeCu is optimised for GaAs chips (having gold pads) and tungsten is optimised for silicon and SiGe chips (having aluminium pads).
5. Launch angle, ϕ.
6. Coaxial connector type: the 3.5 mm Amphenol Precision Connector (APC3.5) is used for operation to 26.5 GHz; the Anritsu K-connector (2.92 mm), for single-mode operation to 46 GHz, is compatible with 3.5 mm connectors; the APC2.4 can be used for measurements up to 50 GHz, while the Anritsu V-connector (1.85 mm), for single-mode operation to 67 GHz, is compatible with 2.4 mm connectors; the Agilent Technologies 1mm connector is used for operation up to 120 GHz, while the Anritsu W-connector (1.1 mm) has a cut-off frequency of either 110 GHz or 116 GHz, depending on the coaxial dielectric used.

If the launch angle is too small, unwanted coupling between the probe and adjacent on-wafer components may occur. For this reason, it is recommended that adjacent components have at least 600 μm of separation for 110 GHz measurements. On the other hand, if the angle is too large there will not be enough skate on the probe pads. It has been found analytically and empirically that the best angle occurs when the horizontal components of the phase velocity for the probe and MMIC transmission lines match one other [44]. Therefore:

$$\phi = \cos^{-1}\left\{\sqrt{\frac{\varepsilon_{eff.probe}}{\varepsilon_{eff.MMIC}}}\right\}$$

where $\varepsilon_{eff.probe}$ = effective permittivity of probe line

and $\varepsilon_{eff.MMIC}$ = effective permittivity of MMIC line

For example, a CPW line on GaAs has $\varepsilon_{eff.MMIC} \approx 6.9$ *at* 76.5 *GHz*

and, therefore, $\phi = 68^{o}$ with *Air Coplanar*^{TM} probes, since $\varepsilon_{eff.probe} = 1$

However, in practice, the launch angle is approximately 20°. This may raise questions as to the possibility of launching unwanted parasitic modes, due to uncompensated velocity mismatches at the RF probe tip, and also fringe fields coupling from the RF probe tip into the wafer.

12.3.2 Prober calibration

During the placement of probes onto an MMIC, there are two mechanisms by which the probe tips become soiled. First, since the probe tip contact's metal-plating is designed to be much harder than the MMIC probe pad ohmic contact's metal, particles of either gold or aluminium will be deposited onto the respective BeCu or tungsten contacts. Second, it is not uncommon for the probe tip contacts to overshoot the unpassivated probe pads and scratch-off some of the Si_2N_3 passivation material surrounding the pads. Without regular cleaning, a build-up of gold/aluminium and Si_2N_3 particles can form around the probe tip contacts. This build-up is likely to degrade the performance of measurements at millimetric frequencies. Therefore, prior to calibrating the measurement system, it is recommended that the probe tips be very gently cleaned. Here, forced-air can be blown onto the probe tip — in a direction parallel to the tip and towards its open contact end — in order to remove any particles. For more stubborn objects, a lint-free cotton bud, soaked in isopropanol (IPA), can be carefully brushed in a direction parallel to the tip and towards its open contact end.

After the probe tips have been inspected for any signs of damage and cleaned, a planarity check must be made between the probes and the ultra-flat surface of the wafer chuck. A contact substrate, consisting of a polished alumina wafer with defined areas of patterned gold, is used to test that all three of the probe tip contacts (e.g. ground-signal-ground) make clear and even markings in the gold. Once this procedure is complete, the probe tip contacts can be cleaned of any residual gold by simply probing onto the exposed, un-metallised, areas of alumina. This is particularly important for tungsten contacts, because tungsten oxidises, and therefore the contact resistance would otherwise increase. However, this is not the case for BeCu contacts as they do not oxidise.

Probe stations use a one-tier calibration procedure, with the standards located either on-wafer or on an impedance standard substrate (ISS). The use of on-wafer standards is by far the best choice. This is because the probe-to-wafer interface can be electromagnetically the same for calibration, verification and all subsequent measurements. Moreover, on-chip launch transition discontinuities (e.g. probe pads and their transmission lines) can be treated as part of the overall measurement system to be calibrated. Ideally, the reference planes within the on-wafer standards should have the same line geometries as those at the on-chip DUT. With a precision ISS, the standards can be fabricated to much tighter tolerances. For example, a pair of 100 Ω resistors are used to implement the CPW 50 Ω load reference impedance. Here, these resistors can be laser-trimmed to achieve an almost exact value of 50 Ω, but at DC only. It should be noted, however, that if a calibration is performed using a 635 μm thick alumina ISS and the verification is performed using 200 μm thick GaAs on-wafer standards (which is a realistic measurement scenario), then problems may be encountered at millimetric frequencies. This is because the probe-to-ISS interface is electromagnetically different to that of the probe-to-wafer interface. As a result, even though the specifications for corresponding calibration and verification standards may be identical, their measured characteristics may differ significantly. For this reason, the UK National Physical Laboratory (NPL) and the US National Institute of Standards and Technology (NIST) are currently developing GaAs ISS wafers with calibration standards and verification components of certified quality [47,48].

There are a number of calibration techniques that are used for on-wafer measurements [44,47–52]. The SOLT technique is not used at upper-microwave frequencies due to the poor quality of planar open standards. For TRL, the reflect standards (either an open or short circuit) must be identical at both ports, but they can be non-ideal and unknown. The TRL technique also requires a minimum of two transmission lines. The reference impedance is taken from the characteristic impedance, Z_0, of these lines [53]. Since a 50 Ω load is not required for TRL calibration, only transmission line standards are needed, and these are easily realisable on-wafer. In practice, in order to cover a useful frequency range, it is necessary to employ a number of different delay line lengths to overcome phase ambiguity at all the measurement frequencies. This means that the probe separation has to be adjusted during the calibration procedure. For many applications such as automated test systems this is a major limitation, and for these applications the line-reflect-match (LRM) calibration [49], developed by Cascade Microtech, is preferred to TRL. The multiple CPW delay lines required with the TRL calibration are effectively replaced by the CPW 50 Ω load, to theoretically represent an infinitely long delay line. This results in the following advantages:

1. an ultra-wideband calibration can be achieved (e.g. DC to 120 GHz)
2. the probes can be set in a fixed position
3. automatic calibration routines can be applied

4. reflections and unwanted modes in long CPW delay lines are avoided
5. a considerable saving of wafer/ISS area can be made.

With SOLT and LRM, the accuracy to which the load is known directly determines the accuracy of the measurement. In other words, perfect models are required for the load impedances. These loads inevitably have some parasitic shunt capacitance (which is equivalent to having negative series inductance), and furthermore, have frequency-dependent resistance due to the 'skin effect'. In addition, with microstrip technology there will be significant series inductance associated with the short and load standards. Cascade's LRRM calibration is a more accurate version of the standard LRM calibration, in which load-inductance correction is incorporated by including an extra reflection standard.

NIST recently released some public domain software on the worldwide web called *MultiCal*™. This software provides a new method for the accurate calibration of VNAs [21–23]. Here, multiple and redundant standards are used to minimise the effects of random errors caused by imperfect contact repeatability. Moreover, with split-band methods (e.g. LRL and TRL), the calibration discontinuities at the frequency break points can be eliminated. With *MultiCal*™-TRL, only the physical lengths of the standards and the DC measurement of the line resistance per unit length (by applying a least-square error fit to the multiple shorted line lengths) are required. The Z_0 of the lines can then be determined and subsequent measurements can be renormalised to the 50 Ω measurement reference impedance.

For the ultimate in ultra-wideband calibration, verification and measurement accuracy, there is strong support for having *MultiCal*™-TRL calibration for frequencies above a few gigahertz (say 1 GHz), combined with LRM for the frequencies below 1 GHz, using on-wafer standards. The LRM's standards should be characterised at DC and at 1 GHz (using *MultiCal*™-TRL); conventional modelling techniques can be used to interpolate the results.

A 2-port probe station traditionally uses a 12-term error model, although a 16-term error model has been introduced that requires five 2-port calibration standards [54]. This more accurate model can correct for poor grounding and the additional leakage paths and coupling effects encountered with open-air probing. With the extremely high levels of accuracy that are possible with modern probe stations, the effects of calibration errors become more noticeable. Calibration errors can result from the following (in the order of greatest significance):

1. probe placement errors – position, pressure and planarity variations
2. degradation with use in the probe tips and the standards' probe pads surface wave effects on calibrations [55]
3. ISS manufacturing variation.s

With the extremely high level of measurement accuracy that can be achieved, the effects of on-chip launch transition discontinuities can be significant above a few gigahertz. So far, it has been assumed that the effects of probe pads and their associated transmission lines have been calibrated out. Here, the on-wafer calibration standards would have the same launch transition discontinuities as the on-chip DUT. However, effective de-embedding techniques can still be performed within the MMIC. If ECMs are to be employed, the foundry that fabricates the MMIC should provide very accurate models for probe pads and transmission lines. The metrologist must use these foundry-specific models to determine the actual measurements of the on-chip DUT. When de-embedding is performed using equivalent circuit modelling, these foundry-specific models can be easily incorporated into conventional frequency-domain simulation software packages.

12.3.3 Measurement errors

Even when the system has been successfully calibrated, measurement errors (or uncertainty) can still occur. Some of the more common sources of errors are as follows:

1. probe placement errors
2. temperature variation between calibration and measurement
3. cable-shift induced phase errors between calibration and measurement
4. radiation impedance changes due to the probes/wafer chuck moving
5. matrix renormalisation not being performed with multiple port MMICs
6. resonant coupling of the probes into adjacent structures [56]
7. low frequency changes in the characteristic impedance and effective permittivity of both microstrip and CPW transmission lines [56]
8. optically-induced measurement anomalies associated with voltage-tunable analogue-controlled MMICs [57].

12.3.4 DC biasing

Depending on the nature and complexity of the device or circuit under test, DC bias can be applied to an MMIC in a number of ways:

1. through the RF probes, via bias-tees in the VNA's test set
2. through single DC needles mounted on probe station positioners
3. with multiple DC needles attached to a DC probe card, which may in-turn be mounted on a positioner.

The DC probe needle has significant inductance, and as a result, provides RF de-coupling for the bias lines that helps to prevent stability problems. However, additional off-chip de-coupling capacitors and resistors can usually be added to the card to further minimise the risk of unwanted oscillations. With bias-tees and DC needles, the maximum DC bias voltage and current are approximately 40 V and 500 mA, respectively. With multiple DC needles, standard in-house DC footprints are recommended wherever possible, in order to provide card re-use. This will reduce measurement costs considerably. There is a limit to the maximum number of needles per card, but ten is typical. One needle is normally required to provide a ground reference.

12.3.5 MMIC layout considerations

The foundry's design guidelines will define a minimum distance between the centres of probe pad vias and the minimum distance from the vias to the edge of the MMIC's active area. Generally, a particular company or institute may standardise on a certain pad size and pitch for a particular probe tip specification. In order to save expensive chip area, probing directly onto via-hole grounds is tempting. However, the probe tip contacts may puncture the gold pads on top of the via-holes, which could damage the probe tips and destroy the MMIC. While on-via probing can be used, in principle, it is likely that the chip would fail a subsequent QA inspection. As a result, when designing MMICs for on-wafer probed measurement, it is important to consult the foundry design guidelines for the probe pad specifications.

The location and orientation of the probe pads must also be considered. If the pads associated with one port are too close to those of another port, the very fragile probe tips are at risk of severe damage should they accidentally touch one another during the probe alignment procedure. The minimum separation distance between probe tips is determined by the design rule on probe pad spacing (typically 250 μm with vias or 200 μm without vias, depending on the thickness of the chip). Moreover, if the spacing between port pads is less than 200 μm, there could be significant measurement errors due to RF crosstalk effects between probes. Finally, if three or four RF probe positioners are attached to the probe station then they will be oriented orthogonal to one another. As a result, the RF probe pads for a 3- or 4-port MMIC must also be orthogonal to one another.

On the MMIC, launch transitions are required to interface between the probes and the DUT. In many cases, the DUT is in the microstrip medium, and so transitions from CPW-to-microstrip must be employed before and after the DUT. With reference to Figure 12.7*a*, microstrip launchers require through-GaAs vias to provide a low inductance earth path from the probe to the MMIC's backside metallisation layer. A microstrip launcher should be long enough for the higher-order evanescent modes, resulting from the CPW-to-microstrip transition, to be sufficiently attenuated and have minimum interaction with the DUT. As a rule of thumb, the microstrip launchers should ideally have a length of three to four times the substrate thickness. With reference to Figure 12.7*b*, when the DUT is in the CPW medium, through-GaAs vias are not required and a matched taper from the probe pads to the DUT is used. Even though this taper is very short, if the 50 Ω characteristic impedance is not maintained throughout the transition, significant parasitic capacitance or inductance can be introduced. In special cases, launchers are not required at all for some devices. One example of this is with a simple FET structure, as shown in Figure 12.7*c*, where two GSG probes are placed directly onto the source-gate-source and source-drain-source pads. This approach eliminates the need for de-embedding the effects of launchers from the measurements, but the effects of the bond pads should still be considered. At this point, it is important to note that for frequencies above a few gigahertz, the equivalent circuit model of a device that has been characterised in one medium (e.g. microstrip or CPW) should only be used in circuits designed in the same medium.

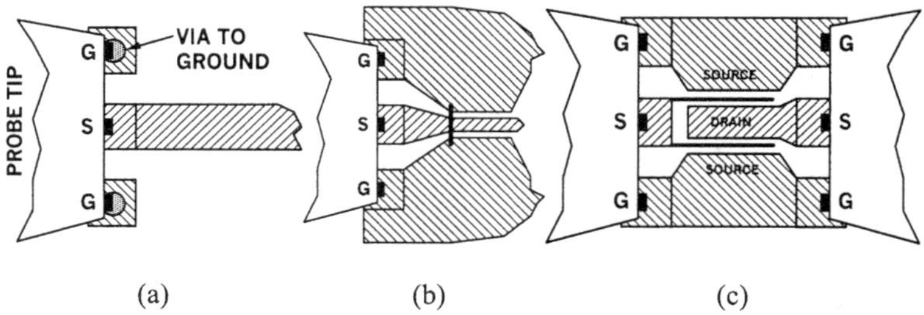

Figure 12.7 Common launcher techniques: (a) microstrip, (b) coplanar waveguide and (c) direct probing onto a FET device

Single devices such as transistors and diodes can be biased through the bias-tees of the network analyser. However, in order to test a complete circuit using a probe station, special consideration has to be given to the layout of the DC bias pads and the design of the bias networks. When using DC needles to bias a circuit, the following points should be considered:

1. The foundry may impose minimum pad sizes and centre-to-centre pitch.
2. For ease of DC probe card fabrication and probe alignment, the DC probe pads should be arranged in a linear array along the edge of the chip's active area, and should be kept away from the RF pads. A common method is to have the RF probe pads on the east and west edges of the chip, and the DC bias pads on the north and/or south edges. If layout constraints suggest that orthogonal RF inputs and outputs would be more convenient, first check that suitable positioners are available.
3. The bias networks of the circuit should be modelled separately to ensure that oscillations will not occur. Off-chip de-coupling capacitors cannot always be placed as near to the chip as they can be in a test fixture.
4. High value resistors can be added on-chip to prevent RF leakage and catastrophic failure resulting from excess forward biasing of diodes and transistors. With varactor diodes, cold-FETs and switching-FETs, a trade-off may have to be made in the value of these bias resistors. If the resistance is too small there may not be enough RF isolation. If the resistance is too high the maximum switching speed may not be reached, due to an excessive R–C time constant. In practice, a minimum resistance value of ~300 Ω should suffice for most applications.

12.3.6 Low-cost multiple DC biasing technique

Conventional DC probe cards may need to be replaced for every new MMIC design, unless standard DC probe footprints can be used. This throwaway approach is very costly, especially when the DC probe cards are supplied by a commercial vendor (as automated and precision manufacturing techniques generally have to be used for aligning multiple needles). Moreover, the cost of the cards increases with the number of needles, as the individual needles are themselves precision-made components.

A flexible, low cost technique has been developed for providing an experimental active filter with multiple DC bias connections [58]. The MMIC is attached to a gold-plated chip carrier using conductive epoxy glue. An array of single-layer microwave capacitors is then attached to the chip carrier in close proximity to the MMIC. BAR-CAPS™, made by Dielectric Labs Inc., are ideal for this purpose since they are available as single-chip strips of three, four or six 100 pF shunt capacitors, each having a probeable area of ~ 650 × 325 μm² and separated by ~ 170 μm. A gold bond wire is then used to connect the MMIC's DC probe pad to its off-chip capacitor.

As an example of this technique, a microphotograph of the experimental MMIC, requiring 15 DC bias lines, is shown in Figure 12.8.

Figure 12.8 Microphotograph of an experimental MMIC with multiple DC biasing using the low-cost technique [58]

It has been found that this low-cost solution has a number of important advantages for use in the R&D laboratory:

1. The high inductance bond wires and off-chip de-coupling capacitors minimise the risk of unwanted oscillations.
2. When designing the MMIC layout, the DC probe pads do not need to be arranged in a linear array along the edges of the chip. This provides greater design layout flexibility.
3. The linear array of off-chip capacitors automatically provides a standard in-house DC footprint, reducing long-term measurement costs considerably.
4. The probeable area of the off-chip capacitors is approximately 15 times larger than that of the MMIC probe pads and the capacitors can withstand greater mechanical forces. As a result, in-house DC probe cards can be made by hand because of the relaxation in manufacturing precision, reducing short-term costs considerably.

12.3.7 Upper-millimetre-wave measurements

The past few years have seen considerable developments in the proposed uses of the millimetric frequency range above 75 GHz for new civil applications; for example, collision avoidance radar at 77 GHz. Also, the 94 GHz band is no longer dominated by military applications. High-resolution radiometric imaging at 94 and 140 GHz has a number of important applications, including aircraft landing systems, finding victims trapped in fires and locating concealed weapons without the use of X-rays. Ultra-high data rate optical communications — using a 'radio-fibre' system at 180 GHz — could transform the way domestic computer networks are distributed. Future EC directives on environmental air pollution monitoring will require cheap high performance terahertz sensors to be mass-produced. Sensors for sub-cellular probing are opening up new areas of medical research. Finally, passive tagging/identification systems are possible, which are both easy to conceal and extremely difficult to forge. With most (if not all) of these applications, monolithic technology will be sought. To this end, there have been major advances in both HEMT and HBT technologies, both of which have attained values of f_{max} greater than 500 GHz [59,60].

Today, vector network analysers are commercially available that can operate in either broadband or banded configurations up to 110 GHz. The Agilent Technologies 8510XF and the Anritsu 360B are examples of two broadband VNAs that are able to measure small-signal *S*-parameters from about 45 MHz to 110 GHz in a single-sweep. Both systems use coaxial cables between the test-sets and the probes.

As frequency increases, the combined losses of all the components between the test-sets' reflectometers and the MMIC under test (e.g. test-set combiners, transmission lines, probes, transitions and connectors) also increase. As a result, the overall system suffers from a reduction in both accuracy and stability [61]. The Anritsu 360B employs two test-sets: one rack-mounted (operating from 40 MHz to 67 GHz) and the other mounted on the probe station (operating from 67 to 110 GHz). Here, a test-set combiner (or forward wave MUX coupler) is used to combine the signals from both test-sets. The drawback with this approach is the considerable losses associated with test-set combiners, which will degrade the effective directivity, source match and frequency tracking of the system at W-band. Ultimately, this will have an impact on the quality of calibrations and the system's ability to hold a calibration in the presence of drift. The Agilent Technologies 8510XF minimises this problem by removing the need for a test-set combiner. Here, ultra-broadband (45 MHz to 110 GHz) directional couplers are utilized to create a single test-set [61].

In order to minimise the losses between the test-set's reflectometer and the MMIC under test, a banded VNA is preferred. This can utilise coaxial cables up to W-band and metal-pipe rectangular waveguides at and/or above W-band. The UK's National Physical Laboratory has recently established a new primary national standard measurement facility for *S*-parameters with rectangular waveguide operating over the frequency range of 75 to 110 GHz, using such a banded VNA system [62]. This facility represents a significant extension to the existing UK national standards for *S*-parameter and impedance measurements [63].

To date, there are still no traceable standards for on-wafer measurements above 75 GHz, from either NPL or NIST. This is due to a multitude of issues (e.g. mechanical precision, multi-moding, radiation effects, dielectric and surface wave propagation, ohmic losses in the dielectric and anomalous skin-effect losses in the conductors) associated with accurate calibration and verification measurements using non-ideal standards. However, there is a great deal of experimental work being undertaken to find the optimum calibration strategy for W-band [64-66].

With the ever-increasing interest in performing on-wafer measurements above 110 GHz, Oleson Microwave Laboratories Inc. can now supply frequency extension modules for the commercial market to include the following waveguide bands: WR-8 for F-band (90-140 GHz) [67]; WR-5 for G-band (140-220 GHz); and WR-3 for H-band (220-325 GHz). Moreover, GGB Industries are now able to supply *Picoprobe*™ on-wafer probes for frequencies up to 220 GHz.

In addition to these commercial systems, the University of Kent has developed an experimental passive on-wafer probing system. Here, ultra-low loss PTFE dielectric waveguides are used to avoid the problem of the skin-effect altogether [68–72]. The dielectric waveguide has been used to implement the multistate reflectometer, interconnecting transmission lines, and even the on-wafer probes. In principle, this system can operate from 118 to 178 GHz [72]. However, the ultimate challenge is to remove all the losses between the test-set's reflectometer and MMIC under test. In an experimental set-up, a full 2-port VNA has been implemented with active probes, enabling *S*-parameter measurements to be made from DC up to 120 GHz [73]. Here, high speed non-linear transmission-line (NLTL)-gated directional time-domain reflectometers (which are essentially directional samplers) were realised using GaAs MMIC technology [74]. More recently, a 70 to 230 GHz VNA has been demonstrated that also employs MMIC reflectometers located on the on-wafer probes [75-76]. The NLTL-based active probes serve as *S*-parameter test-sets for the Agilent Technologies 8510 VNA. Using the Agilent Technologies 8510XF system, good agreement has been demonstrated from 70 GHz to 120 GHz [75].

12.4 Thermal and cryogenic measurements

12.4.1 Thermal measurements

In real-life applications, microwave circuits can be exposed to temperatures other than ambient room temperature (i.e. 23 °C or ~296 K). For example, some components in geostationary orbiting satellites (e.g. within the antenna sub-system) may be periodically exposed to temperatures ranging from −150 °C to +80 °C, depending on the amount of visible sunlight, the levels of localised heat generated within the satellite and the effectiveness of the thermal control sub-system. Also, Gunn diodes can have junction temperatures in excess of +200 °C. At the other extreme, cryogenically-cooled LNAs can operate at −196 °C, with a liquid nitrogen cryogen having a boiling point temperature of 77 K.

During the development of a sub-system, the levels of performance degradation while operating over a predefined temperature range must be known. Therefore, the temperature-dependent characteristics of all the MMIC components that make up a sub-system must be determined. Once the complete sub-system has been assembled, temperature-cycling is performed so that the measured levels of performance degradation can be compared with those predicted during simulation.

The Cascade Microtech Summit 12651 semi-automatic probing system, in conjunction with the *Microchamber*™ enclosure, enables very fast set-up and measurements to be performed up to 110 GHz in a dark, temperature controlled and electromagnetic interference-isolated environment. This thermal probing system [77] can be seen in Figure 12.9. The MMIC under test sits on a temperature controlled wafer chuck, that can be subjected to temperatures ranging from –65 °C to +200 °C or 0 °C to +300 °C. Across these temperature ranges, the parameter values within, say, a FET's equivalent circuit model exhibit a linear temperature dependency. Here, all the resistive and capacitive elements have a positive temperature coefficient, while I_{ds}, g_m and f_T have negative temperature coefficients. Also, as the temperature drops, the gain of an active device can increase significantly. Therefore, to ensure linear operation, and thus avoid oscillation, the input RF power levels need to be reduced accordingly. Also, if the RF probes and cables exhibit large temperature gradients, significant phase changes will be found, even at low microwave frequencies. As a result, an air flow purge is introduced into the chamber in order to minimise the thermal coupling between the chuck and the probe/connector/cables. The air-flow purge also creates a dry, frost-free environment. The system is calibrated for every new wafer chuck temperature setting. An LRRM calibration is used, with the ISS located on a separate thermally isolated stage. The at-temperature calibration procedure can be performed 15 minutes after the chuck temperature has been changed. This short wait corresponds to approximately three thermal time constants for the probe/connector/cable assembly. Since all but the matched load impedance standards are insensitive to temperature, the ISS chuck temperature can be set at –5 °C, for a wafer chuck temperature of –65 °C. This approach results in less than a 1% error in measurements between DC and 65 GHz.

Figure 12.9 Photograph of the Summit 12651 thermal probing system (Courtesy of Cascade Microtech Europe)

As a wafer chuck changes temperature it expands or contracts. For example, the total chuck expansion, from –65 °C to +200 °C, can be about 230 μm. As a result, probe placement errors will become significant. Therefore, at each temperature, the overtravel of the probe tips may need to be adjusted. In addition, as the wafer diameter changes with temperature, there will be small changes in the spacing between devices. Cascade Microtech's Summit series of semi-automatic thermal probe stations include control software that automatically compensates for such changes. This minimises the impact of measurement accuracy.

12.4.2 Cryogenic measurements

Cryogenic hybrid MICs, employing high performance active semiconductor and passive superconductor components, are being more widely used in applications ranging from radio astronomy, to space communications, to medical nuclear magnetic resonance scanners. Therefore, it is important to be able to determine the cryogenic temperature characteristics of these components [3–5,78–82]. At cryogenic temperatures, the noise figures of conventional GaAs transistors are reduced dramatically from their ambient room temperature values. For example, at 10 GHz the measured noise figure of a typical 0.5 × 100 μm MESFET is 0.8 dB at 300 K and only 0.4 dB at 35 K [80]. With HEMT technology, electron mobility can increase by a factor of five when the lattice temperature is reduced from 300 K to 77 K [80], resulting in a considerable improvement in gain and noise performance. Furthermore, measurements made at temperatures as low as 10 K may provide information that can give a unique insight into the physics of experimental devices. Also, in addition to the advances being made in new semiconductor devices, there is considerable interest in the developments of ultra-low loss high temperature superconducting microwave components that currently have to be refrigerated below around 100 K.

The first microwave test fixture to be used in cryogenic measurements was reported in 1976 [3]. The fixture was designed to be immersed in liquid nitrogen (LN_2), which has a boiling point of 77 K. This approach suffers from the problems of poor accuracy and poor repeatability due to the changing temperature gradients exhibited by the cable/connector/launcher assembly, and requires a complicated calibration procedure. Accurate measurements have been reported using a TRL calibrated split-block test fixture mounted on the cold-head of an RMC *Cryosystems*™ LTS-22-IR helium refrigerator [5]. This approach enables small-signal *S*-parameter measurements to be made at 300 K and 77 K.

Cryogenic probe stations have either the MMIC under test and the probes immersed in liquid nitrogen or a liquid cryogen-cooled copper stage with a dry nitrogen vapour curtain. The former approach suffers from poor repeatability (due to varying amounts of LN_2), a short measurement duration (in order to limit the build-up of ice formation), and a limited lifetime due to the degradation of the probes in contact with the LN_2. With the latter approach, accuracy is limited by mechanical stress, caused by the large thermal gradients between the microwave hardware and the MMIC under test. Also, reliability is limited by moisture and the build-up of ice, which increases the wear and tear on manipulators and requires extensive re-planarisation of the mechanical apparatus. Researchers at the University of Illinois have, however, demonstrated the design and operation of a cryogenic vacuum microwave probe station, for the measurement of *S*-parameters from DC to 65 GHz, that minimises the problems of limited accuracy and repeatability [80]. Within a vacuum chamber, the vacuum probe station has high frequency CPW probes connected to cable feeds via a custom bellows and manipulator system. A liquid helium cryogen, with a boiling point temperature of 4.2 K, enables measurements to be performed at temperatures as low as 20 K. The copper stage is continually fed with liquid cryogen, and the system is then left to stand for 15 to 20 minutes in order to achieve temperature equilibrium. Once the at-temperature calibration has been performed, the actual device measurements can be taken for up to four hours before having to recalibrate.

Today, complete on-wafer cryogenic characterisation (from 20 to 300 K) can be performed for *S*-parameters, noise parameters and load-pull measurements [82].

12.5 Experimental field probing techniques

So far only invasive MMIC measurement techniques have been discussed, which generally do not perform internal function and failure analysis. However, one simple technique that can perform such tasks is to realise a coaxial probe with a high impedance tip. Here, a 500 Ω resistor is used to create a potential divider with the 50 Ω oscilloscope. The internal node voltage can be measured without perturbing the operation of the circuit. This technique has been demonstrated on an MMIC power amplifier [83]. Alternatively, non-contacting methods also exist. Again, all the RF ports of the MMIC under test are terminated with matched loads. An RF signal is injected into the MMIC's input port and a micron-level probing system is used to detect the internal signal strength. In the case of non-contacting techniques, different types of fields are detected along transmission lines and at discontinuities. Field probing can detect current crowding, standing waves and unwanted modes of propagation, and *S*-parameters can be determined from time-domain network analysis measurements.

12.5.1 Electromagnetic-field probing

The simplest method of field detection uses a semiconductor diode. At microwave frequencies, however, it becomes difficult to match the diode because its impedance varies with power level. At low power levels, bolometers are traditionally employed for use above 1 GHz. The device is similar to a thin-film resistor, where a high resistivity bismuth film is evaporated onto metallic electrodes. When exposed to microwave radiation, the bolometer absorbs the electromagnetic energy and converts it into heat energy. As the film heats up, its resistivity decreases. Since the bolometer is inherently a square law detector, the measured voltage change across the device is proportional to the change in incident RF power. In practice, however, since the signal levels are so small, the incident microwave signal must be pulsed. This causes the resistance of the bolometer to change at the pulse repetition frequency, which is usually below 100 KHz. With a DC bias current applied, the low frequency voltage signal across the bolometer is applied to a lock-in amplifier that acts as a coherent detector. This technique exhibits a high degree of sensitivity; as an example, a 4 × 5 μm device with a noise equivalent power of 160 $pW/Hz^{1/2}$ has been reported [84]. With the use of conventional probe microfabrication techniques, microbolometers can be employed to detect power levels as low as a few nanowatts along MMIC transmission lines. A microbolometer probe that can be used for microstrip and CPW transmission lines is illustrated in Figure 12.10. With a perfectly symmetrical probe positioned directly above a CPW line, the wanted CPW (or even) mode will be detected and the unwanted slotline (or odd) mode will not. As well as their simple fabrication and calibration, microbolometer probes can be designed to operate in the terahertz frequency range. Unfortunately, the attainable stability and uniformity of the resistive film does not yet appear to be sufficient for the commercial production of these probes.

A more recent development uses a dielectric rod probe, with a thin copper strip at its end face that helps to pick-up the electromagnetic field and couple it to the dielectric waveguide [85]. Using this technique, measured results have been demonstrated between 200 and 220 GHz to show standing wave patterns on a mismatched dielectric waveguide [85].

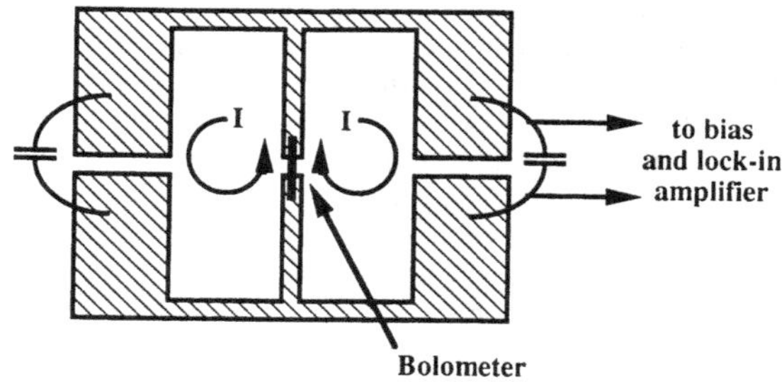

Figure 12.10 Illustration of an electromagnetic-field probe

12.5.2 Magnetic-field probing

The simplest magnetic-field probing technique is to connect a conventional spectrum analyser to a magnetic-field probe. Using wafer probe microfabrication techniques, a miniature magnetic quadrupole antenna can be configured to match the magnetic fields associated with microstrip and CPW transmission lines, as illustrated in Figure 12.11. Placed directly above the transmission line, the lines of magnetic flux will come up through one loop and back down through the other loop. As a result, the induced signals add. From a distance, the probe sees a near uniform magnetic field which induces signals that tend to cancel each other out. In addition to amplitude, phase measurements can also be measured. A reference signal at the same frequency, with a variable amplitude and phase, is combined with the measured signal. The measured phase is equal to the reference phase when the amplitude displayed on the spectrum analyser is at its peak. Therefore, the probe can be used to measure the amplitude and phase of currents at any node within an MMIC.

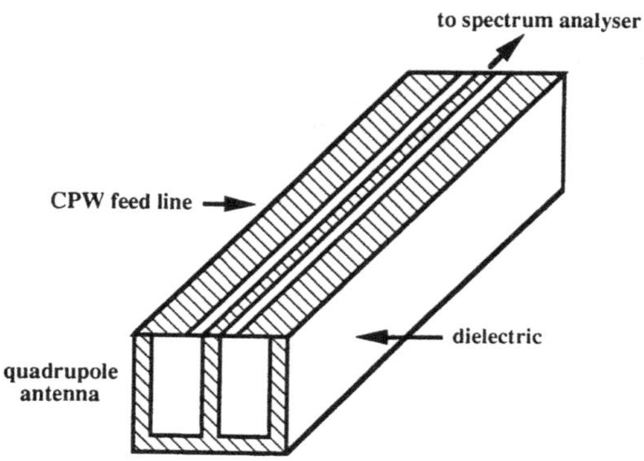

Figure 12.11 Illustration of a magnetic-field probe

An experimental system has been reported that can operate in the 26.5 to 40 GHz frequency range [86]. Here, a 25 to 50 μm separation distance provides sufficient coupling and discrimination, while providing a negligible effect on the MMIC under test. One of the major sources of error is electrostatic pickup. Increasing the width of the loops increases the ratio of magnetic to electric coupling, but it also increases the random radiation picked up from other circuit elements. Reducing the width of the metal conductors reduces capacitive pickup, but increases the conductor's resistance and self-inductance. In practice, an effective method of limiting the errors due to electrostatic pickup is to rotate the probe and average the measurements. This problem can be avoided by having just a single-loop probe [87].

12.5.3 Electric-field probing

The simplest electric-field probing technique is to connect a conventional spectrum analyser to a near electric-field (i.e. capacitive) probe. This technique was first demonstrated on MICs in 1979 [88], but it is still being used today [89]. The probe can be simply realised by removing a small section of the outer screening conductor and dielectric from the end of the analyser's coaxial feed line. Unfortunately, these probes have significant unwanted parasitic reactances at high microwave frequencies, which can severely perturb the operation of the circuit under test, thus causing measurement errors. However, micromachining techniques can be adopted to limit this problem, to realise dipole and monopole antennas [90]. In practice, this technique is only accurate when used with shielded transmission lines. As a result, it is unsuitable for micron-level features found in MMICs.

Over the past decade, a number of alternative electric-field probing techniques have been investigated, with varying degrees of success.

12.5.3.1 Electron-beam probing

The voltage-contrast scanning electron microscope (SEM) was developed in the late 1960's for detecting voltages on the conductor tracks of integrated circuits. A pulsed electron beam stimulates secondary electron emissions from the irradiated surface of metals. For conductors at a negative potential, the secondary electrons have more energy than for conductors at a more positive potential. Commercial SEMs suffer from a poor millivolt potential sensitivity and limited bandwidths of only a few gigahertz [91], although larger bandwidths have been reported [92]. Also, apart from its very high complexity and cost, the electron beam may effect the operation of GaAs MMICs due to charging of deep levels in the GaAs substrate. However, the major advantage of this technique is that the attainable spatial resolution that can be achieved is in the order of a few angstroms.

12.5.3.2 Photo-emissive sampling

Instead of using an electron beam to stimulated secondary electron emissions, another approach uses a high-intensity pulsed laser beam to illuminate the surface of the metals [91]. This time-domain sampling technique offers an improved potential sensitivity and a greatly extended bandwidth. However, as with the SEM, the performance of GaAs MESFETs may be affected by charging of deep level traps.

12.5.3.3 Opto-electronic sampling

Time-domain network analysis can be performed using opto-electronic sampling techniques. Here, electrical pulses can be generated on an MMIC by illuminating DC biased photoconductive switches with a pulsed laser beam. The optical excitation of a photoconductive switch can also perform signal sampling. By comparing the Fourier transforms of the sampled incident and reflected or transmitted waveforms, the complex 2-port *S*-parameters can be determined for the DUT [91,93–99]. Sub-picosecond electrical pulse generation with a photoconductive switch has been reported, enabling terahertz measurement bandwidth [100]. This time-domain opto-electronic sampling technique (also known as photoconductive sampling) requires the DUT to be embedded in a single-chip GaAs test fixture. Each RF port of the DUT is connected to a test structure consisting of a 50 Ω matched load termination, photoconductive switches, DC bias lines and a length of transmission line. These test components are not only wasteful of expensive chip space, but they must also be de-embedded from the measurements. In addition, the fabrication process of the photoconductive switches must be compatible with that of the MMIC under test. However, a DC to 500 GHz measurement system has been demonstrated [99] using this technique.

12.5.3.4 Electro-optic sampling

The most promising electric-field probing technique is electro-optic sampling. A variety of non-centrosymmetric crystals, such as gallium arsenide and indium phosphide, exhibit Pockel's electro-optic effect. The presence of an electric-field will induce small anisotropic variations in the crystal's dielectric constant, and therefore, its refractive index. If a laser beam passes through this material it will experience a voltage-induced perturbation in its polarisation, which is directly proportional to the change in the electric-field strength. As a result, this linear electro-optic effect can be used to provide a non-invasive means of detecting electric fields [91,93–94,101–111].

With internal (or direct) electro-optic probing the laser beam penetrates the GaAs MMIC in a reflection mode, as illustrated in Figure 12.12*a*, giving good beam access and requiring only a single focusing lens [91,93–94,101–103,107–111]. However, optical polishing of the MMIC substrate is required for best results. With front-side probing, the beam is reflected off the back-side ground plane metallisation, adjacent to the circuit conductor. With back-side probing, the beam is reflected off the back of the circuit conductor itself, making this scheme ideal for conventional CPW or coplanar strip lines and slotlines. Today, internal electro-optic sampling can achieve a spatial resolution down to less that 0.5 μm [110].

Centrosymmetric crystals, such as silicon and germanium, do not exhibit the linear electro-optic effect. Therefore, silicon MMICs must employ external (or indirect) electro-optic probing [91, 93, 104-106]. This technique uses an extremely small electric field sensor, consisting of a 40 x 40 μm^2 electro-optic crystal (lithium tantalate) at the end of a fused silica needle, placed in close proximity to the circuit conductor, as shown in Figure 12.12*b*. Sending a laser beam down the needle and measuring the induced change in the refractive index of the crystal from the returning beam can detect the conductor's fringing fields. Since the beam can be focused down to a spot size of 3 to 5 μm in diameter, excellent spatial resolution is achieved. Also, there is no need for MMIC substrate polishing.

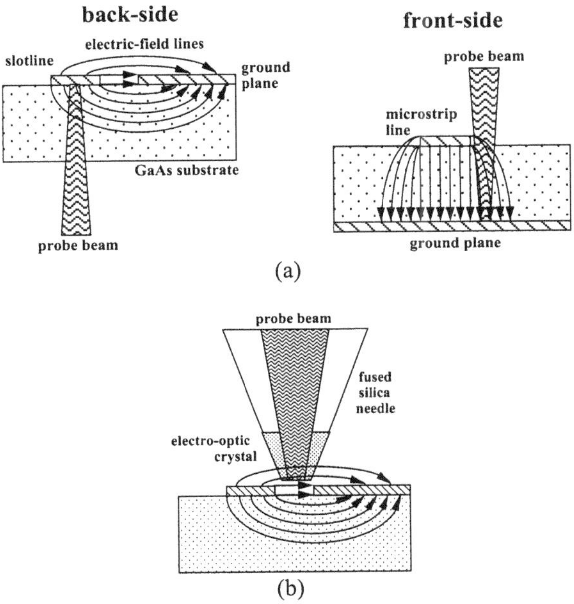

Figure 12.12 Illustration of electric-field probe: (a) internal and (b) external

With electro-optic probing, picosecond optical pulses (generated by a laser with an output power level that is lower than the band-gap energy of the MMIC's semiconductor) pass through the electric-fields associated with the MMIC's circuit conductors. After being passed through a common beam splitter, the incident and return beams are combined, before being passed through a polarising beam splitter. Two photodiodes detect the intensity of the orthogonally polarised components and lock-in amplifiers are then used to determine the electric-field vectors. As a result, internal node voltage measurements can be determined and impressive 2-dimensional mappings of the amplitude [105,107–109] and phase angles [109] of microwave fields within the MMIC can be obtained. Time-domain network analysis can also be performed using electro-optic sampling. Here, picosecond electrical pulses are applied to the input port of the MMIC under test, with the generator connected to the MMIC using traditional invasive techniques. By comparing the Fourier transform of the detected incident and reflected or transmitted waveforms, the complex 2-port *S*-parameters can be determined. To date, a 50 to 300 GHz network analyser has been demonstrated using this technique [106]. More recently, a European consortium (which includes

NPL and the Fraunhofer Institute for Applied Solid State Physics) has developed the first optical instrument capable of testing terahertz circuits and tracing the measurements back to international standards [111].

12.5.3.5 Electrical sampling scanning-force microscopy

A number of non-invasive measurement techniques have been introduced that can perform internal function and failure analysis of MMICs. The electron-beam probing technique is well established and has excellent spatial resolution, but the temporal resolution is limited because of electron transit time effects. Optical probing techniques have a superior temporal resolution, but because of the micron-beam diameters they have a limited spatial resolution. Scanning-force microscopy, in the electrical sampling mode, is a relatively new non-contacting measurement technique that has high spatial, temporal and voltage resolutions [112,113]. Here, an atomically sharp needle is mounted on one end of a cantilever. When the needle is placed a fixed working distance of between 0.1 and 0.5 μm above the MMIC, it will be subjected to attraction or repulsion forces, causing a detectable bending of the cantilever. This very experimental technique has so far demonstrated a spatial resolution of 0.5 μm and a bandwidth of 40 GHz [112].

12.6 Summary

A wide range of techniques has been briefly introduced for the measurement of MMICs. A summary of the main features associated with the most practical invasive techniques is given in Table 12.2. In general, the level of accuracy and repeatability obtainable is proportional to the initial investment costs of the measurement system.

Compared with traditional invasive on-wafer measurement techniques, optical systems have so far demonstrated a lower dynamic range and inferior frequency resolution. In addition, optical techniques have complicated and lengthy calibration procedures. However, with its excellent spatial resolution and extremely wide bandwidth capabilities, electro-optic probing may become commonplace in the not too distant future.

Table 12.2 Comparison of the invasive measurement technologies

	In-house test fixture		Commercial test fixture	On-wafer probe station
calibration	2-tier with ECM de-embedding	1-tier	1-tier	1-tier
accuracy	moderate	high	high	very high
repeatability	moderate	moderate	high	very high
bandwidth	wideband	wideband	wideband	ultra-wideband
flexibility	excellent	poor	poor	poor
cost	very low	low	high	very high

12.7 References

1 LORCH, P.: 'Applications drive the evolution of network analyzers', *Microwaves & RF*, 1994, pp. 79–84

2 LUCYSZYN, S., STEWART, C., ROBERTSON, I. D., and AGHVAMI, A. H.: 'Measurement techniques for monolithic microwave integrated circuits', *IEE Electronics & Communication Engineering Journal*, 1994, pp. 69–75

3 LIECHTI, C. A., and LARRICK, R. B.: 'Performance of GaAs MESFET's at low temperatures', *IEEE Transactions on Microwave Theory and Techniques*, 1976, **MTT-24**, pp. 376–381

4 SMUK, J. W., STUBBS, M. G., and WIGHT, J. S.: 'Vector measurements of microwave devices at cryogenic temperatures', *IEEE MTT-S International Microwave Symposium Digest*, 1989, pp. 1195–1198

5 SMUK, J. W., STUBBS, M. G., and WIGHT, J. S.: 'S-parameter characterization and modeling of three-terminal semiconductive devices at cryogenic temperatures', *IEEE Microwave and Guided Wave Letters*, 1992, **2** (3), pp. 111–113

6 LANE, R.: 'De-embedding device scattering parameters', *Microwave Journal*, 1984, pp. 149–156

7 RIAD, S. M.: 'The deconvolution problem: an overview', *Proceedings of the IEEE*, 1986, **74** (1), pp. 82–85

8 ROMANOFSKY, R. R., and SHALKHAUSER, K. A.: 'Fixture provides accurate device characterization', *Microwaves & RF*, 1991, pp. 139–148

9 LUCYSZYN, S., MAGNIER, V., READER, H. C., and ROBERTSON, I. D.: 'Ultrawideband measurement of multiple-port MMICs using non-ideal test fixtures and a 2-port ANA', *IEE Proceedings A*, 1992, **139** (5), pp. 235–242

10 TIPPET, J. C., and SPACIALE, R. A.: 'A rigorous technique for measuring the scattering matrix of a multiport device with a 2-port network analyzer', *IEEE Transactions on Microwave Theory and Techniques*, 1982, **30**, pp. 661–666

11 DROPKIN, H.: 'Comments on – A rigorous technique for measuring the scattering matrix of a multiport device with a two-port network analyzer', *IEEE Transactions on Microwave Theory and Techniques*, 1983, **MTT-31** (1), pp. 79–81

12 SELMI, L., and ESTREICH, D. B.: 'An accurate system for automated on-wafer characterization of three-port devices, *IEEE GaAs IC Symposium Digest*, 1990, pp. 343–346

13 GOLDBERG, S. B., STEER, M. B., and FRANZON, P. D.: 'Accurate experimental characterization of three-ports', *IEEE MTT-S International Microwave Symposium Digest*, 1991, pp. 241–244

14 STINEHELFER, H. E.: 'Discussion of de-embedding techniques using the time domain analysis', *Proceedings of the IEEE*, 1986, **74**

15 'Introduction to time domain measurements', *HP 8510C Network Analyser: Operating Manual* (Hewlett Packard, USA, 1991)

16 GRONAU, G., and WOLFF, I.: 'A simple broad-band device de-embedding method using an automatic network analyzer with time-domain option', *IEEE Transactions on Microwave Theory and Techniques*, 1989, **MTT-37** (3), pp. 479–483

17 DUMBELL, K. D.: 'TDR for microwave circuits', *Proceedings of IEEE Asia-Pacific Microwave Conference*, 1992, pp. 361–364

18 GRONAU, G.: 'Scattering-parameter measurement of microstrip devices', *Microwave Journal*, 1992, pp. 82–92

19 LU, K., and BRAZIL, T. J.: 'A systematic error analysis of HP8510 time-domain gating techniques with experimental verification', *IEEE MTT-S International Microwave Symposium Digest,* 1993, pp. 1259–1262

20 HJIPIERIS, G.: 'Time and frequency domain measurements', *Proceedings of 11th IEE Training Course on Microwave Measurements*, Great Malvern, May 2000

21 MARKS, R. B.: 'A multiline method of network analyser calibration', *IEEE Transactions on Microwave Theory and Techniques*, 1991, **MTT-39** (7), pp. 1205–1215

22 MARKS, R. B., and WILLIAMS, D. F.: 'Characteristic impedance determination using propagation constant measurements', *IEEE Microwave and Guided Wave Letters*, 1991, **1** (6), pp. 141–143

23 WILLIAMS, D. F., and MARKS, R. B.: 'Transmission line capacitance measurements', *IEEE Microwave and Guided Wave Letters*, 1991, **1** (9), pp. 243–245

24 BURKE, J. J., and JACKSON, R. W.: 'A simple circuit model for resonant mode coupling in packaged MMICs', *IEEE MTT-S International Microwave Symposium Digest*, 1991, pp. 1221–1224

25 SWANSON, D., BAKER, D., and O'MAHONEY, M.: 'Connecting MMIC chips to ground in a microstrip environment', *Microwave Journal*, 1993, pp. 58–54

26 MARCH, S. L.: 'Simple equations characterize bond wires', *Microwaves & RF*, 1991, pp. 105–110

27 NELSON, S., YOUNGBLOOD, M., PAVIO, J., LARSON, B., and KOTTMAN, R.: 'Optimum microstrip interconnects', *IEEE MTT-S International Microwave Symposium Digest*, 1991, pp. 1071–1074

28 IWASAKI, N., KATSURA, K., and KUKUTSU, N.: 'Wideband package using an electromagnetic absorber', *Electronics Letters*, 1993, **29** (10), pp. 875–876

29 WILLIAMS, D. F., PAANANEN, D. W.: 'Suppression of resonant modes in microwave packages', *IEEE MTT-S International Microwave Symposium Digest*, 1989, pp. 1263--1265

30 HEUERMANN, H., and SCHIEK, B.: 'The in-fixture calibration procedure line-network-network-LNN', *Proceedings of 23rd European Microwave Conference*, 1993, pp. 500–503

31 MUNNS, A.: 'Flip-chip solder bonding for microelectronic applications', *Metals and Materials*, 1989, pp. 22–25

32 WARNER, D. J., PICKERING, K. L., PEDDER, D. J., BUCK, B. J., and PIKE, S. J.: 'Flip chip-bonded GaAs MMICs compatible with foundry manufacture', *IEE Proceedings H*, 1991, **138** (1)

33 FELTON, L. M.: 'High yield GaAs flip-chip MMICs lead to low cost T/R modules', *IEEE MTT-S International Microwave Symposium Digest*, 1994, pp. 1707–1710

34 JIN, H., VAHLDIECK, R., MINKUS, H., and HUANG, J.: 'Rigorous field theory analysis of flip-chip interconnections in MMICs using the FDTLM method', *IEEE MTT-S International Microwave Symposium Digest*, 1994, pp. 1711–1714

35 SAKAI, H., OTA, Y., INOUE, K., YOSHIDA, T., TAKAHASHI, K., FUJITA, S., and SAGAWA, M.: 'A novel millimeter-wave IC on Si Substrate using flip-chip bonding technology', *IEEE MTT-S International Microwave Symposium Digest*, 1994, pp. 1763–1766

36 LAU, J. H. (Ed.): 'Flip chip technologies' (McGraw-Hill, 1996)

37 BAUMANN, G., FERLING, D., and RICHTER, H.: 'Comparison of flip chip and wire bond interconnections and the technology evaluation on 51 GHz transceiver modules', *Proceedings of 26th European Microwave Conference*, 1996, pp. 98-100

38 KIM, J., and ITOH, T.: 'A novel microstrip to coplanar waveguide transition for flip-chip interconnection using electromagnetic coupling', *Proceedings of 28th European Microwave Conference*, 1998, pp. 236–240

39 SPIEGEL S. J., and MADJAR, A.: 'Characterization of flip chip bump interconnects', *Proceedings of 28th European Microwave Conference*, 1998, pp. 524–528

40 'Network analysis: Applying the HP 8510B TRL calibration for non-coaxial measurements', Hewlett-Packard Production Note 8510-8, 1987

41 STRID, E., and GLEASON, K.: 'A microstrip probe for microwave measurements on GaAs FET and IC wafers', *GaAs IC Symposium*, paper 31, 1980

42 STRID, E., and GLEASON, K.: 'A DC-12GHz monolithic GaAs FET distributed amplifier', *IEEE Transactions on Microwave Theory and Techniques*, 1982, **MTT-30**, pp. 969–975

43 GODSHALK, E. M.: 'A W-band wafer probe', *IEEE MTT-S International Microwave Symposium Digest*, 1993, pp. 171–174

44 BAHL, I., LEWIS, G., and JORGENSON, J.: 'Automatic testing of MMIC wafers', *International Journal of Microwave and Millimeter-wave Computer-aided Engineering*, 1991, **1** (1), pp. 77–89

45 LIU, J. S. M., and BOLL, G. G.: 'A new probe for W-band on-wafer measurements', *IEEE MTT-S International Microwave Symposium Digest*, 1993, pp. 1335–1338

46 GODSHALK, E. M., BURR, J., WILLIAMS, J.: 'An air coplanar wave probe, *Proceedings of 24th European Microwave Conference*, 1994, pp. 1380–1385

47 BANNISTER, D. J., and SMITH, D. I.: 'Traceability for on-wafer CPW S-parameter measurements', *IEE Colloquium Digest on Analysis, Design and Applications of Coplanar Waveguide*, 1993, pp. 7/1–5

48 PENCE, J. E.: 'Technique verifies LRRM calibrations on GaAs substrates', *Microwaves & RF*, 505–507

49 EUL, H. J., and SCHIEK, B.: 'Thru-match-reflect: one results of a rigorous theory for de-embedding and network analyzer calibration', *Proceedings of 18th European Microwave Conference*, 1988

50 PRADELL, L., CACERES, M., and PURROY, F.: 'Development of self-calibration techniques for on-wafer and fixtured measurements: a novel approach', *Proceedings of 22nd European Microwave Conference*, 1992, pp. 919-924

51 FERRERO, A., and PISANI, U.: 'Two-port network analyzer calibration using an unknown 'thru', *IEEE Microwave and Guided Wave Letters*, 1992, **2** (12), pp. 505–507

52 PURROY, F., and PRADELL, L.: Comparison of on-wafer calibrations using the concept of reference impedance', *Proceedings of 23rd European Microwave Conference*, 1993, pp. 857–859

53 MARKS, R. B. and WILLIAMS, D. F.: 'A general waveguide curcuit theory', *Journal of Research of the National Institute of Standards and Technology*, 1992, **97** (5), 1992, pp. 533–562

54 SILVONEN, K. J.: 'Calibration of 16-term error model', *Electronics Letters*, 1993, **29** (17), pp. 1544–1545

55 GODSHALK, E. M.: 'Wafer probing issues at millimeter wave frequencies', *Proceedings of 22nd European Microwave Conference*, 1992, pp. 925–930

56 MIERS, T. H., CANGELLARIS, A., WILLIAMS, D., and MARKS, R.: 'Anomalies observed in wafer level microwave testing', *IEEE MTT-S International Microwave Symposium Digest*, 1991, pp. 1121–1124

57 LUCYSZYN, S., and ROBERTSON, I. D.: 'Optically induced measurement anomalies with voltage-tunable analog-control MMIC's', *IEEE Transactions on Microwave Theory and Techniques*, 1998, **MTT-46** (8), pp. 1105–1114

58 LUCYSZYN, S., and ROBERTSON, I. D.: 'Monolithic narrow-band filter using ultrahigh-Q tunable active inductors', *IEEE Transactions on Microwave Theory and Techniques*, 1994, **MTT-42** (12), pp. 2617–2622

59 SMITH, P. M., LIU, S.-M. J., KAO, M.-Y., HO, P., WANG, S. C., DUH, K. H. G., FU, S. T., and CHAO, P. C., 'W-band high efficiency InP-based power HEMT with 600 GHz fmax', *IEEE Microwave and Guided Wave Letters*, 1995, **5**, pp. 230–232

60 LEE, Q., MARTIN, S. C., MENSA, D., PULLELA, R., SMITH, R. P., AGARWAL, B., GUTHRIE, J., and RODWELL, M.: 'Deep submicron transferred-substrate heterojunction bipolar transistors', *Proceedings of Device Research Conference*, 1998

61 GIBSON, J.: 'New capabilities for enhancing mm-wave network measurements', *Microwave Journal*, 1998, pp. 86–94

62 FRENCH, G., and RIDLER, N.: 'A primary national standard millimetric waveguide S-parameter measurements', *Microwave Engineering Europe*, 1999, pp. 29–32

63 RIDLER, N. M.: 'A review of existing national measurement standards for RF and microwave impedance parameters in the UK', *IEE Colloquium Digest*, 1999, no. 99/008, pp. 6/1–6

64 KOK, Y.-L., DuFAULT, M., HUANG, T.-W., and WANG, H., 'A calibration procedure for W-band on-wafer testing', *IEEE MTT-S International Microwave Symposium Digest*, 1997, pp. 1663–1666

65 MARKS, R. B.: 'On-wafer millimeter-wave characterization', *Gallium Arsenide and Its Applications Symposium Digest*, 1998, pp. 21–26

66 EDGAR, D. L., ELGAID, K., WILLIAMSON, F. *et al.*: 'W-band on wafer measurements of active and passive devices', *IEE Colloquium Digest*, 1999, pp. 2/1–6
67 ANRITSU Co.: '140 GHz extender modules for vector network analyzers', *Microwave Journal*, 1998, pp. 148–150
68 COLLIER, R. J., and BOESE, I. M.: 'Microwave measurements above 100 GHz', *Proceedings of Microwaves and RF Conference*, 1995, pp. 147–151
69 BOESE, I. M., and COLLIER, R. J.: 'Novel measurement system within 110-170 GHz using a dielectric multistate reflectometer', *Proceedings of 26th European Microwave Conference*, 1996, pp. 806–810
70 BOESE, I. M., COLLIER, R. J, JASTRZEBSKI, A. K., AHMED, H., CLEAVER, J. R., and HASKO, D.: 'An on wafer probe for measurements at 140 GHz', *IEE Colloquium Digest*, 1997, pp. 9/1–7
71 COLLIER, R. J.: 'Measurements of impedance above 110 GHz', *IEE Colloquium Digest*, 1998, pp. 1/1–6
72 BOESE, I. M., and COLLIER, R. J.: 'Measurements on millimeter wave circuits at 140 GHz', *IEE Proceedings – Science, Measurement and Technology*, 1998, **145** (4), pp. 171–176
73 YU, R., REDDY, M., PUSL, J., ALLEN, S., CASE, M., and RODWELL, M.: 'Full two-port on-wafer vector network analysis to 120 GHz using active probes', *IEEE MTT-S International Microwave Symposium Digest*, 1993, pp. 1339–1342
74 RODWELL, M., ALLEN, S., CASE, M., YU, R., BHATTACHARYA, U., and REDDY, M.: 'GaAs nonlinear transmission-lines for picosecond and millimeter-wave applications', *Proceedings of 23rd European Microwave Conference*, 1993, pp. 8–10
75 WOHLGEMUTH, O., RODWELL, M. J. W., REUTER, R., BRAUNSTEIN, J, and SCHLECHTWEG, M.: 'Active probes for network analysis within 70-230 GHz', *IEEE Transactions on Microwave Theory and Techniques*, **47** (12), pp. 2591–2598
76 WOHLGEMUTH, O., AGARWAL, B., PULLELA, R. *et al.*: 'A NLTL-based integrated circuit for a 70-200GHz VNA system', *Microwave Engineering Europe*, 1999, pp. 35–39
77 D'ALMEIDA, D., and ANHOLT, R.: 'Device characterization with an integrated on-wafer thermal probing system', *Microwave Journal*, 1993, pp. 94–105
78 LASKAR, J., KOLODZEY, J.: 'Cryogenic vacuum high frequency probe station', *Journal of Vacuum Science Technology*, 1990, pp. 1161–1165
79 MESCHEDE, H., REUTER, R., ALBERS, J., KRAUS, J., PETERS, D., BROCKERHOFF, W., TEGUDE, F.-J., BODE, M., SCHUBERT, J., and ZANDER, W.: 'On-wafer microwave measurement setup for investigations on HEMT's and high Tc superconductors at cryogenic temperatures down to 20 K', *IEEE Transactions on Microwave Theory and Techniques*, 1992, **MTT-40** (12), pp. 2325–2331
80 LASKAR, J., and FENG, M.: 'An on-wafer cryogenic microwave probing system for advanced transistor and superconductor applications', *Microwave Journal*, pp. 104–114

81 LASKAR, J., LAI, R., BAUTISTA, J. J. *et al.*: 'Enhanced cryogenic on-wafer techniques for accurate InxGa1-xAs HEMT device models', *IEEE MTT-S International Microwave Symposium Digest*, 1994, pp. 1485–1488
82 LASKAR, J., MURTI, M. R., YOO, S. Y., GEBARA, E., and HARRIS, H. M.: 'Development of complete on-wafer cryogenic characterization: S-parameters, noise-parameter and load-pull', *Gallium Arsenide and Its Applications Symposium Digest*, 1998, pp. 33–38
83 WEI, C.-J., TKACHENKO, Y. A., HWANG, J. C. M., SMITH, K. R., and PEAKE, A. H.: 'Internal-node waform analysis of MMIC power amplifiers', *IEEE Transactions on Microwave Theory and Techniques*, 1995, **MTT-43** (12), pp. 3037–3042
84 SCHWARZ, S. E., and TURNER, C. W.: 'Measurement techniques for planar high-frequency circuits', *IEEE Transactions on Microwave Theory and Techniques*, 1986, **MTT-34** (4), pp. 463–467
85 BASU, A., and ITOH, T.: 'A new field-probing technique for millimeter-wave components', *IEEE MTT-S International Microwave Symposium Digest*, 1997, pp. 1667–1670
86 OSOFSKY, S. S., and SCHWARZ, S. E.: 'Design and performance of a non-contacting probe for measurements on high-frequency planar circuits', *IEEE Transactions on Microwave Theory and Techniques*, 1992, **MTT-40** (8), pp. 1701–1708
87 GAO, Y., and WOLFF, I.: 'A new miniature magnetic field probe for measuring three-dimensional fields in planar high-frequency circuits', *IEEE Transactions on Microwave Theory and Techniques*, 1996, **MTT-44** (6), pp. 911–918
88 DAHELE, J. S., and CULLEN, A. L.: 'Electric probe measurements on microstrip', *IEEE Transactions on Microwave Theory and Techniques*, 1980, **MTT-28** (7), pp. 752–755
89 GAO, Y., and WOLFF, I.: 'Electric field investigations on active microwave circuits', *Proceedings of 26th European Microwave Conference*, 1996, pp. 662–664
90 BUDKA, T. P., WACLAWIK, S. D., and REBEIZ, G. M.: 'A coaxial 0.5-18 GHz near electric field measurement system for planar microwave circuits using integrated probes', *IEEE Transactions on Microwave Theory and Techniques*, 1996, **MTT-44** (12), pp. 2174–2182
91 BLOOM, D. M., WEINGARTEN, K. J., and RODWELL, M. J. W.: 'Probing the limits of traditional MMIC test equipment', *Microwaves & RF*, 1987, pp. 101–106
92 KUBALEK, E., and FEHR, J.: 'Electron beam test system for GHz-waveform measurements on transmission-lines within MMIC', *Proceedings of 22nd European Microwave Conference*, 1992, pp. 163–168
93 BIERMAN, H.: 'Improved on-wafer techniques evolve for MMIC testing', *Microwave Journal*, 1990, pp. 44–58
94 LEE, T. T., SMITH, T., HUANG, H. C., CHAUCHARD, E., and LEE, C. H.: 'Optical techniques for on-wafer measurements of MMICs', *Microwave Journal*, 1990, pp. 91–102

95 HUANG, S.-L. L., CHAUCHARD, E. A., LEE, C. H., HUNG, H.-L. A., LEE, T. T., and JOSEPH, T.: 'On-wafer photoconductive sampling of MMICs', *IEEE Transactions on Microwave Theory and Techniques*, 1992, **MTT-40** (12), pp. 2312–2320

96 KIM, J., SON, J., WAKANA, S., NEES, J., WILLIAMSON, S., WHITAKER, J., KWON, Y., and PAVLIDIS, D.: 'Time-domain network analysis of mm-wave circuits based on a photoconductive probe sampling technique', *IEEE MTT-S International Microwave Symposium Digest*, 1993, pp. 1359–1361

97 GOLOB, L. P., HUANG, S. L., LEE, C. H., CHANG, W. H., JONES, K., TAYSING-LARA, M., and DeANNI, T.: 'Picosecond photoconductive switches designed for on-wafer characterization of high frequency interconnects', *IEEE MTT-S International Microwave Symposium Digest*, 1993, pp. 1395–1398

98 ARMENGAUD, L., GERBE, V., LALANDE, M., LAJZEREROWICZ, J., CUZIN, M., and JECKO, B.: 'Electromagnetic study of an electronic sampler for picosecond pulse measurements', *Proceedings of 23rd European Microwave Conference*, 1993, pp. 751–754

99 FRANKEL, M. Y.: '500-GHz characterization of an optoelectronic S-parameter test structure', *IEEE Microwave and Guided Wave Letters*, 1994, **4** (4), pp. 118–120

100 VALDMANIS, J. A., and MOUROU, G.: 'Subpicosecond electrooptic sampling: principles and applications', *IEEE Journal of Quantum Electronics*, 1986, **QE-22**, pp. 69–78

101 BLOOM, D. M., WEINGARTEN, K. J., and RODWELL, M. J. W.: 'Electrooptic sampling measures MMICs with polarized light', *Microwaves & RF*, 1987, pp. 74–80

102 MERTIN, W., BOHM, C., BALK, L. J., and KUBALEK, E.: 'Two-dimensional field mapping in MMIC-substrates by electro-optic sampling technique', *IEEE MTT-S International Microwave Symposium Digest*, 1992, pp. 1443–1446

103 LEE, C. H., LI, M. G., HUNG, H.-L. A., and HUANG, H. C.: 'On-wafer probing and control of microwave by picosecond optical beam', *Proceedings of IEEE Asia-Pacific Microwave Conference*, 1992, pp. 367–370

104 WU, X., CONN, D., SONG, J., and NICKERSON, K.: 'Calibration of external electro-optic sampling using field simulation and system transfer function analysis', *IEEE MTT-S International Microwave Symposium Digest*, 1993, pp. 221–224

105 MERTIN, W., ROTHS, C., TAENZLER, F., and KUBALEK, E.: 'Probe tip invasiveness at indirect electro-optic sampling of MMIC', *IEEE MTT-S International Microwave Symposium Digest*, 1993, pp. 1351–1354

106 CHENG, H., and WHITAKER, J. F.: '300-GHz-bandwidth network analysis using time-domain electro-optic sampling', *IEEE MTT-S International Microwave Symposium Digest*, 1993, pp. 1355–1358

107 HJELME, D. R., YADLOWSKY, M. J., and MICKELSON, A. R.: 'Two-dimensional mapping of the microwave potential on MMIC's using electrooptic sampling', *IEEE Transactions on Microwave Theory and Techniques*, 1993, **MTT-41** (6/7), pp. 1149–1158

108 DAVID, G., REDLICH, S., MERTIN, W. *et al.*: 'Two-dimensional direct electro-optic field mapping in a monolithic integrated GaAs amplifier', *Proceedings of 23rd European Microwave Conference*, 1993, pp. 497–499
109 MERTIN, W., LEYK, A., DAVID, G. *et al.*: 'Two-dimensional mapping of amplitude and phase of microwave fields inside a MMIC using the direct electro-optic sampling technique', *IEEE MTT-S International Microwave Symposium Digest*, 1994, pp. 1597–1600
110 DAVID, G., TEMPEL, R., WOLFF, I., and JAGER, D.: 'Analysis of microwave propagation effects using 2D electro-optic field mapping techniques', *Optical and Quantum Electronics*, 1996, **28**, 1996, pp. 919–931
111 'Maps of electric fields traced back to standards', *Optics and Laser Europe (OLE) Magazine*, 1997, pp. 31–32
112 BOHM, C., ROTHS, C., and KUBALEK, E.: 'Contactless electrical characterization of MMICs by device internal electrical sampling scanning-force microscopy', *IEEE MTT-S International Microwave Symposium Digest*, 1994, pp. 1605–1608
113 MUELLER, U., BOEHM, C., SPRENGEPIEL, J., ROTHS, C., KUBALEK, E., and BEYER, A.: 'Geometrical and voltage resolution of electrical sampling scanning force microscopy', *IEEE MTT-S International Microwave Symposium Digest*, 1994, pp. 1005–1008.

Index

www.ingramcontent.com/pod-product-compliance
Lightning Source LLC
LaVergne TN
LVHW011856200926
848484LV00002B/460

* 9 7 8 0 8 5 2 9 6 7 8 6 7 *